Exact and Fast Algorithms for Mixed-Integer Nonlinear Programming

vorgelegt von
Dipl.-Math. Ambros M. Gleixner
aus Landshut

von der Fakultät II – Mathematik und Naturwissenschaften
der Technischen Universität Berlin
zur Erlangung des akademischen Grades

Doktor der Naturwissenschaften
– Dr. rer. nat. –

genehmigte Dissertation

Promotionsausschuss

Vorsitzender: Prof. Dr. Reinhold Schneider
Berichter: Prof. Dr. Dr. h.c. mult. Martin Grötschel
Prof. Dr. Thorsten Koch
Prof. Dr. Andrea Lodi

Tag der wissenschaftlichen Aussprache: 24. Juni 2015

Berlin 2015
D 83

Bibliografische Information der Deutschen Nationalbibliothek

Die Deutsche Nationalbibliothek verzeichnet diese Publikation in der
Deutschen Nationalbibliografie; detaillierte bibliografische Daten sind
im Internet über http://dnb.d-nb.de abrufbar.

ISBN 978-3-8325-4190-3

Logos Verlag Berlin GmbH
Comeniushof, Gubener Str. 47,
10243 Berlin
Tel.: +49 (0)30 42 85 10 90
Fax: +49 (0)30 42 85 10 92
INTERNET: http://www.logos-verlag.de

Abstract

Mixed-integer nonlinear programming (MINLP) comprises the broad class of finite-dimensional mathematical optimization problems from mixed-integer linear programming and global optimization. The combination of the two disciplines allows us to construct more accurate models of real-world systems, while at the same time it increases the algorithmic challenges that come with solving them. This thesis presents new methods that improve the numerical reliability and the computational performance of global MINLP solvers.

Since state-of-the-art algorithms for nonconvex MINLP fundamentally rely on solving linear programming (LP) relaxations, we address numerical accuracy directly for LP by means of *LP iterative refinement*: a new algorithm to solve linear programs to arbitrarily high levels of precision. The thesis is supplemented by an exact extension of the LP solver SoPlex, which proves on average 1.85 to 3 times faster than current state-of-the-art software for solving general linear programs exactly over the rational numbers. These methods can be generalized to quadratic programming. We study their application to numerically difficult multiscale LP models for metabolic networks in systems biology.

To improve the computational performance of LP-based MINLP solvers, we show how the expensive, but effective, bound-tightening technique called *optimization-based bound tightening* can be approximated more efficiently via feasibility-based bound tightening. The resulting implementation increases the number of instances that can be solved and reduces the average running time of the MINLP solver SCIP by 17–19% on hard mixed-integer nonlinear programs.

Last, we present branching rules that exploit the presence of *nonlinear integer variables*, i.e., variables both contained in nonlinear terms and required to be integral. The new branching rules prefer integer variables when performing spatial branching, and favor variables in nonlinear terms when resolving integer infeasibility. They reduce the average running time of SCIP by 17% on affected instances.

Most importantly, all of the new methods enable us to solve problems which could not be solved before, either due to their numerical complexity or because of limited computing resources.

Acknowledgements

This thesis would not have been possible without the inspiring environment and the technical and financial support I found at Zuse Institute Berlin. For their trust and for this support I wish to thank Martin Grötschel and Thorsten Koch and my colleagues at the Department of Optimization and the administration of the institute. I am indebted to Andreas Bley, without whom I might never have learned about this special place, and to Gary Froyland, who encouraged me during my first steps in discrete optimization at the University of New South Wales. I thank Andrea Lodi for taking an interest in the results of this thesis and agreeing to be a member of my PhD committee, despite his numerous commitments.

I am most grateful to the collaborators and co-authors of the research articles that were written in the course of this thesis. I want to thank Timo Berthold for his creative, critical, and ambitious mind, Dan Steffy for his initiative to study iterative refinement using SoPlex, Benjamin Müller and Stefan Weltge for implementing and improving the OBBT components of SCIP, and Kati Wolter for her thorough testing and profiling of the exact extensions of SoPlex. I wish to thank Dennis Elbrächter, Luca Fabbri, Wei Huang, Leif Naundorf, and Frédéric Pythoud, who supported me during their time as student assistants.

The computational aspect of this thesis builds on the work of many people. I thank Roland Wunderling for creating SoPlex, Tobias Achterberg for creating SCIP, Stefan Vigerske for turning SCIP into an MINLP solver, and the entire team behind the SCIP Optimization Suite for developing this extraordinary code base. In particular, I want to thank Marc Pfetsch for our first debugging sessions in SoPlex, Stefan Heinz for his team spirit, Gregor Hendel for supporting me in the evaluation of the experiments in Chapters 6 and 7 by means of his excellent Ipet tool, and Gerald Gamrath and Matthias Miltenberger for freeing me from a lot of maintenance work on SCIP and SoPlex when I needed it. I am grateful to Ali Ebrahim for providing me with the problem instances which form the basis of the experiments in Chapter 4. Many thanks to Matthias Walter for testing the exact extensions of SoPlex.

The quality of the manuscript was greatly improved by the thorough proofreading of Timo Berthold, Christina Burt, Ali Ebrahim, Gerald Gamrath, Gregor Hendel, Joshua Lerman, Steve Maher, Benjamin Müller, Jonas Schweiger, Felipe Serrano, Dan Steffy, Charles Stewart, Stefan Vigerske, Matthias Walter, and Axel Werner. Thank you for your time and effort. I am particularly indebted to Ralf Borndörfer, Steve Maher, and Marika Neumann for discussing the outline of the introduction.

My research was financially supported by the Berlin Mathematical School[1], the Research Campus MODAL[2] funded by the German Ministry for Research and Education, and the COST Action TD1207[3] funded by the European Union.

Finally, one of the most enriching facets of this time was certainly the opportunity to travel and see the world. I want to express my gratitude to everyone who hosted me during these years, in particular András Frank, Kristóf Bérczi, Erika Kovács, Lászlo Végh, and the EGRES team for many kind invitations to Budapest, Gary Froyland for some Sydney nostalgia after the Discrete Optimization workshop 2010 at the University of New South Wales, Christina Burt, Pascal Van Hentenryck, Adrian Pearce, and the entire NICTA team for their continuing interest in SCIP and for their generous invitations to Melbourne, Katsuki Fujisawa, Yuji Shinano, and Hayato Waki for two wonderful trips to Japan, Yu-Hong Dai, Xin Liu, and Ya-Feng Liu for a very golden week in Beijing, Felipe Serrano and Carolina von Hausen and their families for a memorable time in Chile, Dan Steffy and Huizhu Crystal Wang for Detroit coffee and music, João Pedro Pedroso for the beauty of Porto, Ruth Misener for an inspiring trip to London, Andrea Lodi, Luca Fabbri, and Jonas Schweiger for Bologna's delicacies, Volker Kaibel, Matthias Walter, and Stefan Weltge for Magdeburg's Christmas markets, and Michael Bussieck, Stefan Vigerske, and the GAMS team for the best Ethiopian coffee in Braunschweig. I am grateful to Maike Massierer and James MacLaurin for great times in Basel, Oxford, and Nice, and for continuing to share many common and distinct experiences since our undergraduate year in UNSW's Honours room.

---

[1] http://www.math-berlin.de/
[2] http://www.forschungscampus-modal.de/
[3] http://cost-td1207.zib.de/

Contents

Introduction

Mathematical optimization has become a popular tool for studying steady states of biological systems *in silico*, i.e., via computer models. Mixed-integer linear programming (MIP) is the methodology of choice for many studies on metabolic networks.[4] The use of linear models, however, is to some extent due to limitations in available optimization technology rather than to a perfectly accurate representation of the biological systems at hand. Addressing these limitations is an important step both in the study of biological systems and the future development of mathematical optimization. The following thesis is concerned with the numerical reliability and the computational performance of general-purpose algorithms for the broad class of mixed-integer nonlinear programs.

Mixed-integer nonlinear programming (MINLP) extends the expressiveness of MIP by allowing nonlinear terms in the objective and the constraint functions. To illustrate this using the above example, so-called *ME models* of bacterial growth couple metabolic and regulatory processes nonlinearly: the growth rate limits the *ratio* of reaction rates from both types of processes.[5] Currently these constraints are often linearized by holding the growth rate fixed as a parameter. The actual goal, however, is the computation of the optimal growth rate as a variable of the model, leading to an MINLP formulation with nonlinear constraints.

In mixed-integer linear programming, we have seen tremendous algorithmic progress over the last decades. General-purpose computer implementations, also called *solvers*, for MIP have reached an impressive maturity and become a standard industry tool.[6] In comparison, the range of mixed-integer nonlinear programs that can be solved reliably by general-purpose methods is still rather small. This poses the open question to what extent the advances we have seen in MIP solving can be continued in MINLP.

Note that the difference observed today should not be attributed to the increased theoretical complexity of MINLP alone: extensive computational research in *global optimization*, the discipline concerned with finding globally optimal solutions to non-convex nonlinear programs, only started in the 1990s.[7] By contrast, the seminal work of Dantzig, Fulkerson, and Johnson on the "solution of a large-scale traveling-salesman problem"—maybe the archetypal example of a (combinatorial) integer

---

[4] See, e.g., Zomorrodi et al. (2012).
[5] See Thiele et al. (2012).
[6] See, e.g., Bixby et al. (2000), Bixby and Rothberg (2007), and Koch et al. (2013).
[7] See the historical overview by Liberti (2013).

program—dates back to 1954.[8] With this historical offset in mind, and with the experience gathered in both fields, there is good reason to believe that substantial algorithmic progress is within reach. The increasing attention that MINLP has received in recent years has already led to significant advances both in theory and practice.[9]

For all classes of mathematical programming, the numerical reliability of optimization software is fundamentally important for their use in practice. For nonlinear models, this is often particularly difficult to achieve.[10] Somewhat counterintuitively, in state-of-the-art MINLP solvers numerical instabilities are frequently encountered when solving *linear programming* (LP) relaxations. Because of this, improving the numerical accuracy of LP solvers is both a necessary and effective step towards improving the numerical stability of MINLP solvers.

Also for the biological applications discussed at the beginning, numerical accuracy is highly relevant. One core difficulty in solving the recently developed ME models for bacterial growth is their multiscale nature. This stems from the coupling of metabolism and gene expression, two fundamental processes within a cell that lead to values spanning multiple orders of magnitude, both in the model formulation and the resulting solutions. As a consequence, state-of-the-art LP solvers using double-precision arithmetic frequently fail to produce solutions of sufficient accuracy or even to reliably decide feasibility.

The aim of this thesis is to develop numerically exact algorithms for linear programming and improve the performance of general-purpose algorithms for mixed-integer nonlinear programming. In order to evaluate their computational efficiency and effectiveness, the new methods shall be implemented as part of state-of-the-art solvers and tested on heterogeneous sets of benchmark instances with relevance in practice.

---

[8] See Dantzig, Fulkerson, and Johnson (1954a,b).
[9] See, e.g., Lee and Leyffer (2012).
[10] See, e.g., Domes and Neumaier (2008).

Overview and Main Results

The contribution of this thesis is threefold. First and foremost, we present *LP iterative refinement* (LPIR): a new algorithm to solve linear programs to arbitrarily high levels of precision. LPIR extends the technique of iterative refinement for linear systems of equations by Wilkinson (1963) to the domain of optimization problems, simultaneously correcting violations of feasibility and optimality. Its computational efficiency rests on its mixed-precision nature: the involved part of the computation—the solution of a sequence of related linear programs—is performed in fast floating-point arithmetic, and the cheaper part—essentially matrix-vector multiplications for computing and correcting errors—is performed in exact rational arithmetic. We show that computing solutions up to a precision of 10^{-50}, i.e., far beyond standard floating-point accuracy, incurs only an average slowdown of 3% on numerically easy and 7% on numerically difficult linear programs.[11] Most prominently, this leads to two new algorithms for *exact linear programming over the rational numbers* relying on a proof that a simplex-based implementation of LPIR converges to an optimal basic solution in oracle-polynomial time. The resulting implementation solves linear programs 1.85 to 3 times faster than the current state-of-the-art software.[12] We show that these results can be generalized to *quadratic programming*.

Second, we develop the notion of *Lagrangian variable bounds* for MINLP, which are aggregations of a linear relaxation obtained from dual information and imply bounds on the variables. They can be used to leverage the effect of an expensive, but effective, method called *optimization-based bound tightening* (OBBT). The resulting implementation reduces the average running time by 17–19% on hard mixed-integer nonlinear programs.[13]

Third, we present three new branching rules for MINLP by focusing on decision variables that are both required to be integral and are contained in nonlinear terms of the constraint functions. These are variables that truly combine the two sources of complexity in MINLP: integrality and nonlinearity. In this context we introduce the new concept of a *minimum cover* of a mixed-integer nonlinear program, which leads to a non-standard, discrete measure of nonlinearity. The new branching rules reduce the average running time of LP-based branch-and-bound by 17% on affected instances.[14] Even more importantly, both the new branching rules and the new bound tightening techniques significantly increase the number of instances that can be solved under limited computing resources.

---

[11] This comparison is relative to a standard floating-point LP solver with a tolerance of 10^{-9}. The average increase is calculated via shifted geometric means over a large test set of publicly available benchmark instances excluding trivial instances that can be solved in under two seconds.

[12] This speedup is measured relative to the solver QSopt_ex (Applegate et al., 2007b) over a test set of 492 nontrivial instances on which at least one solver took at least two seconds. On all instances, the speedup factor in shifted geometric mean is 1.85; on the instances where QSopt_ex had to apply extended-precision computation, the speedup factor is three.

[13] These are reductions in shifted geometric mean over instances that can be solved with and without OBBT, but not by both within less than 100 seconds.

[14] These are reductions in shifted geometric mean over instances that can be solved with at least one of two branching rules compared.

The main results of the thesis are:

– An *iterative refinement scheme for linear and quadratic programs*, which computes arbitrarily accurate solutions by repeated calls to a limited-precision floating-point algorithm,

– An *algorithm for exact linear and quadratic programming over the rational numbers* based on the proof that a simplex-based implementation of this refinement scheme converges to an optimal basic solution,

– An *algorithm to reconstruct exact basic solutions of linear and quadratic programs* with rational input data, which only relies on residual errors,

– An application of the above algorithms to so-called *ME models* from systems biology,

– Three enhancements to *optimization-based bound tightening*, which significantly improve the performance of an MINLP solver,

– The notion of a *minimum cover*, a structure that defines a new measure of nonlinearity of a mixed-integer nonlinear program, and

– *Three new branching rules* for MINLP, which significantly improve the performance of LP-based branch-and-bound.

The thesis is organized as follows. Chapter 1 comprises preliminary material that is relevant throughout the thesis. Besides introducing notation and collecting basic mathematical facts for later reference, we discuss benchmarking methodology and give an overview of the test sets of linear and mixed-integer nonlinear programs that are used in the computational experiments of this thesis.

Chapter 2 presents LP iterative refinement. In addition, for infeasible instances, we develop a new method to convert an approximate proof of infeasibility in the form of a Farkas ray into a rigorous and more user-friendly certificate in the form of a box around the origin in which provably no feasible solution exists. This box constitutes an exact infeasibility proof if it extends to infinity or if its radius exceeds the absolute value of the largest bound of a variable.

In Chapter 3, we devise two new algorithms for exact linear programming over the rational numbers. Both rely on the fact that the LPIR scheme converges to an optimal basis if the underlying LP algorithm returns basic solutions. The first algorithm makes use of the fact that this optimal basis uniquely defines an optimal solution that can be computed by a matrix factorization in exact rational arithmetic. The second algorithm exploits the fact that the sequence of increasingly accurate solutions generated by LPIR converges to an optimal solution with bounded denominator; this solution can be reconstructed from a sufficiently advanced iterate using continued fraction approximations. In this context, in order to save refinement steps and running time, we propose a so-called output-sensitive reconstruction algorithm that applies the reconstruction step dynamically to candidates that may be less accurate than suggested by worst-case analysis.

In Chapter 4, we apply LPIR to the *ME models* mentioned above. The multiscale nature of the underlying biological processes results in numerically difficult linear programs, which challenge current floating-point LP software. We demonstrate that LPIR in combination with the infeasibility box algorithm produces highly accurate solutions and decides feasibility of the problems correctly. This facilitates a more reliable analysis of the biological models.

Chapter 5 links the first part of the thesis on linear programming with the second part of the thesis on mixed-integer nonlinear programming. We show that the high-precision and exact linear programming algorithms developed in Chapters 2 and 3 can be generalized to quadratic programs, computing high-precision primal–dual solutions to the first-order necessary Karush–Kuhn–Tucker conditions. Furthermore, we survey basic concepts of mixed-integer nonlinear programming that are fundamental for the contributions in Chapters 6 and 7.

Chapter 6 is concerned with optimization-based bound tightening. In its most basic form, OBBT is a straightforward technique that minimizes and maximizes each variable over a convex relaxation to obtain tighter bounds on the variables. We develop methods to reduce the computational cost of OBBT and exploit dual information to learn valid inequalities. These one-row relaxations approximate the effect of OBBT and can be used to derive further bound tightenings at almost no cost when compared to the effort required by OBBT itself. This provides a cheap version of iterated OBBT during the solving process.

In Chapter 7, we present the new concept of a minimum cover of a mixed-integer nonlinear program, a structure that can be automatically detected and corresponds to a largest linear subproblem. We exploit this in one of three new branching rules, which are rules for subdividing an MINLP problem and constitute a key ingredient of branch-and-bound algorithms. Our computational experiments show significant reductions in average running time and an increase in the number of instances that can be solved within one hour of computation time. Finally, Chapter 8 contains our conclusions.

Publications and Software

The results of this thesis have been published in or submitted to the following peer-reviewed journals and conferences. An early version of iterative refinement for primal and dual feasible linear programs that are solved with the simplex method has been presented at *ISSAC 2012* (Gleixner et al., 2012b). A significantly extended version based on Chapter 2 and parts of Chapter 3 has been accepted for publication in the *INFORMS Journal on Computing* (Gleixner et al., 2015b).

The propagation of Lagrangian variable bounds for mixed-integer nonlinear programs from Section 6.2 first appeared in the proceedings of *CPAIOR 2013* (Gleixner and Weltge, 2013). In conjunction with the new filtering and ordering schemes from Section 6.3 for fast optimization-based bound tightening, an extended computational analysis has been submitted to the *Journal of Global Optimization* (Gleixner et al., 2015a).

The notion of covers for mixed-integer nonlinear programs and their generic detection by solving a vertex covering problem on the co-occurence graph from Chapter 7 has been presented at the *European Workshop on Mixed Integer Nonlinear Programming* (Berthold and Gleixner, 2010) and published in *Mathematical Programming* (Berthold and Gleixner, 2014). Its application as part of the Undercover branching strategy has been presented at *SEA 2013* (Berthold and Gleixner, 2013).

The software implementations that have been developed in the course of this thesis are now part of the LP solver SoPlex[15] and the MINLP solver SCIP.[16] Both are freely available for academic research use as part of the SCIP Optimization Suite. In addition, due to its increased numerical reliability on the multiscale models of bacterial growth studied in Chapter 4, SoPlex has become a standard plug-in in the COBRApy package for constraint-based modeling of biological networks (Ebrahim et al., 2013).[17] Thanks to Hans Mittelmann, SoPlex 2.1 with 80-bit floating-point precision and the new extensions for exact linear programming from Chapters 2 and 3 is easily accessible through an installation on the NEOS Server for Optimization.[18]

---

[15] Zuse Institute Berlin. SoPlex—the Sequential object-oriented simPlex. Available for download under http://soplex.zib.de/.

[16] Zuse Institute Berlin. SCIP—Solving Constraint Integer Programs. Available for download under http://scip.zib.de/.

[17] Ali Ebrahim, Nathan Lewis, and Ronan Fleming. The openCOBRA Project. Available under http://opencobra.github.io/. See also Ebrahim et al. (2013).

[18] NEOS Server: State-of-the-Art Solvers for Numerical Optimization. http://www.neos-server.org/. SoPlex is available under http://www.neos-server.org/neos/solvers/lp:SoPlex80bit/LP.html.

Chapter 1

Preliminaries

This chapter comprises preliminary material that is relevant throughout the thesis. Section 1.1 introduces notation, Section 1.2 collects basic mathematical facts for later reference, Section 1.3 discusses our computational methodology, and Section 1.4 gives an overview of the test sets on which our computational experiments were conducted.

1.1 Notation and Abbreviations

In order to keep the individual chapters as self-contained as possible, most notation will be introduced as it is needed. Here, we only want to mention some common notation and conventions, in which we mainly follow Grötschel, Lovász, and Schrijver (1988).

The sets of real, rational, and integer numbers are denoted by $\mathbb{R}$, $\mathbb{Q}$, and $\mathbb{Z}$, respectively, and their nonnegative versions by $\mathbb{R}_{\geq 0}$, $\mathbb{Q}_{\geq 0}$, and $\mathbb{Z}_{\geq 0}$. For the cardinality of a set S we write $\#S$. For $x \in \mathbb{R}$, the smallest integer greater than or equal to x is written as $\lceil x \rceil$, the largest integer less than or equal to x is written as $\lfloor x \rfloor$.

For a matrix $A \in \mathbb{K}^{m \times n}$ over a generic domain $\mathbb{K}$, we use $A_{\mathcal{I}, \mathcal{J}}$ to denote the submatrix of A with rows from an index set $\mathcal{I} \subseteq \{1, \ldots, m\}$ and columns from an index set $\mathcal{J} \subseteq \{1, \ldots, n\}$. We abbreviate the index set of all columns or rows by '$\cdot$'. Hence, the i-th row vector and j-th column vector are denoted by $A_{i \cdot}$ and $A_{\cdot j}$, respectively. For a vector $b \in \mathbb{K}^n$ we write $b_{\mathcal{J}} := (b_i)_{i \in \mathcal{J}} \in \mathbb{K}^{|\mathcal{J}|}$.

The identity matrix of dimension n is denoted by I_n and the i-th unit vector by e_i. The zero vector and the vector of all ones are written as 0 and $\mathbb{1}$, respectively, their dimension usually being clear from the context. The transpose of a vector or matrix is denoted by a superscript 'T'. Throughout this thesis, all vectors are understood to be column vectors unless otherwise noted. The comparison of vectors $x, y \in \mathbb{K}^n$ is defined componentwise, i.e., $x \leq y$ if and only if $x_i \leq y_i$ for all $i \in \{1, \ldots, n\}$. For two vectors $\ell, u \in \mathbb{K}^n$ we define the box $[\ell, u] := \{x \mid \ell \leq x \leq u\} = \times_i [\ell_i, u_i]$.

7

We use the symbol $\langle \cdot \rangle$ to denote the encoding length of an object. For a list of objects $a, b, c, \ldots$, we write $\langle a, b, c, \ldots \rangle$ for the sum of their encoding lengths.

Many of the abbreviations that are used throughout the thesis refer to specific problem classes, such as "LP" refers to "linear programming". With slight abuse of notation we will use these abbreviations also for the problems of these classes, e.g., "LP" also stands for "linear program". A list of recurring abbreviations and names is given on page 175.

1.2 Mathematical Prerequisites

In this section we summarize basic definitions and well-known facts for later reference. The *maximum norm* of a vector $x \in \mathbb{R}^n$ is defined as

$$\|x\|_\infty := \max_{i=1,\ldots,n} |x_i|. \tag{1.1}$$

The corresponding *row sum norm* of a matrix $A \in \mathbb{R}^{m \times n}$ given by

$$\|A\|_\infty := \max_{i=1,\ldots,m} \sum_{j=1}^{n} |A_{ij}|, \tag{1.2}$$

is compatible with the maximum norm in the sense that

$$\|Ax\|_\infty \leqslant \|A\|_\infty \|x\|_\infty \tag{1.3}$$

for all $A \in \mathbb{R}^{m \times n}$, $x \in \mathbb{R}^n$.

Furthermore, we define the encoding length of an integer $n \in \mathbb{Z}$ as

$$\langle n \rangle := 1 + \lceil \log_2(|n| + 1) \rceil. \tag{1.4}$$

Then, for a rational number p/q with $p \in \mathbb{Z}$ and $q \in \mathbb{Z}_{\geqslant 0}$, $\langle p/q \rangle = \langle p \rangle + \langle q \rangle$. For a vector $v \in \mathbb{Q}^n$, $\langle v \rangle = \sum_i \langle v_i \rangle$ and for a matrix $A \in \mathbb{Q}^{m \times n}$, $\langle A \rangle = \sum_{i,j} \langle A_{ij} \rangle$. Note that $\langle v \rangle \geqslant n$ and $\langle A \rangle \geqslant nm$.

The following lemmas compile some well-known relations involving encoding lengths, norms, and systems of equations.

Lemma 1.1. *For $r_1, \ldots, r_n \in \mathbb{Q}$,*

$$\langle r_1 \cdot \ldots \cdot r_n \rangle \leqslant \langle r_1 \rangle + \ldots + \langle r_n \rangle, \tag{1.5}$$

$$\langle r_1 + \ldots + r_n \rangle \leqslant 2(\langle r_1 \rangle + \ldots + \langle r_n \rangle), \tag{1.6}$$

and for $z_1, \ldots, z_n \in \mathbb{Z}$,

$$\langle z_1 + \ldots + z_n \rangle \leqslant \langle z_1 \rangle + \ldots + \langle z_n \rangle. \tag{1.7}$$

For matrices $A \in \mathbb{Q}^{m \times n}$, $B \in \mathbb{Q}^{n \times p}$,

$$\langle AB \rangle \leqslant 2(p\langle A \rangle + m\langle B \rangle) \tag{1.8}$$

and for a square matrix $D \in \mathbb{Q}^{n \times n}$,

$$|\det D| \leqslant 2^{\langle D \rangle - n^2} - 1 \tag{1.9}$$

and

$$\langle \det D \rangle \leqslant 2\langle D \rangle - n^2. \tag{1.10}$$

Proof. See Grötschel, Lovász, and Schrijver (1988, Lemma 1.3.3, 1.3.4, and Exercise 1.3.5). Note that the factor 2 in (1.6) is best possible. □

Lemma 1.2. *For any matrix* $A \in \mathbb{Q}^{m \times n}$,

$$\|A\|_\infty \leqslant 2^{\langle A \rangle - mn} - mn \leqslant 2^{\langle A \rangle}. \tag{1.11}$$

Proof.

$$\begin{aligned}
\|A\|_\infty &= \max_{i=1,\ldots,m} \sum_{j=1,\ldots,n} |A_{ij}| \\
&\leqslant \sum_{i,j} |A_{ij}| \\
&= \left(\sum_{i,j} (|A_{ij}| + 1) \right) - mn \\
&= 2^{\log_2 \left(\sum_{i,j} (|A_{ij}| + 1) \right)} - mn \\
&\leqslant 2^{\sum_{i,j} \log_2 (|A_{ij}| + 1)} - mn \\
&\leqslant 2^{\left(\sum_{i,j} \langle A_{ij} \rangle \right) - mn} - mn \\
&= 2^{\langle A \rangle - mn} - mn.
\end{aligned}$$

□

Lemma 1.3. *For any nonsingular matrix* $A \in \mathbb{Q}^{n \times n}$, *right-hand side* $b \in \mathbb{Q}^n$, *and solution vector* x *of* $Ax = b$,

$$\langle x_i \rangle \leqslant 4\langle A, b \rangle - 2n^2 \leqslant 4\langle A, b \rangle. \tag{1.12}$$

for all $i = 1, \ldots, n$. *Furthermore,*

$$\langle A^{-1} \rangle \leqslant 4n^2 \langle A \rangle - 2n^4 \leqslant 4n^2 \langle A \rangle. \tag{1.13}$$

Proof. By Cramer's rule, the entries of x are quotients of determinants of submatrices of (A, b). With (1.10) this yields (1.12).

For the second inequality note that the i-th column of A^{-1} is the solution of the system $Ax = e_i$. Using Cramer's rule again and the fact that submatrices of (A, I_n) have size at most $\langle A \rangle$ gives inequality (1.13). □

Lemma 1.4. *For a square matrix* $A \in \mathbb{Q}^{n \times n}$,

$$|\det A| \leqslant \prod_{j=1}^{n} \|A_{j\cdot}\|_2 \leqslant n^{n/2} \prod_{j=1}^{n} \|A_{j\cdot}\|_\infty, \qquad (1.14)$$

where $\|\cdot\|_2$ *is the Euclidean norm. If* A *is nonsingular, the components of any solution vector* x *of* $Ax = b$, $b \in \mathbb{Q}^n$, *have denominator at most*

$$L^n \prod_{j=1}^{n} \|A_{j\cdot}\|_2 \leqslant n^{n/2} L^n \prod_{j=1}^{n} \|A_{j\cdot}\|_\infty, \qquad (1.15)$$

where L *is the least common multiple of the denominators of the entries in* A *and* b.

Proof. (1.14) is the Hadamard inequality in row form plus the fact that $\|x\|_2 \leqslant \sqrt{n}\|x\|_\infty$ for all $x \in \mathbb{R}^n$. For (1.15), note that $Ax = b$ if and only if $LAx = Lb$ and by Cramer's rule, the entries in x have form $\det \tilde{A}/\det(LA)$, where $\tilde{A}$ is a submatrix of (LA, Lb). Since LA, Lb, and $\tilde{A}$ now have integral coefficients, the determinants are integral and the entries in x have denominator at most $|\det(LA)| = L^n|\det A|$. $\qquad\square$

1.3 Experimental Methodology

Throughout the thesis we will analyze the practical performance of our algorithms by computational experiments. Our interest here is not purely *competitive testing* in the sense of determining which algorithm runs fastest, but to analyze the reasons for their behavior by controlled *scientific testing*, see also Hooker (1995). In the following we describe general details concerning our computational methodology.

1.3.1 Hardware and Software

All experiments in this thesis were conducted on a cluster of 64-bit Intel Xeon X5672 CPUs at 3.2 GHz with 12 MB cache and 48 GB main memory. With many modern architectures, these CPUs come with several cores. However, in order to safeguard against a potential mutual slowdown of parallel processes, which can easily distort the measured running times, we ran only one job per node at a time.

As mentioned beforehand, the software implementations used to conduct the experiments described in this thesis are publicly available as part of the solvers SoPlex and SCIP. They are distributed as part of the SCIP Optimization Suite, which is free for academic research use. The specific versions used in the experiments have been archived in order to facilitate reproducibility of the results, as well as the log files of the experiments.

1.3.2 Averaging

In order to evaluate algorithmic performance over a large set of benchmark instances, we have to rely on comparing averages. While in some cases the classical arithmetic average is suitable, it has the property that it is dominated by terms largest in absolute value.

This can create difficulties when a performance measure such as running time, LP iterations, or number of branch-and-bound nodes varies drastically across instances. Hence we often use the *shifted geometric mean*, which provides a measure for relative differences. The shifted geometric mean of values $v_1, \ldots, v_N \geqslant 0$ with shift $s \geqslant 0$ is defined as

$$\left(\prod_{i=1}^{N} (v_i + s) \right)^{1/N} - s. \tag{1.16}$$

While the pure geometric mean (with shift zero) already limits the dominance of large absolute changes in the average, a positive shift helps to avoid an overrepresentation of differences among very small values. See also the discussions by Achterberg (2007, App. A.3), Achterberg and Wunderling (2013), and Hendel (2014, Sec. 3.2).

1.3.3 Statistical Significance

Comparing performance of different algorithms only based on the mean of certain performance measures may be misleading: it does not allow us to draw conclusions on how a change in the mean value is distributed over the test set. A reduction in mean running time may stem from a consistent improvement over the whole test set, or could be the result of a drastic speedup on only few instances while the performance on the majority of instances deteriorates.

Especially for experiments with algorithms that exhibit high performance variability, it is crucial to carefully analyze computational results in this respect. By *performance variability*, we understand the occurrence of considerable changes in the performance of an algorithm caused by small and seemingly insignificant modifications to its implementation. As noted by Danna (2014), this phenomenon can be quite pronounced for state-of-the-art mixed-integer linear programming code. Though this has not been thoroughly studied yet, it is our experience that performance variability looms even larger on mixed-integer nonlinear programs, amongst others because of the effects of branching on continuous variables.

As a consequence, for the computational results presented in Chapters 6 and 7 we use *statistical hypothesis testing* to analyze how consistently a change in performance is distributed across a test set. In the standard setting, we want to compare the running time or the number of branch-and-bound nodes of two algorithms over the same set of instances. Hence we are given two sets of samples $X_1, \ldots, X_N$ and $Y_1, \ldots, Y_N$ that are of same size and paired: X_i and Y_i represent the performance measure of the first and second algorithm, respectively. We want to test whether the shifted geometric means of X and Y are significantly different. Because there is no ground to assume that X or Y adhere to a specific distribution such as the normal distribution, we need to apply nonparametric statistics.

A suitable nonparametric test for the setting described above is the *Wilcoxon signed-rank test*. In the fashion of Hendel (2014, Sec. 3.2) it works as follows. As a first step, we compute the relative difference in shifted performance for each instance,

$$Q_i = \log \frac{Y_i + s}{X_i + s},$$

where s is the same shift as used in (1.16). If we compare running times, for example, then speedups (of the second algorithm with running time Y vs. the first algorithm with running time X) have negative Q values, slowdown positive ones. Instances with identical performance, i.e., with $Q_i = 0$ are removed. Then we sort the remaining instances by the absolute value of their Q_i such that $|Q_1| \leqslant |Q_2| \leqslant \ldots \leqslant |Q_N|$. This allows us to compute the positive and negative rank sums

$$W^+ = \sum_{i:Q_i>0} \frac{i}{\#\{k \mid |Q_k| = |Q_i|\}}$$

and

$$W^- = \sum_{i:Q_i<0} \frac{i}{\#\{k \mid |Q_k| = |Q_i|\}}.$$

Here the denominators are used to split rank contributions for ties (instances with same $|Q_i|$) fairly between W^+ and W^-. (In the common case that all Q_i are distinct, the denominators are one.) In the example of comparing running times, many speedups would lead to a large negative and a small positive rank sum.

Wilcoxon (1945) showed that the difference $W^+ - W^-$ is approximately normally distributed with variance $N(N + 1)(2N + 1)/6$. From the test statistic

$$z = \frac{|W^+ - W^-|}{\sqrt{N(N + 1)(2N + 1)/6}}.$$

we can compute the (approximate) probability, the so-called p-value, that the observed outcome of the experiment holds given the null hypothesis. We always test against the null hypothesis that the setting with better performance in shifted geometric mean is actually not better. This corresponds to a one-sided test. If the p-value is small, the hypothesis is rejected and we may assert that the improvement seen by the shifted geometric mean is significant. Unless stated otherwise, we use a p-value of 5% in our analysis.

We use the implementation of the Wilcoxon signed-rank test available in the SciPy[19] package with parameters `correction=False` (default) and `zero_method= "pratt"`. The Pratt treatment removes anomalies due to samples with no performance difference (Pratt, 1959).

Note that the concept of significance is independent from the magnitude of change in the shifted geometric mean. This is because only the ranks of the logarithmic quotients Q_i count, not their actual value. For example, suppose the running time is reduced by a negligibly small factor, but it is reduced consistently by this same factor on all instances; then this "improvement" would be accepted as highly significant, although it is practically irrelevant.

---

[19] Eric Jones, Travis Oliphant, Pearu Peterson, and others. SciPy: Open source scientific tools for Python. `http://www.scipy.org/`. See also Oliphant (2007) and Millman and Aivazis (2011)

1.4 Test Sets

In the following we give details on the test sets of different problem classes that are used in the thesis.

1.4.1 Linear Programming

For our experiments in Chapters 2 and 3 we collected a large set of publicly available instances from the following sources:

- The Netlib LP test set including the "kennington" folder,[20]

- Hans Mittelmann's benchmark instances,[21]

- Csaba Mészáros's LP collection,[22]

- The LP relaxations of the COR@L mixed-integer linear programming test set,[23] and

- The LP relaxations of the mixed-integer linear programming test sets MIPLIB, MIPLIB 2, MIPLIB 3, MIPLIB 2003, and MIPLIB 2010.[24]

Some instances appeared in several collections. We removed all obvious duplicates and selected the 1,242 primal and dual feasible linear programs. Furthermore, we had to replace blank characters in the column and row names of some MPS files because they could not be parsed by the solver QSopt_ex[25] used in the experiments of Chapter 3.

We had to exclude seven large-scale instances instances[26] with 4,366,648 to 183,263,061 nonzeros in the constraint matrix for which the SoPlex LP solver used in our experiments hit the memory limit of 48 GB when parsing the instance using the exact rational data type. Furthermore, we removed the 33 instances[27] which even the standard, floating-point version of SoPlex could not solve within a time limit of two hours.

Altogether, this left us with 1,202 instances. The number of columns ranges from 3 to 2,277,736, the number of rows from 1 to 656,900, and the constraint matrices

---

[20] University of Tennessee Knoxville and Oak Ridge National Laboratory. Netlib LP Library. `http://www.netlib.org/lp/`, accessed September 2014.

[21] Hans Mittelmann. LP Test Set. `http://plato.asu.edu/ftp/lptestset/`, accessed July 16, 2014.

[22] Csaba Mészáros. LP Test Set. `http://www.sztaki.hu/~meszaros/public_ftp/lptestset/`, accessed July 16, 2014.

[23] Computational Optimization Research At Lehigh. MIP Instances. `http://coral.ie.lehigh.edu/data-sets/mixed-integer-instances/`, accessed June 6, 2011.

[24] Zuse Institute Berlin. MIPLIB—Mixed Integer Problem Library. `http://miplib.zib.de/`, accessed July 16, 2014.

[25] David L. Applegate, William Cook, Sanjeeb Dash, and Daniel G. Espinoza. QSopt_ex. `http://www.dii.uchile.cl/~daespino/ESolver_doc/`.

[26] `cont1_l, hawaiiv10-130, netlarge1,6, pb-simp-nonunif,` and `zib01,02`

[27] `L1_d10_40, Linf_520c, bley_xl1, cdma, cont11, cont11_l, datt256, dbic1, degme, in, karted, mining, nb10tb, neos3, netlarge3, ns1687037, ns1688926, ns1853823, ns1854840, nug15, nug20, nug30, rail02, rail03, rmine21, rmine25, sing161, spal_004, splan1, stat96v2, stat96v3, tp-6,` and `ts-palko`

contain between 6 and 27,678,735 nonzero entries. Detailed problem statistics are given in Table A.1 of the appendix.

Note that the exclusion of the 40 instances does not introduce a bias, since none of the SoPlex versions we test in any of our experiments would be able to solve them. For the comparison against QSopt_ex in Section 3.4, we include them in our evaluation.

1.4.2 Metabolic Networks

In Chapter 4, we study a special class of numerically challenging, multiscale linear programs that model metabolism and macromolecular synthesis of the bacteria *Escherichia coli* K-12 MG1655. To this end, we have been provided with 84 LPs generated by Ali Ebrahim from the Systems Biology Research Group at the University of California, San Diego,[28] some of them infeasible. Each of them has 76,413 variables, 68,726 constraints, and 1,217,886 nonzero entries in the constraint matrix. For more details see Section 4.2.

1.4.3 Mixed-Integer Nonlinear Programming

For our experimental evaluation of the bound tightening and branching algorithms in Chapters 6 and 7, we used the publicly available benchmark instances from the MINLPlib2.[29] MINLPlib2 includes amongst others instances from the first MINLPlib (Bussieck et al., 2003), from the nonlinear programming library GLOBALLib,[30] and from the recent CMU-IBM initiative minlp.org[31] and puts a focus on models that are relevant in practice. At the time of the experiments, the collection contained 1,357 instances. Of these, 1,279 instances were available in OSiL format and could be parsed by our testing framework SCIP.

We removed instances that could be solved already during the presolving phase of SCIP or that were linearized after presolving, since they are not relevant for the algorithms under investigation. In our analyses, we distinguished instances with and without integer variables after presolving and evaluated results on further subgroups of particular interest. For more details we refer to the computational studies in Sections 6.4, 7.3, and 7.4.

---

[28] Systems Biology Research Group, University of California, San Diego. http://systemsbiology.ucsd.edu/.

[29] Stefan Vigerske. MINLP Library 2. http://gamsworld.org/minlp/minlplib2/.

[30] GAMS Global World. GLOBAL Library: A Collection of Nonlinear Programming Models. http://www.gamsworld.org/global/globallib.htm

[31] Carnegie Mellon University. CMU-IBM Cyber-Infrastructure for MINLP. http://www.minlp.org/.

Chapter 2

Iterative Refinement for Linear Programming

Most fast linear programming (LP) solvers available today use floating-point arithmetic, which can lead to numerical errors. Although such implementations are effective at computing perfectly satisfactory approximate solutions for a wide range of instances, there are situations when they give unreliable results, or when extended-precision or exact solutions are desirable. In this chapter, we show how such limited-precision algorithms can be used as oracles in an iterative refinement procedure in order to compute high-precision solutions for linear programs.

To motivate this work we want to mention such applications as the multiscale ME models in systems biology discussed in Chapter 4. In addition, high-precision LP solutions are important in algorithms for exact integer programming, with applications in the verification of computer chips or the execution of combinatorial auctions, see also the discussion of Cook et al. (2013). Moreover, linear and integer programming have been used to establish theoretical results in a number of recent papers (Hicks and McMurray Jr., 2007; Buchheim et al., 2008; Bulutoglu and Kaziska, 2010; de Oliveira Filho and Vallentin, 2010; Burton and Ozlen, 2012; Held et al., 2012). Like the prominent example of the proof for the Kepler conjecture by Hales (2005), approximate floating-point solutions often have to be post-processed in order to obtain provably correct results (Obua and Nipkow, 2009). Such procedures might benefit from the availability of exact or high-precision solutions.

The chapter is organized as follows. In Section 2.1, we give an introduction to the basics of linear programming and highlight the limitations of modern floating-point implementations. Our main contribution—an iterative refinement scheme for primal and dual feasible linear programs—is presented in Section 2.2. For clarity, we present its basic description for problems given in standard form with equality constraints and lower bounds on the variables; Section 2.3 explains how to handle inequalities and general bounds.

Section 2.4 discusses how infeasible and unbounded problems can be addressed via homogenization. In this context, we develop an algorithm to transform an approximate Farkas proof of infeasibility into a rigorous infeasibility box centered at the origin, so providing a provable and easily interpretable answer to the user of an LP solver. Finally, we describe an integrated refinement algorithm for linear programs with an a priori unknown status of primal and dual feasibility. The computational results in Section 2.5 analyze the effectiveness of the procedure using an implementation based on the LP solver SoPlex.

The general idea of iteratively refining linear programming solutions is joint work with Daniel E. Steffy and Kati Wolter (Gleixner et al., 2012b). The running-time analysis of Sections 2.2.2 and 2.2.3, the extension of the method to infeasible and unbounded LPs of general form discussed in Sections 2.3 and 2.4, and the implementation presented in Section 2.5 are original contributions of this thesis. A condensed version of this chapter including the experiment described in Section 3.2.1 is available as preprint (Gleixner et al., 2015b).

2.1 Introduction to Linear Programming

This section summarizes some fundamental facts about linear programming we will need later. For readers familiar with the topic, Sections 2.1.1 and 2.1.2 will serve as an introduction to our notation and a reference. For a thorough treatment of the theory of linear programming, many textbooks are available, see for instance Dantzig (1963), Chvátal (1983), or Vanderbei (1996).

2.1.1 Basic Concepts

A *linear program* in its most general form is given as

$$\min\{c^\mathsf{T} x \mid L \leqslant Ax \leqslant U, \ell \leqslant x \leqslant u\} \tag{2.1}$$

where $x \in \mathbb{R}^n$ is the vector of variables, $c \in \mathbb{R}^n$ is the objective function vector, $\ell \in (\mathbb{R} \cup \{-\infty\})^n$ and $u \in (\mathbb{R} \cup \{+\infty\})^n$ are the vectors of lower and upper bounds on the variables, respectively, $A \in \mathbb{R}^{m \times n}$ is the constraint matrix, and the vectors $L \in (\mathbb{R} \cup \{-\infty\})^m$ and $U \in (\mathbb{R} \cup \{+\infty\})^m$ contain the left-hand and right-hand sides of the constraints (also called rows), respectively. By abuse of notation, we will use the abbreviation LP for the class of problems as well as for an individual problem instance. For the sake of clarity, our presentation of the main ideas will first use the simpler form

$$\min\{c^\mathsf{T} x \mid Ax = b, x \geqslant \ell\} \tag{2.2}$$

with equality constraints, $b = L = U \in \mathbb{R}^m$, and with finite lower bounds and no upper bounds on the variables. Without loss of generality, A has full row rank and $n \geqslant m$.

The feasible region $F := \{x \in \mathbb{R}^n \mid Ax = b, x \geqslant \ell\}$ of (2.2) is called a polyhedron, see Schrijver (1986) for details on polyhedral theory. If F is nonempty and bounded, then (because F is closed) an *optimal* solution is guaranteed to exist, i.e., an $x^* \in F$ with $c^\mathsf{T} x^* \leqslant c^\mathsf{T} x$ for all $x \in F$. In this case, we also call the LP itself optimal.

Problem (2.2) is called *unbounded* if it contains solutions with arbitrarily low objective function value. This is the case if and only if the LP is feasible and the so-called *recession cone* of F,

$$\text{rec}(F) := \{r \in \mathbb{R}^n \mid x + \lambda r \in F \text{ for all } x \in F, \lambda \geqslant 0\}$$
$$= \{r \in \mathbb{R}^n \mid Ar = 0, r \geqslant 0\}, \tag{2.3}$$

contains a ray with negative objective function value, i.e., $r \in \text{rec}(F)$ such that $c^T r < 0$. We will refer to elements of the recession cone as *unbounded rays*. If F is empty, then problem (2.2) is called *infeasible*. This is characterized by the following version of the Farkas lemma.

Theorem 2.1 (Solvability of Linear Equations with Bounded Variables). *The system* $Ax = b, x \geqslant \ell$ *has a solution* x *if and only if there does not exist a vector* $y \in \mathbb{R}^m$ *such that* $y^T A \leqslant 0$ *and* $y^T(b - A\ell) > 0$.

The vector y here is called a *Farkas ray* or *Farkas proof* for the infeasibility of (2.2). It corresponds to an unbounded ray of the so-called *dual LP* of (2.2),

$$\max\{b^T y + \ell^T z \mid A^T y + z = c, z \geqslant 0\} \tag{2.4}$$

where $z = c - A^T y$ is the vector of *dual slacks*. In an optimal solution, the dual slacks are often called *reduced costs*. They can be associated with the variables and the reduced cost of variable x_i in an optimal solution then quantifies by how much the optimal value of (2.2) would decrease if the lower bound ℓ_i were relaxed by one (and no other constraints are blocking).

Because z is uniquely determined by the vector of dual multipliers y, we will sometimes speak of a dual solution y when we actually mean a solution (y, z) of the dual LP (2.4). The original problem (2.2) is then often called the *primal LP*.

Our discussion on infeasibilty and unboundedness above shows that if either of (2.2) or (2.4) has an optimal solution, then so does the other. Because of this strong connection between primal and dual, we often speak of a primal–dual solution x, y. We do not necessarily assume that they are primal or dual feasible. However, if x is a feasible solution of the primal (2.2) and (y, z) is a feasible solution of the dual LP (2.4), then their objective values bound each other, since

$$c^T x = (A^T y + z)^T x = y^T A x + z^T \ell + z^T (x - \ell)$$
$$= b^T y + \ell^T z + z^T (x - \ell) \geqslant b^T y + \ell^T z. \tag{2.5}$$

This inequality is tight if and only if $z^T(x - \ell) = 0$, which proves the following fundamental optimality condition.

Theorem 2.2 (Complementary Slackness). *Let* x *and* (y, z) *be feasible solutions to the primal LP (2.2) and its dual (2.4), respectively, then* x *is optimal for (2.2) and* (y, z) *is optimal for (2.4) if and only if*

$$x_i = \ell_i \text{ or } y^T A_{\cdot i} = c_i \tag{2.6}$$

holds for all $i \in \{1, \ldots, n\}$.

The value

$$\gamma(x, y) := (x - \ell)^\mathsf{T}(c - A^\mathsf{T}y), \tag{2.7}$$

called the *duality gap*, equals the difference between primal and dual objective function with value $c^\mathsf{T}x - (b^\mathsf{T}y + \ell^\mathsf{T}z)$ and is a measure for the total violation of complementary slackness.

Although linear programs are continuous optimization problems, they have a discrete structure given by the basic solutions. *Basic solutions* are solutions uniquely determined by $n - m$ variables held fixed at their bounds. The so-called *basis* $\mathcal{B}$, $\mathcal{B} \subseteq \{1, \ldots, n\}$, $|\mathcal{B}| = m$, is formed by the indices of the remaining, unfixed variables. This transforms the constraints $Ax = b$ to

$$A_{.\mathcal{B}}x_{\mathcal{B}} = b - \sum_{i \notin \mathcal{B}} A_{.i}\ell_i \tag{2.8}$$

where $A_{.\mathcal{B}} \in \mathbb{R}^{m \times m}$ is the submatrix of A formed by the columns with index in $\mathcal{B}$, called the *basis matrix*.

We are only interested in *regular bases*, i.e., bases $A_{.\mathcal{B}}$ with full rank. Then (2.8) uniquely defines a primal solution by $x_{\mathcal{B}} = A_{.\mathcal{B}}^{-1}(b - \sum_{j \notin \mathcal{B}} A_{.j}\ell_j)$ and $x_i = \ell_i$ for $i \notin \mathcal{B}$. Furthermore, the system

$$(A_{.\mathcal{B}})^\mathsf{T}y = c_{\mathcal{B}} \tag{2.9}$$

provides us with a unique dual solution $y \in \mathbb{R}^m$. Hence, via (2.8) and (2.9) each regular basis uniquely determines a primal–dual solution. By construction, this solution satisfies complementary slackness. As a result, a basic primal–dual solution x, y is optimal if and only if both x is primal feasible, i.e., if $x \geqslant \ell$, and y is dual feasible, i.e., $A^\mathsf{T}y \leqslant c$. Geometrically, the vertices of the polyhedral feasible region F are precisely the primal feasible basic solutions.

Alternatively, when an LP is given in inequality form $\min\{c^\mathsf{T}x \mid Ax \geqslant b\}$, $A \in \mathbb{Q}^{m \times n}$, basic solutions correspond to an index set $\mathcal{S} \subseteq \{1, \ldots, m\}$ of n linearly independent rows that are tight at their right-hand side. In this case, the primal–dual solution can be computed by solving with the basis matrix $A_{\mathcal{S}.}$. This is also referred to as a *row basis*.

2.1.2 Algorithms

Today, the algorithms most used in practice to solve linear programs are variations of the *simplex algorithm* or the *interior point method*. While economically motivated linear programming problems were described as early as 1939 by Kantorovich (1960), the (primal) simplex algorithm developed by Dantzig in 1947 was the first computational procedure to solve linear programs. For an account of the origins of linear programming see, for instance, Dantzig (1963, Chap. 2).

The primal simplex algorithm traverses vertices of F with increasing objective function value until an optimal vertex, certified by dual feasibility, is reached. In many cases a computationally more efficient variant is the dual simplex algorithm independently described by Beale (1954) and Lemke (1954), which follows dual feasible basic solutions until primal feasibility is reached. Although all known versions of the simplex method exhibit exponential worst-time complexity, they prove to be very efficient in practice. One of its outstanding features is the capability of hot starting from an almost feasible basis after slight modifications of the objective function or bounds. Here, the term *hot starting* is used to describe a warm start that also reuses internal matrix factorizations. This is effectively harnessed, for instance, in branch-and-cut algorithms for integer programming.

In 1979, Khachiyan noted that LPs can be solved in polynomial-time by the so-called *ellipsoid method* (Khachiyan, 1979, 1980). Although practically inefficient, this method became an important theoretical tool. For details on the ellipsoid method and implications in combinatorial optimization, see Grötschel, Lovász, and Schrijver (1988).

The first practical polynomial-time algorithm was an interior point algorithm described by Karmarkar (1984). It resembles methods pioneered by Fiacco and McCormick (1968) for nonlinear programming, who are sometimes credited as the true, though to some extent unknowing inventors of the interior point method—their interest was neither much in linear programming nor in complexity theory, a field that was only emerging at that time (Shanno, 2012). In a sense complementary to the simplex, an interior point algorithm produces a sequence of solutions that are both primal and dual feasible where the violation of complementary slackness converges to zero. Modern versions typically outperform the simplex method computationally when solving an LP from scratch, among other reasons because the underlying linear algebra routines can be efficiently parallelized on modern hardware. For the simplex method, no comparably efficient parallelization schemes have been developed so far. However, the recent works of Hall and Huangfu (2012) and Lubin et al. (2013) show promising results in this direction.

Note that in general the solutions produced by interior point methods are not basic, and current warm-starting strategies for interior point methods are not as effective as those for the simplex method. These disadvantages often make interior point methods less suitable for use in branch-and-cut algorithms for integer programming. So-called *crossover* algorithms for obtaining a basic solution from an interior point solution, from which one can continue with a simplex algorithm, are sequential procedures similar to the simplex method—see Bixby and Saltzman (1994) for an example.

2.1.3 Numerics

In contrast to our initial definition of linear programming problems over the field of real numbers, most LP solvers operate with the limited-precision *floating-point arithmetic* that is available on modern computer hardware, i.e., with a finite subset of the rational numbers. In this section we will discuss the implications of this limitation for LP solvers. We refer to the recent article of Klotz (2014) for a detailed analysis of numerical issues in floating-point LP solvers with a focus on implementations of the simplex algorithm.

The de facto standard for floating-point computation on modern hardware is double precision. This is specified by the IEEE Microcomputer Standards Committee (2008) as a 64-bit data type that can hold the binary representation of numbers

$$\pm 1.q_1 q_2 q_3 \ldots q_{52} \cdot 2^e$$

where $q_i \in \{0, 1\}$ and $e \in \{-1022, -1021, \ldots, 1023\}$. This gives 15 to 17 significant decimal digits. Special bit patterns are reserved for zero, plus and minus infinity, and the result of undefined arithmetic operations ("not-a-number").

The 128-bit floating-point data type defined by the IEEE standard is unfortunately not widely supported by current hardware. Hence the term *quad(ruple) precision*, although strictly speaking it designates a 128-bit data type, in practice often implies the 80-bit extended-precision format supported for instance by the common x86-family processors for the purpose of numerically accurate exponentiation of double-precision numbers. This provides eleven more bits to the significand—63 instead of 52 as in double-precision—so (only) three to four more significant decimal digits.

The speed of hardware-supported floating-point computations comes with the disadvantages of potential numerical errors. The result of an arithmetic operation on floating-point numbers is not necessarily floating-point representable and these errors, although small for each operation, may accumulate over a longer sequence of operations.

A typical routine that accumulates rounding errors in a floating-point LP solver is the solution of linear systems of equations such as system (2.8) as part of the simplex algorithm. Here the right-hand side is already the result of inexact floating-point operations. The factor by which this residual error may increase when propagated to the solution vector $x_{\mathcal{B}}$ can be estimated by the *condition number* of the basis matrix $A_{\cdot\mathcal{B}}$, see, for instance, Golub and van Loan (1983) for more details.

When solving LPs in practice with the simplex algorithm, one commonly encounters basis matrices with condition numbers of 10^6 and above. Although condition numbers give only worst-case estimates, this shows that—with 15 to 17 significant decimal digits in double-precision—we can in general not expect to obtain solutions with errors far below 10^{-9}. Note that this still holds if the discrete information given by the basis corresponds to an optimal solution.

As a consequence, floating-point LP solvers have to relax their feasibility and optimality conditions by a small tolerance $\varepsilon > 0$. They promise to compute an approximate solution $\tilde{x}$, $\tilde{y}$ for (2.2) such that $\|A\tilde{x} - b\|_\infty \leqslant \varepsilon$, $\tilde{x} \geqslant \ell - \varepsilon\mathbb{1}$, $\tilde{y}^\mathsf{T}A \leqslant c + \varepsilon\mathbb{1}$, and $\gamma(x, y) \leqslant \varepsilon$, where the tolerance in each condition can be chosen independently and may be relative to the input data.

Note that in theory this corresponds to a relaxation of the feasible regions of the primal *and* the dual LP and is not a relaxation of the original problem in the classical sense. In practice, LP solvers do not try to solve any relaxations, but compute approximate solutions with small infeasibilities. However, $\tilde{x}$, $\tilde{y}$ is exactly feasible and optimal for a *perturbation*:

Theorem 2.3. *Let $\tilde{x}$, $\tilde{y}$ be an approximate primal–dual solution to an LP of form* $\min\{c^\mathsf{T}x \mid Ax = b, x \geqslant \ell\}$, *then $\tilde{x}$, $\tilde{y}$ is exactly feasible and optimal for the perturbed* *problem* $\min\{\tilde{c}^\mathsf{T}x \mid Ax = \tilde{b}, x \geqslant \tilde{\ell}\}$ *with*

$$\tilde{c}_i := \begin{cases} \tilde{y}^\mathsf{T}A_{\cdot i} & \text{if } c_i - \tilde{y}^\mathsf{T}A_{\cdot i} < 0 \text{ or } c_i - \tilde{y}^\mathsf{T}A_{\cdot i} < \tilde{x}_i - \ell_i, \\ c_i & \text{otherwise,} \end{cases}$$

$$\tilde{\ell}_i := \begin{cases} \tilde{x}_i & \text{if } \tilde{x}_i < \ell_i \text{ or } \tilde{x}_i - \ell_i \leqslant c_i - \tilde{y}^\mathsf{T}A_{\cdot i}, \\ \ell_i & \text{otherwise,} \end{cases}$$

for $i = 1, \ldots, n$, *and* $\tilde{b} := A\tilde{x}$.

Proof. Primal, dual feasibility, and complementary slackness hold by construction. $\square$

Intuitively, we hope that any approximate solution with small violations will be close to an exactly feasible and optimal solution. Although this may hold in practice for most applications, the following simple example shows that in general this expectation is false.

Example 2.4. The zero solution is feasible for $\{x_1 - \delta^2 x_2 = -\delta, x_1, x_2 \geqslant 0\}$ within a tolerance of $\delta > 0$, but has Euclidean distance at least $1/\delta$ from any exactly feasible solution.

As mentioned above, for basic solutions, the condition number of the basis matrix measures the sensitivity to errors in the input and the floating-point operations locally at this solution. In order to quantify the sensitivity of an entire LP, Renegar (1994) and Freund and Vera (1999) consider the *distance to ill-posedness* defined as the minimum perturbation to the input data A, b, ℓ, c such that the LP changes its status— optimal, primal infeasible, or dual infeasible. The *condition measure* of an LP, defined to be inversely proportional to this distance to ill-posedness, is a generalization of the condition number for linear systems of equations that can be used to compute bounds on the feasible region and the change of the optimal value of an LP with respect to changes in the input. Ordóñez and Freund (2003) report that a large portion of the optimal LPs from the Netlib benchmark set[32] have infinite condition measure, i.e., an arbitrarily small perturbation of the input renders them primal or dual infeasible. This is largely explained by linear dependencies and redundant constraints, which are often removed by preprocessing steps of modern LP solvers. Nevertheless, this shows that numerical issues are not only a theoretical concern, but do appear in practice.

We want to emphasize here that for the majority of applications and instances, modern floating-point solvers will provide perfectly satisfactory, sufficiently accurate results (Dhiflaoui et al., 2003; Koch, 2004). However, there are cases when their answers are unreliable, or when higher-precision or even exact solutions are required. (Steffy, 2011, pp. 4) underlines the importance of handling floating-point computation with care by mentioning two tragic accidents due to faulty floating-point software:

---

[32] University of Tennessee Knoxville and Oak Ridge National Laboratory. Netlib LP Library. http://www.netlib.org/lp/.

There are several documented tragic errors involving floating-point computation. During the first US gulf war patriot missiles were used to intercept SCUD missiles, the software controlling missiles was based on floating-point computations. Repeated use of the number 1/10 in the code, which is not representable exactly as a base-2 floating point [sic] number, led to miscalculations that accumulated to form significant errors. On Feb 25, 1991, as a direct result of this miscalculation, a patriot missile failed to intercept an incoming Iraqi SCUD missile; it was off target by more than 0.6 kilometers and resulted in the death of 28 US soldiers [...]. In a later incident, the 1996 launch of the European Ariane 5 Rocket ended in failure when it went out of control and exploded 37 seconds into its flight path. The explosion was due to a software error caused by improper handling of a floating-point calculation; the software converted a 64-bit floating-point number to a 16-bit signed integer causing an overflow and system crash. The rocket and its cargo were worth an estimated 360 million USD [...].

These sound like naïve errors, still they show the relevance of research on numerically accurate computation, even more so for software as complex as LP solvers that are mostly employed as black-box subroutines.

We conclude this introduction with a description of *iterative refinement*, which is a well-known technique for improving numerical accuracy when solving linear systems of equations. Many LP solvers apply iterative refinement when solving linear systems, for instance, with the basis matrix and its transpose in the simplex method (Maes, 2013).

2.1.4 Iterative Refinement for Linear Systems of Equations

Given a system $Mx = r$, $M \in \mathbb{Q}^{n \times n}$, $r \in \mathbb{Q}^n$, iterative refinement constructs a sequence of increasingly accurate solutions $x_1, x_2, \ldots$, by first computing an approximate solution x^1, with $Mx_1 \approx r$. Then for $k \geqslant 2$, a refined solution $x_k \leftarrow x_{k-1} + \hat{x}$ is computed where $\hat{x}$ satisfies $M\hat{x} \approx \hat{r}$ and is a correction of the error $\hat{r} = r - Mx_{k-1}$ observed from the solution at the previous iteration, see Algorithm 2.1 for a summary.

Algorithm 2.1: Iterative Refinement for Linear Systems of Equations

---

    input: $Mx = r$, $M \in \mathbb{Q}^{n \times n}$, $r \in \mathbb{Q}^n$, termination tolerance $\varepsilon > 0$
    output: approximate solution $x^* \in \mathbb{Q}^n$
1 **begin**
2   get x_1 with $Mx_1 \approx r$
3      $\hat{r} \leftarrow r - Mx_1$
4   **foreach** $k \leftarrow 1, 2, \ldots$ **do**
5   **if** $\|\hat{r}\| \leqslant \varepsilon$ **then** return $x^* \leftarrow x_k$
6   get $\hat{x}$ with $M\hat{x} \approx \hat{r}$
7          $x_{k+1} \leftarrow x_k + \hat{x}$
8          $\hat{r} \leftarrow r - Mx_{k+1}$

This procedure can either be applied with *fixed precision* where all operations are performed using the same level of precision, or in *mixed precision* where the computation of the residual errors $\hat{r}$ and the addition of the correction—lines 3, 7, and 8—are computed with a higher level of precision than the system solves. For more details, see Wilkinson (1963) and Golub and van Loan (1983). Algorithm 2.1 can be applied to irrational input data if arbitrarily precise rational approximations of M and r can be computed.

However, if M and r are rational, then iterative refinement can even be used for computing exact solutions: after a sufficiently accurate solution has been constructed, the exact rational solution vector can be recovered using Diophantine approximation, see also Section 3.1.4. This idea was first described by Ursic and Patarra (1983) and improved upon by Wan (2006), Pan (2011), and Saunders et al. (2011).

2.2 An Iterative Refinement Scheme for Linear Programs

As mentioned above, iterative refinement is already applied by many LP solvers to improve their numerical robustness. Intensifying this use of iterative refinement inside floating-point LP solvers may yield more accurate solutions. We will take this idea one step further and show how iterative refinement can not only be applied to linear systems of equations, but to an entire LP.

Much of the previous work in the literature has focused on solving LPs with rational input data and our discussion will also concentrate on rationally defined problems. However, we note that there are some applications where it is meaningful and interesting to solve, or bound the optimal value of, LPs whose input data are defined by irrational numbers, or are given as interval enclosures. One advantage of the iterative refinement developed here is that it can easily be applied to irrational input data as long as rational approximations can be readily computed.

2.2.1 The Basic Algorithm

Our method solves a sequence of LPs, each one computing a correction of the previous to build a high-precision primal–dual solution. This strategy will simultaneously refine both the primal and dual solutions by adjusting the primal feasible region and the objective function of the LP to be solved. It is based on the following theorem, which formally holds for all positive scaling factors Δ_P, Δ_D. As will become clear soon, we are interested in the case when $\Delta_P, \Delta_D \gg 1$.

Theorem 2.5 (LP Refinement). *Suppose we are given an LP in form*

$$\min\{c^\mathsf{T}x \mid Ax = b, x \geqslant \ell\}, \tag{P}$$

then for $x^* \in \mathbb{R}^n$, $y^* \in \mathbb{R}^m$, and scaling factors $\Delta_P, \Delta_D > 0$, consider the transformed problem

$$\min\{\Delta_D \hat{c}^\mathsf{T}x \mid Ax = \Delta_P \hat{b}, x \geqslant \Delta_P \hat{\ell}\} \tag{$\hat{P}$}$$

where $\hat{c} = c - A^\mathsf{T}y^*$, $\hat{b} = b - Ax^*$, and $\hat{\ell} = \ell - x^*$. Then for any $\hat{x} \in \mathbb{R}^n$, $\hat{y} \in \mathbb{R}^m$ the following hold:

1. $\hat{x}$ *is primal feasible for* $\hat{P}$ *within an absolute tolerance* $\varepsilon_P \geqslant 0$ *if and only if* $x^* + \frac{1}{\Delta_P}\hat{x}$ *is primal feasible for* P *within* ε_P/Δ_P.

2. $\hat{y}$ *is dual feasible for* $\hat{P}$ *within an absolute tolerance* $\varepsilon_D \geqslant 0$ *if and only if* $y^* + \frac{1}{\Delta_D}\hat{y}$ *is dual feasible for* P *within* ε_D/Δ_D.

3. $\hat{x}, \hat{y}$ *violate complementary slackness for* $\hat{P}$ *by at most* $\varepsilon_S \geqslant 0$ *if and only if* $x^* + \frac{1}{\Delta_P}\hat{x}, y^* + \frac{1}{\Delta_D}\hat{y}$ *violate complementary slackness for* P *by at most* $\varepsilon_S/(\Delta_P\Delta_D)$.

4. $\hat{x}, \hat{y}$ *is an optimal primal–dual solution for* $\hat{P}$ *if and only if* $x^* + \frac{1}{\Delta_P}\hat{x}, y^* + \frac{1}{\Delta_D}\hat{y}$ *is optimal for* P.

5. $\hat{x}, \hat{y}$ *is a basic primal–dual solution of* $\hat{P}$ *associated with basis* $\mathcal{B}$ *if and only if* $x^* + \frac{1}{\Delta_P}\hat{x}, y^* + \frac{1}{\Delta_D}\hat{y}$ *is a basic primal–dual solution for* P *associated with basis* $\mathcal{B}$.

Proof. For primal feasibility, point 1, we must check that the violation of variable bounds and equality constraints is simply scaled by $1/\Delta_P$:

$$\left(x^* + \frac{1}{\Delta_P}\hat{x}\right) - \ell = \frac{1}{\Delta_P}(\hat{x} - \Delta_P(\ell - x^*)) = \frac{1}{\Delta_P}(\hat{x} - \Delta_P\hat{\ell})$$

and

$$A\left(x^* + \frac{1}{\Delta_P}\hat{x}\right) - b = \frac{1}{\Delta_P}(\Delta_P A x^* + A\hat{x} - \Delta_P b) = \frac{1}{\Delta_P}(A\hat{x} - \Delta_P\hat{b}).$$

For dual feasibility, point 2, we check the reduced costs,

$$c - A^\mathsf{T}\left(y^* + \frac{1}{\Delta_D}\hat{y}\right) = \frac{1}{\Delta_D}(\Delta_D c - \Delta_D A^\mathsf{T} y^* - A^\mathsf{T}\hat{y}) = \frac{1}{\Delta_D}(\Delta_D\hat{c} - A^\mathsf{T}\hat{y}).$$

Using this, point 3 on complementary slackness follows from the definition of the duality gap via

$$\gamma_P\left(x^* + \frac{1}{\Delta_P}\hat{x}, y^* + \frac{1}{\Delta_D}\hat{y}\right) = \left(\left(x^* + \frac{1}{\Delta_P}\hat{x}\right) - \ell\right)^\mathsf{T}\left(c - A^\mathsf{T}\left(y^* + \frac{1}{\Delta_D}\hat{y}\right)\right)$$

$$= \frac{1}{\Delta_P\Delta_D}(\hat{x} - \Delta_P\hat{\ell})^\mathsf{T}(\Delta_D\hat{c} - A^\mathsf{T}\hat{y})$$

$$= \gamma_{\hat{P}}(\hat{x}, \hat{y})/(\Delta_P\Delta_D).$$

where γ_P and $\gamma_{\hat{P}}$ represent the duality gaps of the problems P and $\hat{P}$, respectively. And since a solution is optimal if and only if it is primal and dual feasible and satisfies the complementary slackness conditions, these first points entail point 4.

Finally, a solution is basic if there is a regular basis $\mathcal{B}$ such that the nonbasic variables—the variables with index $i \notin \mathcal{B}$—are at their bounds and the basic variables have zero reduced cost. In the following equivalence, the left-hand equations are the conditions for $\hat{x}, \hat{y}$ to solve $(\hat{P})$, while the right-hand equations are the conditions for $x^* + \frac{1}{\Delta_P}\hat{x}, y^* + \frac{1}{\Delta_D}\hat{y}$ with respect to the same basis for (P):

$$\hat{x}_i = \Delta_P \ell_i \Leftrightarrow x_i^* + \frac{\hat{x}_i}{\Delta_P} = \ell_i \text{ for all } i \notin \mathcal{B},$$

$$A_{.\mathcal{B}}^\mathsf{T}\hat{y} = \Delta_D \hat{c}_{\mathcal{B}} \Leftrightarrow A_{.\mathcal{B}}^\mathsf{T}\left(y^* + \frac{1}{\Delta_D}\hat{y}\right) = c_{\mathcal{B}}.$$

This proves point 5. □

This theorem can be viewed in two complementary ways. From a numerical perspective, $(\hat{P})$ is formed by replacing the right-hand side b, the bounds on the variables ℓ, and the objective function vector c by the corresponding residual errors in an approximate solution x^*, y^*. This is similar to a refinement step in Algorithm 2.1, but additionally the residual errors are magnified by the scaling factors Δ_P and Δ_D. Points 1 to 3 state the improved accuracy of the corrected solution $x^* + \frac{1}{\Delta_P}\hat{x}, y^* + \frac{1}{\Delta_D}\hat{y}$ if $\hat{x}, \hat{y}$ is an approximate solution to $(\hat{P})$.

Geometrically, problem $(\hat{P})$ is the result of applying the affine transformation $x \mapsto \Delta_P(x - x^*)$ to the primal and $y \mapsto \Delta_D(y - y^*)$ to the dual solution space of (P). Theorem 2.5 summarizes the straightforward one-to-one correspondence between solutions of the original problem (P) and the transformed problem $(\hat{P})$. Graphically, the primal transformation zooms in on the reference solution x^*—by first shifting the reference solution x^* to the origin, then scaling the problem by a factor of Δ_P. The dual transformation tilts the objective function to become the vector of reduced cost. This is illustrated by the following examples.

Example 2.6 (Primal LP Refinement). Consider the LP on two variables

$$\min\{x_1 + x_2 \mid 2x_1 + x_2 \geqslant 3, x_1 + 2x_2 \geqslant 3, x_1 + x_2 \geqslant 2 + 10^{-6}, x_1, x_2 \geqslant 0\}$$

with an approximate solution $x^* = (1,1)^\mathsf{T}$ as depicted in the Figure 2.1a. (We use the inequality form without slack variables here for better visualization.) Note that the constraint $x_1 + x_2 \geqslant 2 + 10^{-6}$ is indistinguishable from $x_1 + x_2 \geqslant 2$ and the tiny violation of 10^{-6} is invisible on this scale.

Shifting the problem such that the reference solution is centered at the origin gives the shifted LP in Figure 2.1b. After scaling the primal space by $\Delta_P = 10^6$, we obtain the transformed problem

$$\min\{x_1 + x_2 \mid 2x_1 + x_2 \geqslant 0, x_1 + 2x_2 \geqslant 0, x_1 + x_2 \geqslant 1, x_1, x_2 \geqslant -10^6\}$$

in Figure 2.1c, denoted by $(\hat{P})$ in Theorem 2.5.

Here, the infeasibility of the initial solution is apparent. An LP solver might return the solution $\hat{x} = (-1, 2)^\mathsf{T}$ instead, which corresponds to the corrected, exactly feasibly solution $x^* + \frac{1}{\Delta_P}\hat{x} = (1 - 10^{-6}, 1 + 2 \cdot 10^{-6})$ for the original problem.

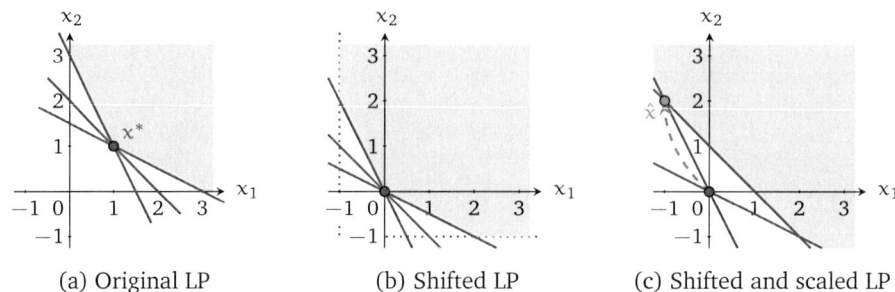

(a) Original LP   (b) Shifted LP   (c) Shifted and scaled LP

Figure 2.1: Two-variable example for primal LP refinement

Example 2.7 (Dual LP Refinement). Consider the LP on two variables

$$\min\{x_1 + (1 - 10^{-6})x_2 \mid x_1 + x_2 = 2, x_1, x_2 \geqslant 0\}$$

with an approximate solution $x^* = (2, 0)^\mathsf{T}$ as shown in Figure 2.2a. Note that any point on the line $x_1 + x_2 = 2$ looks optimal because on this scale the objective function is not distinguishable from $x_1 + x_2$. With dual multiplier $y^* = 1$ the reduced cost vector is $\hat{c} = (0, -10^{-6})^\mathsf{T}$. The solution x^*, y^* satisfies complementary slackness, but is dual infeasible and slightly suboptimal.

After replacing the objective function with the reduced cost vector and scaling it by $\Delta_\mathrm{D} = 10^6$, we obtain the transformed LP with objective function $-x_2$ in Figure 2.2b. Now, the initial solution is seen to be clearly suboptimal and any LP solver should return the solution $\hat{x} = (2, 0)^\mathsf{T}$ instead, which—because we did not transform the primal in this example—is already the corrected, now optimal solution for the original problem.

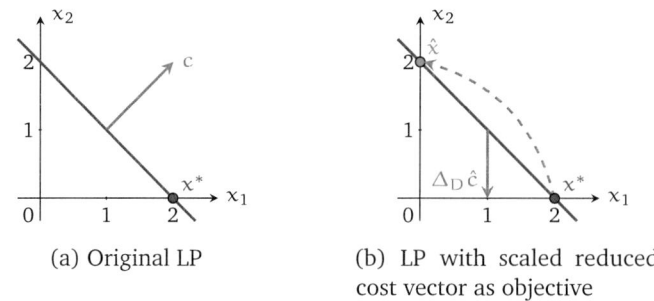

(a) Original LP   (b) LP with scaled reduced
                          cost vector as objective

Figure 2.2: Two-variable example for dual LP refinement

Alternatively, we can view this dual refinement as a primal-refinement step on the one-dimensional dual LP

$$\max\{2y \mid y \leqslant 1, y \leqslant 1 - 10^{-6}\}.$$

Shifting by the approximate dual solution $y^* = 1$ and scaling by $\Delta_D = 10^6$ yields the transformed problem

$$\max\{2y \mid y \leqslant 0, y \leqslant -1\}.$$

This gives the dual corrector $\hat{y} = -1$ and the corrected dual solution $y^* + \frac{1}{\Delta_D}\hat{y} = 1 - 10^{-6}$.

Iterating this refinement step gives the scheme outlined in Algorithm 2.2. First, the LP is solved approximately, producing an initial primal–dual solution x_k, y_k for $k = 1$. Then, the primal and dual residual errors are computed and used to check whether termination tolerances for primal and dual feasibility and complementary slackness have been reached. If not, then the transformed problem $(\hat{P})$ is set up as in Theorem 2.5. The scaling factors are chosen as the inverse of the maximum primal and dual violations in order to normalize right-hand side, lower bounds, and objective function vector. Additionally, we limit the increase of the scaling factors from round to round by the incremental scaling factor α to ensure that we do not scale by infinity if one of the violations drops to zero. Eventually, the transformed LP is solved approximately to obtain a solution $\hat{x}, \hat{y}$, which is used to refine the accuracy of the candidate solution x_k, y_k. This process is repeated until the required accuracy is reached. All operations are performed in exact rational arithmetic unless otherwise noted.

For clarity of presentation, the basic scheme above only considers optimal, i.e., primal and dual feasible LPs. In Section 2.4, we will give an extension that handles also potentially infeasible or unbounded LPs whose status is a priori unknown.

To our knowledge, such an iterative refinement algorithm for linear programming has not been described in the literature. However, we want to mention the presentation of Saunders and Tenenblat (2006) on warm starting interior point methods for convex quadratic programs via a "zoom strategy" that also shifts the problem and scales quadratic terms in the objective function. Furthermore, we believe that some aspects of our approach might have been used in software packages, although most likely not with extended precision or rational arithmetic. In particular, we have heard that some interior point solvers may have experimented with the idea of replacing the objective function of an LP by its reduced cost vector and resolving the problem to improve some of its numerical properties—this would correspond to performing a single dual-refinement step (Ladányi, 2011).

Remark 2.8 (Importance of Scaling). Classical iterative refinement for linear systems as studied by Wilkinson (1963) does not scale the residual errors on the right-hand side. Instead it exploits the fact that floating-point arithmetic is more accurate close to zero and so solving the linear system in floating-point naturally yields an error that is relative in magnitude to the right-hand side. This does not hold for linear programming since LP solvers work with fixed, absolute tolerances. If we did not scale the right-hand side, bounds, and reduced-cost values in the objective function of the transformed problem $\hat{P}$, a standard floating-point LP solver would consider the zero solution optimal within its tolerances. This would result in a zero corrector solution $\hat{x}, \hat{y}$ and no increase in precision. Furthermore, we need scaling for the

---

Algorithm 2.2: Iterative Refinement for a Primal and Dual Feasible LP

---

input: rational, primal and dual feas. LP $\min\{c^\mathsf{T}x \mid Ax = b, x \geq \ell\}$, termination
           tolerances $\varepsilon_\mathrm{P}, \varepsilon_\mathrm{D}, \varepsilon_\mathrm{S} > 0$

params: incremental scaling limit $\alpha > 1$

output: primal–dual solution $x^* \in \mathbb{Q}^n, y^* \in \mathbb{Q}^m$

1 **begin**

2     $\Delta_{\mathrm{P},1} \leftarrow 1, \Delta_{\mathrm{D},1} \leftarrow 1$ `/* initial solve */`

3   get $\bar{A}, \bar{b}, \bar{\ell}, \bar{c} \approx A, b, \ell, c$ in precision of the LP solver

4   solve $\min\{\bar{c}^\mathsf{T}x \mid \bar{A}x = \bar{b}, x \geq \bar{\ell}\}$ approximately

5     $x_1, y_1 \leftarrow$ approximate primal–dual solution returned

6   **for** $k \leftarrow 1, 2, \dots$ **do**   `/* refinement loop */`

7         $\hat{b} \leftarrow b - Ax_k$ `/* 1. compute violations */`

8         $\hat{\ell} \leftarrow \ell - x_k$

9         $\hat{c} \leftarrow c - A^\mathsf{T}y_k$

                                                      `/* 2. check termination */`

10        $\delta_{\mathrm{P},k} \leftarrow \max\{\max_{j=1,\dots,m} |\hat{b}_j|, \max_{i=1,\dots,n} \hat{\ell}_i\}$

11        $\delta_{\mathrm{D},k} \leftarrow \max\{0, \max\{-\hat{c}_i \mid i = 1, \dots, n\}\}$

12        $\delta_{\mathrm{S},k} \leftarrow |\sum_{i=1,\dots,n} -\hat{\ell}_i\hat{c}_i|$

13   **if** $\delta_{\mathrm{P},k} \leq \varepsilon_\mathrm{P}$ and $\delta_{\mathrm{D},k} \leq \varepsilon_\mathrm{D}$ and $\delta_{\mathrm{S},k} \leq \varepsilon_\mathrm{S}$ **then**

14   **return** $x^* \leftarrow x_k, y^* \leftarrow y_k$

                                   `/* 3. solve transformed problem */`

15        $\Delta_{\mathrm{P},k+1} \leftarrow 1/\max\{\delta_{\mathrm{P},k}, (\alpha\Delta_{\mathrm{P},k})^{-1}\}$

16        $\Delta_{\mathrm{D},k+1} \leftarrow 1/\max\{\delta_{\mathrm{D},k}, (\alpha\Delta_{\mathrm{D},k})^{-1}\}$

17   get $\bar{b}, \bar{\ell}, \bar{c} \approx \Delta_{\mathrm{P},k+1}\hat{b}, \Delta_{\mathrm{P},k+1}\hat{\ell}, \Delta_{\mathrm{D},k+1}\hat{c}$ in precision of the LP solver

18   solve $\min\{\bar{c}^\mathsf{T}x \mid \bar{A}x = \bar{b}, x \geq \bar{\ell}\}$ approximately

19        $\hat{x}, \hat{y} \leftarrow$ approximate primal–dual solution returned

                                      `/* 4. perform correction */`

20        $x_{k+1} \leftarrow x_k + \frac{1}{\Delta_{\mathrm{P},k+1}}\hat{x}$

21        $y_{k+1} \leftarrow y_k + \frac{1}{\Delta_{\mathrm{D},k+1}}\hat{y}$

---

same reason as for the iterative refinement scheme for linear systems of Wan (2006): because we want to compute solutions that have a higher precision than can be represented in the precision of the floating-point solver.

In the following we will discuss further details of Algorithm 2.2 that concern convergence properties, the role of the underlying LP solver, the arithmetic precision, and the treatment of general bounds on the variables and inequality constraints.

2.2.2 Convergence

In the iterative refinement scheme of Algorithm 2.2, the approximate LP solver is treated as a black-box oracle. This permits an implementation where the LP solver is accessed through an interface and has the advantage that it allows substitution whenever an application benefits from a specific LP algorithm. The following basic assumption suffices to obtain a sequence of increasingly accurate solutions.

Assumption 2.9 (Accuracy of the LP Oracle). *There exist constants ε, $0 \leqslant \varepsilon < 1$, and $\sigma \geqslant 0$ such that for all $A \in \mathbb{R}^{m \times n}$, $b \in \mathbb{R}^m$, and $c, \ell \in \mathbb{R}^n$ for which*

$$\min\{c^\mathsf{T}x \mid Ax = b, x \geqslant \ell\} \tag{2.2}$$

is primal and dual feasible, the LP solver returns an approximate primal–dual solution $\bar{x} \in \mathbb{Q}^n, \bar{y} \in \mathbb{Q}^m$ that satisfies

1. $\|A\bar{x} - b\|_\infty \leqslant \varepsilon$,

2. $\bar{x} \geqslant \ell - \varepsilon \mathbb{1}$,

3. $c - A^\mathsf{T}\bar{y} \geqslant -\varepsilon \mathbb{1}$, *and*

4. $|\gamma(\bar{x}, \bar{y})| \leqslant \sigma$

when it is given the LP

$$\min\{\bar{c}^\mathsf{T}x \mid \bar{A}x = \bar{b}, x \geqslant \bar{\ell}\} \tag{$\bar{\mathrm{P}}$}$$

where $\bar{A} \in \mathbb{Q}^{m \times n}$, $\bar{c}, \bar{\ell} \in \mathbb{Q}^n$, and $\bar{b} \in \mathbb{Q}^m$ are A, c, ℓ, and b rounded to the working precision of the LP solver.

This assumption can be combined with Theorem 2.5 to give the following simple convergence result.

Corollary 2.10 (Convergence of Iterative LP Refinement). *Suppose we are given a primal and dual feasible LP as in (2.2) and an LP oracle which staisfies Assumption 2.9 with constants ε and σ. Let $x_k, y_k, \Delta_{P,k},$ and $\Delta_{D,k}$, $k = 1, 2, \ldots$, be the sequences of primal–dual solutions and scaling factors produced by Algorithm 2.2 with incremental scaling limit $\alpha > 1$, and let $\tilde{\varepsilon} := \max\{\varepsilon, 1/\alpha\}$. Then for all k,*

1. $\Delta_{P,k}, \Delta_{D,k} \geqslant 1/\tilde{\varepsilon}^{k-1}$,

2. $\|Ax_k - b\|_\infty \leqslant \tilde{\varepsilon}^k$,

3. $x_k - \ell \geqslant -\tilde{\varepsilon}^k \mathbb{1}$,

4. $c - A^\mathsf{T}y_k \geqslant -\tilde{\varepsilon}^k \mathbb{1}$, *and*

5. $|\gamma(x_k, y_k)| \leqslant \sigma\tilde{\varepsilon}^{2(k-1)}$.

Hence, Algorithm 2.2 terminates after at most

$$\max\{\log(\varepsilon_P)/\log(\tilde{\varepsilon}), \log(\varepsilon_D)/\log(\tilde{\varepsilon}), \log(\varepsilon_S\varepsilon/\sigma)/2\log(\tilde{\varepsilon})\} \tag{2.10}$$

approximate LP solves.

Proof. We prove all points together by induction over k. For k = 1 they hold trivially.

Consider $k + 1$ for $k \geqslant 1$. Because points 2–4 hold for x_k, y_k, their violations satisfy $\delta_{P,k}, \delta_{D,k} \leqslant \tilde{\varepsilon}^k$. Because point 1 holds for k, $\alpha \Delta_{P,k} \geqslant \alpha/\tilde{\varepsilon}^{k-1} \geqslant 1/\tilde{\varepsilon}^k$; analogously, $\alpha \Delta_{D,k} \geqslant 1/\tilde{\varepsilon}^k$. Hence, in lines 15 and 16 of the algorithm we have

$$\Delta_{P,k+1}, \Delta_{D,k+1} \geqslant 1/\tilde{\varepsilon}^k \tag{2.11}$$

proving point 1.

If we let $x^* = x_k$, $y^* = y_k$, $\Delta_P = \Delta_{P,k+1}$, and $\Delta_D = \Delta_{D,k+1}$ in Theorem 2.5, then $(\hat{P})$ is the exact shifted and scaled LP before rounding it to the working precision of the LP solver. $(\hat{P})$ is primal and dual feasible because it is simply an affine transformation of the original LP (P) (Theorem 2.5).

By Assumption 2.9, the LP solver returns $\hat{x}, \hat{y}$ that are primal and dual feasible for $(\hat{P})$ within absolute tolerance $\varepsilon \leqslant \tilde{\varepsilon}$. By Theorem 2.5, the corrected solution x_{k+1}, y_{k+1} violates primal and dual feasibility for the original LP by at most $\tilde{\varepsilon}/\Delta_{P,k+1}$ and $\tilde{\varepsilon}/\Delta_{D,k+1}$, respectively. By (2.11), this is at most $\tilde{\varepsilon}^{k+1}$, proving points 2, 3, and 4 for $k + 1$.

Finally, by Assumption 2.9, the violation of complementary slackness $|\gamma(\hat{x}, \hat{y})|$ in $(\hat{P})$ is at most σ. Using again Theorem 2.5 with $\varepsilon_S = \sigma$ we get

$$|\gamma(x_{k+1}, y_{k+1})| \leqslant \sigma/(\Delta_{P,k+1}\Delta_{D,k+1}) \overset{(2.11)}{\leqslant} \sigma\tilde{\varepsilon}^{2k}, \tag{2.12}$$

proving point 5 for $k + 1$.

Assuming slowest convergence gives

$$\tilde{\varepsilon}^k \leqslant \varepsilon_P, \tilde{\varepsilon}^k \leqslant \varepsilon_D, \text{ and } \sigma\tilde{\varepsilon}^{2k} \leqslant \varepsilon_S, \tag{2.13}$$

in the termination conditions in line 13 of the Algorithm 2.2, which is equivalent to the stated bound on the number of refinement rounds. □

Several remarks on the reasonableness of Assumption 2.9 are in order. As Klotz (2014) points out for the CPLEX[33] LP code, state-of-the-art LP solvers typically use an absolute definition for their tolerance requirements. However, because limited floating-point precision is by construction relative, an LP solver based only on floating-point computation will in general not be able to return solutions within absolute tolerances—certainly not the fast LP solvers we have in mind for practical applications. Otherwise, this would permit the following, much simpler approach. In Theorem 2.5, choose $x^*, y^* = 0$ and scaling factors $\Delta_P = \Delta_D = N$ arbitrarily large and solve $(\hat{P})$ for a solution $\bar{x}, \bar{y}$ as guaranteed by Assumption 2.9. Then $\bar{x}/N, \bar{y}/N$ would violate primal and dual feasibility by at most ε/N and complementary slackness by at most σ/N^2. One refinement step would suffice to reach arbitrary precision.

Because of this, the primal and dual scaling factors in Algorithm 2.2 are limited by the inverse of the maximum primal and dual violations, respectively. As a result, the largest entries in the right-hand side, lower bounds, and objective function vector

---

[33] IBM.   CPLEX   Optimizer.   http://www.ibm.com/software/commerce/optimization/
    cplex-optimizer/.

of the transformed LP that correspond to violations of primal and dual feasibility become at most one in absolute value. For these variables and constraints, a relative tolerance requirement implies the absolute tolerance requirement of Assumption 2.9. However, note that when in x_k, y_k the lower bound is already satisfied for a variable then its lower bound in the transformed LP may have a large absolute value; and if its reduced cost is already nonnegative then the same holds for the transformed objective function coefficient.

Although there is no guarantee that a floating-point solver will produce solutions that are accurate to within an absolute tolerance, with our scaling strategy we expect that a modern LP solver will yield satisfactory results for most LPs in practice. However, in some cases LPs may be so poorly conditioned that the floating-point LP solver produces meaningless results. In such a case performing extended-precision computations within the solver may be necessary. A minor modification could be done to Algorithm 2.2 to incrementally boost the working precision of the LP solver when needed similarly as employed by Applegate et al. (2007b), confer Section 3.1.3 later.

Note that if the floating-point solver encounters numerical difficulties, it may not only return a solution with large violations, but even incorrectly conclude infeasibility or unboundedness. Section 2.4 describes a robust implementation of the iterative refinement scheme that, to the extent possible, tries to cope with such violations of Assumption 2.9.

2.2.3 Oracle-Polynomial Running Time

According to Corollary 2.10, the number of calls to the LP oracle until reaching the termination tolerances ε_P, ε_D, and ε_S is linear in their encoding length. It is also easy to check that the number of arithmetic operations executed per iteration is polynomial in the input size: it is dominated by the matrix vector products in lines 7 and 9. The topic of this section is to show that also the encoding length of the numbers computed during the algorithm is bounded by a polynomial in the input size.

From this it follows that iterative refinement for linear programming has polynomial running time as an oracle algorithm in the sense that it can be implemented on an oracle Turing machine such that it terminates after a polynomial number of steps in the encoding length of the input, counting calls to the oracle as one step. For a compact introduction to the complexity of oracle algorithms we refer to Grötschel, Lovász, and Schrijver (1988, Sec. 1.2).

The crucial point here are the correction steps in lines 20 and 21, which read $x_{k+1} \leftarrow x_k + \frac{1}{\Delta_{P,k+1}} \hat{x}$ and $y_{k+1} \leftarrow y_k + \frac{1}{\Delta_{D,k+1}} \hat{y}$. By the addition, the size of x_{k+1} and y_{k+1} may be as large as $2(\langle x_k \rangle + \langle \Delta_{P,k+1} \rangle \langle \hat{x} \rangle)$ and $2(\langle y_k \rangle + \langle \Delta_{D,k+1} \rangle \langle \hat{y} \rangle)$, respectively, if the numbers involved have arbitrary denominators. A factor of two per refinement round could lead to an exponential growth. This can be avoided if we guarantee that the solutions of the LP oracle and the scaling factors have denominators with small common multiple. It is standard to assume that oracles return answers of polynomial size in the input, see Grötschel, Lovász, and Schrijver (1988, General Assumption 1.2.1). The following assumption goes beyond that and requires bounded floating-point solutions with a common base. It is naturally satisfied by the limited-precision solvers we have in mind.

Assumption 2.11 (Output of the LP Oracle). *There exist natural numbers B and C such that each number in the solutions returned by the LP oracle has form n/B^C with $n \in \mathbb{Z}$, $|n| \leqslant C$.*

Correspondingly, we round the primal and dual scaling factor to powers of base B, i.e., lines 15 and 16 in Algorithm 2.2 are implemented as

$$\Delta_{P,k+1} \leftarrow B^{\lceil -\log_B \max\{\delta_{P,k}, (\alpha \Delta_{P,k})^{-1}\}\rceil}$$
$$\Delta_{D,k+1} \leftarrow B^{\lceil -\log_B \max\{\delta_{D,k}, (\alpha \Delta_{D,k})^{-1}\}\rceil} \tag{2.14}$$

Because the scaling factors keep their order of magnitude, this does not affect convergence. Since they are rounded up, the proof of Corollary 2.10 continues to hold. Then we can prove the following.

Theorem 2.12. *The iterative refinement procedure of Algorithm 2.2 with an LP oracle satisfying Assumptions 2.9 and 2.11 and modification (2.14) runs in oracle-polynomial time in the size of the input A, b, ℓ, c, ε_P, ε_D, and ε_S.*

Proof. The initial rounding of the constraint matrix and the computation of the residuals takes $O(nm)$ elementary operations. The computation of the maximum violation and checking of the termination criteria is of order $O(n+m)$. The computation of the scaling factors takes constant effort and the correction of the primal–dual solution vectors at the end of each iteration is of order $O(n + m)$.

It remains to be shown that the intermediate numbers do not grow too much. Let $K = \max\{\log(1/\varepsilon_P), \log(1/\varepsilon_D), \log(1/\varepsilon_S)\}$, then by Corollary 2.10, the maximum number of iterations is linear in K. From (2.14) it follows that the scaling factors increase by at most a factor of $\beta := B^{\lceil \log_B \alpha \rceil}$ at each iteration. Hence, the scaling factors $\Delta_{P,k}, \Delta_{D,k}$ are bounded by β^{k-1}, and

$$\max\{\langle \Delta_{P,k} \rangle, \langle \Delta_{D,k} \rangle\} \leqslant 1 + \log_2 \beta^{k-1} \leqslant K \lceil \log_B \alpha \rceil \log_2 B \in O(K).$$

Note that this uses the fact that the scaling factors are integers, because the encoding length of rational numbers cannot be bounded by their magnitude.

Next we show by induction over k that the entries in the refined primal and dual solution vectors x_k, y_k have form $p/(\Delta_{P,k}B^C)$ with $|p| \leqslant C(\beta^k - 1)$. Here B and C are the constants from Assumption 2.11, which guarantees that $\hat{x}_i = \hat{n}_i/B^C$ for $|\hat{n}_i| \leqslant C$. For $k = 1$, Assumption 2.11 ensures that x_1, y_1 have the claimed form. Suppose now that $(x_k)_i = p/(\Delta_{P,k}B^C)$ with $|p| \leqslant C(\beta^k - 1)$, then for $k + 1$,

$$\begin{aligned}
(x_{k+1})_i &= (x_k)_i + \hat{x}_i/\Delta_{P,k+1} \\
&= p/(\Delta_{P,k}B^C) + \hat{n}_i/(\Delta_{P,k+1}B^C) \\
&= (p\tfrac{\Delta_{P,k+1}}{\Delta_{P,k}} + \hat{n}_i)/(\Delta_{P,k+1}B^C)
\end{aligned}$$

and

$$|p\tfrac{\Delta_{P,k+1}}{\Delta_{P,k}} + \hat{n}_i| \leqslant |p|\beta + C$$
$$\leqslant C(\beta^k - 1)\beta + C$$
$$\leqslant C(\beta^{k+1} - 1)$$

because $\beta \geqslant 2$. Hence,

$$\langle x_k \rangle \leqslant \langle C(\beta^k - 1) \rangle + \langle \Delta_{P,k} B^C \rangle$$
$$\leqslant \langle C \rangle + K\langle \beta \rangle + \langle \Delta_{P,k} \rangle + C\langle B \rangle \in O(K).$$

The same holds for y_k.

According to the formulas of Lemma 1.1, the sizes of the entries in $\hat{b}$, $\hat{\ell}$, and $\hat{c}$ are bounded by $4(\langle b \rangle + \langle A \rangle + \langle x_k \rangle)$, $2(\langle \ell \rangle + \langle x_k \rangle)$, and $4(\langle c \rangle + \langle A \rangle + \langle y_k \rangle)$, respectively. Hence, so are $\langle \delta_{P,k} \rangle$ and $\langle \delta_{D,k} \rangle$, and $\langle \delta_{S,k} \rangle \leqslant 2(\langle \hat{\ell} \rangle + \langle \hat{c} \rangle)$. As shown above, the scaling of the vectors in line 17 only increases their size by $O(K)$. To summarize, the sizes of the number appearing during the course of the algorithm are of order $O(K^2 + \langle A, b, \ell, c \rangle K)$, i.e., polynomial in the size of the input. $\qquad\square$

While the above discussion was theoretically motivated, the following section is concerned with the cost of the arithmetic operations in practice.

2.2.4 Arithmetic Precision

Like many iterative refinement procedures for linear systems, Algorithm 2.2 is a mixed-precision procedure. This has the advantage that the most involved part—LP solving—is executed in fast floating-point arithmetic. Expensive rational arithmetic is only used for computing and scaling the new objective function, right-hand side, and bounds of the transformed LP, which amounts to two matrix-vector multiplications and a constant multiple of $n + m$ elementary operations per refinement round. Hence the number of elementary operations performed in rational arithmetic grows linearly with the number of nonzeros of the constraint matrix.

Still, rational arithmetic remains computationally expensive. Compared to floating-point arithmetic, its cost is not constant per operation, but depends on the encoding length of the numbers involved. Hence its cost increases as the encoding length of the corrected solution grows with increased accuracy. In this respect, the technique of rounding the scaling factors to powers of the same base as used by the underlying floating-point solver is also practically important for a fast implementation. Otherwise, the potential exponential growth in the size of the numbers would prevent the refinement from reaching very high levels of accuracy. Similar ideas have been used by Wan (2006) for solving linear systems of equations.

Finally we wish to remark that we use exact rational arithmetic in order to allow the computation of arbitrarily precise solutions. If the goal is merely to reach a certain fixed level of accuracy then it may be possible to replace rational by (sufficiently high) extended-precision arithmetic. This is potentially faster, but may at the same time require a more careful analysis and implementation.

2.3 Handling LPs of General Form

This section describes the technical adaptations necessary to handle LPs in the general form that is used in most practical implementations of LP solvers such as in the SoPlex LP solver, which is the basis for our computational study presented in Section 2.5.

2.3.1 Variables with General Bounds

For clarity of presentation we have so far only considered lower bounds on the variables. Variables with additional upper bounds, i.e., $\ell \leqslant x \leqslant u, \ell, u \in \mathbb{R}^n$, require only a few modifications in Algorithm 2.2. For the primal refinement we need to compute $\hat{u} \leftarrow u - x_k$ in line 8 and consider it in the calculation of the primal violation line 10,

$$\delta_{P,k} \leftarrow \max\{ \max_{j=1,\ldots,m} |\hat{b}_j|, \max_{i=1,\ldots,n} \hat{\ell}_i, \max_{i=1,\ldots,n} -\hat{u}_i \}.$$

In line 17, we also need to compute and round the transformed upper bound vector $\bar{u} \leftarrow \Delta_{P,k+1}\hat{u}$.

The dual refinement is only affected in the calculation of the dual scaling factor. With upper bounds, the dual LP contains separate reduced cost multipliers for the lower and the upper bound of each variable and the dual LP (2.4) reads

$$\max\{b^\mathsf{T} y + \ell^\mathsf{T} z^\ell - u^\mathsf{T} z^u \mid A^\mathsf{T} y + z^\ell - z^u = c, z^\ell, z^u \geqslant 0\}. \tag{2.15}$$

The reduced cost vector $z = c - A^\mathsf{T} y$ is split into $z = z^\ell - z^u$, where z^ℓ is associated with the lower bounds and z^u with the upper bounds. The duality gap (2.7), our measure for the violation of complementary slackness, becomes

$$\gamma(x, y) := (x - \ell)^\mathsf{T} z^\ell + (u - x)^\mathsf{T} z^u. \tag{2.16}$$

If all variables have finite lower and upper bounds then each dual solution y gives a feasible solution to (2.15) if we let z^ℓ be the positive and z^u be the negative part of z. However, this ignores its connection to the primal solution. Suppose some variable x_i with domain $[0, U]$ is tight at its lower bound, $x_i = 0$, but its reduced reduced cost $z_i = -10^{-9}$ is slightly negative. Setting $z_i^\ell = 0$ and $z_i^u = 10^{-9}$ would guarantee dual feasibility, but the variable would contribute $U \cdot 10^{-9}$ to the duality gap.

A more suitable definition of the dual violation in line 11 is obtained if we distribute the reduced cost value not according to its sign, but associate it with the bound which is closest to the primal solution value via

$$\begin{aligned}\delta_{D,k} \leftarrow \max\{0, &\max\{-\hat{c}_i \mid i = 1, \ldots, n, (x_k)_i \leqslant (\ell_i + u_i)/2\}, \\ &\max\{\hat{c}_i \mid i = 1, \ldots, n, (x_k)_i > (\ell_i + u_i)/2\}\}\end{aligned} \tag{2.17}$$

where $\hat{c}$ is the reduced cost vector computed as before in line 9. The according violation of complementary slackness is calculated as

$$\delta_{S,k} \leftarrow \Big| \sum_{i:(x_k)_i \leqslant (\ell_i + u_i)/2} -\hat{\ell}_i \hat{c}_i + \sum_{i:(x_k)_i > (\ell_i + u_i)/2} \hat{u}_i \hat{c}_i \Big| \tag{2.18}$$

in line 12. This definition is also meaningful if one of the bounds is $\pm\infty$. If the variable is free we include the absolute value of its reduced cost in the dual violation and exclude it from the violation of complementary slackness. This has the same effect as splitting the variable into its positive and negative part.

Remark 2.13. If the floating-point solver returns basic solutions, one alternative is to directly use the basis information in the definition of dual feasibility. This is described in more detail in Section 2.5.1 together with our extension of the LP solver SoPlex, which implements the revised simplex algorithm.

2.3.2 Inequality Constraints

For clarity of presentation, we have so far only considered equality constraints. In the case of an LP with ranged rows

$$\min\{c^\mathsf{T}x \mid L \leqslant Ax \leqslant U, \ell \leqslant x \leqslant u\} \tag{2.1}$$

with some $L_i \neq U_i$, the primal refinement only requires a small adjustment of Algorithm 2.2. In line 7, both the left-hand and right-hand side vector has to be shifted via $\hat{L} \leftarrow L - Ax_k$ and $\hat{U} \leftarrow U - Ax_k$, and the computation of the primal violation in line 10 must include $\max_{j=1,\ldots,m} \hat{L}_j$ and $\max_{j=1,\ldots,m} -\hat{U}_j$.

The dual-refinement step, however, does not allow for an equally straightforward generalization. The dual LP (2.4) now reads

$$\max\{L^\mathsf{T}y^L - U^\mathsf{T}y^U + \ell^\mathsf{T}z^\ell - u^\mathsf{T}z^u \mid A^\mathsf{T}(y^L - y^U) + z^\ell - z^u = c,$$
$$y^L, y^U, z^\ell, z^u \geqslant 0\}, \tag{2.19}$$

i.e., the dual vector is split into $y = y^L - y^U$, where y^L is associated with the left-hand sides and y^U with the right-hand sides. If entries in L and U are $\pm\infty$, the corresponding dual variables are left out. The duality gap (2.7) becomes

$$\gamma(x,y) := (x - \ell)^\mathsf{T}z^\ell + (u - x)^\mathsf{T}z^u + (Ax - L)^\mathsf{T}y^L + (U - Ax)^\mathsf{T}y^U. \tag{2.20}$$

Similar to the reduced costs for variables with lower and upper bounds in Section 2.3.1, the dual multipliers should not be distributed to y^L and y^U according to their sign, but according to the row activity of the primal solution, i.e., the dual violation in line 11 should be computed as

$$\begin{aligned}
\delta_{D,k} \leftarrow \max\{0, &\max\{-\hat{c}_i \mid i = 1,\ldots,n, (x_k)_i \leqslant (\ell_i + u_i)/2\}, \\
&\max\{\hat{c}_i \mid i = 1,\ldots,n, (x_k)_i > (\ell_i + u_i)/2\}, \\
&\max\{-(y_k)_j \mid j = 1,\ldots,m, A_{j\cdot}x \leqslant (U_j + L_j)/2\}, \\
&\max\{(y_k)_j \mid j = 1,\ldots,m, A_{j\cdot}x > (U_j + L_j)/2\}\},
\end{aligned} \tag{2.21}$$

and the violation of complementary slackness in line 12 as

$$\begin{aligned}
\delta_{S,k} \leftarrow \Bigg| &\sum_{i:(x_k)_i \leqslant (\ell_i + u_i)/2} -\hat{\ell}_i \hat{c}_i + \sum_{i:(x_k)_i > (\ell_i + u_i)/2} \hat{u}_i \hat{c}_i \\
&+ \sum_{j:A_{j\cdot}x \leqslant (U_j + L_j)/2} -\hat{L}_j(y_k)_j + \sum_{j:A_{j\cdot}x > (U_j + L_j)/2} \hat{U}_j(y_k)_j \Bigg|.
\end{aligned} \tag{2.22}$$

These modifications would allow us to execute Algorithm 2.2 formally, but further adjustments are necessary for convergence. In the case of equality constraints, $Ax = b$, replacing the objective function vector by the reduced cost vector $c - A^Ty_k$ is an equivalent transformation because it amounts to subtracting the constant offset $(A^Ty_k)^Tx = (Ax)^Ty_k = b^Ty_k$ from the objective function value of any primal solution x. This argument falls short if the row activity Ax may vary between L and U. In order to compensate for this, the activity of each row must be considered in the objective function with its dual multiplier as objective coefficient.

The naïve solution is to call Algorithm 2.2 for the reformulated LP with slack variables

$$\min\{c^Tx + 0^Ts \mid Ax - s = 0, \ell \leqslant x \leqslant u, L \leqslant s \leqslant U\}, \qquad (2.23)$$

where for notational simplicity we write slacks also for equality constraints. While dual feasibility is identical for (2.23) and (2.1) this does not hold for primal feasibility. If x, s is primal feasible for (2.23) with maximum violation of ε_P then x is primal feasible for (2.1) with maximum violation of $2\varepsilon_P$ because the violation of $L \leqslant s \leqslant U$ and $Ax - s = 0$ may add up. Additionally, because for an approximate primal solution the constraints do not hold exactly, i.e., only $Ax \approx s$ we should check complementary slackness in terms of (2.20) to meet the termination conditions in the original LP.

We obtain a slightly more involved implementation if we note that it suffices to introduce slack variables in the floating-point solver and work with the original LP in the main algorithm. The first approximate LP solve in line 4 can be performed without slack variables. During the refinement loop in line 18 we then solve

$$\min\{\bar{c}^Tx + \bar{y}^Ts \mid \bar{A}x - s = 0, \bar{\ell} \leqslant x \leqslant \bar{u}, \bar{L} \leqslant s \leqslant \bar{U}\} \qquad (2.24)$$

where $\bar{y} \approx \Delta_{D,k+1}y_k$, $\bar{L} \approx \Delta_{P,k+1}\hat{L}$ and $\bar{U} \approx \Delta_{P,k+1}\hat{U}$. The approximate solution values of the slack variables are ignored. As before, only $\hat{x}, \hat{y}$ are used for the subsequent correction.

Besides the tiny advantage this approach gives in avoiding overhead in the data structures and directly checking the termination conditions of the original problem (for instance, we do not need to reduce ε_P by two, a priori), there also may be performance improvements. If the floating-point solver already uses slack variables internally and provides access to set their objective coefficients to a value different than zero, or if it has another more efficient way of solving the slack formulation, this becomes possible. We exploit this in the implementation described in Section 2.5.1. We conclude this section with a discussion of an alternative, overly simplistic strategy for handling inequality constraints that we have experimented with.

Remark 2.14 (Naïve Treatment of Inequality Constraints). The slack formulation of LP (2.24) could be simplified by substituting and removing the slack variables exactly as a modern LP solver would do in its presolving phase. This would result in

$$\min\{(\bar{c} + \bar{A}^T\bar{y})^Tx \mid \bar{L} \leqslant \bar{A}x \leqslant \bar{U}, \bar{\ell} \leqslant x \leqslant \bar{u}\}, \qquad (2.25)$$

where the objective function vector $\bar{c} + \bar{A}^T\bar{y} \approx \Delta_{D,k+1}c$. From this observation one could think of using a refinement step where the objective function is not replaced by the scaled reduced cost vector, but by $\Delta_{D,k+1}(c - \sum_{j:L_j=u_j} y_j A_{j\cdot}^\top)$, i.e., we only subtract the row vectors corresponding to equality constraints.

The same objective function is obtained when forcing the dual multipliers of all non-equality constraints to zero after the correction step and so in theory Theorem 2.5 could be applied. In practice, however, this algorithm fails to converge because in general Assumption 2.9 does not hold. If ranged rows with large dual multipliers are ignored in the new objective function, it may contain coefficients of large absolute value and a floating-point solver will in general not be able to satisfy an absolute dual feasibility tolerance below one.

If they are considered, i.e., if we do replace the objective function by the scaled reduced cost vector, then as explained above we would need to fix the non-equality constraints with a nonzero dual multiplier to one of their sides. As part of the simplex-based implementation described in Section 2.5.1, we had also experimented with such a scheme that fixes each ranged row with nonzero dual multiplier in the transformed problem, either to its left-hand side if finite and the multiplier is positive or to its right-hand side if finite and the multiplier is negative, and forces infeasible multipliers to zero. This ensures that replacing the objective function by the (scaled) reduced cost vector is an equivalent affine transformation for this heuristically restricted LP.

If the dual multiplier changes sign after the correction step, the fixing was relaxed and the multiplier reset to zero. If the transformed LP becomes infeasible, this is performed for all fixed rows.

For cases when the initial LP solve returned a basis that was almost optimal, the approach worked well. For more complicated cases, however, this strategy seemed to increase the number of LP solves, sometimes drastically, because too many unfixings were necessary.

2.4 Refining Infeasible and Unbounded LPs

In Examples 2.6 and 2.7, we illustrated how the refinement scheme can be viewed as zooming in to an approximate primal and dual feasible solution. For infeasible or unbounded LPs there exists no such reference solution that could be refined. In these cases, our task is to construct a high-precision certificate of the infeasibility or unboundedness. In the following we will describe how this is achieved via auxiliary reformulations of the LP and how it can be integrated into one solving loop. Our presentation will switch between the equality form

$$\min\{c^\top x \mid Ax = b, x \geqslant \ell\} \tag{2.2}$$

and the general form

$$\min\{c^\top x \mid L \leqslant Ax \leqslant U, \ell \leqslant x \leqslant u\} \tag{2.1}$$

of an LP—the first used for clarity of presentation, the second because the technical details are crucial for an implementation within a full-fledged LP solver as described in Section 2.5.

2.4.1 Testing Feasibility

As a consequence of Theorem 2.1, the feasibility of an LP given in equality form (2.2) can be decided by solving the auxiliary LP

$$\max\{(b - A\ell)^\mathsf{T}y \mid A^\mathsf{T}y \leqslant 0, (b - A\ell)^\mathsf{T}y \leqslant 1\}. \tag{2.26}$$

The last inequality on the objective function ensures boundedness, and that if the optimal objective value is nonzero, it is equal to one. Feasible solutions to (2.26) with positive objective value serve as infeasibility certificates to the LP (2.2) and are often referred to as *Farkas proofs*. Because the zero solution is trivially feasible, this LP is primal and dual feasible and iterative refinement can be applied to compute an arbitrarily accurate Farkas proof.

In order to integrate this most seamlessly into one refinement scheme for optimal, infeasible, or unbounded LPs as described below in Section 2.4.4, we will see that it is more suitable to consider the dual of (2.26), which reads

$$\min\{\tau \mid A\xi + (b - A\ell)\tau = (b - A\ell), \xi, \tau \geqslant 0\}. \tag{2.27}$$

Substituting $1 - \tau$ for τ gives the more natural formulation

$$\max\{\tau \mid A\xi - (b - A\ell)\tau = 0, \xi \geqslant 0, \tau \leqslant 1\}, \tag{2.28}$$

which we refer to as the *feasibility LP*. The following lemma summarizes how solving this LP gives either a primal feasible solution or a Farkas proof of infeasibility for (2.2).

Lemma 2.15 (Auxiliary Feasibility LP). *Suppose we are given an LP in equality form (2.2), then the following hold.*

1. *The auxiliary LP (2.28) is primal and dual feasible.*

2. *The original LP (2.2) is feasible if and only if the auxiliary LP (2.28) has an optimal objective value of one.*

3. *If the optimal objective value of (2.28) is less than one and y^* is an optimal dual solution vector for (2.28) (associated to the equality constraints), then y^* is a Farkas proof for the infeasibility of (2.2).*

4. *If (ξ^*, τ^*), $\tau^* > 0$, is an approximate optimal solution of (2.28) that violates primal feasibility by at most ε_P, then $x^* = \frac{1}{\tau^*}\xi^* + \ell$ is a feasible solution for (2.2) within tolerance ε_P/τ^*.*

Proof. 1. The zero vectors are feasible in the primal and the dual of (2.28). 2. A primal solution x^* for (2.2) gives the primal solution $(x^* - \ell, 1)$ for (2.28), and $(\xi^*, 1)$ can be mapped back as $\xi^* + \ell$. 3. The dual of (2.28) is $\min\{z \mid A^\mathsf{T}y \leqslant 0, (b - A\ell)^\mathsf{T}y + z = 1, z \geqslant 0\}$. If its optimal value is less than one then for an optimal dual solution y^*, $(b - A\ell)^\mathsf{T}y^* = 1 - z^* > 0$, but $A^\mathsf{T}y^* \leqslant 0$. 4. $\|A(\frac{1}{\tau^*}\xi^* + \ell) - b\|_\infty = \|A\xi^* - (b - A\ell)\tau^*\|_\infty/\tau^*$ and $\xi^* \geqslant -\varepsilon_P \Rightarrow \frac{1}{\tau^*}\xi^* + \ell \geqslant \ell - \varepsilon_P/\tau^*$. □

As point 4 shows, when applying iterative refinement to the feasibility LP we have to adjust our termination criterion for primal feasibility. When a primal violation of $\delta_{P,k}$ is achieved in Algorithm 2.2, we may terminate if $\tau_k > 0$ and $\delta_{P,k}/\tau_k \leqslant \varepsilon_P$. If $\tau_k \approx 0$, the dual solution gives an approximate Farkas proof. Because a Farkas proof remains valid when multiplied by an arbitrary positive scalar, absolute tolerance requirements are meaningless. We will address this question in the next section.

Remark 2.16 (Feasibility LP in General Form). For an LP given in general form (2.1), the slightly more technical formulation of the feasibility LP becomes

$$\max\{\tau \mid L - At - w \leqslant A\xi - w\tau \leqslant U - At - w,$$
$$\ell - t \leqslant \xi \leqslant u - t, \tau \leqslant 1\} \tag{2.29}$$

with shift vectors $t = t(\ell, u) \in \mathbb{Q}^n$,

$$t(\ell, u)_i := \begin{cases} \ell_i & \text{if } \ell_i > 0, \\ u_i & \text{if } u_i < 0, \\ 0 & \text{otherwise,} \end{cases} \tag{2.30}$$

and $w = w(L, U, \ell, u) \in \mathbb{Q}^m$,

$$w(L, U, \ell, u)_j := \begin{cases} L_j - A_j.t(\ell, u) & \text{if } L_j - A_j.t(\ell, u) > 0, \\ U_j - A_j.t(\ell, u) & \text{if } U_j - A_j.t(\ell, u) < 0, \\ 0 & \text{otherwise,} \end{cases} \tag{2.31}$$

to ensure that the zero solution is primal feasible. In order to minimize the overhead in transforming the original LP into the this form, we do not shift variables that already contain zero within their bounds and we do not homogenize the bounds, since then these would become additional constraints involving the auxiliary variable τ.

This generalized feasibility LP cannot be derived simply by a twofold dualization as was the case for the standard form (2.28), but it can be verified that points 1 to 3 of Lemma 2.15 still hold. Point 4, however, becomes slightly more technical. If (ξ^*, τ^*), $\tau^* \approx 1$, is an approximate optimal solution of the feasibility LP that violates primal feasibility by at most ε_P, then $x^* = \frac{1}{\tau^*}\xi^* + t$ is a feasible solution for the original LP within tolerance $M\frac{\max\{1-\tau^*,0\}}{\tau^*} + \varepsilon_P/\tau^*$, where M is the maximum of $\|\ell - t\|_\infty$, $\|u - t\|_\infty$, $\|L - At - w\|_\infty$, and $\|U - At - w\|_\infty$. The first term goes to zero as τ^* goes to one. If τ is bounded away from one, we will not converge to a feasible solution and have to apply infeasibility detection.

While for (2.28) the only possible optimal values are zero and one, this is not the case for this general form of the feasibility LP. Consider, for instance, the system

$$\{x = 1 + \varepsilon, 0 \leqslant x \leqslant 1\},$$

which is infeasible for $\varepsilon > 0$. The corresponding feasibility LP is

$$\max\{\tau \mid \xi - (1 + \varepsilon)\tau = 0, 0 \leqslant \xi \leqslant 1\}.$$

Its optimal solution is $(\xi, \tau) = (1, 1/(1 + \varepsilon))$ and its objective value $1/(1 + \varepsilon)$ is arbitrarily close to one as ε approaches zero.

2.4.2 Approximate Proofs of Infeasibility

Similar to Theorem 2.1, an exact Farkas proof for an LP of general form (2.1) consists of a vector of dual multipliers $y = y^L - y^U \in \mathbb{Q}^m$ for the rows and a "reduced cost" vector $z = z^\ell - z^u \in \mathbb{Q}^n$ of multipliers for the bound constraints, all $y^L, y^U, z^\ell, z^u \geqslant 0$, such that

$$A^\mathsf{T} y + z = 0 \tag{2.32}$$

and

$$L^\mathsf{T} y^L - U^\mathsf{T} y^U + \ell^\mathsf{T} z^\ell - u^\mathsf{T} z^u > 0 \tag{2.33}$$

hold. An approximate Farkas proof may violate both these conditions. For instance, if some y_j is only slightly positive although constraint j has no left-hand side, i.e., $L_j = -\infty$, then the left-hand side of (2.33) is already at $-\infty$.

This may be corrected by setting such entries in y to zero. Also, the first equation can be enforced by adjusting z to be equal to $-A^\mathsf{T} y$. This, however, may increase the violation of (2.33) or create it in the first place. In particular, if some $\ell_i = -\infty$ or $u_i = \infty$, there is no guarantee that $z_i \leqslant 0$ and $z_i \geqslant 0$, respectively. Even if iterative refinement applied to the feasibility LP produces Farkas proofs with smaller and smaller violations, this is not guaranteed to also produce a reliable certificate of infeasibility.

In the following, we introduce the concept of an *infeasibility box* that does not prove "approximate" infeasibility of the entire LP, but establishes exactly proven infeasibility within restricted bounds. As noted by Neumaier and Shcherbina (2004), even if a Farkas proof is invalid in the classical sense of (2.32) and (2.33), the vector of dual multipliers y by aggregation gives the valid inequality

$$(y^\mathsf{T} A)x \geqslant L^\mathsf{T} y^L - U^\mathsf{T} y^U \tag{2.34}$$

which we call a *Farkas cut*. Its right-hand side is finite if L, U are finite or inconsistent entries in y have been adjusted to zero if necessary. If we can show that (2.34) is violated by all points x with $\ell \leqslant x \leqslant u$, then the LP is proven infeasible. In Neumaier and Shcherbina (2004) it was observed that interval arithmetic can be used to compute a lower bound on the left-hand side of (2.34), and if this is below the right-hand side value it produces a certificate of infeasibility. However, in the case that some entries of ℓ or u are not finite, this approach may fail, even if the approximate Farkas proof is very accurate.

We extend this notion by describing a method that will, given an approximate Farkas proof y, work backwards to determine a domain in which no feasible solution x can exist. We compute the largest value R such that (2.34) is violated by all points x with $\|x\|_\infty < R$, i.e., that the feasible region (of the one-row relaxation given by the Farkas cut) intersected with the infeasibility box $\{x \mid -R\mathbb{1} < x < R\mathbb{1}\}$ is empty.

If the Farkas cut is written in general form $d^\mathsf{T} x \geqslant d_0$, this largest R is computed by the mapping $\rho : \mathbb{Q}^n \times \mathbb{Q} \to \mathbb{Q}_{\geqslant 0} \cup \{\infty\}$ defined as

$$\rho(d, d_0) := \begin{cases} 0 & \text{if } d_0 \leqslant 0, \\ d_0/\|d\|_1 & \text{if } d \neq 0 \text{ and } d_0 > 0, \\ \infty & \text{if } d = 0 \text{ and } d_0 > 0. \end{cases} \tag{2.35}$$

If $\rho(d, d_0) = 0$, then the infeasibility box is empty; $\rho(d, d_0) = \infty$ implies that the full LP is successfully proven infeasible.

This kind of answer is both mathematically sound and helpful to users of an LP solver in practice. It allows them to conclude that feasible solution vectors must have large entries in absolute value, which might or might not be viable for the application at hand. The dimensions of this infeasibility box may be more comprehensible to an end user than, for example, the relative feasibility of a normalized Farkas proof.

Algorithm 2.3 describes the complete procedure for computing the infeasibility box. We assume that all arithmetic operations are executed in rational arithmetic. Throughout the algorithm, $d^T x \geqslant d_0$ is a valid inequality for the LP and $R = \rho(d, d_0)$. We begin by sorting the bounds of the variables as will be needed later, labeled by their negative or positive index to distinguish whether they correspond to a lower or upper bound.

The first loop adjusts the approximate Farkas proof given and gradually builds up the aggregated constraint row by row. Meanwhile, we have a valid inequality and if the domain of x is covered by the interior of the infeasibility box, we can terminate and conclude global infeasibility in line 12. To decide this we use the maximum entry in the bound list.

The second loop tries to incorporate the bounds on the variables in order to increase the size of the infeasibility box further. At this point, $d = A^T y^*$, so $-d$ corresponds to the vector z in (2.32) and (2.33) of dual multipliers for the bound constraints. Suppose d_0 is positive and d contains more than one nonzero entry. For some $d_i < 0$, adding the inequality $-d_i x_i \geqslant -d_i \ell_i$ to $d^T x \geqslant d_0$ increases the value of $\rho(d, d_0)$ if and only if

$$\rho(d - d_i e_i, d_0 - d_i \ell_i) > \rho(d, d_0) \Leftrightarrow \frac{d_0 - d_i \ell_i}{\sum_{q \neq i} |d_q|} > \frac{d_0}{\sum_q |d_q|}$$

$$\Leftrightarrow (d_0 - d_i \ell_i) \sum_q |d_q| > d_0 \sum_{q \neq i} |d_q|$$

$$\Leftrightarrow d_0 |d_i| - d_i \ell_i \|d\|_1 > 0$$

$$\Leftrightarrow -\ell_i < \rho(d, d_0). \tag{2.36}$$

Similarly, if we add an upper bound with multiplier $d_i > 0$, i.e., $-d_i x_i \geqslant -d_i u_i$, we know that

$$\rho(d - d_i e_i, d_0 - d_i u_i) > \rho(d, d_0) \Leftrightarrow u_i < \rho(d, d_0). \tag{2.37}$$

This motivates the special sorting at the beginning of the algorithm that sorts tuples B^k by their first component; these tuples store variable upper bounds or *negations* of variable lower bounds in their first component, and the corresponding variable index in their second component.

At the end of the first loop, we may have $d_0 \leqslant 0$ and $R = 0$. As long as this holds, including positive lower bounds and negative upper bounds, i.e., negative values B_1^k in the first components of the tuples, will increase d_0, although not necessarily R. Once the B_1^k become positive, d_0 starts to decrease; hence, if it is still not positive, we have to terminate without success in line 16. Otherwise, we continue to include bound constraints in increasing order as long as the criteria given by (2.36) and (2.37), are satisfied, respectively. This is checked in line 15. Additionally, as in the first loop, if all bounds are finite then we can conclude global infeasibility as soon as they are strictly within the infesibility box in line 21.

Algorithm 2.3: Infeasibility Box Computation

---

input: rational inequality system $L \leqslant Ax \leqslant U, \ell \leqslant x \leqslant u$,
         approximate Farkas proof $\bar{y} \in \mathbb{Q}^m$
output: $R \geqslant 0$ such that $\|x\|_\infty \geqslant R$ for all feasible solutions x

1 begin
2   **for** $i \leftarrow 1, 2, \ldots, n$ **do**
3          $B^i \leftarrow (-\ell_i, -i), B^{i+n} \leftarrow (u_i, i)$
4   sort $B^1, \ldots, B^{2n}$ by nondecreasing first component
5      $R, y^*, d, d_0 \leftarrow 0$
6   **for** $j \leftarrow 1, 2, \ldots, m$ **do**   /* aggregate rows */
7   **if** $(L_j > -\infty$ and $\bar{y}_j > 0)$ or $(U_j < \infty$ and $\bar{y}_j < 0)$ **then**
8              $y_j^* \leftarrow \bar{y}_j$
9   **if** $y_j^* > 0$ **then** $d_0 \leftarrow d_0 + y_j^* L_j$ **else** $d_0 \leftarrow d_0 + y_j^* U_j$
10             $d \leftarrow d + y_j^* (A_{j.})^\mathsf{T}$
11             $R \leftarrow \rho(d, d_0)$
12   **if** $R > B_1^{2n}$ **then return** $R \leftarrow \infty$

13   **for** $k \leftarrow 1, 2, \ldots, 2n$ **do**   /* include bound constraints */
14         $i \leftarrow B_2^k$
15   **if** $B_1^k \geqslant R$ **then**
16   **return** R
17   **else if** $(i < 0$ and $d_{|i|} < 0)$ or $(i > 0$ and $d_i > 0)$ **then**
18   **if** $i < 0$ **then** $d_0 \leftarrow d_0 - d_{|i|}\ell_{|i|}$ **else** $d_0 \leftarrow d_0 - d_i u_i$
19             $d_{|i|} \leftarrow 0$
20             $R \leftarrow \rho(d, d_0)$
21   **if** $R > B_1^{2n}$ **then return** $R \leftarrow \infty$

22   **return** R

Remark 2.17 (Infeasibility Box with Slack Variables). Instead of skipping ranged rows with infeasible dual multipliers in line 7, we could (implicitly or explicitly) introduce slack variables. Then Algorithm 2.3 would effectively compute a radius R such that $\|x\|_\infty \geqslant R$ *and* $\|Ax\|_\infty \geqslant R$ must hold for any feasible solution. We decided against this in our implementation because a certificate on the bounds of the variables alone is easier to interpret by the user.

It remains to be explained how the computation of an infeasibility box interacts with iterative refinement of the feasibility LP. Intuitively, the size of the infeasibility box should grow as the approximate Farkas proof given as input becomes more accurate. This becomes most clear when we look at the simple case of a system of equations without bounds, $A x = b$, with the feasibility LP $\max\{\tau \mid A\xi - b\tau = 0, \tau \leqslant 1\}$.

Let ξ_k, τ_k, y_k be a sequence of more and more accurate primal–dual solutions produced by Algorithm 2.2. From the proof of Theorem 2.5 it follows that not only the total violation of complementary slackness γ, but also the individual violation with respect to τ goes to zero, i.e.,

$$(1 - b^\mathsf{T} y_k)(1 - \tau_k) \to 0. \tag{2.38}$$

If τ_k does not converge to one, i.e., $\tau_k < C < 1$ (indicating infeasibility of the system), then $b^\mathsf{T} y_k \to 1$. At the same time, the dual violation $\|0 - A^\mathsf{T} y_k\|_\infty$ goes to zero. If we apply Algorithm 2.3 to y_k, then after the first loop we will have $d = A^\mathsf{T} y_k$ and $d_0 = b^\mathsf{T} y_k$. Hence, $R = d_0/\|d\|_\infty$ grows towards infinity with increasingly accurate iterates y_k.

This makes it natural to interleave the infeasibility box computation with the iterative refinement applied to the feasibility LP. The resulting procedure is summarized in Algorithm 2.4 for an LP in general form. We refer to Remark 2.16 for the technical background. The description of the algorithm is rather technical and is not a prerequisite for later sections.

It will typically be called after a normal floating-point solve that claimed infeasibility, hence we assume an approximate Farkas proof as input that is tested right at the beginning with Algorithm 2.3. If the computed radius of the infeasibility box is below the termination threshold, we continue to construct the feasibility LP by shifting bounds and sides. As a first reference point we choose the all zero solution for the primal and the given approximate Farkas proof for the dual solution.

The refinement loop works as in Algorithm 2.2 except when checking termination (which is skipped for the first artificial solution). To check primal feasibility in the original LP we use the formula from Remark 2.16. If tolerance ε_P is not satisfied, this is either because (ξ_k, τ_k) is not yet sufficiently accurate for the feasibility LP or because the optimal value of the feasibility LP is less than one. In the former case we continue with the next refinement step, in the latter case we run Algorithm 2.3 on y_k. The latter is checked by $\tau_k < 1 - \delta_{S,k+1}$ because careful calculation shows that for y_k the left-hand term in (2.33), which needs to be positive, equals $1 - \tau_k - \gamma(\xi_k, \tau_k, y_k)$. It is important that we do not use a fixed threshold to compare τ_k against 1 since—as shown in Remark 2.16—the optimal value of the feasibility LP for infeasible LPs may be arbitrarily close to one.

Remark 2.18 (Infeasibility Box Arithmetic). The infeasibility box algorithm as described uses rational arithmetic in order to obtain provable results. To save computation, R can be updated efficiently instead of recomputing it from scratch. In line 19 of Algorithm 2.3, for instance, only one component of d is removed. Lines 11 and 12 may be dropped. Using the sparsity of the constraint matrix A and of vectors y^*, d is also critical.

Still, rational arithmetic may in some cases be too expensive. Alternatively, Algorithm 2.3 could be implemented using interval arithmetic (Moore et al., 2009), which

---

Algorithm 2.4: Feasibility Test via Iterative Refinement

---

input: rational inequality system $L \leqslant Ax \leqslant U, \ell \leqslant x \leqslant u$,
              approximate Farkas proof $y^* \in \mathbb{Q}^m$,
              termination tolerances $\varepsilon_P, \varepsilon_R > 0$

params: incremental scaling limit $\alpha > 1$

output: primal solution $x^* \in \mathbb{Q}^n$ feasible within ε_P or infeasibility box radius $R \geqslant \varepsilon_R$

1 **begin**

                                                 `/* test initial proof */`

2     $R \leftarrow$ Algorithm 2.3 for $L \leqslant Ax \leqslant U, \ell \leqslant x \leqslant u$ and y^*

3   **if** $R \geqslant \varepsilon_R$ **then return** R

4     $t \leftarrow t(\ell, u), w \leftarrow w(L, U, \ell, u)$ `/* setup feas. LP */`

5     $\tilde{\ell} \leftarrow \ell - t, \tilde{u} \leftarrow u - t, \tilde{L} \leftarrow L - At - w, \tilde{U} \leftarrow U - At - w$

6     $M \leftarrow \|(\tilde{\ell}, \tilde{u}, \tilde{L}, \tilde{U})\|_\infty$

7   get $\bar{A} \approx A$ in precision of the LP solver

8     $\Delta_{P,1} \leftarrow 1, \Delta_{D,1} \leftarrow 1$

9     $(\xi_1, \tau_1, y_1) \leftarrow (0, 1, y^*)$

10   **for** $k \leftarrow 1, 2, \ldots$ **do**   `/* 1. compute residuals */`

11        $\hat{L} \leftarrow \tilde{L} - A\xi_k + w\tau_k, \hat{U} \leftarrow \tilde{U} - A\xi_k + w\tau_k$

12        $\hat{\ell} \leftarrow \tilde{\ell} - \xi_k, \hat{u} \leftarrow \tilde{u} - \xi_k$

13        $\hat{c} \leftarrow -y_k^\mathsf{T} A, \hat{c}_0 \leftarrow 1 - y_k^\mathsf{T} w$

14   compute violations $\delta_{P,k}, \delta_{D,k}, \delta_{S,k}$

15   **if** $k \geqslant 2$ **then**   `/* 2a. test feasibility */`

16   **if** $M\frac{\max\{1-\tau_k, 0\}}{\tau_k} + \delta_{P,k}/\tau_k \leqslant \varepsilon_P$ **then**

17   **return** $x^* \leftarrow \xi_k/\tau_k + t$

18   **else if** $\tau_k < 1 - \delta_{S,k}$ **then**   `/* 2b. test infeasibility */`

19                $R \leftarrow$ Algorithm 2.3 for $L \leqslant Ax \leqslant U, \ell \leqslant x \leqslant u$ and y_k

20   **if** $R \geqslant \varepsilon_R$ **then return** R

                                      `/* 3. solve transformed problem */`

21        $\Delta_{P,k+1} \leftarrow 1/\max\{\delta_{P,k}, (\alpha\Delta_{P,k})^{-1}\}$

22        $\Delta_{D,k+1} \leftarrow 1/\max\{\delta_{D,k}, (\alpha\Delta_{D,k})^{-1}\}$

23   get $\bar{L}, \bar{U}, \bar{\ell}, \bar{u}, \bar{u}_0, \bar{c}, \bar{c}_0 \approx \Delta_{P,k+1}\hat{L}, \Delta_{P,k+1}\hat{U}, \Delta_{P,k+1}\hat{\ell}, \Delta_{P,k+1}\hat{u},$
                       $\Delta_{P,k+1}(1 - \tau_k), \Delta_{D,k+1}\hat{c}, \Delta_{D,k+1}\hat{c}_0$ in prec. of the LP solver

24   solve $\max\{\bar{c}^\mathsf{T}\xi + \bar{c}_0\tau \mid \bar{L} \leqslant \bar{A}x \leqslant \bar{U}, \bar{\ell} \leqslant \xi \leqslant \bar{u}, \tau \leqslant \bar{u}_0\}$

25        $\hat{\xi}, \hat{\tau}, \hat{y} \leftarrow$ approximate primal–dual solution returned

                                        `/* 4. perform correction */`

26        $\xi_{k+1} \leftarrow \xi_k + \frac{1}{\Delta_{P,k+1}}\hat{\xi}, \tau_{k+1} \leftarrow \tau_k + \frac{1}{\Delta_{P,k+1}}\hat{\tau}, y_{k+1} \leftarrow y_k + \frac{1}{\Delta_{D,k+1}}\hat{y}$

---

is faster and still yields proven results, if also a potentially smaller infeasibility box.

2.4.3 Testing Unboundedness

A certificate of primal unboundedness consists of a feasible primal solution vector and an unbounded direction of improving objective function value in the recession cone (2.3). The former can be computed as described above. For the latter we apply the above feasibility test—assuming an LP in general form (2.1)—to the system

$$
\begin{aligned}
A_{j\cdot}v &\leqslant 0 \text{ for all } j \text{ with } U_j < \infty, \\
A_{j\cdot}v &\geqslant 0 \text{ for all } j \text{ with } L_j > -\infty, \\
v &\leqslant 0 \text{ for all } i \text{ with } u_i < \infty, \\
v &\geqslant 0 \text{ for all } i \text{ with } \ell_i > -\infty, \\
c^T v &= -1.
\end{aligned}
\tag{2.39}
$$

Since the primal violation of an unbounded ray is not scale-invariant, we should normalize the violation of primal feasibility by the objective function decrease, i.e., given an approximate unbounded direction v^* we should use the violation of $v^*/|c^T v^*|$ for checking the primal termination tolerance. This guarantees a maximum increase of the primal violation by ε_P as we decrease the objective function by one unit along the ray.

2.4.4 An Integrated Refinement Algorithm

After discussing how iterative refinement is applied to optimal, infeasible, and unbounded LPs separately, this section will show how these techniques are integrated into one algorithm to handle LPs of an a priori unknown status. A crucial property of such an algorithm is its ability to cope with incorrect answers of the underlying floating-point LP solver, since numerically challenging problems for which floating-point solvers return inconsistent results are one of the prime motivations for applying iterative refinement.

Figure 2.3 gives a flowchart of the integrated algorithm. We start by applying Algorithm 2.2 to the given LP. If the floating-point solver returns approximately optimal solutions at each call, the refinement steps will yield increasingly accurate solutions and we return a solution meeting the termination tolerances. If the floating-point solver concludes infeasibility or unboundedness for the initial or one of the transformed LPs, then the refinement loop is interrupted to test this claim. Here we implicitly rely on the fact that the transformed LP is infeasible or unbounded if and only if the original LP is infeasible or unbounded, respectively, which is a consequence of Theorem 2.5.

In the case of floating-point infeasibility, we apply the feasibility test of Algorithm 2.4 as described in Section 2.4.2. Using the primal formulation of the feasibility LP (2.28) instead of the dual formulation keeps the overhead of transforming the original LP low. We only need to add one column for the auxiliary variable τ to the constraint matrix, then modify the objective function and some of the bounds and sides. The feasibility test either concludes infeasibility or results in an approximately feasible solution. In the latter case we reject the floating-point solver's claim of infeasibility and start the iterative refinement algorithm again for the original LP; running the floating-point solver with a variety of settings, we hope that one of them succeeds in this second try.

In the case that the floating-point solver claims unboundedness, we first apply iterative refinement to (2.39) to compute a primal ray. If this fails we restart the refinement of the original LP with modified floating-point parameters. If it succeeds we continue testing feasibility. If we obtain a primal feasible solution we conclude unboundedness. Otherwise, we return (primal and dual) infeasibility.

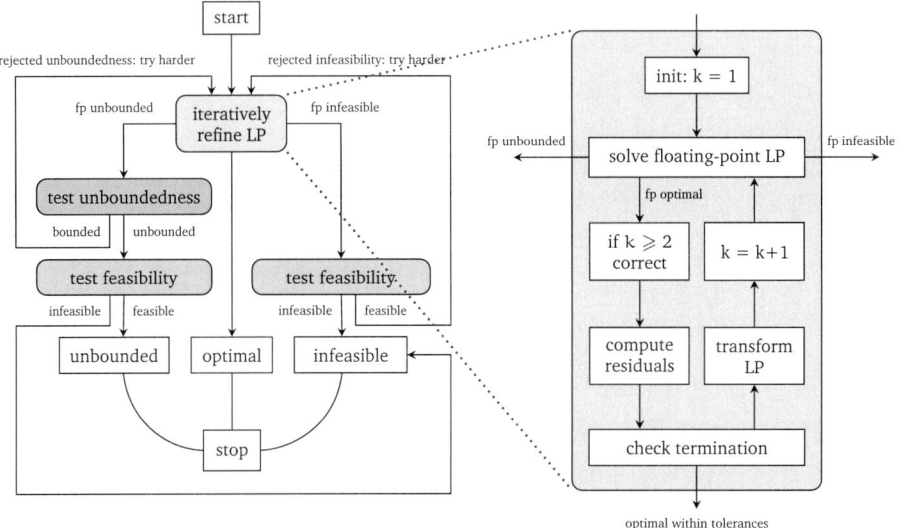

Figure 2.3: Iterative refinement for optimal, unbounded, and infeasible LPs

Remark 2.19 (Handling Floating-Point Failures). Our description above implicitly assumes that the floating-point solver always returns the answer infeasible, unbounded, or an approximate solution that is optimal within the specified tolerances. In practice, this may not hold. The solver could return an approximate solution with high violations, fail to converge, or terminate due to a singularity in a factorization procedure, to name only a few possibilities. One way to cope with this could be to increase the working precision of the solver to 128-bit or higher as is done in the incremental precision boosting procedure of the exact rational LP solver QSopt_ex (Applegate et al., 2007b), see also Section 3.1.3. This could also be applied when the floating-point solver repeatedly finds infeasibility or unboundedness, although this has been tested and rejected before.

Remark 2.20 (Ill-conditioned Problems). Some LPs are formulated in such a way that a small perturbation may change the feasibility status of the problem. For such LPs, there may simultaneously exist solutions that are feasible within a small feasiblity tolerance as well as highly accurate Farkas proofs. LPs near or on this boundary of feasibility and infeasibility are called ill-conditioned, or ill-posed, confer Section 2.1.3.

Thus, concerning the question of feasibility, if our algorithm produces an LP solution that is accurate to within some small tolerance, this does not guarantee that the

LP is feasible. However, in such cases one may at least conclude that a small perturbation of the LP would be feasible as stated in Theorem 2.3. Since our algorithm first attempts to find and refine a feasible solution, it is in a sense biased toward finding approximate feasible solutions if they exist, even if approximate Farkas proofs also exist.

2.5 Computational Study

As we have pointed out, Assumption 2.9 on the accuracy of the floating-point solutions returned will not hold in general. We hope that in practice it is satisfied for the LPs constructed during iterative refinement. To analyze the performance of the algorithm, we have implemented it as part of the LP solver SoPlex[34], which is based on the revised simplex method (Wunderling, 1996) and is freely available in source code under an academic license.

One of the motivations for using a simplex solver as the basis for our experiments instead of an interior point solver is the fact that the sequence of LPs solved by means of the iterative refinement procedure are highly similar in that they all share the same constraint matrix and their solution spaces are affine transformations of each other. Hence the solution refined by the last LP solve gives a starting point for the next LP solve and its basis information carries over as shown by Theorem 2.5. While modern implementations of interior point methods today are reported to be faster on average for solving LPs from scratch, the unmatched hot-starting capabilities of the simplex method promise greater gains in performance when a larger number of refinement rounds are applied.

In the following, we will discuss implementation details and present computational results.

2.5.1 Implementation

Starting with Version 1.7, released in July 2012, we extended SoPlex with functionality to read, store, check, and process LPs and LP solutions in rational precision. Our implementation is based on GMP, a free library for arbitrary precision arithmetic.[35]

Rational Arithmetic. We created a C++ wrapper class around the `mpq_t` objects of GMP's C interface, not only for convenience within SoPlex's C++ code, but because it enabled us to make additional performance improvements through non-local changes to the code:

– A pool to store `mpq_t` objects when they run out-of-scope and reuse them once new rational objects are needed, reducing expensive memory management operations and calls to `mpq_init`,

– One-time allocation of constants zero, one, and negative one,

---

[34] Zuse Institute Berlin. SoPlex—the Sequential object-oriented simPlex. Available for download under `http://soplex.zib.de/`.

[35] GMP. The GNU Multiple Precision Arithmetic Library. `http://www.gmplib.org/`.

- Using the sign function when comparison operators are called with argument zero,

- Skipping redundant addition and subtraction of zero and calling negation instead of subtracting a value from zero,

- Skipping multiplication and division by one, negating instead of multiplying or dividing by negative one, returning zero directly instead of multiplying by zero explicitly, and

- Avoiding redundant assignments by checking if values are already equal.

Each of these points led to a slight improvement in speed, even the last four points, which only involve trivial operations. Additionally, we link to the EGlib package[36] by Daniel G. Espinoza and Marcos Goycoolea, which improves memory allocation procedures for GMP by a slab pool management similar to the one described by Bonwick (1994). This is functionally complementary to our mpq_t pool.

LP and Basis Representation.   As described in Section 2.3.2, the original LP in rational precision is stored in general form (2.1). Instead of introducing explicit slack variables in the floating-point LP, which would have to be considered during pricing and ratio testing routines, we directly modify the objective coefficients of the implicit slack variables, i.e., the unit vectors that enter the basis matrix for rows not forced to be tight at one of their sides. This applies to the so-called column-oriented computational form—"column representation" in SoPlex speak—, which is used in textbook descriptions and most implementations of the simplex algorithm.

Alternatively, SoPlex implements a row-oriented computational form—"row representation" in SoPlex speak—as described for example by Nazareth (1987). For LPs with many more rows than columns, $m \gg n$, this can be beneficial because it works with a basis matrix of dimension $n \times n$, hence smaller than the $m \times m$ basis matrix of the column-oriented form with slack variables. The objective coefficients of the slack variables in column representation here take the role of row multipliers. They would need to be considered when solving for the dual solution with the row-oriented basis matrix. Currently the transformed LPs during iterative refinement can only be solved using column representation.

Basic Solutions.   The basis information is first used for computing the maximum dual violation $\delta_{D,k}$ in line 11 of Algorithm 2.2. The reduced cost of a variable is considered infeasible if either it is positive but the basis status of the column is *not* nonbasic (i.e., not fixed) at the lower bound or it is negative but the basis status is *not* nonbasic at the upper bound. Similarly, the dual multiplier of a row is considered infeasible if either it is positive but the basis status of the column is *not* nonbasic at its left-hand side or it is negative but the basis status is *not* nonbasic at its right-hand side. It is faster to check this than to calculate whether the solution value of the variable is closer to its lower or upper bound by rational arithmetic as described in

---

[36] EGlib. Efficient General Library. http://www.dii.uchile.cl/~daespino/EGlib_doc/main.html.

Section 2.3.1. The violation of complementary slackness is not tracked and line 12 is skipped since basic solutions, by construction, satisfy complementary slackness.

Second, we hot start the LP solver in line 18 from the previous basis, reusing the factorization of the basis matrix. (This is possible since the constraint matrix is not modified.) Note that unless the primal and dual scaling factors are limited by the incremental scaling limit α, the corresponding basic solution is guaranteed to be primal and dual infeasible for the scaled LP with a maximum violation of one each. Whereas textbook simplex algorithms usually require primal or dual feasibility to warm start the primal or dual simplex algorithm, respectively, SoPlex, like most other competitive simplex implementations, allow for arbitrary regular starting bases. It uses an initial shift of either bounds and sides or objective function coefficients to make the basic solution artificially primal or dual feasible and continues to switch between primal and dual simplex pivots until the shifts are reduced to zero and an optimal basis is reached (Wunderling, 1996, Sec. 1.4).

Third, we have to consider basis information in the correction step. Suppose in the floating-point solution $\hat{x}$ some variable is nonbasic at one of its bounds, $\hat{x}_i = \bar{\ell}_i$, say. Then the corrected solution value $(x_{k+1})_i$ may differ from ℓ_i by $(\bar{\ell}_i/\Delta_{D,k+1}) - \ell_i$. This is due to the necessary rounding step in line 17. We correct this by setting $\hat{x}_i$ directly to ℓ_i in this case, which has the added advantage of avoiding one multiplication and addition in rational arithmetic. This adjustment of the corrector solution is on a tiny order of magnitude and should not compromise Assumption 2.9—if it does then the floating-point solutions are almost certainly already unreliable.

We also noticed that in some cases it can cause numerical difficulties if the primal violation decreases much faster than the dual violation and as a result the primal scaling factor increases much faster than the dual scaling factor. If basic variables are forced out of the basis because the basis is still slightly dual infeasible, they then become fixed at one of their bounds. These will typically have a large value if the primal scaling factor is large and may challenge the numerics of the floating-point solver. In our implementation, we limit the primal scaling factor by the dual scaling factor.

Robust Floating-Point Solves. For instances with particularly bad numerical properties, SoPlex may fail to solve one of the floating-point LPs—running into a singular basis, aborting due to cycling, returning infeasible or unbounded for the auxiliary LPs when testing feasibility or unboundedness, or returning infeasible or unbounded although these results have been rejected earlier in the integrated solving loop displayed in Figure 2.3. If this happens, we try again with a series of different parameter settings until successful: forcing an increased Markowitz threshold, turning simplification on/off, turning matrix scaling on/off, tightening/relaxing tolerances, switching ratio test, switching pricing strategy, and solving from scratch in between. If this still does not help, we terminate without reaching the desired tolerances.

Updating Residual Vectors. In line 7 of Algorithm 2.2 for $k \geqslant 2$, we can use the relation $x_k = x_{k-1} + \frac{1}{\Delta_{P,k}}\hat{x}$ in order to compute $b - Ax_k$ by subtracting $\frac{1}{\Delta_{P,k}}A\hat{x}$ from the previous $\hat{b}$. If the corrector solution $\hat{x}$ is sparser than the corrected solution x_k

then this update is likely to be faster than recomputing $\hat{b}$ from scratch. The same holds for computing the new objective function in line 9. We already store the differences $\frac{1}{\Delta_{P,k+1}}\hat{x}$ and $\frac{1}{\Delta_{D,k+1}}\hat{y}$ as sparse vectors when correcting the primal and dual solution in lines 20 and 21.

This is a heuristic minimization of the number of arithmetic operations performed in rational arithmetic since we do not account for a potentially unequal distribution of nonzeros over the constraint matrix. However, it promises a good approximation. Especially when the basis between two subsequent floating-point solves does not change much or at all, which is typical after a few refinement rounds, the primal update vectors will contain only entries for basic variables; as a result, they will typically be much sparser than the corrected solution.

2.5.2 Experimental Setup

The goal of our experiments was to answer the following questions: Does the iterative refinement procedure always converge and how fast does it converge, both in terms of the number of refinements and the time spent? How much time is spent in the refinement phase compared to the initial floating-point solve? How expensive are the rational arithmetic operations performed? Are there differences between numerically easy and difficult instances? Does it always converge to an optimal basis and how many refinements are needed for this?

To this end, we performed the plain floating-point solve of standard SoPlex without iterative refinement with a primal and dual tolerance of 10^{-9} and compared it to iterative refinement runs with primal and dual tolerances of 10^{-50} and 10^{-250}. We will denote these settings by SoPlex_9, SoPlex_{50}, and SoPlex_{250}, respectively. In all cases, we parsed LP files exactly and compute violations in rational arithmetic. To check feasibility and optimality of the final bases, we used the basis verification tool perPlex.[37]

While by default, SoPlex uses an absolute feasibility tolerance of 10^{-6}, in our experiments, we have tightened it to 10^{-9}. Generally, using a stricter tolerance will return higher-precision corrector solutions. At the same time too strict of a tolerance can lead to a numerical breakdown, which is why we did not use a value of 10^{-12}. (We note that the analogous tolerance in the commercial LP solver CPLEX is adjustable by the user to no smaller than 10^{-9}.) LP presolving and matrix scaling were applied when solving LPs from scratch (initially and possibly after numerical difficulties), but not when hot starting from an advanced basis.

Hardware. The experiments were conducted on a cluster of 64-bit Intel Xeon X5672 CPUs at 3.2 GHz with 12 MB cache and 48 GB main memory. In order to safeguard against a potential mutual slowdown of parallel processes, we ran only one job per node at a time. We used a time limit of two hours per instance for each SoPlex and perPlex run.

---

[37] Thorsten Koch. perPlex LP feasiblity and optimality verificator. `http://www.zib.de/koch/perplex`.

Software. We used the SoPlex developer version 2.0.1.1, which implements the iterative refinement algorithm with features and parameters as described above. SoPlex was compiled with GCC 4.8.2 and linked to GMP 5.1.3 and EGlib 2.6.20.

Instances. We used the linear programming test set described in Section 1.4.1. It consists of 1,202 publicly available, primal and dual feasible LPs collected from different sources, some of which are well-known to be numerically challenging.

2.5.3 Computational Results

Let us first look at the behavior of iterative refinement on two individual instances, `momentum3` from MIPLIB 2003 and `world` from Mészáros's test set. These are nontrivial instances with 949,495 nonzeros and 164,470 nonzeros in their constraint matrix, respectively. Table 2.1 shows how the primal and dual violations progressed with time and number of simplex iterations elapsed. Because we are mainly interested in the order of magnitude of the violations, we only report the precision as the rounded negative base-10 logarithm.

Instance `momentum3` showed steady convergence. Each iteration added between seven and 16 orders of magnitude to the precision of the solution, such that after 18 refinement rounds primal and dual violations below 10^{-250} were reached. The initial floating-point solve with 46,184 iterations consumed the largest part of the running time. Only during the first refinement LP one more simplex iteration was performed. After that the basis remained unchanged. The refinement phase only incurred an overhead of 4.2% in running time and was dominated by rational arithmetic.

The hot start proved very efficient and the simplex time for solving the refinement LPs was negligible. The time consumed by rational arithmetic grew only slightly as the precision of the solution improved, from 0.3 seconds for the first refinement to 0.5 seconds for the last refinement. This was largely due to the positive effect of rounding the scaling factors to powers of two. In an earlier implementation without this feature, the last refinements consumed almost two seconds. When refining further, this slowdown had become increasingly pronounced.

The observed distribution of running time is typical and was similar for the instance `world`. Numerically, however, it seemed to be more challenging to SoPlex and the convergence of iterative refinement was slower. Simplex pivots, although comparatively few, were performed up to refinement round twelve until the final basis was reached. In the rounds where pivots were performed, the dual violation did not decrease, meaning that the floating-point corrector solutions returned by SoPlex must have exhibited high absolute violations in dual multipliers or reduced costs. This violation of Assumption 2.9 is explained as follows. In the transformed LP, non-basic variables with (dual) feasible reduced cost may have large objective coefficients, because these do not limit the dual scaling factor. If they are pivoted into the basis, which may happen, e.g., when basic variables with bound violations are forced to leave the basis, their reduced cost should be zero. Solving the linear system (2.9) for the dual solution vector, however, only yields a precision that is relative with respect to the right-hand side, i.e., the objective coefficients.

Table 2.1: Progress of iterative refinement for `momentum3` and `world`

R — number of refinements
iter — number of simplex iterations elapsed
t — time elapsed (in seconds)
t_{rat} — time spent on rational arithmetic (cumulative, in seconds)
δ_P — maximum primal violation (rounded negative $\log_{10}$)
δ_D — maximum dual violation (rounded negative $\log_{10}$)

	momentum3					world				
R	iter	t	t_{rat}	δ_P	δ_D	iter	t	t_{rat}	δ_P	δ_D
0	46184	190.4	0.3	8	10	70204	131.5	0.1	8	11
1	46185	191.0	0.6	23	17	70256	131.7	0.2	17	12
2	46185	191.3	1.0	37	33	70282	131.9	0.4	22	13
3	46185	191.7	1.3	51	48	70287	132.1	0.5	23	13
4	46185	192.0	1.7	65	65	70289	132.3	0.7	23	14
5	46185	192.4	2.1	80	78	70292	132.4	0.9	24	13
6	46185	192.8	2.4	93	93	70292	132.6	1.0	24	29
7	46185	193.2	2.8	108	107	70315	132.8	1.2	34	13
8	46185	193.6	3.2	121	120	70315	133.0	1.4	34	29
9	46185	194.0	3.6	136	134	70319	133.2	1.6	40	13
10	46185	194.4	4.0	150	149	70319	133.4	1.7	40	29
11	46185	195.1	4.7	165	164	70319	133.6	1.9	40	45
12	46185	195.6	5.1	178	177	70320	133.8	2.1	55	15
13	46185	196.0	5.6	193	191	70320	134.0	2.3	70	30
14	46185	196.4	6.0	207	205	70320	134.2	2.5	85	46
15	46185	196.9	6.4	222	220	70320	134.4	2.7	100	61
16	46185	197.4	6.9	235	234	70320	134.6	2.9	115	77
17	46185	197.9	7.4	250	248	70320	134.8	3.1	130	93
18	46185	198.4	7.9	264	262	70320	135.0	3.3	146	107

As we can see in Table 2.1, after the first refinement round without pivots (round four), the dual violation started to decrease below 10^{-29}. In the next round, however, new pivots occurred, and the maximum dual violation even fell back to 10^{-13}. Only from round 13 onward both primal and dual precision improved continuously, in each round reducing the violations by about 15 orders of magnitude. Because of the setbacks during the refinement rounds with simplex pivots, the dual precision lagged behind the primal precision. This could have been even more pronounced, but because we limited the primal scaling factor by the dual scaling factor in our implementation, the primal precision stalled during some rounds. After 29 refinement rounds (not shown in the table) a maximum violation below 10^{-250}, comparable to the precision on `momentum3` after 18 rounds, was reached.

General Results.   We move on to discuss the results over the entire test set. Detailed results for each instance can be found in Table A.2 of the appendix. Out of the 1,202 instances in our test set, SoPlex$_{50}$ and SoPlex$_{250}$ converged successfully to the specified tolerance on 1,195 instances. On three instances,[38] they timed out because the floating-point solver could not solve the first refinement LP within the time limit. For three instances, the initial floating-point solve (equivalent to SoPlex$_9$) incorrectly claimed unboundedness and for one instance it incorrectly claimed infeasibility.[39] In all cases, SoPlex$_{50}$ and SoPlex$_{250}$ rejected these claims successfully using the feasibility and unboundedness tests described in Section 2.4, but after starting to refine the original LP again, floating-point SoPlex failed to return an approximately optimal solution even when run with different settings. Furthermore, for five of the 1,195 instances[40] the floating-point solver claimed one of the intermediate refined LPs infeasible. SoPlex$_{50}$ and SoPlex$_{250}$ rejected these claims and continued to converge to their target tolerance.

Results for the Floating-Point Solves. For four instances SoPlex$_9$ claimed a wrong status,[41] and for 99 instances it returned a numerical solution that exhibited violations *above* 10^{-9}, though most of those only slightly. However, for the instance de063155 from Mészáros's "problematic" test set, which features constraint coefficients with absolute values ranging from approximately 10^{-7} to 10^{12} SoPlex$_9$ even returned a completely meaningless solution violating primal and dual feasibility by almost 10^3 and 10^8, respectively. This instance is solved correctly by SoPlex$_{50}$ and SoPlex$_{250}$ in seven and 20 refinements, respectively.

For 1,024 instances perPlex verified that the basis returned by SoPlex$_9$ was indeed optimal. For 95 instances, the SoPlex$_9$ basis was detected as primal infeasible, for 30 instances as dual infeasible, and for six instances as both primal and dual infeasible. On the remaining instances, perPlex hit a time or memory limit, or—as for the instance neos-619167—it could not handle free nonbasic variables correctly. Hence, our test set seems to contain a not negligible number of numerically challenging instances, certainly to SoPlex$_9$. Although for the majority of instances, we can confirm the conclusion that Dhiflaoui et al. (2003) and Koch (2004) have drawn based on the Netlib test set, namely that floating-point LP solvers often succeed in returning optimal bases, we have to relativize this finding for more than 14% of the instances.

Results for Iterative Refinement. As mentioned above, on all 1,195 instances, both SoPlex$_{50}$ and SoPlex$_{250}$ converged to the specified tolerance. On all instances that perPlex could handle, it verified that the final basis returned was primal and dual feasible. For 61 instances, they even terminated with an exactly optimal numeric solution with a zero primal and dual violation.

For the vast majority of instances, the basis after the initial floating-point solve was already the final basis. On 93 and 30 instances, one and two refinement rounds, respectively, were needed to reach the final (optimal) basis. Only for the three in-

---

[38] neos-954925, neos-956971, and neos-957143

[39] de063157, 130, and stat96v1 were claimed to be unbounded, neos-1603965 was claimed to be infeasible by SoPlex$_9$

[40] fome11,12,13, rail01, and shs1023

[41] de063157, 130, and stat96v1 were claimed to be unbounded, neos-1603965 was claimed to be infeasible by SoPlex$_9$

Table 2.2: Computational comparison of iterative refinement and pure floating-point performance (results aggregated by number of refinement rounds to final basis)

R_0 — number of refinements to final basis
N — number of instances in this R_0-class
R — number of refinements (arithmetic mean)
iter — number of simplex iterations (shifted geometric mean)
t — total running time (shifted geom. mean in seconds)
Δt — relative running time w.r.t. SoPlex_9

		SoPlex_9		SoPlex_{50}				SoPlex_{250}			
R_0	N	iter	t	R	iter	t	Δt	R	iter	t	Δt
0	326	19291.8	20.0	3.0	19291.8	20.6	1.03	16.4	19291.8	22.7	1.14
1	26	16936.8	20.6	3.9	17918.2	22.3	1.08	17.8	17918.2	25.1	1.22
2	2	79895.4	90.3	4.5	79964.6	92.8	1.03	17.5	79964.6	99.0	1.10
11	1	58340.0	100.4	14.0	58534.0	103.8	1.03	26.0	58534.0	104.2	1.04
12	1	70204.0	131.2	15.0	70320.0	132.1	1.01	29.0	70320.0	137.6	1.05
$\geqslant 1$	30	20531.5	25.7	4.6	21562.4	27.6	1.07	18.5	21562.4	30.7	1.19

stances de063155, mod2, and world (from Table 2.1), three, eleven, and twelve rounds, respectively, were performed until the basis did not change in the following refinement rounds. This happened already when using SoPlex_{50}, so the additional rounds performed by SoPlex_{250} essentially only refine the primal and dual solutions to the linear systems defined by the basis matrix. Furthermore, for the five instances for which feasibility was tested, few refinements with actual pivots were performed.

Average Performance Comparison.   In order to compare performance of the three runs on subsets of instances with varying numerical difficulty, we categorized the instances according to the number of refinement rounds performed until the final basis was reached, denoted by R_0. We excluded the five instances where the infeasibility test was triggered, because their run actually amounts to the refinement of two LPs. We excluded instances that were solved in under two seconds by each algorithm.

For the resulting 356 instances, Table 2.2 reports the average number of refinement rounds, simplex iterations, and the average running time over these subsets. Because simplex iterations and running times vary across instances, we use shifted geometric means, see also Section 1.3, with shifts of two seconds and 100 simplex iterations. For the number of refinements we report arithmetic averages.

For SoPlex_{50}, most notably, the average running time increases by only 3% for the easy subset $R_0 = 0$ and by only 7% for the set $R_0 \geqslant 1$ of instances with at least one refinement round with simplex pivots. This is reflected in the small number of additional simplex iterations that are performed during the refinement rounds. Furthermore, the average number of refinement rounds is often smaller than expected. To achieve a maximum violation of 10^{-250} by floating-point solves with tolerance 10^{-9}, one would have to estimate approximately $250/9 - 1 \approx 26.8$ refinement rounds. By contrast, even the class $R_0 \geqslant 1$ shows only 18.5, indicating that most floating-point solves—in particular the final ones without simplex pivots—return higher-precision solutions. This is slightly less pronounced for SoPlex$_{50}$.

The overhead in the running time of SoPlex$_{250}$ beyond SoPlex$_{50}$ stems only from refining the primal and dual solution to the linear systems defined by the basis matrix and its transpose. We note that the increase in time of SoPlex$_{250}$ to SoPlex$_9$ is still very small on average—14% for $R_0 = 0$ and 19% for $R_0 \geqslant 1$. Implementing the more sophisticated techniques recently developed by Wan (2006), Pan (2011), or Saunders et al. (2011) may close this gap even further.

2.6 Conclusion

In this chapter we have presented a new algorithm to solve linear programs to high levels of precision. It extends the idea of iterative refinement for linear systems of equations by Wilkinson (1963) to the domain of optimization problems by simultaneously correcting primal and dual residual errors. Algebraically, it builds up an increasingly accurate solution by solving a sequence of LPs, which differ only in the bounds of the variables, the sides of the constraints, and the objective function coefficients, to fixed precision. Geometrically, it can be viewed as zooming further and further into the area of interest around the refined solution. While it is designed to work with an arbitrary LP oracle, it combines particular well with the hot-starting capabilities of the simplex method.

For infeasible and unbounded instances iterative refinement can be used to compute high-precision certificates via an auxiliary reformulation of the LP. In this context, we have developed an algorithm to convert an approximate Farkas proof into a rigorous infeasibility box centered at the origin that helps users to analyze the domains in which feasible solutions can or cannot exist.

For a simplex-based implementation we demonstrated it to be efficient in practice: On a large test set of publicly available benchmark instances, computing solutions up to a precision of 10^{-50} incurred an average slowdown of only 3% on numerically easy and 7% on numerically difficult LPs. In addition we saw that the basis corresponding to the refined solution was always optimal.

With classical iterative refinement the algorithm shares the limitation that it breaks down when the LP is too ill-conditioned for the underlying floating-point routine. This could be overcome by increasing the working precision of the underlying floating-point LP solver whenever necessary in a similar fashion as in the exact LP solver QSopt_ex by Applegate et al. (2007b) discussed in Section 3.1.3.

As a final remark, we note that some applications may require extended-precision solutions, but not need exact solutions. Examples include the multiscale models of metabolic networks from Chapter 4. In such cases iterative refinement can be used to meet this demand without requiring the amount of time taken by an exact LP solver, giving it a competitive advantage. At the same time, an obvious question is how iterative refinement can be used to advance the state-of-the-art in exact linear programming. This is the topic of the next chapter.

Chapter 3

Exact Linear Programming over the Rational Numbers

The development and discussion of the iterative refinement procedure in Chapter 2 concentrated on computing a sequence of numeric solutions to a linear program that gave smaller and smaller violations. This chapter will show how this technique can be applied to solve linear programs with rational input data *exactly* and obtain solutions with a zero violation.

Section 3.1 offers a survey of previous work and related research. In Sections 3.2 and 3.3, we discuss two techniques that can be combined efficiently with iterative refinement to solve LPs exactly—the exact solution of the linear systems of equations defining a basic solution and Diophantine approximation. We prove that they yield oracle-polynomial time algorithms. The computational performance of the algorithms is compared in Section 3.4. Section 3.5 contains our conclusions.

3.1 Background and Previous Work

There is a trivial method of solving LPs with rational input data exactly, which is to apply a simplex algorithm and perform all computations in (exact) rational arithmetic. An example is the solver exlp by Masashi Kiyomi.[42] Implementations of this approach can also be found in packages for discrete computational geometry. Two examples are the vertex enumeration packages cdd+[43] (Fukuda and Prodon, 1996) and lrs[44] based on the reverse search algorithm (Avis and Fukuda, 1992).

Because of the higher computational cost of rational arithmetic, this approach becomes prohibitively slow for large instances. The cdd+ manual states that its LP algorithm works well for problems with up to one hundred variables, while the number of inequality constraints may be significantly larger. This reflects the typical

---

[42] Masashi Kiyomi. exlp, an exact LP solver. http://members.jcom.home.ne.jp/masashi777/exlp.html, accessed September 2014.

[43] Komei Fukuda. cdd/cdd+. http://www.inf.ethz.ch/personal/fukudak/cdd_home/, accessed September 2014.

[44] David Avis. lrs. http://cgm.cs.mcgill.ca/~avis/C/lrs.html, accessed September 2014.

use cases in computational geometry and is not targeted towards large-scale real-world applications.

More precisely, as Espinoza (2006) demonstrates computationally, the running time of the naïve approach is correlated not so much with the size of an LP, but rather with the encoding length of the basic solutions traversed by the simplex algorithm. We can see this by observing that the cost of rational arithmetic scales with the size of denominators and numerators involved. In the following, we will discuss more involved techniques that have been suggested in the literature.

3.1.1 Edmonds' Q-Pivoting

Edmonds (1994) shows that—essentially as a consequence of Cramer's rule—the inverse of a basis matrix can be represented as matrix of integer coefficients divided by a common denominator and that this representation can be efficiently updated when pivoting from basis to basis. This is improved by Edmonds and Maurras (1997) and Azulay and Pique (1998) and is also exploited in the vertex enumeration package lrs.

The advantage of this technique, compared to computing a basis inverse with rational coefficients, is that the arithmetic operations are performed on multi-precision integers, which is generally faster than rational arithmetic. This is demonstrated by the computational study of Azulay and Pique (2001) on relatively small instances with up to 517 constraints and 1,028 variables from Netlib.[45] Unfortunately, their underlying implementation of the simplex algorithm is not sophisticated enough to draw firm conclusions on the overall potential of this approach.

3.1.2 Gärtner's Mixed-Precision Simplex Algorithm

With a focus on LPs from computational geometry, Gärtner (1999) developed an implementation of the simplex algorithm that combines inexact and exact arithmetic. This implementation uses fast floating-point arithmetic in heuristic components only in order to determine the order of basic solutions that the algorithm traverses, the so-called pricing step. The exactness of the algorithm is guaranteed by an exact handling of the linear algebra operations involving the basis matrix and its inverse via Edmonds' Q-matrices.

The performance results on applications from computational geometry with few variables show that the implementation clearly outperforms a version where exact arithmetic is used for all computations. A comparison with the LP solver CPLEX[46] 4.0.9 on the three Netlib instances with highest variable-to-constraint ratio (which is favorable for Gärtner's code) makes clear that the pivoting speed may still be up to a factor of 10 000 slower when compared to a floating-point solver. However, to a large extent this may be due to the use of dense Q-matrices versus CPLEX's sparse LU

---

[45] University of Tennessee Knoxville and Oak Ridge National Laboratory. Netlib LP Library. http://www.netlib.org/lp/.

[46] IBM.   CPLEX   Optimizer.   http://www.ibm.com/software/commerce/optimization/cplex-optimizer/.

factorization, and so it is not possible to draw any conclusions on how small the performance overhead due to the exact computation can be made. The implementation is publicly available as part of the computational geometry package CGAL.[47]

3.1.3 QSopt_ex's Incremental Precision Boosting

Kwappik (1998) and Dhiflaoui et al. (2003) note that the basis information returned by the simplex algorithm can be exploited to check and verify whether the basis corresponds to an optimal solution—even if the basis was returned by an inexact solver. It suffices to solve the linear systems of equations (2.8) and (2.9) in rational arithmetic, because basic solutions satisfy complementary slackness by construction and hence primal and dual feasibility certify optimality. They used this in order to verify that most bases returned by CPLEX on a selection of the Netlib test set correspond to an optimal solution.

This result was improved by Koch (2004) who could compute optimal bases for all Netlib instances using CPLEX 8.0 with adjusted parameter settings. He provided the basis verification software perPlex[48] implementing a straightforward LU factorization in rational arithmetic. Koch also experimented with the LP solver SoPlex[49] which failed to return optimal basis for five instances. When increasing the floating-point precision from double to 128-bit, SoPlex then returned optimal bases for these LPs as well.

Applegate, Cook, Dash, and Espinoza (2007b) systematized this approach and extended the simplex-based LP solver QSopt to the exact rational LP solver QSopt_ex.[50] If an optimal basis is not identified by the double-precision subroutines, QSopt_ex performs more simplex pivots using increased levels of precision until the exact rational solution is identified. A simplified version of this procedure is summarized by Algorithm 3.1. Whenever possible, the LP solve in line 4 is warm started with a basis $\mathcal{B}$ computed at a previous iteration. Analogous ideas are used for infeasible and unbounded LPs, see Espinoza (2006).

QSopt_ex is often very effective at finding exact solutions quickly, especially when the double-precision LP subroutines are able to find an optimal LP basis. However, in cases when extended-precision computations are used to identify the optimal basis, or when the rational systems of equations solved to compute the rational solution are difficult, solution times can increase significantly. We will refer to this strategy of iteratively increasing the working precision for the simplex algorithm as *incremental precision boosting*.

---

[47] CGAL—Computational Geometry Algorithms Library. http://www.cgal.org/.

[48] Thorsten Koch. perPlex LP feasiblity and optimality verificator. http://www.zib.de/koch/perplex.

[49] Zuse Institute Berlin. SoPlex—the Sequential object-oriented simPlex. Available for download under http://soplex.zib.de/.

[50] David L. Applegate, William Cook, Sanjeeb Dash, and Daniel G. Espinoza. QSopt_ex. http://www.dii.uchile.cl/~daespino/ESolver_doc/.

Algorithm 3.1: Incremental Precision Boosting

---

    input: rational, primal and dual feas. LP $\min\{c^\mathsf{T}x \mid Ax = b, x \geqslant \ell\}$

    output: optimal primal–dual solution $x^* \in \mathbb{Q}^n, y^* \in \mathbb{Q}^m$,

                 optimal basis $\mathcal{B} \subseteq \{1, \dots, n\}$

1 **begin**

2   **for** $p \leftarrow$ double, $128, 256, \dots$, rational **do**

3   get $\bar{A}, \bar{b}, \bar{c}, \bar{\ell} \approx A, b, c, \ell$ in precision p

4   solve $\min\{\bar{c}^\mathsf{T}x \mid \bar{A}x = \bar{b}, x \geqslant \bar{\ell}\}$ in precision p

5   get basis $\mathcal{B}$ returned as optimal

6          $x^*_{\mathcal{B}} \leftarrow A^{-1}_{.\mathcal{B}}(b - \sum_{i \notin \mathcal{B}} A_{.i}\ell_i)$ in rational arithmetic

7          $y^* \leftarrow (A^\mathsf{T}_{.\mathcal{B}})^{-1}c_{\mathcal{B}}$ in rational arithmetic

8   **if** $x^*_{\mathcal{B}} \geqslant \ell_{\mathcal{B}}$ and $A^\mathsf{T}y^* \leqslant c$ **then**

9   **foreach** $i \notin \mathcal{B}$ **do** $x^*_i \leftarrow \ell_i$

10   return $x^*, y^*, \mathcal{B}$

3.1.4 Diophantine Approximation Algorithms

One fundamental technique for computing exact rational solutions to linear systems of equations or inequalities is to "round" a sufficiently accurate approximate solution to a nearby solution with small denominators. Informally speaking, the rationale behind these methods is the following. Approximate solutions that are obtained by a numerical algorithm tend to have a complicated rational representation with higher and higher denominators as their accuracy increases. However, we may know from the problem that there exist nearby exact rational solutions with a simpler representation and that the complicated rational representaion is merely "noise" in the approximate solution caused by the remaining tiny errors. In this case we can try to remove this noise by identifying the closest rational solution within a certain bound on the denominators.

In one dimension, i.e., for reconstructing a single rational number, the formal basis for these techniques is the following theorem found similarly in Schrijver (1986, Cor. 6.3b) or Wan (2006).

Theorem 3.1. *For* $\alpha \in \mathbb{Q}$, $M > 0$ *there exists at most one rational number* p/q *such that* $|p/q - \alpha| < 1/2Mq$ *and* $1 \leqslant q \leqslant M$. *There exists a polynomial-time algorithm to test whether this number exists and if so, to compute this rational number.*

The proof makes use of continued fraction approximations and the resulting algorithm—essentially the extended Euclidean algorithm—runs in polynomial time (in the encoding length of α and M) and is very fast in practice. The problem solved by this theorem is known as the *Diophantine approximation* problem. This technique is sometimes referred to as "rounding" since for $M = 1$ above it corresponds to rounding to the nearest integer.

Theorem 3.1 implies that if there is an unknown rational number p/q, whose denominator q is bounded by M, then if we compute an approximation α that satisfies $|p/q - \alpha| < 1/2M^2$, then we may recover the exact rational number p/q in polynomial time. Since basic solutions of linear programs correspond to solutions of linear systems of equations, see (2.8) and (2.9), one can derive an a priori bound on the size of the denominators of any basic solution using Cramer's rule and the Hadamard inequality, as in Lemma 1.4. It can be shown, for instance, that all basic primal–dual solutions of a rational LP $\min\{c^\mathsf{T}x \mid Ax = b, x \geq \ell\}$, $A \in \mathbb{Q}^{m \times n}$, have entries with denominator bounded by

$$H := n^{m/2}L^n \prod_{j=1}^{m} \|A_{j\cdot}\|_\infty \qquad (3.1)$$

where L is the least common multiple of the denominators of the entries in A, b, ℓ, and c. Therefore, upon computing an approximation of an optimal basic solution vector where each component is within $1/2H^2$ of the exact value, we may apply Theorem 3.1 componentwise to recover the exact solution vector.

QSopt_ex makes use of this and tries to round the approximate primal–dual solution it has computed in line 4 of Algorithm 3.1 to an exact rational solution before continuing to compute the exact basic solution from scratch in lines 6 and 7 by a rational LU factorization. Cook and Steffy (2011) showed that solutions of these sparse systems of linear equations can often be accelerated further by alternative strategies such as the DLCM method used by Kaltofen and Saunders (1991) and Chen and Storjohann (2005) for modular reconstruction of rational vectors and an output-sensitive version of Wan's algorithm (Wan, 2006).

Output-sensitive algorithms, as used by Chen and Storjohann (2005), address the fact that a priori bounds on the accuracy of an approximate solution that guarantee success of the reconstructions routines are usually weak. Because denominators in the actual solutions are in many cases much smaller than guaranteed by these worst-case bounds, output-sensitive algorithms heuristically apply reconstruction to less accurate solution candidates and check the result. If the reconstructed solution is not yet feasible, the approximate solution is refined further.

The DLCM method is designed for reconstructing a vector of rational numbers with denominators that share many common factors—a typical situation for solution vectors of linear systems. Instead of reconstructing the components separately one by one, when reconstructing the i-th entry, DLCM takes into account the denominators of the previous $i - 1$ reconstructed entries. It computes their least common multiple and forces the denominator of the i-th entry to be a multiple of this. This strategy reduces the amount of computational work and has the advantage that it aborts faster when reconstruction is attempted from a solution with insufficient accuracy. As one disadvantage, DLCM requires the bound M in Theorem 3.1 to hold for the least common multiple of the denominators of the reconstructed vector, not only for the individual denominators. As stated above, DLCM is designed for the case when this overhead is small.

A different technique for reconstructing rational vectors, one that is computationally more expensive but works under milder assumptions, is based on polynomial-time *lattice reduction* algorithms as pioneered by Lenstra, Lenstra, and Lovász (1982) and recently improved by Nguyen and Stehlé (2009). It rests on the following theorem.

Theorem 3.2 (Lenstra et al. (1982), Prop. 1.39). *There exists a polynomial-time algorithm that, given positive integers* n, M *and* $\alpha \in \mathbb{Q}^n$, *finds integers* $p_1, \ldots, p_n, q$ *for which*

$$|p_i/q - \alpha_i| \leqslant 1/Mq \text{ for } i = 1, \ldots, n,$$
$$1 \leqslant q \leqslant 2^{n(n+1)/4} M^n.$$

This has prominently been applied by Grötschel, Lovász, and Schrijver (1988) in the following fashion.

Lemma 3.3 (Grötschel et al. (1988), Lem. 6.2.9). *Suppose we are given a polyhedron* P *in* $\mathbb{R}^n$ *with facet complexity at most* φ *and a point* $\alpha \in \mathbb{R}^n$ *within Euclidean distance at most* $2^{-6n\varphi}$ *of* P. *If integers* $p_1, \ldots, p_n, q$ *satisfy*

$$\|p - \alpha q\|_2 \leqslant 2^{-3\varphi} \text{ and} \tag{3.2}$$
$$1 \leqslant q < 2^{4n\varphi}$$

then $\frac{1}{q}p$ *is in* P.

Here the facet complexity of P is the smallest number $\varphi \geqslant n$ such that P is defined by a list of inequalities each having encoding length at most φ. A solution $\frac{1}{q}p$ satisfying (3.2) can be computed in polynomial time by the lattice reduction algorithm behind Theorem 3.2. Current lattice reduction algorithms are, however, significantly slower than the continued-fraction based techniques discussed above. The main advantage of Theorem 3.2 is that it only requires α to be close to an exact solution $x \in P$, while Theorem 3.1 additionally needs the entries of x to have small denominators. For more details on rational reconstruction algorithms, see, e.g., von zur Gathen and Gerhard (1999).

3.1.5 Ellipsoid and Interior Point Methods

While our focus is mostly on computational methods, we note that the idea of using an approximate solution as a starting point in finding an exact solution also appears in more theoretically targeted research. Grötschel, Lovász, and Schrijver (1988) give exact polynomial-time algorithms for solving LPs defined by rational data using variants of the ellipsoid method. These produce smaller and smaller ellipsoids enclosing an optimal solution such that eventually the simultaneous Diophantine approximation explained in the previous section can be applied to recover an exact rational solution from the center of the ellipsoid.

The same methods could be applied in the context of interior point algorithms in order to convert an approximate, sufficiently advanced solution along the central path to an exact rational solution. However, the original algorithm of Karmarkar (1984) moves from the approximate solution to a nearby basis solution that matches it as closely as possible and checks this for optimality. This technique is in some sense related to QSopt_ex's incremental precision boosting with the difference that the basis defining the linear systems solved is not given by a simplex algorithm executed in higher and higher precision but obtained from a more and more advanced solution along the central path.

The variant of Renegar (1988) instead identifies an optimal face of the feasible region from approximately tight inequalities and performs a projection step via solving a system of linear (normal) equations. He gives a detailed analysis and discussion of these ideas and also references the method developed by Edmonds (1967) for solving linear systems of equations in polynomial time. Note that these methods rely on the explicit description of the LP, and cannot be applied in the oracle setting of the ellipsoid method considered by Grötschel, Lovász, and Schrijver (1988).

3.1.6 Rigorous Bounds

Finally, we want to mention a related line of research that has focused on computing rigorous bounds on the objective function value of an LP. These approaches are based on the use of interval arithmetic, an extension of traditional arithmetic operations where rigorous upper and lower bounds on computed values are maintained, as an alternative to using exact representations or approximate floating-point representations. For details on interval arithmetic we refer to the recent book of Moore et al. (2009) and the extensive list of references therein.

In particular, some recent work along this line has proven effective for efficiently computing rigorous bounds on the objective value of LPs without the use of exact rational arithmetic. These rigorous bounds can be useful when solving mixed-integer linear programming problems exactly because algorithms such as branch-and-cut may use bounds on the objective value of an LP relaxation in order to prune nodes of a search tree. See Jansson (2004), Neumaier and Shcherbina (2004), and Steffy and Wolter (2013) for more details.

Althaus and Dumitriu (2009) describe an involved method to certify exact feasibility of an LP also using interval arithmetic. First, they iteratively detect and remove redundant constraints using rational arithmetic. Second, they compute an approximate primal solution in floating-point arithmetic and use an approach of Higham (1987) to compute safe error bounds around this solution such that all remaining constraints are satisfied for all solution vectors within these bounds. While this does not produce a high-precision solution, it is able to certify that an exactly feasible solution exists. The algorithm successfully certifies feasibility for all but three LPs from the Netlib test set. They note that—for an LP in equality form (2.2)—the same technique can be applied to the dual system

$$A^\mathsf{T} y \leqslant 0, (b - A\ell)^\mathsf{T} y = 1 \qquad (3.3)$$

to certify infeasibility by the existence of a Farkas proof. However, they do not report computational experience on the effectiveness of this approach.

One advantage of interval arithmetic techniques is that they may also be applied to problems with irrational input data without a large overhead. For example, one component of Thomas Hales' proof of the Kepler conjecture relies on the solution of many LPs, some of which had irrational input data arising from the geometric structure of the problem (Hales, 2005). Valid bounds on those LPs sufficient to establish the conjecture were obtained using interval methods (Obua and Nipkow, 2009).

3.2 Exact Bases via Iterative Refinement

In the previous section we have already seen two techniques that lend themselves to extending an iterative refinement algorithm for LPs towards computing exact solutions—Diophantine approximation and exact rational solution of the linear systems of equations defining a primal–dual basic solution. The latter technique and its combination with iterative refinement is the topic of this section. We start with a simple experiment to demonstrate the potential of computational improvement that lies in iterative refinement.

3.2.1 A First Experiment

A first idea to accelerate exact LP solving is to warm start the QSopt_ex solver from the advanced starting basis obtained after several rounds of iterative refinement. This only requires linking two "out-of-the-box" libraries that are readily available. We evaluated the performance impact of this strategy and describe our computational results in the following.

The design of this experiment is joint work with Daniel E. Steffy and Kati Wolter and appeared in an early version in Gleixner et al. (2012b). The results of this section are contained in the preprint version of Chapter 2 (Gleixner et al., 2015b).

Experimental Setup.   We continued from the experiments described in Section 2.5 for primal and dual feasible LPs and selected the 1,195 instances for which both SoPlex$_9$ and SoPlex$_{50}$ returned basic solutions. We used the same hardware and software with addition of QSopt_ex 2.5.10.

We warm started QSopt_ex from the basis returned by SoPlex$_{50}$ and measured the total running time of both runs. Comparing this to the plain QSopt_ex performance would be biased, however, since the underlying implementation of the floating-point simplex solvers in SoPlex and QSopt_ex is significantly different. (As just one example, QSopt_ex does not implement presolving techniques.) Hence, as a meaningful point of reference we used the performance of QSopt_ex when warm started from the basis returned by SoPlex$_9$.

All selected instances could be solved by SoPlex$_9$ and SoPlex$_{50}$ in under two hours. We enforced a time limit of two hours for each QSopt_ex run.

Results.   1,166 instances could be solved by both versions. Nine instances could be solved only when warm starting from the advanced basis returned by SoPlex$_{50}$ (with running times of between 108 and 2,026 seconds); on these instances when starting from SoPlex$_9$'s basis QSopt_ex hit the time limit of two hours.[51] 20 instances could

---

[51] fome13 (in 328.6 seconds), mod2 (in 108.1 seconds), momentum3 (in 2,025.5 seconds), ofi (in 432.2 seconds), sgpf5y6 (in 228.7 seconds), shs1023 (in 625.4 seconds), watson_1 (in 379.8 seconds), watson_2

not be solved by either version. Detailed results for each instance can be found in Table A.3 of the appendix.

Table 3.1 gives an aggregated comparison over the 307 instances that were solved by both versions, but were nontrivial in the sense at least one version took more than two seconds. It compares the total number of iterations taken (by SoPlex and QSopt_ex), the total running time, and additionally states the running times of SoPlex and QSopt_ex individually. We report shifted geometric means as explained in Section 1.3, using a shift of two seconds and 100 simplex iterations. (Note that it is hence correct that the values in columns $t_{9/50}$ and t_{ex} do not exactly add up to the t-values.) In order to distinguish numerically easy from difficult instances, we again categorized them by the number of refinement rounds (R_0) needed by SoPlex$_{50}$ in order to reach the final basis. (22 instances needed one refinement round and one instance[52] took two refinements.)

On the instances for which iterative refinement did not change the basis, the QSopt_ex performance is necessarily identical. It never increased the simplex precision beyond 64-bit except for the instance maros-r7, for which it started pivoting despite the optimality of the starting basis and performed to precision boosts to 192-bit. The total running time on these instances increases by only 1% on average, corresponding to the small overhead of iterative refinement.

By contrast, for the 23 instances for which iterative refinement affected the final basis the performance gain is drastic: the total running time is reduced to only 18%, i.e., by a factor 5.5 on average. (Note that this does not even involve the nine instances on which SoPlex$_9$+QSopt_ex timed out.) This improvement cannot be explained by the reduction in simplex iterations alone. Most importantly, warm starting from the basis returned by SoPlex$_{50}$, no precision boosts were performed and so the number of computationally expensive pivots performed by QSopt_ex in extended precision was reduced to zero.

Conclusion. In this simple experiment we observed a marginal slowdown on numerically easy instances and an extraordinary speedup on more challenging LPs. Furthermore, nine more instances could be solved within the time limit of two hours. The next section presents a more systematic analysis of this approach.

3.2.2 Convergence to Optimal Bases

From Theorem 2.5 we know that if the corrector solution $\hat{x}, \hat{y}$ to the transformed problem $\hat{P}$ is basic, then also the corrected solution is basic with respect to the same basis. Note, however, that the underlying LP solver returns only approximate solutions which will in general not respect the definition of a basic solution in exact arithmetic. (While the nonbasic variables might be fixed exactly to their bounds if they are representable in the working precision of the solver, in general the equality constraints will be violated and the reduced costs on the basic variables will not evaluate exactly to zero.)

(in 1,538.1 seconds), and world (in 136.6 seconds).
[52] momentum2

Table 3.1: Computational comparison of QSopt_ex's performance warm started from bases returned by SoPlex$_9$ and SoPlex$_{50}$ (results aggregated by number of refinements to final basis)

R_0 — number of refinements to final basis
N — number of instances in this R_0-class
iter — number of simplex iterations by SoPlex+QSopt_ex (shifted geom. mean)
B — number of precision boosts in QSopt_ex (total over all LPs in this R_0-class)
t_9/t_{50} — running time of SoPlex$_9$/SoPlex$_{50}$ (shifted geom. mean in seconds)
t_{ex} — running time of QSopt_ex (shifted geom. mean in seconds)
t — total running time (shifted geom. mean in seconds)
Δt — relative total running time w.r.t. SoPlex$_9$+QSopt_ex

		SoPlex$_9$+QSopt_ex				SoPlex$_{50}$+QSopt_ex					
R_0	N	iter	B	t_9	t_{ex}	t	iter	B	t_{50}	t_{ex}	t	Δt
0	284	22306.5	2	20.7	3.1	27.2	22306.5	2	21.2	3.1	27.5	1.01
1	22	23812.9	13	18.1	110.2	168.7	16423.4	0	19.4	4.6	31.0	0.18
2	1	51469.0	1	18.8	240.8	259.6	45950.0	0	19.7	2.6	22.2	0.09
$\geqslant 1$	23	24625.4	14	18.1	114.0	171.9	17176.3	0	19.4	4.5	30.6	0.18

Still, we can ask whether and under which conditions the sequence of bases is guaranteed to becomes optimal in a finite number of refinements. We will show that this holds if the LP oracle satisfies an extension of Assumption 2.9 such that not only the residual error of the numerical solution, but also the deviation from the basis information is sufficiently bounded. In this case, the number of refinements is polynomial in the size of the input.

The proof relies on the fact that there are only finitely many non-optimal basic solutions and that their uniquely determined infeasibilities are hence bounded. This is formalized by the following lemma.

Lemma 3.4 (Minimum Infeasibilities of Basic Solutions). *Suppose we are given an LP as in* (2.2),
$$\min\{c^\mathsf{T}x \mid Ax = b, x \geqslant \ell\},$$
with rational data $A \in \mathbb{Q}^{m \times n}$, $b \in \mathbb{Q}^m$, and $\ell, c \in \mathbb{Q}^n$. Then for any basic primal–dual solution x, y the following hold:

1. *Either x is (exactly) primal feasible or its maximum primal violation is at least $1/2^{4\langle A,b\rangle+5\langle \ell\rangle+2n^2+4n}$.*

2. *Either y is (exactly) dual feasible or its maximum dual violation is at least $1/2^{4\langle A,c\rangle+2n^2+4n}$.*

Proof. Suppose x, y is a basic primal–dual solution with respect to some basis $\mathcal{B}$. By standard arguments, we will show that the size of the entries in x and y is bounded by a polynomial in $\langle A, b, \ell, c \rangle$ and that all possible violations can be expressed as differences of rational numbers with bounded denominator (or zero).

Let $N = \{1, \ldots, n\} \setminus \mathcal{B}$, let $B = A_{\cdot\mathcal{B}}$ be the corresponding (square, non-singular) basis matrix and $N = A_{\cdot\mathcal{N}}$ the matrix formed by the nonbasic columns. Then the primal solution is given as solution of

$$\underbrace{\begin{pmatrix} B & N \\ 0 & I_{n-m} \end{pmatrix}}_{=:\tilde{B} \in \mathbb{Q}^{n \times n}} \begin{pmatrix} x_{\mathcal{B}} \\ x_{\mathcal{N}} \end{pmatrix} = \underbrace{\begin{pmatrix} b \\ \ell_{\mathcal{N}} \end{pmatrix}}_{\tilde{b} \in \mathbb{Q}^n} \tag{3.4}$$

and the dual vector y is determined by

$$\tilde{B}^\top \begin{pmatrix} y \\ z \end{pmatrix} = c \tag{3.5}$$

together with the vector $z \in \mathbb{Q}^{n-m}$ containing the dual slacks of the nonbasic variables.

First, primal infeasibilities can only stem from violations of the bound constraints on basic variables since the equality constraints and the nonbasic bounds are satisfied by construction. Hence, they are of the form $|x_i - \ell_i|$ for some $i \in \mathcal{B}$. From Lemma 1.3 applied to (3.4) we know that the entries in $x_{\mathcal{B}}$ have size at most $4\langle \tilde{B}, \tilde{b} \rangle - 2n^2$. Because

$$\langle \tilde{B}, \tilde{b} \rangle = \langle B \rangle + \langle N \rangle + \langle 0 \rangle + \langle I_{n-m} \rangle + \langle b \rangle + \langle \ell_{\mathcal{N}} \rangle$$
$$= \langle A, b, \ell_{\mathcal{N}} \rangle + (n+1)(n-m)$$
$$\leqslant \langle A, b, \ell \rangle + (n+1)(n-m),$$

it follows that all nonzero entries of $x_{\mathcal{B}}$ are of form p/q, $p \in \mathbb{Z}$, $q \in \mathbb{Z}_{\geqslant 0}$, with

$$q \leqslant 2^{4\langle A,b,\ell \rangle + 4(n+1)(n-m) - 2n^2} \leqslant 2^{4\langle A,b,\ell \rangle + 2n^2 + 4n}.$$

The entries in ℓ can be written as p'/q', $p' \in \mathbb{Z}$, $q' \in \mathbb{Z}_{\geqslant 0}$, with $q' \leqslant 2^{\langle \ell \rangle}$. Combining this we know that for all nonzero primal violations

$$|x_i - \ell_i| = \left| \frac{p}{q} - \frac{p'}{q'} \right| = \frac{|pq' - p'q|}{qq'}$$
$$\geqslant 1/(2^{4\langle A,b,\ell \rangle + 2n^2 + 4n} \cdot 2^{\langle \ell \rangle}) = 1/2^{4\langle A,b \rangle + 5\langle \ell \rangle + 2n^2 + 4n}.$$

Second, note that dual infeasibilities are precisely the (absolute values of the) negative entries of z in (3.5). Again from Lemma 1.3 we have

$$\langle z_j \rangle \leqslant 4\langle \tilde{B}, c \rangle - 2n^2$$
$$\leqslant 4\langle A, c \rangle + 4(n+1)(n-m) - 2n^2$$
$$\leqslant 4\langle A, c \rangle + 2n^2 + 4n$$

for all $j \in \mathcal{N}$, and so any nonzero dual violation is at least $1/2^{4\langle A,c \rangle + 2n^2 + 4n}$. $\qquad\square$

The following theorem states the main convergence result.

Theorem 3.5 (Polynomial Convergence to Optimal Bases). *Suppose we are given an LP as in (2.2),*

$$\min\{c^\mathsf{T}x \mid Ax = b, x \geqslant \ell\},$$

and a sequence of primal–dual solutions $\hat{x}_k, \hat{y}_k$ *with associated bases* $\mathcal{B}_k$ *such that*

1. $\|A\hat{x}_k - b\|_\infty \leqslant \varepsilon^k$,

2. $(\hat{x}_k)_i - \ell_i \geqslant -\varepsilon^k$ *for all* $i \in \{1, \dots, n\}$,

3. $c_i - \hat{y}_k^\mathsf{T} A_{\cdot i} \geqslant -\varepsilon^k$ *for all* $i \in \{1, \dots, n\}$,

4. $|(\hat{x}_k)_i - \ell_i| \leqslant \varepsilon^k$ *for all* $i \notin \mathcal{B}_k$, *and*

5. $|c_i - \hat{y}_k^\mathsf{T} A_{\cdot i}| \leqslant \varepsilon^k$ *for all* $i \in \mathcal{B}_k$

for $k = 1, 2, \dots$ *and* $0 < \varepsilon < 1$.

Then there exists $K \in O((m^2 \langle A \rangle + \langle b, \ell, c \rangle + n^2)/\log_2(1/\varepsilon))$ *such that the basis* $\mathcal{B}_k$ *is optimal for all* $k \geqslant K$.

Remark 3.6. Points 1 to 3 require that the primal and dual violations of $\hat{x}_k, \hat{y}_k$ converge to zero precisely as is guaranteed by Corollary 2.10 for the sequence of numeric solutions produced by iterative refinement. Additionally, we assume with points 4 and 5 that the numeric solutions become "more and more basic" in the sense that the deviation of the nonbasic variables from their bounds and the absolute value of the reduced costs of basic variables converges to zero at the same rate as the primal and dual violations. With Theorem 3.8 below we will prove that iterative refinement produces solutions that satisfy these conditions if the underlying floating-point solver returns approximate basic solutions.

Proof of Theorem 3.5. We first give a short outline of the idea of the proof. Define $\tilde{x}_k, \tilde{y}_k$ to be the exact basic solution vectors corresponding to $\mathcal{B}_k$, let $\mathcal{N}_k = \{1, \dots, n\} \setminus \mathcal{B}_k$, and let $B^k = A_{\cdot \mathcal{B}_k}$ and $N^k = A_{\cdot \mathcal{N}_k}$.

The proof will use points 4 and 5 to show that the convergence assumptions of points 1 to 3 on $\hat{x}_k, \hat{y}_k$ hold similarly for the basic solutions $\tilde{x}_k, \tilde{y}_k$. From this we know that for sufficiently large k, the primal and dual violations of $\tilde{x}_k, \tilde{y}_k$ drop below the minimum infeasibility thresholds stated in Lemma 3.4 and from then on $\mathcal{B}_k$ must be optimal. In the following we derive the threshold K stated in the theorem from which on this holds.

By construction, the basic solution $\tilde{x}_k$ satisfies the equality constraints $Ax = b$ exactly. For the violation of the lower bounds, we first show that $\hat{x}_k$ and $\tilde{x}_k$ converge towards each other.

Claim 1. For $k = 1, 2, \dots$, $\|\hat{x}_k - \tilde{x}_k\|_\infty \leqslant 2^{4m^2\langle A \rangle + 1}\varepsilon^k$.

For the nonbasic variables we have

$$\|(\hat{x}_k)_{\mathcal{N}_k} - (\tilde{x}_k)_{\mathcal{N}_k}\|_\infty = \|(\hat{x}_k)_{\mathcal{N}_k} - \ell_{\mathcal{N}_k}\|_\infty \leqslant \varepsilon^k$$

by point 4. For the basic variables we have

$$
\begin{aligned}
\|(\hat{x}_k)_{\mathcal{B}_k} - (\tilde{x}_k)_{\mathcal{B}_k}\|_\infty &\leqslant \|(B^k)^{-1}\|_\infty \|B^k((\hat{x}_k)_{\mathcal{B}_k} - (\tilde{x}_k)_{\mathcal{B}_k})\|_\infty \\
&= \|(B^k)^{-1}\|_\infty \|\underbrace{B^k(\hat{x}_k)_{\mathcal{B}_k}}_{= A\hat{x}_k - N^k \hat{x}_k} - (b - N^k \ell_{\mathcal{N}_k})\|_\infty \\
&\leqslant \|(B^k)^{-1}\|_\infty \big(\underbrace{\|A\hat{x}_k - b\|_\infty}_{\leqslant \varepsilon^k \text{ by point 1}} + \|N^k\|_\infty \underbrace{\|(\hat{x}_k)_{\mathcal{N}_k} - \ell_{\mathcal{N}_k}\|_\infty}_{\leqslant \varepsilon^k \text{ by point 4}}\big) \\
&\leqslant \|(B^k)^{-1}\|_\infty (\|N^k\|_\infty + 1)\varepsilon^k.
\end{aligned}
$$

By Lemma 1.2 and Lemma 1.3,

$$
\|(B^k)^{-1}\|_\infty \leqslant 2^{\langle (B^k)^{-1}\rangle} \leqslant 2^{4m^2\langle B^k\rangle}
$$

and

$$
\|N^k\|_\infty + 1 \leqslant 2^{\langle N^k\rangle} + 1 \leqslant 2^{\langle N^k\rangle + 1},
$$

hence

$$
\|(B^k)^{-1}\|_\infty (\|N^k\|_\infty + 1) \leqslant 2^{4m^2\langle B^k\rangle} \cdot 2^{\langle N^k\rangle + 1} \leqslant 2^{4m^2\langle A\rangle + 1},
$$

proving the claim. As a consequence, $\tilde{x}_k$ satisfies the lower bound constraints in the limit.

Claim 2. For $k = 1, 2, \ldots$ and $i \in \{1, \ldots, n\}$, $(\tilde{x}_k)_i - \ell_i \geqslant -2^{4m^2\langle A\rangle + 2}\varepsilon^k$.

This follows from

$$
\begin{aligned}
(\tilde{x}_k)_i - \ell_i &= (\tilde{x}_k)_i - (\hat{x}_k)_i + (\hat{x}_k)_i - \ell_i \\
&\geqslant -\|(\tilde{x}_k)_i - (\hat{x}_k)_i\|_\infty + (\hat{x}_k)_i - \ell_i \\
&\geqslant -2^{4m^2\langle A\rangle + 1}\varepsilon^k - \varepsilon^k \geqslant -2^{4m^2\langle A\rangle + 2}\varepsilon^k.
\end{aligned}
$$

For dual feasibility, we first show that the dual solutions $\hat{y}_k$ and $\tilde{y}_k$ converge towards each other.

Claim 3. For $k = 1, 2, \ldots$, $\|\hat{y}_k - \tilde{y}_k\|_\infty \leqslant 2^{4m^2\langle B^k\rangle}\varepsilon^k$.

This follows from point 5 via

$$
\begin{aligned}
\|\hat{y}_k - \tilde{y}_k\|_\infty &\leqslant \|((B^k)^\mathsf{T})^{-1}\|_\infty \|(B^k)^\mathsf{T}(\hat{y}_k - \tilde{y}_k)\|_\infty \\
&= \|((B^k)^\mathsf{T})^{-1}\|_\infty \underbrace{\|(B^k)^\mathsf{T}\hat{y}_k - c_{\mathcal{B}_k}\|_\infty}_{=\max\{|c_i - \hat{y}_k^\mathsf{T}A_{\cdot i}|, i \in \mathcal{B}_k\}} \\
&\leqslant \|((B^k)^\mathsf{T})^{-1}\|_\infty \varepsilon^k
\end{aligned}
$$

and

$$
\|((B^k)^\mathsf{T})^{-1}\|_\infty \leqslant 2^{\langle((B^k)^\mathsf{T})^{-1}\rangle} \leqslant 2^{4m^2\langle(B^k)^\mathsf{T}\rangle} \leqslant 2^{4m^2\langle B^k\rangle}
$$

using Lemma 1.2 and Lemma 1.3. As a consequence, also the $\tilde{y}_k$ satisfy dual feasibility in the limit.

Claim 4. For $k = 1, 2, \ldots$ and $i \in \{1, \ldots, n\}$, $c_i - \tilde{y}_k^\mathsf{T}A_{\cdot i} \geqslant -2^{4m^2\langle A\rangle+1}\varepsilon^k$.

For the basic variables, $c_i - \tilde{y}_k^\mathsf{T}A_{\cdot i} = 0 \geqslant -2^{4m^2\langle A\rangle}\varepsilon^k$. For $i \in \mathcal{N}_k$,

$$
\begin{aligned}
|(\hat{y}_k - \tilde{y}_k)^\mathsf{T}A_{\cdot i}| &\leqslant \|A_{\cdot i}\|_\infty \|\hat{y}_k - \tilde{y}_k\|_\infty \\
&\leqslant \|N^k\|_\infty 2^{4m^2\langle B^k\rangle}\varepsilon^k \\
&\leqslant 2^{\langle N^k\rangle} 2^{4m^2\langle B^k\rangle}\varepsilon^k \\
&\leqslant 2^{4m^2\langle B^k\rangle + \langle N^k\rangle}\varepsilon^k \\
&= 2^{4m^2\langle A\rangle}\varepsilon^k.
\end{aligned}
$$

This proves the claim via

$$
\begin{aligned}
c_i - \tilde{y}_k^\mathsf{T}A_{\cdot i} &= c_i - \hat{y}_k^\mathsf{T}A_{\cdot i} + (\hat{y}_k - \tilde{y}_k)^\mathsf{T}A_{\cdot i} \\
&\geqslant c_i - \hat{y}_k^\mathsf{T}A_{\cdot i} - |(\hat{y}_k - \tilde{y}_k)^\mathsf{T}A_{\cdot i}| \\
&\geqslant -\varepsilon^k - 2^{4m^2\langle A\rangle}\varepsilon^k \\
&\geqslant -2^{4m^2\langle A\rangle+1}\varepsilon^k.
\end{aligned}
$$

Claim 5. $\tilde{x}_k, \tilde{y}_k$ is optimal for $k \geqslant K \in O((m^2\langle A\rangle + \langle b, \ell, c\rangle + n^2)/\log_2(1/\varepsilon))$.

From Claim 2 and 4, the maximum violation of primal and dual feasibility of $\tilde{x}_k, \tilde{y}_k$ is $2^{4m^2\langle A\rangle+2}\varepsilon^k$ and $2^{4m^2\langle A\rangle+1}\varepsilon^k$, respectively. These drop below the thresholds from Theorem 3.4 as soon as

$$
2^{4m^2\langle A\rangle+2}\varepsilon^k < 1/2^{4\langle A,b\rangle+5\langle\ell\rangle+2n^2+4n}
$$

$$
\Leftrightarrow \varepsilon^k < 1/2^{4m^2\langle A\rangle+2+4\langle A,b\rangle+5\langle\ell\rangle+2n^2+4n}
$$

$$
\Leftrightarrow k > \frac{4m^2\langle A\rangle + 4\langle A, b\rangle + 5\langle\ell\rangle + 2n^2 + 4n + 2}{\log_2(1/\varepsilon)} =: K_P
$$

and

$$2^{4m^2\langle A\rangle+1}\varepsilon^k < 1/2^{4\langle A,c\rangle+2n^2+4n}$$

$$\Longleftrightarrow \varepsilon^k < 1/2^{4m^2\langle A\rangle+1+4\langle A,c\rangle+2n^2+4n}$$

$$\Longleftrightarrow k > \frac{4m^2\langle A\rangle + 4\langle A,c\rangle + 2n^2 + 4n + 1}{\log_2(1/\varepsilon)} =: K_D.$$

The resulting threshold $K := \max\{K_P, K_D\} + 1$ has the order claimed. From Theorem 3.4, $\tilde{x}_k, \tilde{y}_k$ must be primal and dual feasible for $k \geq K$. Since they are basic, hence complementary slack, they are also optimal. □

It remains to be shown that the sequence of solutions produced by iterative refinement actually satisfies the conditions of Theorem 3.5. For this we need the following extension of Assumption 2.9.

Assumption 3.7 (LP Oracle for Approximate Basic Solutions). *The LP Oracle satisfies Assumption 2.9 with constants ε, $0 \leq \varepsilon < 1$, and $\sigma \geq 0$ such that for all $A \in \mathbb{R}^{m\times n}$, $b \in \mathbb{R}^m$, and $c, \ell \in \mathbb{R}^n$ for which*

$$\min\{c^\mathsf{T}x \mid Ax = b, x \geq \ell\} \tag{2.2}$$

is primal and dual feasible, it returns an approximate primal–dual solution $\bar{x} \in \mathbb{Q}^n, \bar{y} \in \mathbb{Q}^m$ and a basis $\mathcal{B}$ that satisfy

1. *$|\bar{x}_i - \ell_i| \leq \varepsilon$ for all $i \notin \mathcal{B}$, and*

2. *$|c_i - \bar{y}^\mathsf{T}A_{\cdot i}| \leq \varepsilon$ for all $i \in \mathcal{B}$*

when it is given the LP

$$\min\{\bar{c}^\mathsf{T}x \mid \bar{A}x = \bar{b}, x \geq \bar{\ell}\} \tag{$\bar{P}$}$$

where $\bar{A} \in \mathbb{Q}^{m\times n}$, $\bar{c}, \bar{\ell} \in \mathbb{Q}^n$, and $\bar{b} \in \mathbb{Q}^m$ are A, c, ℓ, and b rounded to the working precision of the LP solver.

Then we can prove that conditions 4 and 5 of Theorem 3.5 are satisfied, i.e., that the refined solutions become "more and more basic".

Theorem 3.8 (Iterative Refinement of Basic Solutions). *Suppose we are given a primal and dual feasible LP as in (2.2) and an LP oracle which satisfies Assumption 3.7 with constants ε and σ. Let $x_k, y_k, \mathcal{B}_k, \Delta_{P,k}$, and $\Delta_{D,k}$, $k = 1, 2, \ldots$, be the sequences of primal–dual solutions, bases, and scaling factors produced by Algorithm 2.2 with incremental scaling limit $\alpha > 1$, and let $\tilde{\varepsilon} := \max\{\varepsilon, 1/\alpha\}$.*

Then for all k,

   1. $|(x_k)_i - \ell_i| \leqslant \tilde{\varepsilon}^k$ *for all* $i \notin \mathcal{B}_k$, *and*

   2. $|c_i - y_k^\mathsf{T} A_{\cdot i}| \leqslant \tilde{\varepsilon}^k$ *for all* $i \in \mathcal{B}_k$.

Proof. We prove both points together by induction over k. For $k = 1$, they hold directly because Assumption 3.7 is satisfied for the initial floating-point solution and $\varepsilon \leqslant \tilde{\varepsilon}$.

Consider $k + 1$ for $k \geqslant 1$ and let $\hat{x}, \hat{y}, \hat{\mathcal{B}}$ be the last approximate solution returned by the LP solver. Then for point 1 we have

$$
\begin{aligned}
|(x_{k+1})_i - \ell_i| &= \left| \left((x_k)_i + \frac{\hat{x}_i}{\Delta_{P,k+1}} \right) - \ell_i \right| \\
&= |\hat{x}_i + \Delta_{P,k+1} \underbrace{((x_k)_i - \ell_i)}_{=-\hat{\ell}_i}| / \underbrace{\Delta_{P,k+1}}_{\geqslant \tilde{\varepsilon}^k \text{ by Cor. 2.10}} \\
&\leqslant \underbrace{|\hat{x}_i - \Delta_{P,k+1} \hat{\ell}_i|}_{\leqslant \varepsilon \text{ by Ass. 3.7 for } i \notin \hat{\mathcal{B}} = \mathcal{B}_{k+1}} / \tilde{\varepsilon}^k \\
&\leqslant \varepsilon \tilde{\varepsilon}^k \\
&\leqslant \tilde{\varepsilon}^{k+1}
\end{aligned}
$$

for all $i \notin \mathcal{B}_{k+1}$. For point 2 we get similarly

$$
\begin{aligned}
|c_i - y_{k+1}^\mathsf{T} A_{\cdot i}| &= \left| c_i - \left(y_k + \frac{\hat{y}}{\Delta_{D,k+1}} \right)^\mathsf{T} A_{\cdot i} \right| \\
&= |\Delta_{D,k+1} \underbrace{(c_i - y_k^\mathsf{T} A_{\cdot i})}_{=\hat{c}_i} - \hat{y}^\mathsf{T} A_{\cdot i}| / \underbrace{\Delta_{D,k+1}}_{\geqslant \tilde{\varepsilon}^k \text{ by Cor. 2.10}} \\
&\leqslant \underbrace{|\Delta_{D,k+1} \hat{c}_i - \hat{y}^\mathsf{T} A_{\cdot i}|}_{\leqslant \varepsilon \text{ by Ass. 3.7 for } i \in \hat{\mathcal{B}} = \mathcal{B}_{k+1}} / \tilde{\varepsilon}^k \\
&\leqslant \varepsilon \tilde{\varepsilon}^k \\
&\leqslant \tilde{\varepsilon}^{k+1}
\end{aligned}
$$

for all $i \in \mathcal{B}_{k+1}$, completing the induction step. $\qquad\square$

Remark 3.9. The bound on the number of refinements may seem surprisingly high when considering that the best-known iteration complexity for interior point methods is $O(\sqrt{n+m}\langle A, b, \ell, c \rangle)$ (Renegar, 1988) and that in each refinement round we solve an entire LP. One reason for this difference is that iterative refinement converges only linearly as proven in Corollary 2.10, while interior point algorithms are essentially a form of Newton's method, which allows for superlinear convergence. The motivation for our interest in iterative refinement is not in improving worst-case complexity, but in its practical efficiency for obtaining exact solutions. The convergence results of this section, however, provide an important theoretical underpinning of the algorithms developed.

3.2.3 Iterative Refinement with Basis Verification

As already seen in the experiments of Section 2.5 and 3.2.1, the worst-case bound of Theorem 3.5 is only of theoretical interest. In practice, an optimal basis is typically arrived at after very few refinements. Hence, we do not want to rely on bounds computed a priori, but check the optimality of the basis early. A natural idea is to perform an exact rational solve as soon as the basis has not changed for a specified number of refinements.

This strategy is applied by Algorithm 3.2. It takes a stalling threshold L as parameter and computes an exact primal–dual solution corresponding to the current basis as soon as the LP solver returned this basis for the last L refinements. This is controlled by a counter k^*, which is set to the current refinement round in line 13 when the basis changes. The optimality of the basis is checked tested after L more refinements without updates to k^*. (L may be zero, in which case we check each basis.) If the basis turns out to be primal or dual infeasible, we set k^* to infinity in order to skip the checks of the basis until the LP solver returns a new candidate.

It is possible that the refined numeric solution x_k, y_k itself may turn out to be optimal in exact arithmetic. In our experiments, we have experienced this only for very few simple LPs, but because it is cheap, we compute and check the residual errors before an exact solve. Furthermore the residuals are a necessary ingredient to setting up the modified LP for the next refinement round if the basis is not detected to be optimal. The following corollary summarizes our findings.

Corollary 3.10 (Iterative Refinement with Basis Verification). *Algorithm 3.2 with an LP oracle satisfying Assumption 3.7 terminates with an optimal solution in oracle-polynomial running time.*

Proof. Corollary 2.10 and Theorem 3.8 prove that the sequence of basic solutions $x_k, y_k, \mathcal{B}_k$ satisfies the conditions of Theorem 3.5. Hence $\mathcal{B}_k$ is optimal after a polynomial number of refinements. According to Theorem 2.12, this runs in oracle-polynomial time. As proven by Edmonds (1967), the linear systems in lines 15 and 16 can be solved in polynomial time, which is performed at most once per refinement round. $\square$

Remark 3.11 (Replacing Approximate Solutions by Infeasible Basic Solutions). Optionally, if the basis verification fails, one could update the numeric solution x_k, y_k to the exact basic solution x^*, y^* —even if the infeasibilities in x^*, y^* are higher than in x_k, y_k. The motivation for this might be to generate the next modified LP in such a way as to maximally exhibit the infeasibilities of the current basis and force the floating-point LP solver to move to a new basis.

However, this would come with the disadvantage that x^*, y^* may have a more complicated rational representation. It would affect the proof of Theorem 2.12 that the x_k, y_k have polynomial encoding length. In practice it may slow down the rational arithmetic operations because the numbers then fail to have denominators in a common base.

Algorithm 3.2: Iterative Refinement with Basis Verification

---

input: rational, primal and dual feas. LP $\min\{c^\mathsf{T}x \mid Ax = b, x \geqslant \ell\}$
params: stalling threshold $L \geqslant 0$
output: optimal primal–dual solution $x^* \in \mathbb{Q}^n, y^* \in \mathbb{Q}^m$

1 **begin**
2   initialize Algorithm 2.2 with tolerances $\varepsilon_P, \varepsilon_D, \varepsilon_S = 0$
3     $\mathcal{B}_0 \leftarrow \emptyset$
4     $k^* \leftarrow \infty$

5   **for** $k \leftarrow 1, 2, \dots$ **do**   `/* refinement loop */`
6   perform next refinement step in Algorithm 2.2
7        $x_k, y_k \leftarrow$ refined numeric solution
8        $\mathcal{B}_k \leftarrow$ approximate basis from LP solver
9        $\delta_{P,k}, \delta_{D,k}, \delta_{S,k} \leftarrow$ residual errors (primal, dual, comp. slack.)

10   **if** $\delta_{P,k}, \delta_{D,k}, \delta_{S,k} = 0$ **then**   `/* check termination */`
11   return $x^* \leftarrow x_k, y^* \leftarrow y_k$

12   **if** $\mathcal{B}_k \neq \mathcal{B}_{k-1}$ **then**   `/* mark new basis candidate */`
13           $k* \leftarrow k$

14   **if** $k \geqslant k^* + L$ **then**   `/* check current basis */`
15           $x^*_{\mathcal{B}_k} \leftarrow A^{-1}_{\cdot \mathcal{B}_k}(b - \sum_{i \notin \mathcal{B}_k} A_{\cdot i} \ell_i)$ in rational arithmetic
16           $y^* \leftarrow (A^{-1}_{\cdot \mathcal{B}_k})^\mathsf{T} c_{\mathcal{B}_k}$ in rational arithmetic
17   **if** $x^*_i \geqslant \ell_i$ *for all* $i \in \mathcal{B}_k$ and $(y^*)^\mathsf{T} A_{\cdot i} \leqslant c_i$ *for all* $i \notin \mathcal{B}_k$ **then**
18   **foreach** $i \notin \mathcal{B}_k$ **do** $x^*_i \leftarrow \ell_i$
19   return x^*, y^*
20   **else**
21              $k^* \leftarrow \infty$ `/* do not factorize basis again */`

3.3 From Approximate to Exact Solutions

The LP algorithm developed in the previous section relies solely on the optimality of the basis information to construct an exact solution. Except for the computation of the residual vectors it does not make use of the more and more accurate solutions produced. In this section, we discuss a conceptually different class of techniques that exploit the approximate solutions as starting points in order to move from them to an exact optimal solution.

3.3.1 Convergence to an Optimal Solution

Until now the convergence of the residual errors to zero was sufficient for our results and we did not have to address the question whether the sequence of solutions x_k, y_k itself converges to a limit point. If the solutions returned by the LP oracle are not bounded—which can happen if the set of optimal solutions is unbounded—then this does not hold as can be seen in the following example.

Example 3.12. Consider the degenerate LP

$$\min\{x_1 - x_2 \mid x_1 - x_2 = 0, -x_1 + x_2 = 0, x_1, x_2 \geqslant 0\}.$$

One can show that iterative refinement produces the sequence of primal–dual solutions

$$\big((x_k)_1, (x_k)_2\big) = \big(2^k + 2^{-k-1}, 2^k - 2^{-k-1}\big)$$
$$\big((y_k)_1, (y_k)_2\big) = \big(2^k + 2^{-1} + 2^{-3k-1}, 2^k - 2^{-1} - 2^{-3k-1}\big)$$

for $k = 1, 2, \ldots$, if for the k-th transformed problem (built with scaling factors $\Delta_{P,k} = \Delta_{D,k} = 2^k$)

$$\min\{-1/2^{2k}x_1 + 1/2^{2k}x_2 \mid x_1 - x_2 = -1, -x_1 + x_2 = +1,$$
$$x_1 \geqslant -2^{2k-1} - 1/2, x_2 \geqslant -2^{2k-1} + 1/2\},$$

the LP solver returns the approximate solution $\big((\hat{x}_k)_1, (\hat{x}_k)_2\big) = \big(2^{2k} - 1/4, 2^{2k} + 1/4\big)$ and $\big((\hat{y}_k)_1, (\hat{y}_k)_2\big) = \big(2^{2k} - 7 \cdot 2^{-2k-4}, 2^{2k} + 7 \cdot 2^{-2k-4}\big)$. Their primal violation ($= 2^{-k}$), dual violation ($= 2^{-3k}$), and their violation of complementary slackness ($= 2^{-4k}$) all go to zero, but the iterates themselves go to infinity.

This is obviously an artificial construction. If the solutions returned by the LP solver are bounded by a constant, $\|\hat{x}_k\|_\infty, \|\hat{y}_k\|_\infty \leqslant C$, say, then

$$x_k = \sum_{\kappa=1}^{k} \frac{1}{\Delta_{P,\kappa}} \hat{x}_\kappa$$

is a Cauchy sequence, because for any $k, k' \geqslant K$, the distance $\|x_k - x_{k'}\|_\infty$ is at most $C \sum_{\kappa=K+1}^{\infty} \varepsilon^\kappa = \varepsilon^{K+1}/(1 - \varepsilon)$, where ε is the rate of convergence from Corollary 2.10. Hence, it is guaranteed to converge to a finite limit point $\tilde{x}$; equally, y_k converges to a limit $\tilde{y}$. Note that the boundedness of the corrector solutions is an immediate consequence when either Assumption 2.11 holds, stating that the LP solver returns limited-precision floating-point solutions, or Assumption 3.7 holds that it returns approximately basic solutions.

However, without any further assumptions, $\tilde{x}$ and $\tilde{y}$ may have irrational components. Even if they are rational, we still have no general guarantee on their size or the size of their denominators. On the other hand, if the LP solver returns approximately basic solutions as in Assumption 3.7 then we know that $\tilde{x}$ and $\tilde{y}$ are basic solutions. (This follows from Claims 1 and 3 in the proof of Theorem 3.5.) Hence, their size is polynomial in the size of the input.

In Section 3.3.2 we will briefly discuss methods to compute an exact solution when the limit point is irrational. Sections 3.3.3 and 3.3.4 focus on the case when iterative refinement converges to rational limit points, ideally with polynomially bounded denominators, for which more efficient methods using continued fraction approximations are available.

3.3.2 Interior Point Projections and Lattice Reduction

One general method for converting a high-precision into an exact optimal solution is the one applied by Renegar (1988) as last step of his interior point algorithm. For an LP given in inequality form, $\min\{c^\mathsf{T}x \mid Ax \geqslant b\}$, and the last interior point iterate x_k he identifies all approximately tight rows as j with $A_j.x_k \leqslant 2^{-C\langle A,b,c\rangle}$ for some large constant C. Subsequently, if $\mathcal{S}$ is the index set of an (inclusion-wise) maximal linearly independent subset of these rows and $P = A_{\mathcal{S}.}$ is the matrix formed by these rows (computed by Gaussian elimination), then the orthogonal projection of x_k onto $Px = b_{\mathcal{S}}$,

$$x^P := x_k - P^\mathsf{T}(PP^\mathsf{T})^{-1}(Px_k - b_{\mathcal{S}}), \tag{3.6}$$

is optimal if x_k is sufficiently advanced.

This can be interpreted as a generalization of the case where x_k is already approximately basic. If $\mathcal{S}$ is chosen as the index set of the n tight rows of a row basis (confer Section 2.1), then (3.6) collapses to the solution of the standard basis system. From this observation, we could generalize Algorithm 3.2 to handle LP oracles without basis information by adding the detection of $\mathcal{S}$ in line 8 and by replacing lines 15, 16, and 17 by projection steps for the primal and dual solution and an extended optimality check. As before, this verification procedure has to be performed in rational arithmetic.

A conceptually different technique is simultaneous Diophantine approximation as employed by Grötschel, Lovász, and Schrijver (1988) in the context of the ellipsoid method and outlined in Section 3.1.4. To apply Lemma 3.3, we need a bound on the facet complexity of both the feasible region $P = \{x \mid Ax \geqslant b\}$ and the optimal hyperplane $\{x \mid c^\mathsf{T}x = C\}$, where C is the optimal value of the LP. This can be computed as $\varphi = 5\langle A, b, c\rangle$, confer Schrijver (1986, Theorem 10.3). Then, after performing iterative refinement until a sufficiently accurate solution x_k within distance $2^{-6n\varphi}$ of the optimal hyperplane is reached, performing the lattice reduction algorithm behind Theorem 3.2 to $\alpha = x_k$ gives an optimal primal solution $x^* = \frac{1}{q}p$.

The same can be applied to the refined dual solution y_k (considered as primal solution to the dual LP) to obtain a proof of optimality y^*. This can be used to apply this reconstruction in an output-sensitive fashion, i.e., for early iterates x_k for which Lemma 3.3 would not yet guarantee the optimality of the reconstructed x^*. In this case the reconstructed dual solution y^* can be checked for verification—if this fails, refinement is continued. This also resolves the issue that in fact we only know a bound on the residual error of the current solution x_k, y_k, but not on its error, i.e., its actual distance to the optimal face. However, since we know that it *does* converge to a point on the optimal face at a linear rate of convergence, we are guaranteed that the reconstructed solution is exactly optimal after at most $O(n\langle A, b, c\rangle)$ refinements. We may hope that it becomes optimal earlier.

3.3.3 Reconstruction of Rational Limit Points

If the limit point $\tilde{x}, \tilde{y}$ of the sequence of approximate solutions x_k, y_k is rational itself, then under certain conditions the more efficient algorithms behind Theorem 3.1

can be applied. Suppose we know *a priori* a bound M on the denominators in the limit. Then we can compute $\tilde{x}, \tilde{y}$ from an approximate solution x_k, y_k satisfying $\|(x_k, y_k) - (\tilde{x}, \tilde{y})\|_\infty < 1/2M^2$ by rounding the entries to the largest convergent in the continued fraction expansion that has denominator at most M. If the size of M is small, i.e., polynomial in the input size, then iterative refinement produces a sufficiently accurate solution after a polynomial number of refinements.

We need to address the following difficulties. In general, rational solutions may have arbitrarily large denominator. Even if we restrict ourselves to basic solutions, which have denominators of polynomial size, then classical Hadamard type bounds are often very weak. This has been demonstrated by Abbott and Mulders (2001) for random square matrices and by Steffy (2011) for the special case of selected basis matrices from linear programs. For the Hadamard bound H from (3.1) that holds for *all* basis matrices of an LP, this situation must be even more pronounced. Tighter bounds that are reasonably cheap to compute are—to the best of our knowledge—not available. Computing an approximate solution with error below $1/2H^2$ before applying the rounding procedure can thus be unnecessarily expensive. We will design an output-sensitive algorithm in the spirit of Chen and Storjohann (2005) that attempts to reconstruct exact solution vectors during early rounds of refinement, i.e., from approximate solutions with error above $1/2H^2$, using a bound on the denominators smaller than H. The correctness of these heuristically reconstructed solution can be tested exactly using rational arithmetic.

As a second difficulty, Theorem 3.1 requires us to bound the error of the solution, i.e., the distance $\|(x_k, y_k) - (\tilde{x}, \tilde{y})\|_\infty$. During the iterative refinement procedure, however, we only know bounds on the residual errors that converge to zero. For bounding the error from the residual in the fashion of $\|x_k - \tilde{x}\|_\infty \leqslant C \|Ax_k - b\|_\infty$ only weak bounds on C are known, even for basic solutions. Again, assuming the worst case would mostly lead to an excessively large number of refinements. Both issues are addressed by Algorithm 3.3. We now give a description of it, followed by a proof of correctness in Lemma 3.13.

The algorithm is an extension of the basic iterative refinement for linear programs, Algorithm 2.2, interleaved with attempts at rational reconstruction. For $k = 1$, the algorithm starts with the first floating-point solve to obtain the initial approximate solution and the corresponding residual errors. Unless the solution is exactly optimal, we enter the rounding routine. The working bound M_k on the denominator is computed such that $1/2M_k^2$ equals $\beta^k \max\{\delta_{P,k}, \delta_{D,k}, \delta_{S,k}\}$, which gives an estimate for the error in the solution that is motivated as follows. If reconstruction attempts fail, the term β^k, governed by parameter $\beta > 1$, keeps growing exponentially such that we eventually obtain a true bound on the error. Initially, however, β^k is small in order to account for the many cases where the residual is a good proxy for the error.

We first round the entries of the primal vector x_k to the closest rational number with denominator at most M_k. This is achieved by running, for each component, the extended Euclidean algorithm in order to compute the largest convergent in the continued fraction expansion that has denominator at most M_k. It is denoted by "round_to_denom$((x_k)_i, M_k)$". Then we check primal feasibility before proceeding to the dual vector. If feasibility and optimality could be verified in rational arithmetic, the rounded solution is returned as optimal. Otherwise, we compute the next refinement round after which reconstruction should be tried again. Because computing continued fraction approximations becomes increasingly expensive as the encoding length of the approximate solutions grows, rational reconstruction is executed at a geometric frequency governed by parameter f such that the cumulative effort spent in this part of the algorithm grow only linearly in the number of refinement rounds. The following lemma shows that the algorithm computes an exact optimal solution to a primal and dual feasible LP if the approximate solutions converge to a basic solution.

Lemma 3.13 (Output-Sensitive Reconstruction of Basic Limit Points). *Suppose that the sequence x_k, y_k in Algorithm 3.3 converges to a basic solution $\tilde{x}, \tilde{y}$ associated with basis $\mathcal{B}$ and each entry in $\tilde{x}, \tilde{y}$ has denominator at most $\tilde{q}$.*
    If there exist constants $0 < \varepsilon < 1$ and $C \geqslant 1$ such that for all k

1. *$\max\{\delta_{P,k}, \delta_{D,k}, \delta_{S,k}\} \leqslant C\varepsilon^k$,*

2. *$|(\hat{x}_k)_i - \ell_i| \leqslant C\delta_{P,k}$ for all $i \notin \mathcal{B}$, and*

3. *$|c_i - \hat{y}_k^T A_{\cdot i}| \leqslant C\delta_{D,k}$ for all $i \in \mathcal{B}$,*

then Algorithm 3.3 with parameters $\beta < 1/\varepsilon$ and $f \geqslant 1$ terminates with an optimal solution in $O(\max\{\langle \tilde{q} \rangle, m^2 \langle A \rangle\})$ refinements.

Remark 3.14 (Implicit Basis Information). Note that neither the basis nor the limit point nor the bound on the denominator must be known a priori in Algorithm 3.3. Their existence suffices, which is, for instance, guaranteed by Theorems 3.5 and 3.8 under Assumption 3.7 that the LP oracle returns approximately basic solutions.

Proof of Lemma 3.13. In order to apply Theorem 3.1 we first need to choose k sufficiently large such that the bound on the denominator M_k exceeds $\tilde{q}$,

$$1/\sqrt{2\beta^k \max\{\delta_{P,k}, \delta_{D,k}, \delta_{S,k}\}} \geqslant \tilde{q}$$
$$\Longleftrightarrow \beta^k \max\{\delta_{P,k}, \delta_{D,k}, \delta_{S,k}\} \leqslant 1/2\tilde{q}^2$$

which holds by point 1 if $C\beta^k \varepsilon^k \leqslant 1/2\tilde{q}^2$, i.e., if

$$k \geqslant K_1 := (2\log_2 \tilde{q} + \log_2 C + 1)/\log_2(1/\beta\varepsilon).$$

Second, we need to bound the errors in the primal and dual solution by their residuals. A simple bound can be computed as in the proofs of Claim 1 and 3 for Theorem 3.5 and yields

$$\|\tilde{x} - x_k\|_\infty \leqslant 2^{4m^2\langle A\rangle + 1} C\delta_{P,k}, \text{ and}$$
$$\|\tilde{y} - y_k\|_\infty \leqslant 2^{4m^2\langle A\rangle + 1} C\delta_{D,k}.$$

For $k \geqslant K_1$ we have that

$$2^{4m^2\langle A\rangle + 1} C \max\{\delta_{P,k}, \delta_{D,k}\} < 1/2M_k\tilde{q}$$

if

$$2^{4m^2\langle A\rangle+1} C \max\{\delta_{P,k}, \delta_{D,k}\} < 1/2M_k^2$$

$$\Leftrightarrow 2^{4m^2\langle A\rangle+1} C \max\{\delta_{P,k}, \delta_{D,k}\} < \beta^k \max\{\delta_{P,k}, \delta_{D,k}, \delta_{S,k}\}.$$

This in turn holds if $2^{4m^2\langle A\rangle+1} C < \beta^k$, which is equivalent to

$$k > (4m^2\langle A\rangle + \log_2 C + 1)/\log_2 \beta =: K_2.$$

Hence the errors drop below $1/2M_k\tilde{q}$ for $k > K^* := \max\{K_1, K_2\}$.

All in all, for $k > K^*$, Theorem 3.1—applied with $M = M_k$ and $\alpha = (x_k)_i$ or $(y_k)_j$—guarantees that the continued fraction expansion in lines 13 and 16 returns the optimal solution $\tilde{x}, \tilde{y}$. If $f > 1$ the algorithm skips the call to the reconstruction routines during some refinement rounds. By construction of the relative increase of k^* by a factor of f in line 19, however, it is guaranteed that they are executed at least once during the refinement rounds $K^* + 1, K^* + 2, \ldots, \lceil fK^*\rceil + 1$—unless an exactly optimal solution was found before. Hence the algorithm terminates after at most $\lceil fK^*\rceil + 1$ refinement rounds, which has the order claimed. □

Several comments are in place. First, in our simplex-based implementation discussed in Section 3.4, conditions 2 and 3 are naturally satisfied with $C = 1$ because nonbasic variables are always exactly fixed to their bound and nonzero reduced costs on the basic variables are treated as violations. Second, in our experience, the number of refinements is never dominated by the term $m^2\langle A\rangle$. The denominators in the reconstructed solution are the decisive factor. Third, we already mentioned that the basis does not need to be known explicitly. Accordingly, Algorithm 3.3 may even return an optimal solution x^*, y^* that is different from $\tilde{x}, \tilde{y}$ if for an early iterate x_k, y_k is successful. In this case, x^*, y^* is not guaranteed to be a basic solution unless we explicitly discard solutions that are not basic in the optimality checks.

Remark 3.15 (Basic Limit Points via Objective Perturbation). If the limit point is not a basic solution, Algorithm 3.3 could be applied as a heuristic, but without guarantee of termination. One possible way to enforce convergence to a basic solution is a perturbation strategy as applied in Grötschel, Lovász, and Schrijver (1981). By adding Q^{-i} to the objective coefficient of the i-th variable, where $Q > 1$ is chosen sufficiently large, ties between multiple optimal vertices are broken, and it is ensured that the perturbed LP has a unique optimal and basic solution (which is also optimal for the unperturbed problem). For practical efficiency, certainly a dynamic scheme would be called for that applies perturbations only to variables with small reduced cost and adjusts the value of Q gradually over the refinements.

3.3.4 Exploiting Common Factors in Vector Reconstruction

As mentioned already in Section 3.1.4, Cook and Steffy (2011) have investigated the solution of sparse systems of linear equations by iterative refinement combined with rational reconstruction of the solution vector. They demonstrated that using the

---

Algorithm 3.3: Iterative Refinement with Rational Reconstruction

---

input: rational, primal and dual feas. LP $\min\{c^\mathsf{T}x \mid Ax = b, x \geqslant \ell\}$
params: geometric reconstruction frequency $f \geqslant 1$,
              error correction factor $\beta > 1$
output: optimal primal–dual solution $x^* \in \mathbb{Q}^n, y^* \in \mathbb{Q}^m$

1 **begin**
2 $\quad$ initialize Algorithm 2.2 with tolerances $\varepsilon_P, \varepsilon_D, \varepsilon_S = 0$
3 $\quad$ $k^* \leftarrow 0$

4 $\quad$ **for** $k \leftarrow 1, 2, \dots$ **do**   /* refinement loop */
5 $\quad\quad$ perform next refinement step in Algorithm 2.2
6 $\quad\quad$ $x_k, y_k \leftarrow$ refined numeric solution
7 $\quad\quad$ $\delta_{P,k}, \delta_{D,k}, \delta_{S,k} \leftarrow$ residual errors (primal, dual, complementarity)

8 $\quad\quad$ **if** $\delta_{P,k}, \delta_{D,k}, \delta_{S,k} = 0$ **then**   /* check termination */
9 $\quad\quad\quad$ **return** $x^* \leftarrow x_k, y^* \leftarrow y_k$
10 $\quad\quad$ **else if** $k \geqslant k^*$ **then**   /* try reconstruction */
11 $\quad\quad\quad$ $M_k \leftarrow 1/\sqrt{2\beta^k \max\{\delta_{P,k}, \delta_{D,k}, \delta_{S,k}\}}$
12 $\quad\quad\quad$ **for** $i \leftarrow 1, \dots, n$ **do**
13 $\quad\quad\quad\quad$ $x_i^* \leftarrow$ round_to_denom$((x_k)_i, M_k)$

14 $\quad\quad\quad$ **if** x^* *feasible in rational arithmetic* **then**
15 $\quad\quad\quad\quad$ **for** $j \leftarrow 1, \dots, m$ **do**
16 $\quad\quad\quad\quad\quad$ $y_j^* \leftarrow$ round_to_denom$((y_k)_j, M_k)$

17 $\quad\quad\quad\quad$ **if** x^*, y^* *optimal in rational arithmetic* **then**
18 $\quad\quad\quad\quad\quad$ **return** x^*, y^*

19 $\quad\quad\quad$ $k^* \leftarrow \lceil fk \rceil$ /* counter for next reconstruction */

DLCM method instead of simple componentwise reconstruction yields an average decrease of 60% in running time. Since their test instances were derived from basis matrices of linear programs, similar results can be expected for the reconstruction routines in Algorithm 3.3.

The according adaption of Algorithm 3.3 to use the DLCM strategy is straightforward. The loops in lines 12, 13, 14, and 16 over the components of x_k and y_k are each replaced by the vector-wise DLCM routines. For details we refer to Steffy (2011).

Note that in theory we could perform one DLCM reconstruction on the entire primal–dual solution vector x_k, y_k. This is not necessarily beneficial because the efficiency of DLCM relies on the denominators of the components sharing common factors, and in general it is not clear to what extent this holds true across x_k and y_k. Hence, we perform two separate DLCM sweeps over x_k and y_k.

3.4 Computational Study

In the following we will analyze the computational performance of the exact linear programming algorithms developed in this chapter. For our theoretical discussions above we have made the assumptions that the LP is primal and dual feasible and given in the form of equality constraints and lower bounds on the variables. Hence, in Section 3.4.1 we first describe how they can be applied to LPs of general form that may be infeasible or unbounded. Section 3.4.2 provides details of our implementation and Section 3.4.4 presents the computational analysis.

3.4.1 Exact Solution of General Linear Programs

Since Algorithms 3.2 and 3.3 are extensions of the iterative refinement algorithm, it is natural to handle infeasible and unbounded LPs the same way as described in Section 2.4. We can set up auxiliary LPs that are artificially primal and dual feasible and have optimal solutions from which a Farkas proof or a primal feasible solution and a primal unbounded ray can be derived. Solving these reformulations exactly yields exact proofs of infeasibility or unboundedness, or allows these claims to be rejected. Hence we can follow the integrated solving loop shown in Figure 2.3 with the only adaption that each LP is solved exactly. This is achieved by solving the basis system exactly or performing rational reconstruction during the step "check termination". When testing feasibility, the infeasibility box algorithm from Section 2.4.2 can be applied optionally to try and convert the approximate Farkas proof to an exact proof of infeasibility.

LPs of general form with ranged rows and arbitrary bounds on the variables require no adaption in the rational reconstruction routines since these only consider the solution vectors. The standard procedure to solve the basis system exactly when inequalities are present involves adding slack variables for basic rows or, when using a row basis matrix, unit rows for nonbasic variables. A slightly more efficient approach might be to use the so-called *reduced basis matrix* first applied by Zoutendijk (1970) and Powell (1975) and recently reconsidered by Wunderling (2012) and Gleixner (2012) for the simplex algorithm. The reduced basis matrix is the submatrix of the constraint matrix corresponding to basic columns and nonbasic rows. Since it is smaller in dimension than either the column and row basis matrices it could lead to fewer operations in rational arithmetic during the exact solves. Since it is a separate routine, how the exact solves are implemented does not interact with the main algorithm.

3.4.2 Implementation

We implemented Algorithms 3.2 and 3.3 by extending the iterative refinement procedure of the LP solver SoPlex described in Section 2.5.1. Rational reconstruction and the exact solution of the basis systems can be activated independently by parameters. If both are active and called during the same refinement round, reconstruction is performed first, since we expect it is typically much faster. Their execution can be controlled by changing the reconstruction frequency f and the stalling threshold L for the rational factorization. Their default values also used in the experiments described below are f = 1.2 and L = 2.

Rational LU Factorization.   For solving the basis systems exactly, we use a rational factorization of the standard basis matrix with slack columns for basic rows. To this end, we modified the implementation of the floating-point LU factorization of SoPlex and replaced all data types and arithmetic operations from floating-point to rational arithmetic. This was facilitated by the arithmetic operators provided by the wrapper class around the rational GMP data type described in Section 2.5.1.

We had to adjust one implementation trick that used tiny nonzeros to mark non-zero entries of sparse vector that become zero. Furthermore, we keep the Markowitz threshold fixed at its default value of 0.01. This is possible since rational arithmetic can handle numbers with arbitrarily small absolute value safely. The Markowitz threshold tries to balance stability and sparsity when choosing the pivot order, see Markowitz (1957); in the floating-point factorization it is dynamically increased in the case that the factorization is judged instable. If a time limit is specified, this is checked regularly during factorization and solves because compared to the floating-point implementation these procedures can exhibit huge running times.

DLCM Reconstruction.   For rational reconstruction, we have integrated and adapted the code of Daniel Steffy [53] for our purposes. We changed the computation of the bound on the denominator to use the residual errors as in line 11 of Algorithm 3.3. The error correction factor β was set to 1.1. Furthermore, since we hold nonbasic variables fixed exactly at one of their bounds, we skip the corresponding entries of the primal vector during reconstruction.

3.4.3 Experimental Setup

The goal of our experiments was to analyze individually and compare against each other the computational performance of Algorithms 3.2 and 3.3 in the implementation described above. We will denote these by SoPlex$_{fac}$ and SoPlex$_{rec}$, respectively. In particular we were interested in the following questions. How many instances can be solved by each algorithm and how do they compare in running time? How much of the solving time is consumed by rational factorization and rational reconstruction, and how often are these routines called? How many refinements are needed until reconstruction succeeds, and are the reconstructed solutions always basic? In addition, we compared both algorithms against the current state-of-the-art, the incremental precision boosting procedure of Algorithm 3.1 as implemented by the solver QSopt_ex.

Hardware.   The experiments were conducted on a cluster of 64-bit Intel Xeon X5672 CPUs at 3.2 GHz with 12 MB cache and 48 GB main memory. In order to safeguard against a potential mutual slowdown of parallel processes, we ran only

---

[53] Daniel E. Steffy. Dense Iterative Refinement Solver Version 1.1, `https://files.oakland.edu/users/steffy/web/rational/`, accessed January 2015.

one job per node at a time. We used a time limit of two hours per instance for each SoPlex and QSopt_ex run.

Software. We used the SoPlex developer version 2.0.1.1, which implements the algorithms described above. For details on the implementation of iterative refinement, see Section 2.5. SoPlex was compiled with GCC 4.8.2 and linked to the external libraries GMP 5.1.3, EGlib 2.6.20, and zlib 1.2.8 for reading compressed instance files. Additionally, we compare SoPlex with QSopt_ex 2.5.10.[54]

Instances. We used the 1,202 LP instances described in detail in Section 1.4.1, which also served as test set for our previous experiments in Section 2.5.

3.4.4 Computational Results

The experimental results of the SoPlex$_{fac}$, SoPlex$_{rec}$, and QSopt_ex runs for each instance can be found in detail in Table A.4 in the appendix. For all solvers we state the number of simplex iterations and their total running time. Additionally, for SoPlex$_{fac}$, we list the number of refinement rounds, the number of rational factorizations, and their time including the time for the rational triangular solves and the subsequent optimality test; for SoPlex$_{rec}$, we give the number of refinement rounds, the number of calls to rational reconstruction, and their time, again including the optimality test. For QSopt_ex we also state the maximum level of precision it used.

The columns "dlcm" for SoPlex$_{fac}$ and SoPlex$_{rec}$ state the size of the least common multiple of the denominators in the solution, computed as rounded base-10 logarithm. (To be precise, we compute this size separately for the primal and the dual solution and state their maximum, which is typically the size in the primal.) This value indicates how "complicated" the representation of the exact solution is and is relevant for the analysis of the reconstruction solver below.

The average performance of SoPlex$_{fac}$ and SoPlex$_{rec}$ on the instances solved by both is compared in Table 3.2. Besides averaging over all instances, we consider subsets of LPs with increasing difficulty. The classes "$\leqslant$ T" refer to the set of instances that are easy for *both* algorithms in the sense that *both* could solve them by time T. The classes "$\geqslant$ T" denote the set of instances that are hard in the sense that at least one solver took more than time T. It is important to consider both running times in the definition of these sub sets in order to avoid a systematic bias towards one solver, see, e.g., the discussion in Achterberg and Wunderling (2013). An aggregated comparison against QSopt_ex is given in Table 3.3. For average times and simplex iterations we always report shifted geometric means as explained in Section 1.3, using a shift of two seconds and 100 simplex iterations.

Solved Instances. Of the 1,202 instances, 1,186 instances could be solved exactly by both SoPlex$_{fac}$ and SoPlex$_{rec}$. Five instances could only be solved by rational factor-

---

[54] David L. Applegate, William Cook, Sanjeeb Dash, and Daniel G. Espinoza. QSopt_ex. http://www.dii.uchile.cl/~daespino/ESolver_doc/.

ization,[55] three instances only by the reconstruction solver.[56] On cont1, both timed out. As in the experiments of Section 2.5, the iterative refinement process suffered from insufficient performance of the floating-point solver on the same seven instances. The four time outs of SoPlex$_{fac}$ (that were not caused by the floating-point LP solves) were due to the rational factorization in three cases[57], while for one instance[58] the time limit was hit during the subsequent rational triangular solve.

Results for Rational Factorization.   For most LPs (1,123 out of 1,191 solved), SoPlex$_{fac}$ performs exactly one rational factorization. Only for seven instances two factorizations were necessary. These were the five LPs[59] for which floating-point infeasibility occured during the refinement and the feasibility LP had to be solved additionally, and the infamous instances mod2 and world, for which eleven and twelve refinement rounds, respectively, are performed before reaching the final (optimal) basis.

Notably, there are 61 instance where no factorization is necessary because the approximate solution is exactly optimal. This is explained by the fact that the "dlcm" size of this solution is very small. In almost all cases it is one, i.e., the solution is integral, except for the two instances air03 with "dlcm" size two and neos-1171448 with "dlcm" size three. This is also the reason that the averages in column "#fac" of Table 3.2 are slightly below one. As can be seen this occurs also for LPs with longer running times, because the simplicity of the solution does not imply short running times of the floating-point simplex.

As can be seen in Table 3.2, on average the time for rational factorization and triangular solves t$_{fac}$ is small compared to the total solving time. Also in absolute values, t$_{fac}$ is small for the vast majority of instances—for 901 instances it is below 0.1 seconds. Only for 17 instances, t$_{fac}$ is above seven seconds—for 14 instances between 21 and 412 seconds and for three instances as high as 4,923 (nug08-3rd), 3,548 (self), and 5,936 seconds (gen2). In these cases, the time for rational factorization and triangular solves typically constitutes more than 90% of the running time. (These instances are also the reason why we did not exclude LPs with small running times of the floating-point solver from our test set, because LPs that can be solved quickly *in an approximate sense* may still be difficult to solve *exactly*.)

Results for Rational Reconstruction.   To analyze the reconstruction routines, consider the "dlcm" column of Table A.4. Roughly speaking, in order to reconstruct a solution with "dlcm" size x, we need to reach an approximate solution with error approximately 10^{-2x} or less, see Theorem 3.1. As can be seen, this value mostly correlates with the number of refinement rounds, as expected from the theory of

---

[55] ofi (in 414.9 seconds), self (in 3700.2 seconds), stat96v4 (in 1811.2 seconds), stat96v5 (in 681.5 seconds), and watson_1 (in 370.2 seconds)

[56] cont4 (in 5508.7 seconds), gen4 (in 2016.9 seconds) and rat7a (in 2339.7 seconds)

[57] cont1, cont4, and gen4

[58] rat7a

[59] fome11,12,13, rail01, and shs1023

Table 3.2: Computational comparison of exact SoPlex with rational factorization vs. rational reconstruction (results aggregated by running time)

N — number of instances in this R_0-class
R — number of refinements (arithmetic mean)
#fac — number of rational factorizations (arithmetic mean)
#rec — number of rational reconstructions (arithmetic mean)
t_{fac} — time for rational factorization and LU solves (shifted geom. mean in seconds)
t_{rec} — time for rational reconstructions (shifted geom. mean in seconds)
t — total running time (shifted geom. mean in seconds)
Δt — relative running time of $SoPlex_{rec}$ w.r.t. $SoPlex_{fac}$

Instances	N	SoPlex$_{fac}$				SoPlex$_{rec}$				
		R	#fac	t_{fac}	t	R	#rec	t_{rec}	t	Δt
all	1186	2.1	0.95	0.21	2.8	68.3	6.74	1.26	5.2	1.82
$\geqslant 0.1s$	876	2.1	0.97	0.29	4.6	92.2	8.48	1.88	9.2	2.01
$\geqslant 1s$	591	2.3	0.98	0.43	8.8	135.1	11.04	3.28	21.8	2.47
$\geqslant 10s$	311	2.4	0.99	0.83	24.1	241.6	15.47	9.15	101.9	4.22
$\geqslant 100s$	161	2.7	0.98	1.40	42.8	384.9	19.70	22.95	340.3	7.95
$\geqslant 1000s$	28	2.0	0.86	4.70	2128.1	178.4	11.00	12.11	2730.6	1.28
$\leqslant 0.1s$	320	1.8	0.90	0.00	0.032	1.0	1.84	0.01	0.030	0.95
$\leqslant 1s$	597	1.9	0.92	0.01	0.177	2.0	2.47	0.02	0.182	1.03
$\leqslant 10s$	875	1.9	0.94	0.02	0.658	6.8	3.64	0.11	0.773	1.18
$\leqslant 100s$	1025	2.0	0.95	0.07	1.411	18.6	4.70	0.37	1.907	1.35
$\leqslant 1000s$	1158	2.1	0.96	0.15	2.176	65.7	6.64	1.15	4.211	1.93

Diophantine approximation. Nevertheless, there are several instances where reconstruction succeeds with fewer refinements than predicted.

The large "dlcm" size for the solutions of the five instances that could only be solved by $SoPlex_{fac}$ also explains the time out of $SoPlex_{rec}$. The number of refinements that could be performed within the time limit did not suffice to produce an approximate solution of sufficiently high accuracy.

The "dlcm" columns for $SoPlex_{fac}$ and $SoPlex_{rec}$ contain the same value for almost all instances solved by both because both algorithms typically return the same basic solution. Only for seven instances[60] did $SoPlex_{fac}$ and $SoPlex_{rec}$ return solutions of different "dlcm" size. Notably, these are exactly the instances for which the reconstructed solution is not basic.

Factorization vs. Reconstruction. As already mentioned, there are five instances that $SoPlex_{fac}$ solves within the time limit that $SoPlex_{rec}$ does not, as opposed to just three instances where $SoPlex_{rec}$ solves within the time limit and $SoPlex_{fac}$ does not. Table 3.2 shows further that $SoPlex_{fac}$ is significantly faster than $SoPlex_{rec}$ on the subsets "$\leqslant 1s$", "$\leqslant 10s$", "$\leqslant 100s$", and "$\leqslant 1000s$". Only for the simple, quickly

---

[60] complex, deter3, gesa2, haprp, model9, n3703, and noswot

solved instances in class "$\leqslant 0.1$s" is SoPlex$_{rec}$ faster, by 5%. Column "Δt" reports the relative time of SoPlex$_{rec}$ vs. SoPlex$_{fac}$ for the discrete filtering thresholds 0.1, 1, 10, 100, and 1,000 seconds. Figure 3.1 plots this average speedup over the classes "$\geqslant T$s" (solid line) and "$\leqslant T$s" (dashed line) for all filtering times T from zero to the longest running time in the test. Here the solid line shows, on one hand, that the average speedup factor of SoPlex$_{fac}$ vs. SoPlex$_{rec}$ is as large as 8.4. On the other hand, we can observe that it drops below one for large filtering times. This is due to the two instances nug08-3rd and gen2, for which the rational factorization takes 4,924 and 5,935 seconds, whereas SoPlex$_{rec}$ solves each in under 2,000 seconds.

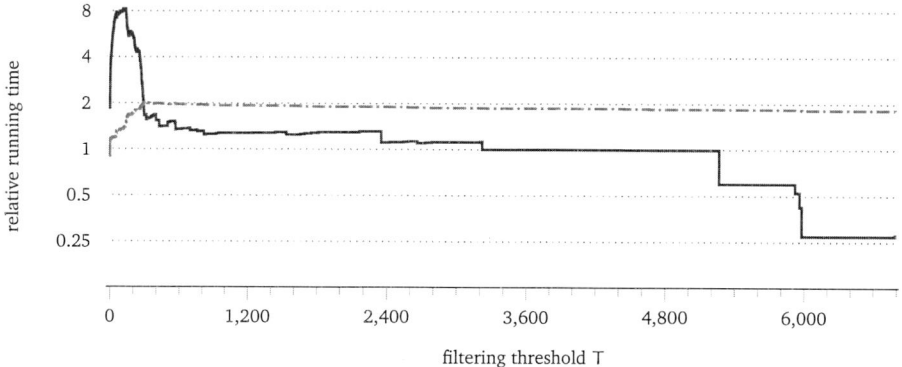

Figure 3.1: Average running time of SoPlex$_{rec}$ relative to SoPlex$_{fac}$ filtered by time T vs. T (in seconds): solid line over instances for which at least one run took time T ("$\leqslant$ T"), dashed line over instances solved within time T by both ("$\geqslant$ T")

Furthermore, as our previous experiments suggested, the number of refinements for SoPlex$_{fac}$ is small, because the final, optimal basis is almost always reached by the second round. By contrast, SoPlex$_{rec}$ takes many more refinements on average, which is highly correlated with the total running time. As can be seen from column "t_{rec}", the strategy of calling rational reconstruction at a geometric frequency succeeds in keeping down t_{rec} as the number of refinements increases.

Table 3.2 does not report the average number of simplex iterations because it remains the same for both runs. This is not surprising since neither rational factorization nor reconstruction affects the refinement procedure and we know from our previous experiments that the final basis is reached in the first refinement rounds.

SoPlex vs. QSopt_ex. Both SoPlex$_{fac}$ and SoPlex$_{rec}$ could solve more instances than QSopt_ex within the available time and memory resources: 1,158 out of the 1,202 instances were solved by all three solvers; QSopt_ex could solve five further instances where both SoPlex versions timed out, while SoPlex$_{rec}$ solved 31 and SoPlex$_{fac}$ even 33 instances for which QSopt_ex did not succeed.

For 32 of the 44 instances not solved by QSopt_ex this was due to the time limit; while the other twelve instances it terminated because it could not allocate enough

Table 3.3: Computational comparison of exact SoPlex with rational factorization or rational reconstruction vs. QSopt_ex (results aggregated by max. precision of QSopt_ex)

prc — max. precision used by QSopt_ex
N — number of instances in this prc-class
iter — number of simplex iterations (shifted geom. mean)
t — total running time (shifted geom. mean in seconds)
Δt — relative total running time w.r.t. QSopt_ex

		QSopt_ex		SoPlex$_{fac}$			SoPlex$_{rec}$		
prc	N	iter	t	iter	t	Δt	iter	t	Δt
all	492	8025.7	15.6	9740.6	8.5	0.54	9740.6	24.2	1.55
64	324	8368.3	16.1	11683.7	11.3	0.70	11683.7	14.8	0.92
128	163	7217.1	13.9	6757.2	4.3	0.31	6757.2	58.5	4.21
192	5	16950.9	72.5	10763.4	20.5	0.28	10763.4	134.6	1.86
$\geqslant 128$	168	7403.6	14.6	6851.7	4.6	0.32	6851.7	59.9	4.10

memory on the 48 GB machine. SoPlex never ran out of memory. This can partially be explained by the fact that, in addition to the original LP in rational precision, QSopt_ex has to keep the extended-precision LP in the main memory, while SoPlex only requires the floating-point rounding of the LP in double precision: for seven instances, QSopt_ex ran out of memory while boosting the precision or solving an extended-precision LP;[61] for five instances, this happened already when reading the LP or solving in double precision.[62]

To compare performance, we first counted the number of instances an algorithm "won" in the sense that it solved the instance in no more than 5% of the time taken by the fastest algorithm: QSopt_ex won 324 times, SoPlex$_{fac}$ 702 times, and SoPlex$_{rec}$ won 569 times; on the instances that took at least two seconds by one of the solvers, QSopt_ex won 116 times, SoPlex$_{fac}$ 372 times, and SoPlex$_{rec}$ 197 times. Average running times and simplex iterations are compared in Table 3.3 on basis of the 492 instances that could be solved by all three algorithms, but were sufficiently nontrivial that one of the solvers took at least two seconds. We state averages over all instances and over the instances for which QSopt_ex boosted the precision beyond 64-bit.

It can be seen that SoPlex$_{fac}$ clearly outperformed the other two algorithms. Overall, it was a factor of 1.85 faster than QSopt_ex. On the instances where QSopt_ex found the optimal solution after the double-precision solve (line "64"), SoPlex$_{fac}$ was 30% faster than QSopt_ex despite an increase in the average number of simplex iterations by 40%. On these instances, SoPlex$_{rec}$ was 8% faster than QSopt_ex, while in general it took 55% more time. When QSopt_ex had to boost the working precision of the floating-point solver (line "$\geqslant 128$"), SoPlex$_{fac}$ was faster than QSopt_ex by a factor of three.

---

[61] `buildingenergy`, `ds-big`, `ofi`, `stat96v5`, `sgpf5y6`, `shs1023`, and `watson_1`
[62] `ivu06-big`, `mspp16`, `ns1663818`, `rail2586`, and `4282`

3.5 Conclusion

In this chapter, we developed two new algorithms for exact linear programming over the rational numbers based on iterative refinement. To this end, we proved that iterative refinement with an LP oracle that returns approximately basic solutions, such as a standard floating-point simplex implementation, is guaranteed to converge to an optimal basis in polynomial time, supporting the experimental results observed in Chapter 2.

The first algorithm exploits this by computing exact solutions from the basis information accompanying the refined numeric solutions. In practice, this can be done by a LU factorization and two triangular solves in rational arithmetic. The second algorithm uses continued fraction approximations to "round" the refined solutions to an exact solution with small denominator. For this, we designed an output-sensitive reconstruction algorithm that relies only on the residual errors available during iterative refinement.

Extending the iterative refinement implementation of Chapter 2, we analyzed the computational performance of both algorithms within the SoPlex LP solver. On our test set, the rational factorization approach solved slightly more instances and was about 46% faster on average. However, we identified several instances that could be solved by rational reconstruction only or much faster than by the solver using a rational factorization; we also found that the reconstruction approach was slightly faster for those LPs with very short running times. This raises the question how to combine both techniques most efficiently into a hybrid algorithm. An immediate idea would be to perform a rational factorization only after reconstruction has failed for a minimum number of refinements. However, the critical instances on which reconstruction wins typically have solutions with large denominators and require a high number of refinements. Putting the factorization on hold in the meantime would incur a major slowdown on the majority of instances. Hence, currently the most promising hybridization seems to be a straightforward parallelization: whenever iterative refinement reaches a basis candidate that is assumed to be optimal, start the rational factorization in a background thread and continue with refinement and reconstruction in the foreground. Although the competition of the two threads for cache usage may lead to a small slowdown of both algorithms, this would help safeguard against cases where rational factorization does not succeed. A deterministic result could be achieved by accepting only reconstructed solutions that conform to the basis that was given to the factorization thread.

Finally, we compared the iterative refinement based algorithms against the current state-of-the-art approach, the incremental precision boosting procedure implemented by the solver QSopt_ex. We found that SoPlex (using the rational factorization strategy) is 1.85 to 3 times faster on our test set and solves more instances within given time and memory restrictions. However, the advantage of incremental precision boosting is its capability to handle extremely ill-conditioned LPs by increasing the working precision of the floating-point solver when necessary. In order to harness the strengths of both approaches, incremental precision boosting can be integrated into iterative refinement quite naturally: whenever the underlying floating-point solver encounters numerical difficulties and fails to return a satisfactory approximate solution, boost the precision of the floating-point solver to the next level. This would help to handle instances on which iterative refinement failed in our experiments because SoPlex's double-precision simplex broke down. (In our test set, these made up seven out of 1,202 instances, of which QSopt_ex could solve six.) At the same time, for the vast majority of instances no precision boosts would be necessary, retaining the drastic performance benefits of iterative refinement.

Chapter 4

Accurate Optimization over Multiscale Metabolic Networks

Over many years, linear programming has become a popular tool in computational systems biology for analyzing steady-state behavior of metabolic networks, see, e.g., the recent survey of Zomorrodi et al. (2012). Examples include genome-scale models for bacterial growth that incorporate both metabolism and gene expression. As input, they take the nutritional environment and produce predictions for the cell's maximum growth rate, inputs and outputs, metabolic fluxes, and gene expression levels. One core difficulty is that these models describe events at multiple scales that span several orders of magnitude. As a result, standard linear programming solvers using double precision struggle to produce solutions of sufficient accuracy or even to reliably decide feasibility.

The subject of this chapter is the robust and accurate solution of multiscale linear programs that model these genome-scale biological systems. Section 4.1 gives a short overview of the biological models and current state-of-the-art techniques. In Section 4.2 we study the accuracy of solutions produced by standard floating-point solvers in double and 80-bit extended precision. We then compare this accuracy to that achieved by the iterative refinement procedure developed in the previous chapters. These experiments are based on sequences of LPs solved during a binary search for the maximum growth rate of Escherichia coli in four different nutritional environments, which have been generated using the COBRApy package for constraint-based modeling of biological networks.[63] Section 4.3 contains our conclusions.

4.1 Background

In the following we give a short overview of the basic methodology relevant to the computational experiments in Section 4.2.

---

[63] Ali Ebrahim, Nathan Lewis, and Ronan Fleming. The openCOBRA Project. Available under http://opencobra.github.io/. See also Ebrahim et al. (2013).

4.1.1 Flux Balance Analysis

A metabolic network can be viewed as a directed hypergraph. The compounds $X_1, \ldots, X_m$ of the system form the nodes of the network; the reactions

$$\underline{s}_{1,i}X_1 + \ldots + \underline{s}_{m,i}X_m \longrightarrow \bar{s}_{1,i}X_1 + \ldots + \bar{s}_{m,i}X_m$$

for $i = 1, \ldots, n$, can be modeled as a directed hyperedge pointing from the reactants on the left-hand side to the products on the right-hand side. The so-called stoichiometric coefficients $\underline{s}_{k,i}$ and $\bar{s}_{k,i}$ quantify the participation of each compound in the reactions. We refer to Michal and Schomburg (2012) for more details.

While in reality the state of the system varies over time and is most accurately modeled dynamically, much insight can be gained by inspecting steady-state solutions. This is the approach of *flux balance analysis*. At a steady state, mass balance must hold for each compound, i.e., the amount consumed and produced by all reactions must be equal. This can be written as the flow balance constraint

$$\underline{s}_{k,1}v_1 + \ldots + \underline{s}_{k,n}v_n = \bar{s}_{k,1}v_1 + \ldots + \bar{s}_{k,n}v_n,$$

for each compound X_k. Here v_i is the *flux* through reaction i, which can be informally described as the number of times the reaction is executed per time unit. In general, reactions are reversible and negative fluxes correspond to the associated backward reaction. In a specific environment, lower and upper bounds on the fluxes may be available, $\ell_i \leqslant v_i \leqslant u_i$. Using the so-called *stoichiometric matrix* $S \in \mathbb{R}^{m \times n}$, which is defined via $S_{k,i} = \bar{s}_{k,i} - \underline{s}_{k,i}$, the mass balance constraints at a steady state can be expressed in matrix form as $Sv = 0$.

Interesting systems typically feature many more reactions than compounds and exhibit no unique steady-state solution. Flux balance analysis aims to identify interesting steady states that maximize a biological objective of the system at hand. An example is the maximal production of biomass by a cell that can serve as a proxy for cell growth. To this end, the biological objective is formulated as a linear function $c^T v$ of the flux vector and maximized by solving the linear program

$$\max\{c^T v \mid Sv = 0, \ell \leqslant v \leqslant u\}.$$

The coefficient c_i quantifies the contribution of reaction i to the objective for all $i \in \{1, \ldots, n\}$.

One limitation of the approach is that it does not compute concentrations of the compounds in a steady-state solution. Its clear advantage, given the growing availability of detailed data at the genome scale, is that it can harness the computational performance of state-of-the-art linear programming solvers to study large systems and test a multitude of scenarios, which is currently not viable with more complex models. As a result, a variety of advanced techniques based on classic flux balance analysis have been developed in recent years. For further details we refer to Orth et al. (2010) and Lewis et al. (2012).

4.1.2 Integrated Models for Metabolism and Gene Expression

A prominent application for the flux balance analysis methodology has been the study of the genotype–phenotype relationship at the genome scale, i.e., the correlation between the genes inside a cell and the resulting biochemical reactions that can be observed. One recent advancement in this field is the generation of models that integrate the processes of *metabolism* and *gene expression*. By gene expression we mean the process of so-called macromolecular synthesis, during which the genetic code is read, leading to gene products such as proteins that regulate the functioning of a cell, in particular its metabolic reactions. We refer to the textbook by Alberts et al. (2002) for more details on the biological background.

Integrated models for metabolism and gene expression have been pioneered by Thiele et al. (2012) for *Escherichia coli* and applied to *Thermotoga maritima* by Lerman et al. (2012). Given a specified nutritional environment, these models predict the cell's inputs and outputs, metabolic fluxes, and gene expression levels at its maximum possible growth rate. They are also called *ME models* (for *metabolism and gene expression*) as opposed to previous *M models* that account only for metabolic reactions.

An inherent difficulty is that the described events of metabolism and macromolecular synthesis (gene expression) occur at multiple scales that span several orders of magnitude. ME models use *coupling constraints* of the form

$$r_{\min} \leqslant \frac{v_2}{v_1} \leqslant r_{\max}$$

with positive constants, $r_{\min}$ and $r_{\max}$, in order to link related fluxes of reactions from both parts of the model. Since flux variables are nonnegative, these constraints can be linearized as $r_{\min}v_1 - v_2 \leqslant 0$ and $v_2 - r_{\max}v_1 \leqslant 0$ and appended to the mass balance constraints. At the end, a linear program of the form

$$\max\{c^\mathsf{T}v \mid Sv = 0, Cv \leqslant 0, \ell \leqslant v \leqslant u\} \tag{4.1}$$

is solved where $Cv \leqslant 0$, $C \in \mathbb{R}^{m' \times n}$ contains the linearized coupling constraints.

A further refinement of this methodology was developed by O'Brien et al. (2013) that does not rely on the maximization of a biomass function as an artificial proxy for bacterial growth. They take into account that the coefficients of the matrix C actually depend on the growth rate μ of the bacterium and that (4.1) is a linear program only when the growth rate is fixed. Consequently, they determine the maximal growth rate directly by a binary search over μ, for each fixed growth rate solving (4.1) to decide its feasibility. In this case, the objective function may be set to zero. If it remains unchanged, the optimal objective value function value can be used to check consistency of the results: at the optimal growth rate it should be close to zero, since no excessive biomass is produced.

4.1.3 Increased Demand for Numerically Reliable LP Solvers

The multiscale nature of the recently developed ME models poses a particular challenge to the numerics of standard floating-point solvers. In these models, quantities at multiple scales are encountered in two different ways. First, the coefficients of the constraint matrix span many orders of magnitude (from 10^{-6} to 10^5 in the instances used for our experiments in Section 4.2). The basis matrices traversed when applying the simplex method often have large condition numbers (as large as 10^{16} in our experiments). Second, optimal steady-state solutions can contain tiny flux values (10^{-12} and smaller) that bear biological significance and need to be reliably differentiated from zero, while others are significantly larger. With the 15 to 17 significant decimal digits of double-precision arithmetic, it is no surprise that standard LP solvers struggle to produce solutions of sufficient accuracy or even to reliably decide feasibility. The smallest feasibility tolerance that can be set inside the CPLEX simplex solver, for example, is 10^{-9}.[64]

In response, systems biologists have started to employ such exact linear programming solvers as the QSopt_ex package of Applegate et al. (2007b) used in the studies of Lerman et al. (2012) and O'Brien et al. (2013). Because of the large overhead in running time, however, this limits the number of simulations that can be conducted. Hence, researchers have sought alternatives methods in order to improve the numerical stability of floating-point solvers. Sun et al. (2013) developed an ad-hoc lifting scheme that reduces the value of large matrix coefficients by introducing auxiliary variables prior to the application of an LP solver. While this seemed to improve the stability of the CPLEX simplex implementation, it had limited effect on the barrier algorithm and does not fully address the issue of tiny, but meaningful positive flux values encountered in steady-state solutions.

An alternative direction is the use of extended-precision arithmetic. The 80-bit version of the SoPlex[65] floating-point solver is employed in the study of O'Brien et al. (2013). ME models have been one of the motivating applications for Ma and Saunders (2015) to extend the primal simplex implementation of the MINOS[66] solver to quadruple (128-bit) precision. They use a three phase approach that can be viewed as a limited version of the incremental precision boosting applied by Applegate et al. (2007b) in the QSopt_ex solver, only without rational factorization. This is achieved by first solving the LP in double precision. The quad-precision solver is then warm started from the resulting basis. A final run with matrix scaling deactivated is performed to ensure that constraint violations are below the specified tolerances on the original, unscaled LP.

In the following, we will analyze how the iterative refinement procedure of Chapter 2 performs on multiscale ME models and evaluate whether it constitutes a useful addition to the toolbox of systems biologists.

---

[64] IBM. CPLEX Optimizer. `http://www.ibm.com/software/commerce/optimization/cplex-optimizer/`.

[65] Zuse Institute Berlin. SoPlex—the Sequential object-oriented simPlex. Available for download under `http://soplex.zib.de/`.

[66] Bruce Murtagh and Michael Saunders. MINOS, Modular In-core Nonlinear Optimization System. `http://www.sbsi-sol-optimize.com/asp/sol_products_minos_desc.htm`.

4.2 Iterative Refinement for Metabolic Networks

Chapters 2 and 3 have demonstrated the computational benefits of the iterative refinement approach for linear programming on a large test set of heterogeneous benchmark instances. In this section, we focus specifically on the multiscale ME models introduced above, which have been described as particularly numerically challenging in the literature.

The basis of the experiments are a set of 84 LPs generated from an ME model of *Escherichia coli* K-12 MG1655, which have been provided to us by Ali Ebrahim from the Systems Biology Research Group at the University of California, San Diego.[67] The LPs are divided into four sets of 21 LPs each, where each set models a different nutritional environment: "Acetate", "D-Glucose", "Fumarate", and "Pyruvate". The 21 LPs from each set form a sequence parametrized by the growth rate μ. They stem from a binary search for the maximal growth rate μ^* such that the corresponding LP is feasible, similar to the method employed by O'Brien et al. (2013). Hence, by construction, each of the four sequences partition into a set of feasible LPs for $\mu \leqslant \mu^*$ and a set of infeasible LPs for $\mu > \mu^*$. They feature 76,413 variables, 57,659 equality constraints, and 11,067 inequality constraints. The 1,217,886 nonzero entries in each constraint matrix have absolute values that span eleven orders of magnitude from $2.8 \cdot 10^{-6}$ to $17,878$; the variables are bounded between zero and 1,000.

4.2.1 Experimental Setup

The first goal of our experiments was to rigorously analyze the numerical accuracy of the solutions returned by floating-point LP solvers, in terms of the objective function value, the error in the solution space, and the correct detection of feasibility. To this end, we used CPLEX as one of the most mature, numerically robust, and fast LP solvers available today, as well as the 80-bit version of SoPlex in order to investigate the effect of extended-precision arithmetic. To evaluate the error, we used the exact solving routines of SoPlex that were developed in Chapter 3 in order to compute exact objective function values and the distance of the floating-point solutions to the set of exact optimal solutions.

Second, we tested the performance of the iterative refinement procedure from Chapter 2—in particular, to evaluate the computational cost it takes to obtain higher accuracy. Third, the infeasible instances provide an ideal test set to analyze the effectiveness of the infeasibility box algorithm developed in Section 2.4.2.

We conducted our experiments as follows. We started by solving each LP using the dual simplex of CPLEX to a precision of 10^{-9} (later referred to as CPLEX$_9$), either giving an approximate floating-point solution v^* or reporting the problem as infeasible. Next, we solved the LP exactly using SoPlex with iterative refinement and rational LU factorization as denoted by SoPlex$_{fac}$ in Chapter 3, resulting in an exact solution $\tilde{v}$ unless infeasibility was proven. If both LPs were feasible, we computed the deviation of the objective function value as the relative difference $|c^\mathsf{T} v^* - c^\mathsf{T} \tilde{v}| / c^\mathsf{T} \tilde{v}$,

---

[67] Systems Biology Research Group, University of California, San Diego. `http://systemsbiology.ucsd.edu/`, accessed March 2015.

where c is the objective function vector of the LP. Note that the exact objective function value is always positive.

Since the exact solution is not necessarily uniquely determined, considering the maximum difference $\|v^* - \tilde{v}\|_\infty$ as the error of the floating-point solution is inaccurate. See also the study of alternate optimal solutions by Thiele et al. (2010). Only looking at the residual errors given by violation of bounds, constraints, and reduced cost values also seemed unsatisfactory because optimal basis matrices for these numerically sensitive LPs may exhibit large condition numbers and hence amplify the residual errors significantly. Hence, we set up an auxiliary LP to compute the distance of v^* to the set of optimal solutions. Instead of adding an objective function cut-off, this is achieved most efficiently by fixing variables with positive reduced cost to their upper bound, variables with negative reduced cost to their lower bound, and converting "$\leqslant$" inequality constraints with negative dual multiplier to an equality constraint. By LP duality, the resulting feasible region is exactly the set of optimal solutions. We computed the minimum distance of v^* to this optimal hyperplane by adding an auxiliary variable τ to be minimized and the linear constraints $\tau + v_i \geqslant v_i^*$ and $\tau - v_i \geqslant -v_i^*$ for each $i = 1, \dots, n$ in order to model $\tau = \|v - v^*\|_\infty$. Solving this LP exactly gives the true error in the solution space.

We performed the same procedure three times: once for CPLEX$_9$ as described above; once for the floating-point simplex implementation of SoPlex in 80-bit extended precision, setting the primal and dual tolerances to 10^{-12}, which we denote by SoPlex$_{12}$; and once for SoPlex with iterative refinement to a primal and dual termination tolerance of 10^{-25}, which we denote by SoPlex$_{25}$. In addition to the total running time of each solver, we measured the running time spent in the rational factorization of the exact SoPlex run and—for infeasible instances—the running time of the infeasibility box algorithm.

Parameter Settings and Further Details. We observed the most stable performance of CPLEX when using the "numerical emphasis" meta setting. Additionally, we experimented with turning matrix scaling to aggressive and deactivating it because the latter is sometimes advocated in order to avoid infeasibilities after unscaling. In preliminary tests we also tried CPLEX's barrier algorithm without crossover, but its numerical accuracy was inferior to that of the simplex. This result is consistent with that reported by Sun et al. (2013).

We applied iterated geometric scaling (allegedly similar to CPLEX's aggressive scaling) for SoPlex$_{12}$ and for the floating-point LPs solved during iterative refinement and SoPlex$_{fac}$. For increased numerical stability, 80-bit precision was not only used during SoPlex$_{12}$, but also for the floating-point LPs solved by the exact SoPlex$_{fac}$ and by SoPlex$_{25}$. Best performance of SoPlex$_{25}$ was obtained by solving the underlying floating-point LPs to a tolerance of 10^{-6}.

In preliminary experiments we observed that on our test set CPLEX outperformed SoPlex's floating-point solver by a large margin. This seemed mainly due to stronger presolving reductions. Hence, to focus the analysis on the numerical aspects rather than on the efficiency of the underlying simplex implementations, we warm started all SoPlex runs from the final CPLEX$_9$ basis. (Note that this is also possible for infeasible

LPs by running CPLEX again without presolving in case it detects infeasibility.) The time of the CPLEX run is included in the reported running times for the SoPlex runs.

Finally, in order to avoid an unfair information bias towards the exact LP solver or iterative refinement, we parsed the instance files as double-precision floating-point data, which can be stored accurately by CPLEX and SoPlex.

Hardware and Software. The experiments were conducted on a cluster of 64-bit Intel Xeon X5672 CPUs at 3.2 GHz with 12 MB cache and 48 GB main memory. In order to safeguard against a potential mutual slowdown of parallel processes, we ran only one job per node at a time. We imposed a time limit of two hours for each call to a solver. We used CPLEX version 12.6.0.0 and SoPlex version 2.0.1.2. SoPlex was compiled with GCC 4.8.2 and linked to the external libraries GMP 5.1.3 and EGlib 2.6.20.

4.2.2 Computational Results

Detailed results of the experiment are given in Table A.5 in the appendix. For each nutritional environment, the LPs are ordered by growth rate μ. Note that this is not the order in which they were generated during the binary search. We first focus on the "Acetate" LPs, because they appear to be the numerically most challenging to all solvers.

Exact Solutions. The exact solver $SoPlex_{fac}$ classified the ten instances with growth rate up to $\mu^* = 1.0319224$ as feasible and the remaining eleven LPs as infeasible. The optimal objective function values of the feasible LP decreases from approximately $4.2 \cdot 10^{-4}$ to $5.3 \cdot 10^{-8}$ as μ increases. This is consistent with the semantics of the model since the objective is formulated to express excess biomass production at the given growth rate—with increasing growth rate, more of the produced biomass is consumed by the replication process, until the optimal growth rate is reached where the excessive biomass production should be zero. Technically, a zero objective function could be used since only the feasibility status of the LP affects the result of the binary search. The objective function was set by the modelers to verify the consistency of the results.

The largest portion of the running time—up to 61% on average—is consumed by the rational LU factorization and the triangular solves. Only for two LPs (the infeasible LPs at $\mu = 1.1293750$ and $\mu = 1.5025$) is more than half the running time spent in solving the floating-point LPs.

In order to measure the dual degeneracy of the LPs, we counted the number of nonbasic variables with a zero reduced cost plus the number of nonbasic constraints with a zero dual multiplier. Relative to the number of variables, this was between 2.1 and 2.8%, hence the LPs seem to be mildly degenerate and contain multiple optimal solutions.

Floating-Point Solutions. CPLEX performed best when using aggressive scaling. With matrix scaling turned off, we observed higher condition numbers in the order of 10^3 and slightly increased errors in the objective function value and the primal solution. While scaling did not generally affect running time, there were a few instances in which running time was drastically increased by a factor of up to 22 when turning off scaling. Hence, for better readability, Table A.5 reports only the results for the parameter setting with aggressive scaling.

The first observation is that CPLEX incorrectly classified three infeasible instances as feasible, returning the status CPX_STAT_OPTIMAL. Not surprisingly, these are the first three LPs with a growth rate slightly above μ^*. By construction, these are highly ill-conditioned in the sense that they are just at the border between feasibility and infeasibility and slight perturbations to the input data can easily change their status, confer Renegar (1994) and Freund and Vera (1999) discussed in Section 2.1.3. For all other instances, the status is decided correctly.

On the feasible LPs, the distance to the optimal hyperplane is consistently around 10^{-8}, but the relative deviation in the objective function value increased with μ, from $2.1 \cdot 10^{-4}$ to as large as 1.7. This comes from the fact that the absolute deviation is always around 10^{-8}, while the correct objective value decreases with μ. Especially considering the large condition numbers of the final basis matrices, with order of 10^{13} to 10^{16}, the observed performance of CPLEX should still be considered as quite robust.

Using extended precision, the 80-bit version $SoPlex_{12}$ decided feasibility correctly for all LPs. Furthermore, it achieved notably smaller deviations in the objective function and the solutions are as close as 10^{-11} to the optimal hyperplane—three to four orders of magnitude smaller than the $CPLEX_9$ solutions, which quite accurately matches the extra significant digits of 80-bit versus 64-bit precision. However, $SoPlex_{12}$ failed to return a solution twice, once terminating due to cycling, once due to singularity of the basis matrix. The increase in running time compared to $CPLEX_9$ was up to a factor of 2.1 on the feasible instances and a factor of 1.13 for the infeasible instances. (The large running time for $\mu = 1.5025$ is due to CPLEX, which in just this case took a drastically higher number of iterations when we started it again with presolving turned off in order to create the basis information for warm starting $SoPlex_{12}$.)

Iterative Refinement and the Infeasibility Box Algorithm. Iterative refinement yielded the most accurate results by far. On the feasible LPs, the error in the objective function value and the distance of the solution to the optimal hyperplane ranged between 10^{-18} and 10^{-16}. By comparison, Figure 4.1 plots the relative deviation of the objective function and the relative distance to the optimal hyperplane for the approximate solutions returned by $CPLEX_9$, $SoPlex_{12}$, and $SoPlex_{25}$.

Additionally, we counted the number of times a variable had a zero solution value when the solution returned by the exact solver was nonzero. To distinguish zero from nonzero flux reliably is crucial in the context of the biological application. (Although this analysis is not precise due to the presence of dual degeneracy—we have not performed a flux variability analysis as in the study of Mahadevan and Schilling (2003)—it still gives an indication of the reliability of a solver in this respect.) Notably, the solution of $SoPlex_{25}$ *never* showed zero flux where the exact solution had a

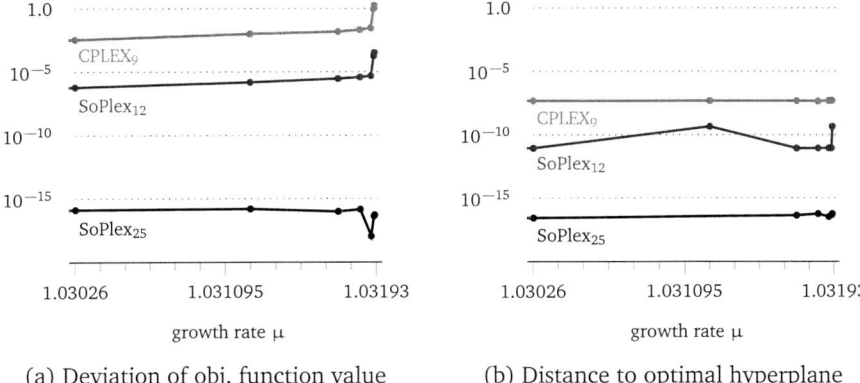

(a) Deviation of obj. function value   (b) Distance to optimal hyperplane

Figure 4.1: Accuracy of approximate solutions to feasible multiscale ME models versus growth rate μ for $\mu \in [1.0302637, 1.0319224]$

nonzero flux value; for CPLEX$_9$ this was the case on average 1,572 times, for SoPlex$_{12}$ 191 times.

The numerical difficulty of the test set can also be seen on the number of refinement rounds performed until the final basis was reached (R_0 in the notation of Chapters 2 and 3): on eight of the ten feasible "Acetate" instances, SoPlex$_{25}$ took two rounds of refinement, only on the remaining two LPs was one round sufficient; this indicates that the initial basis was never optimal. On three infeasible LPs, the initial floating-point solve returned an "optimal" solution and one round of refinement was necessary to trigger the transformation to the feasibility LP.

The overhead in running time of SoPlex$_{25}$ compared to SoPlex$_{12}$ was always below 53% for the feasible LPs and up to 315% on the infeasible LPs. This difference was largely due to the time of the infeasibility box computation, which on average consumed 33% of the running time. It successfully converted the approximate Farkas proofs into exact ones, but its computational cost shows the need for improving its implementation. Currently it uses rational arithmetic in a rather naïve way. We believe that this can be significantly improved by using interval arithmetic, see Remark 2.18. Note that for the first five infeasible LPs $\mu = 1.0319233, \ldots, 1.0319641$, it is called twice—once (unsuccessfully) for the Farkas proof from the original LP, once (successfully) for the refined Farkas proof from the infeasibility LP. For the four remaining infeasible LPs, the first Farkas proof could already be successfully transformed to an infeasibility box covering the upper bounds of 1,000 on the flux variables. Over these four instances, the running time is significantly reduced, also for the reason that no refinement is performed.

Remaining Test Sets. The results for the three nutritional environments *D-Glucose*, *Fumarate*, and *Pyruvate* are similar. Except for the fact that SoPlex$_{12}$ terminated more often due to cycling (on seven out of 63 instances), they seem to be slightly less numerically challenging in the sense that none of the solvers, including CPLEX$_9$, decided feasibility incorrectly.

4.3 Conclusion

This chapter was concerned with the accurate solution of multiscale linear programming models that have recently been developed by systems biologists in order to study bacterial growth. We applied the exact linear programming solver developed in Chapter 3 to exactly decide the feasibility status of these often ill-conditioned linear programs and rigorously analyze errors of approximate solutions. For the latter, we computed the exact distance of solutions delivered by approximate solvers to the set of optimal solutions; this value provides a meaningful measure of the error in the solution space not only when the exact solution is unique, but also when multiple optima exist due to dual degeneracy. We confirmed that state-of-the-art double-precision floating-point solvers may struggle to decide feasibility correctly and often report zero solution values for variables that may be nonzero in an exact solution. This is critical for the reliable interpretation of the results in the context of the application. We saw that using 80-bit extended-precision gives a slight improvement.

While the most accurate results are delivered by an exact LP solver, this may not always be viable or desirable due to running time requirements. This shows the advantages of the iterative refinement procedure of Chapter 2, which delivered consistently small errors of up to order 10^{-16} and did so without the need for solving rational systems of equations, which would have taken at least 3.5 times as long as the iterative refinement solver itself. On infeasible LPs that were modeled with an overestimated growth rate, we could test the efficacy of the infeasibility box algorithm developed in Section 2.4.2. It successfully converted approximate Farkas proofs into exact proofs of infeasibility. For less ill-conditioned LPs, it even eliminated the need for the iterative refinement algorithm in further refinements, thereby improving its computational performance.

Chapter 5

From Linear to Mixed-Integer Nonlinear Programming

While the previous chapters are centered around *linear* programming, the following chapters will present new algorithms for nonconvex mixed-integer *nonlinear* programming (MINLP). Section 5.1 gives an introduction to basic concepts and algorithms in MINLP, which we will rely on in the subsequent chapters. We try to point out the role of LP solving in algorithms for MINLPs, in particular when nonconvex nonlinearities are present.

In Section 5.2 we address the natural question of whether the iterative refinement methodology from Chapters 2 and 3 can be applied to nonlinear optimization problems. We show that the algorithms can be generalized in order to compute high-precision and exact solutions for the class of *quadratic programs*. Quadratic programming (QP) is an important class of problems relevant in practice. One example are variations of the LP models for metabolic networks from Chapter 4 that feature a convex quadratic objective, see Segrè et al. (2002). Moreover, QP serves as a basic subroutine of the sequential quadratic programming method for solving general nonlinear programs.

5.1 Introduction to Mixed-Integer Nonlinear Programming

Mixed-integer nonlinear programming studies the large class of finite-dimensional mathematical programs whose constraints and objective function may be general nonlinear functions and whose variables may be required to be integer. Its combination of the special cases *mixed-integer linear programming* and *nonlinear programming* makes it attractive both as a flexible modeling tool for a wide range of academic and industrial applications and an area of research rich in computationally challenging optimization problems. In the following, we will introduce notation, give a short overview of the different problem classes, and discuss complete algorithms and available implementations.

101

5.1.1 Basic Concepts

Formally, a *mixed-integer nonlinear program* is an optimization problem of the form

$$\min\{f(x) \mid x \in \mathcal{X}, x_i \in \mathbb{Z} \text{ for } i \in \mathcal{I}\} \tag{5.1}$$

where $f : \mathbb{R}^n \to \mathbb{R}$ is the objective function, $\mathcal{X} \subseteq \mathbb{R}^n$, and $\mathcal{I} \subseteq \{1, \ldots, n\}$ is the index set of variables required to take integral values. By abuse of notation, we will use the abbreviation MINLP for the class of problems as well as for an individual problem instance, likewise for the subclasses introduced below. The feasible region $\mathcal{X}$ is specified by a list of linear and nonlinear constraints and bounds on the variables,

$$\mathcal{X} := \{x \in \mathbb{R}^n \mid g_k(x) \leqslant 0, k = 1, \ldots, m, x \in [\ell, u]\}, \tag{5.2}$$

where ℓ and u are the vectors of lower bounds $\ell_i \in \mathbb{R} \cup \{-\infty\}$ and upper bounds $u_i \in \mathbb{R} \cup \{+\infty\}$ and $[\ell, u] = \times_i [\ell_i, u_i]$. In general, the constraint functions $g_k : [\ell, u] \to \mathbb{R}$ (and hence $\mathcal{X}$) may be nonconvex. We do not make any assumptions about the smoothness of the nonlinear functions, since there are relevant applications with nonsmooth components such as piecewise linear functions. Algorithms that are based on polyhedral relaxations often permit the presence of nondifferentiable functions, see Section 5.1.2.

In this thesis, as in much of the MINLP literature, we are interested in problems for which the objective and constraint functions are known analytically. However, we want to mention the classes of black-box, grey-box, and derivative-free optimization problems for which model functions are only given by function evaluations. These have gained growing attention with the increasing employment of complex simulation software in modeling real-world processes, see Floudas and Gounaris (2009, Sec. 6) for a review on recent research efforts in this direction.

In particular we are interested in the class of *factorable programming problems* as introduced by McCormick (1976). These are MINLPs with constraints and objective function recursively defined by

$$g_k(x) = \sum_{p=1}^{k-1} T_p^k(g_p(x)) + \sum_{p=1}^{k-1} \sum_{q=1}^{p} V_{q,p}^k(g_p(x)) \cdot W_{q,p}^k(g_q(x)) \tag{5.3}$$

where $g_1(x) = x_1, \ldots, g_n(x) = x_n$, and T_p^k, $V_{p,q}^k$, and $W_{p,q}^k$ are functions of a single variable such as square root, logarithm, sine, etc., for which the convex hull of the graph is known. In other words, each nonlinear term is arrived at by a (finite) series of well-understood unary operations or the addition or subtraction of variables.

This algorithmically motivated focus on factorable problems still comprises a large class of mathematical programs. Important special cases are

- *Linear programming* problems (LPs), the focus of the preceding chapters,

- *Mixed-integer linear programming* problems (MIPs), defined by purely linear constraints and objective function,

– *Mixed-integer quadratically constrained quadratic programs* (MIQCQPs), when all constraints and objective function are quadratic, i.e., of the form

$$x^\top Q x + q^\top x + q_0 \tag{5.4}$$

where $Q \in \mathbb{R}^{n \times n}$ symmetric, $q \in \mathbb{R}^n$, and $q_0 \in \mathbb{R}$, and

– *Quadratically constrained programs* (QCPs) as a special case of MIQCQPs with linear objective function and no integrality restrictions, and

– Factorable *nonlinear programming* problems (NLPs), i.e., factorable MINLPs without integrality restrictions.

A wide range of applications from the fields of biomedical and biological engineering, combinatorial optimization, computational chemistry, computational geometry, finance, process networks, and transportation can be adressed as MINLPs, see, e.g., the surveys of Tawarmalani and Sahinidis (2002), Burer and Letchford (2012, Sec. 2), and the references in Misener (2012) and Vigerske (2013). Particular examples on which we have worked are optimal water network operation (Gleixner et al., 2012a) and mine production scheduling with bilinear stockpiling constraints (Bley et al., 2012).

The diversity of mixed-integer nonlinear programming is matched by its complexity. Detailed discussions on the hardness of deciding feasibility and computing or approximating optimal solutions of MINLPs under specific assumptions are found in Hochbaum (2007), Hemmecke et al. (2010), Köppe (2012), and Locatelli and Schoen (2013, Chap. 2). Here, we only want to illustrate the complexity by looking at some fundamental facts.

There is provably no algorithm to decide whether a given polynomial has an integer root (Matiyasevich, 1970), rendering mixed-integer nonlinear programming *incomputable* in general. (This already holds for the quadratic case.) Fortunately, this negative worst-case result ceases to hold, for instance, as soon as all variables are bounded. Then, mixed-integer nonlinear programming is $\mathcal{NP}$-hard in general and one can investigate special cases that can be solved in polynomial time. Note that when continuous variables are present, exact solutions may not have polynomial encoding length and efficient solvability must be investigated as the efficient approximability of an optimal solution.

The most fundamental distinction concerning computational difficulty, both in theory and in practice, runs not so much along the classes listed above, but between *convexity* and *nonconvexity*. Nonconvexity comes in two shades: first, as the integrality conditions $x_i \in \mathbb{Z}$ for all $i \in \mathcal{I}$, and second, as nonconvex terms appearing in the objective and the constraints. Although theoretically related via the equivalence $x_i \in \{\ell_i, \ell_i + 1, \ldots, u_i\} \Leftrightarrow \prod_{j=0}^{u_i - \ell_i}(x_i - \ell_i - j) = 0$, they are addressed very differently. The solution of nonconvex NLPs without integer variables is the classical domain of *global optimization*; if all constraint functions and the objective function are convex (over their domain), we speak of a *convex MINLP*.

Both classes remain hard. By contrast, continuous convex optimization problems mostly allow for polynomial-time solution algorithms under mild assumptions.

Examples are the ellipsoid method built on an efficient membership oracle with some boundedness conditions (e.g., Grötschel et al., 1988), or interior point methods with so-called self-concordant barrier functions (Nesterov and Nemirovskii, 1994). However, this line is blurred by polynomial-time solvable nonconvex problems such as minimizing a quadratic function over a sphere (Ye, 1992) and $\mathcal{NP}$-hard convex problems such as copositive programming (Bomze et al., 2000): although the set of copositive matrices is convex, membership cannot be tested efficiently; and while they admit self-concordant barrier functions, these cannot be evaluated efficiently.

In this thesis, we are most interested in the computational aspects of mixed-integer nonlinear programming relevant to solving practical applications. The next section will give an overview of the algorithmic techniques employed in today's state-of-the-art solvers.

5.1.2 Algorithms

The following gives an overview over global solution algorithms for convex and nonconvex MINLPs proposed in the literature, inspired by the more detailed surveys of Grossmann (2002), Tawarmalani and Sahinidis (2002), Bonami et al. (2012), Belotti et al. (2013), Vigerske (2013, Sec. 6.1), and Locatelli and Schoen (2013, Chap. 4 and 5).

Since many of the methods employ related techniques, we put an emphasis on their central building blocks rather than on complete descriptions of specific algorithms. For convenience of presentation, we will assume MINLPs with linear objective function,

$$\min\{c^\mathsf{T}x \mid g_k(x) \leqslant 0, k = 1, \ldots, m,$$
$$x \in [\ell, u], \tag{5.5}$$
$$x_i \in \mathbb{Z} \text{ for } i \in \mathcal{I}\}$$

where $c \in \mathbb{R}^n$, since we can rewrite the nonlinear objective function f in (5.1) as the constraint $f(x) - x_0 \leqslant 0$ and minimize the additional auxiliary variable x_0. We will be concerned with algorithms that are—at least on a high level—indifferent to this reformulation.

Convex Mixed-Integer Nonlinear Programming.   For MINLPs with all constraint functions g_k convex, two important properties are commonly exploited. First, locally optimal solutions to the NLP relaxation

$$\min\{c^\mathsf{T}x \mid g_k(x) \leqslant 0, k = 1, \ldots, m, x \in [\ell, u]\}, \tag{5.6}$$

obtained by dropping the integrality requirements, are globally optimal for this relaxation and thus provide a valid dual bound on the optimal objective function value of the MINLP. Under conditions of sufficient smoothness, (5.6) can be solved efficiently in practice, as well as the NLP relaxation with fixed integer variables

$$\min\{c^\mathsf{T}x \mid g_k(x) \leqslant 0, k = 1, \ldots, m, x \in [\ell, u], x_i = x_i^* \text{ for } i \in \mathcal{I}\}, \tag{5.7}$$

for an integer feasible $x^* \in [\ell, u]$, and the feasibility NLP

$$\min\{s \mid g_k(x) \leqslant s, k = 1, \ldots, m, x \in [\ell, u], s \geqslant 0, x_i = x_i^* \text{ for } i \in \mathcal{J}\}, \qquad (5.8)$$

with added slack variable s.

Second, if g_k is convex and differentiable and $x^* \in [\ell, u]$, then

$$\nabla g_k(x^*)^\mathsf{T}(x - x^*) + g_k(x^*) \leqslant g(x) \qquad (5.9)$$

holds for all x, where $\nabla g_k(x^*)$ is the gradient vector of g_k at x^*. Hence,

$$\nabla g_k(x^*)^\mathsf{T} x \leqslant \nabla g_k(x^*)^\mathsf{T} x^* - g_k(x^*) \qquad (5.10)$$

is a valid linear inequality, sometimes called *gradient cut* or *outer approximation cut*. (For nondifferentiable constraints this can be generalized using subgradients.)

If x^* violates the constraint, i.e., $g_k(x^*) > 0$, then (5.10) cuts off the point x^*, which can be used in separation algorithms built on a linear relaxation. Figure 5.1 illustrates this for the simple constraint $10(x_1^2 + x_2^2) - 4 \leqslant 0$ linearized at $x^* = (-1/2, -3/4)$.

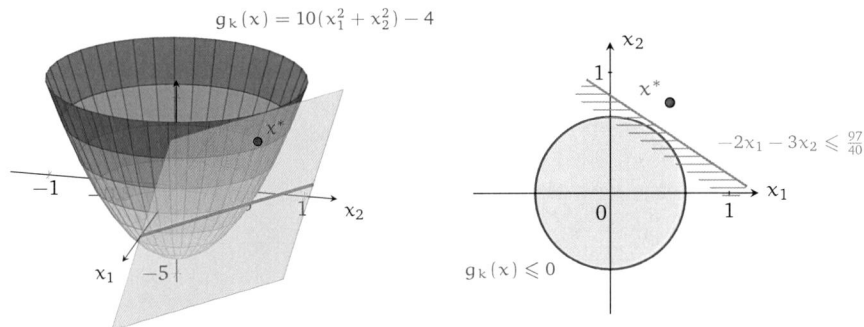

(a) Epigraph and tangent hyperplane   (b) Cut in x-space

Figure 5.1: Illustration of an outer approximation cut

These linearization techniques have been combined in various ways with a branch-and-bound search to obtain global solution algorithms, which we list by their first appearance in the literature:

- *NLP-based branch-and-bound*, noted already by Dakin (1965) and first thoroughly investigated by Gupta (1980) and Gupta and Ravindran (1985),

- *Generalized Benders decomposition* (Geoffrion, 1972), which iteratively solves and augments a master problem on the integer variables by Benders cuts derived from the dual solution to the nonlinear subproblem (5.7) or (5.8),

– *Outer approximation* (Duran and Grossmann, 1986), which iteratively solves and augments a mixed-integer linear programming relaxation on the original variables by outer approximation cuts at solutions to the NLPs (5.7) or (5.8),

– *Hybrid LP/NLP-based branch-and-bound*, first proposed by Quesada and Grossmann (1992) and improved by Bonami et al. (2008) and Abhishek et al. (2010), where the solution of NLPs and the generation of linearizations is integrated into the MIP branch-and-bound search, and the

– *Extended cutting plane method* of Westerlund and Pettersson (1995), which in its simplest form iteratively solves and augments a mixed-integer linear programming relaxation on the original variables by outer approximation cuts at MIP solutions.

For detailed descriptions we refer to the original papers and the surveys of Grossmann (2002), Bonami et al. (2012), and (Belotti et al., 2013, Sec. 3). Recent developments include QP-Diving, a branch-and-bound algorithm of Mahajan et al. (2012) that solves quadratic programming relaxations for bounding; a study of branching rules by Bonami et al. (2013); and valid inequalities generated as a byproduct of strong branching, see Kılınç et al. (2014).

While classical outer approximation cuts (5.10) do not distinguish between integer and continuous variables, several lines of research have focused on making use of the integrality information for generating tighter relaxations. Belotti et al. (2013, Sec. 4) mention in particular *perspective cuts* (Frangioni and Gentile, 2006; Günlük and Linderoth, 2008), *disjunctive cuts* (Stubbs and Mehrotra, 1999; Ceria and Soares, 1999), and cutting planes exploiting the presence of so-called *second-order cone constraints* (Drewes, 2009; Atamtürk and Narayanan, 2010).

Some algorithms allow for generalizations to functions with convex level sets, called *quasi-convex*, or to *pseudo-convex* functions, which have unique global minima, e.g., the extended cutting plane method (Pörn and Westerlund, 2000). Furthermore, there are nonconvex MINLPs for which a problem-specific analysis can still exhibit some convex characteristics. Fügenschuh and Humpola (2013) study an example concerned with nonlinear network flows going back to Maugis (1977) and Collins et al. (1978).

For general nonconvex problems, however, algorithms for convex MINLPs can be used only as heuristic procedures since they may discard optimal solutions due to invalid dual bounds. In order to alleviate this effect, see the work of Viswanathan and Grossmann (1990) for "soft" outer approximation constraints with penalized slack variables and of Bragalli et al. (2012), who use multistart NLP solves and heuristically relaxed dual bounds for pruning branch-and-bound nodes in an application to water network design. In order to find provably optimal solutions more conservative techniques are required.

Piecewise-Linear Approximations and Relaxations.   A classic idea to handle nonlinear functions is the idea of replacing them by a piecewise-linear approximation.

This was proposed as early as 1957 by Markowitz and Manne as a motivation to develop algorithms for integer programs, since models with piecewise linear functions are naturally modeled as MIPs using binary variables to distinguish which of the linear pieces is active within each interval.

Geißler et al. (2012) give an excellent overview of this approach with computational examples. Its appeal is apparent. The resulting models can be solved by readily available state-of-the-art solvers for mixed-integer linear programming, which have improved drastically in performance over the last decades (e.g., Koch et al., 2011). This yields globally optimal solutions within error bounds that can be specified a priori. One disadvantage is that it usually produces slightly infeasible solutions. In order to guarantee small error bounds, a large numbers of auxiliary binary variables can be required, particularly in higher dimensions, which may affect the scalability of the approach.

As they note, instead of an *approximation*, piecewise linear functions can equally be used to underestimate and overestimate the original nonlinearities and obtain a valid *relaxation*. Leyffer et al. (2008) describe a global optimization algorithm termed *branch-and-refine*, which is a branch-and-bound search based on a dynamically refined piecewise-linear relaxation, interleaved with fixing integer variables and solving NLP (5.7). They emphasize that this can be integrated with the spatial branch-and-bound algorithm explained below.

Finally, the recent paper of Rovatti et al. (2014) addresses piecewise-linear approximations of multivariate functions. Their "optimistic" modeling allows the value of the approximation to be determined *a posteriori* by the optimization of the MIP model instead of *a priori*.

Spatial Branch-and-Bound. For *separable* nonconvex NLPs, i.e., for problems where nonconvexities only depend on a single variable, Falk and Soland (1969) and Soland (1971) described a branch-and-bound algorithm that converges to a globally optimal solution if the nonconvex functions are lower semi-continuous. This exploits the fact that for most univariate functions it is easy to compute their *convex envelopes*, i.e., the convex hull of their epigraph, over a finite interval. However, separability is a quite restrictive condition, and although in theory it can be artificially established by introducing auxiliary variables (McCormick, 1972), this is not practical.

The first global optimization algorithm for a general class of *nonseparable* nonconvex NLPs was developed by McCormick (1976). The methodology described in his seminal paper is still the foundation for how today's state-of-the-art MINLP solvers handle nonconvex nonlinearities. As may be expected, all of its aspects have seen major advances, extensions, and structure-specific specializations.

His first observation is that most problems of practical interest are built from *factorable* functions as defined by (5.3). Second, he gives explicit formulas for convex underestimators and concave overestimators of the product and the composition of two bounded, real-valued functions. Assuming that for functions of a single variable, convex underestimators and concave overestimators are readily available, this makes it possible to recursively construct a convex relaxation that can be used to compute valid dual bounds.

Third, because these relaxations become tighter as the bounds on the variables shrink, it can be proven that recursively partitioning the feasible region by *branching* on a variable contained in constraints that are violated by the relaxation solution yields a convergent algorithm. Precisely, if each time a subpartition with lowest dual bound is selected for branching then the algorithm either terminates with a global minimizer or every accumulation point of the series of relaxation solutions is a global minimizer. Hence, convergence is only established in the limit and the algorithm terminates if the gap between the objective value of the best feasible solution and the lower bound drops below a specified tolerance $\varepsilon > 0$. We say that the algorithm is *approximately complete* and solves the MINLP to ε-*global optimality*. Necessary conditions for the convergence of a tree search algorithm are studied in detail by Horst and Tuy (1990).

Figure 5.2 depicts the bilinear constraint $x_1 x_2 \leqslant -1/4$ as a simple example. It illustrates that an outer approximation cut does not yield a valid relaxation and how the convex underestimators become tighter when they are computed over subpartitions of the domain.

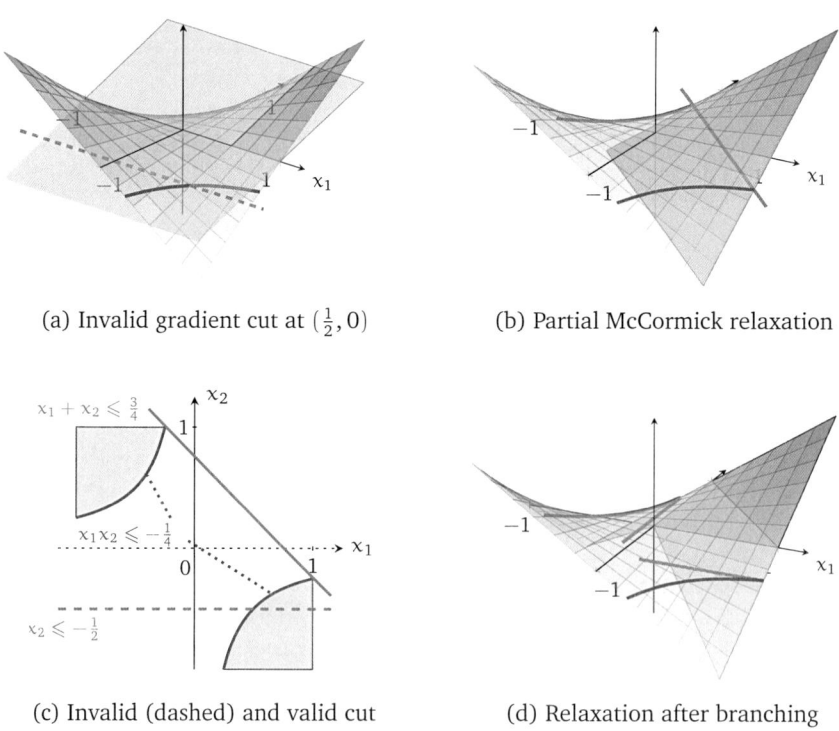

(a) Invalid gradient cut at $(\frac{1}{2}, 0)$ (b) Partial McCormick relaxation

(c) Invalid (dashed) and valid cut   (d) Relaxation after branching

Figure 5.2: Illustration of the McCormick relaxation of $x_1 x_2 \leqslant -\frac{1}{4}$ over $[-1, 1]^2$

Whereas McCormick's original algorithm generates convex underestimators in the original space, today's implementations mostly follow the approach of Smith (1996) and Smith and Pantelides (1999), who propose the introduction of auxiliary variables and *reformulate* each factorable nonlinearity recursively into elementary constraints consisting of bilinear terms or univariate functions with known convex envelope. This increases the dimension of the relaxation, but it eases implementation and may even yield tighter bounds if the same auxiliary variable is used for identical terms appearing in different constraints. Although convex envelopes are in general nonlinear, spatial branch-and-bound algorithms today often use a linear relaxation built from outer approximation cuts. This is advocated, e.g., by Tawarmalani and Sahinidis (2005), who point out the advantages in performance and stability of LP solvers over NLP solvers. The analysis of convex envelopes is only one stream of research for generating tight lower bounds. For a detailed overview of this wide subject we refer to the recent overviews of Belotti et al. (2013, Chap. 5) and Locatelli and Schoen (2013, Chap. 4).

This general approach is called *spatial* branch-and-bound because it potentially branches on the space of continuous variables. If the domain is not partitioned into hyperrectangles, but instead more complex geometric objects, the term *geometric* branching is common, see Locatelli and Schoen (2013, Chap. 5) for an overview. When integer variables are present, then spatial branch-and-bound must be interleaved with branching on integrality restrictions. In this case, well-known branching rules from the MIP literature can be applied. In Chapter 7, provide a review on branching for MINLP and develop new strategies that try to overcome the separation of spatial and integer branching.

Bound Tightening and Interval Arithmetic. One specific characteristic of convex envelopes as well as some other lower bounding techniques for MINLPs is that their tightness depends on the size of the underlying domain. Because of this, *bound tightening* procedures, not yet mentioned by McCormick, have become a crucial supplementary technique in modern solvers for nonconvex MINLPs since they help to tighten the relaxation. A more detailed overview on bound tightening techniques can be found in Chapter 6, where we also extend the well-known idea of *optimization-based bound tightening* by exploiting duality.

Some bound tightening techniques rely on *interval arithmetic*, which is an extension of traditional arithmetic that can be used to compute the range of nonlinear expressions over interval domains of the variables involved. As shown already by Hansen (1979, 1980), box-constrained nonconvex NLPs can be solved to global optimality by a branch-and-bound search based purely on interval arithmetic used to test for monotonicity and convexity, and to compute local optima. In general, however, bounds from interval arithmetic alone are too weak to obtain fast algorithms. For details on interval arithmetic we refer to the recent book of Moore et al. (2009) and the extensive list of references therein.

5.1.3 Solvers

General-purpose solvers for mixed-integer linear programs have reached an exceptional level of maturity over the last decades (e.g., Bixby et al.; Koch et al.; Achterberg and Wunderling, 2000; 2011; 2013). The situation for MINLPs, at least today, is not comparable, which is naturally explained by the larger complexity and diversity of the problem class. Even encoding general nonlinear expressions requires a significantly increased effort compared to linear constraints, let alone the numerically stable implementation of global optimization algorithms. (See Liberti and Maculan (2006) for discussions on practical implementations of global optimization algorithms, in particular Chapter 8.) On top of that, as Liberti (2013) points out, extensive computational research in global optimization only began in the 1990s; by contrast, the first research efforts in mixed-integer linear programming date back to the 1950s, see, e.g., Dantzig, Fulkerson, and Johnson (1954a,b).

However, the last years have seen much activity not only in theoretical research, but also in practical implementations for solving MINLPs to global optimality. A comprehensive survey of available solver software including a historical overview is given by Bussieck and Vigerske (2010). For a more practically oriented survey of MINLP tools targeted at finding high-quality solutions, see D'Ambrosio and Lodi (2013). We limit ourselves to discuss new developments since 2010 concerning global solvers for nonconvex MINLPs. The numerous activities over this short period may be taken as an indication of the great interest in this field.

GloMIQO and ANTIGONE. In 2013, Misener and Floudas released the new MINLP solver ANTIGONE[68] as extension of their GloMIQO[69] solver for MIQCQPs available since 2012 (Misener and Floudas, 2013, 2014). ANTIGONE reformulates a factorable MINLP into elementary nonlinearities and emphasizes the detection of special structures useful in constructing tight relaxations. It performs spatial branch-and-bound based on *MIP* relaxations, which are solved by CPLEX. This has the advantage that the core solver needs to address only the nonlinear nonconvexities and can, to a large extent, delegate the handling of integralities to the MIP solver.

MIP Features in BARON. The importance of using state-of-the-art MIP technology when integrality restrictions are present has also been recognized by other solvers such as the established BARON[70] solver developed by Tawarmalani and Sahinidis (2005) and co-workers. Classically, BARON employs an LP relaxation constructed from a reformulation of a given factorable MINLP into elementary nonlinearities. With Version 14 available since 2014 BARON also includes MIP presolving and cutting plane techniques and partially solves MIP relaxations for computing dual bounds and primal solution candidates, see Kılınç and Sahinidis (2014). Furthermore, for

---

[68] Ruth Misener and Christodoulos A. Floudas. ANTIGONE: Algorithms for coNTinuous / Integer Global Optimization of Nonlinear Equations. `http://helios.princeton.edu/ANTIGONE/`.

[69] Ruth Misener and Christodoulos A. Floudas. GloMIQO: Global Mixed-Integer Quadratic Optimizer. `http://helios.princeton.edu/GloMIQO/`.

[70] Mohit Tawarmalani and Nikolaos V. Sahinidis. BARON. Branch-And-Reduce Optimization Navigator. `http://archimedes.cheme.cmu.edu/?q=baron`.

infeasible problems BARON offers the functionality to compute an irreducible inconsistent subset of constraints, see Puranik and Sahinidis (2014).

Advances in Couenne. The open-source solver Couenne[71] developed by Pietro Belotti and co-workers has been an active platform for computational research, see Belotti et al. (2009). Recent advances include the sparsification of so-called SDP cuts (Qualizza et al., 2012) and a feasibility pump heuristic for nonconvex MINLPs (Berthold, 2014). It utilizes the MIP solver Cbc[72] to implement an LP-based spatial branch-and-bound algorithm.

Nonconvex MIQP with CPLEX. While previously CPLEX[73] could only solve convex MIQCQPs nonconvex bilinear terms with binary variables via a linear reformulation, CPLEX 12.6, released in late 2013, also computes globally optimal solutions to nonconvex MIQPs, i.e., MINLPs with linear constraints and a potentially nonconvex quadratic objective function. It automatically chooses between the original formulation or a factorized eigenvalue reformulation and applies LP-based spatial branch-and-bound (Bliek and Bonami, 2014).

The MINOTAUR Toolkit. The MINOTAUR[74] project developed by Ashutosh Mahajan and co-workers addresses the fact that due to the high diversity of MINLP applications, there is often not one dominating algorithmic approach. Rather than one complete solver, MINOTAUR provides a toolkit of data structures and interfaces to enable faster implementation of specific MINLP algorithms. Currently, NLP-based branch-and-bound for convex MINLPs and a rudimentary spatial branch-and-bound for nonconvex MIQCQPs are provided (Mahajan, 2014).

SCIP as MINLP Solver. Over the last few years, the constraint integer programming framework SCIP[75] developed at Zuse Institute Berlin has been extended to solve nonconvex MINLPs to global optimality. We explain SCIP's approach in more detail because it serves as the basis for our computational experiments.

SCIP was originally developed by Tobias Achterberg in order to integrate algorithms from constraint programming, satisfiability solving, and mixed-integer linear programming into one coherent branch-and-price-and-cut algorithm (Achterberg, 2007, 2009). Despite its greater generality, it proved to be competitive with state-of-the-art MIP code and still provides the fastest academically developed MIP solver. Its flexibility allowed the extension first to nonconvex MIQCQPs (Berthold, Heinz, and Vigerske, 2012b) and subsequently to general factorable MINLPs (Vigerske, 2013).

---

[71] COIN-OR. Couenne, an exact solver for nonconvex MINLPs. `https://projects.coin-or.org/Couenne`.

[72] COIN-OR. Cbc, Coin branch-and-cut. `https://projects.coin-or.org/Cbc`.

[73] IBM. CPLEX Optimizer. `http://www.ibm.com/software/commerce/optimization/cplex-optimizer/`.

[74] Ashutosh Mahajan, Sven Leyffer, Jeffrey T. Linderoth, James Luedtke, and Todd Munson. MINOTAUR: a Toolkit for Solving Mixed-Integer Nonlinear Optimization. `http://wiki.mcs.anl.gov/minotaur`.

[75] Zuse Institute Berlin. SCIP—Solving Constraint Integer Programs. Available for download under `http://scip.zib.de/`.

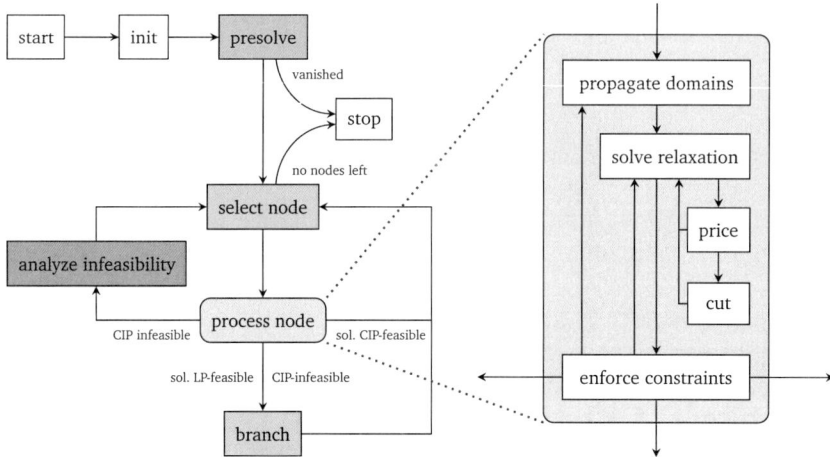

Figure 5.3: Flowchart of the main solving loop of SCIP from Berthold et al. (2012b)

Figure 5.3 shows the main solving loop of SCIP, into which the handling of non-linear constraints has been integrated seamlessly. During presolving, SCIP reformulates factorable nonlinear constraints into elementary nonlinearities by introducing auxiliary variables. It tries to detect special mathematical structures and convexity and applies feasibility- and optimality-based bound tightening procedures. The bounding step solves an LP relaxation built from convex envelopes for well-known univariate or concave functions, outer approximation cuts for convex constraints, and the McCormick relaxation of bilinear terms. All of these are dynamically separated, for pure NLPs also at a solution of the NLP relaxation. The McCormick relaxation is generated without artificial variables by adding up violated facets for each bilinear term in a quadratic constraint. The separation rounds are iterated with bound tightening. Branching is first performed on integer variables with fractional value in the LP solution, subsequently on variables contained in violated nonconvex constraints. Primal heuristics that attempt to find high-quality solutions are applied at various points during the solution process, see Berthold (2014) for details. If an integer feasible solution candidate is available, the integer variables are fixed and the remaining NLP (5.7) on the space of continuous variables is solved to local optimality.

The strength of SCIP lies in the deep integration of these nonlinear components into the main solving loop and hence with the state-of-the-art techniques that are available for constraint integer programs in general and for MIPs in particular. Two examples of such techniques are MIP cutting planes, which tighten the relaxation using integrality information, and conflict analysis, which tries to obtain short certificates of infeasibility from infeasible nodes and uses them to reduce the size of the search tree. This improves the performance on problems where the discrete part poses the main challenge.

Additionally, it enables the user to combine classical models for MINLPs with more general types of constraints found in constraint programming. Examples include constraints from scheduling applications (e.g., Berthold et al., 2010) or pseudo-boolean optimization (e.g., Berthold et al., 2009).

5.2 Iterative Refinement for Quadratic Programming

A natural question is how the iterative refinement procedure developed in Chapter 2 can be applied to nonlinear optimization problems. As explained in Section 5.1.2, many algorithms rely on the solution of linear programs. In this case, a simple option is to solve these linear programs at higher precision via iterative refinement.

In this section, we go one step further and show that iterative refinement can be generalized to *quadratic programs*: it can be adapted to compute high-precision primal–dual solutions to the first-order necessary Karush–Kuhn–Tucker conditions. This holds without assuming convexity. For the important case that the problem is convex, in which the Karush–Kuhn–Tucker conditions also suffice for local and global optimality, iterative refinement yields a high-precision global optimum. If the input data is rational and all constraints are linear, exact solutions can be computed using the techniques developed in Chapter 3 for linear programs.

5.2.1 Quadratic Programming

A *quadratic program* is an optimization problem with linear constraints and quadratic objective function. We will consider the general form

$$\min\{\tfrac{1}{2}x^\mathsf{T}Qx + c^\mathsf{T}x \mid Ax = b, x \geqslant \ell\} \tag{5.11}$$

where $x \in \mathbb{R}^n$ is the vector of variables, $\ell \in \mathbb{R}^n$ is the vector of their lower bounds, $A \in \mathbb{R}^{m \times n}$ is the constraint matrix of full row rank m, and $b \in \mathbb{R}^m$ is the right-hand side vector. The linear part of the objective function is defined by $c \in \mathbb{R}^n$ and the quadratic part by a symmetric square matrix $Q \in \mathbb{R}^{n \times n}$, which is the Hessian matrix of the objective function. By abuse of notation, we will use the abbreviation QP both for the class of problems and for an individual problem instance.

Solving QPs to global optimality is in $\mathcal{NP}$ (Vavasis, 1990); it is already $\mathcal{NP}$-hard when Q has only one negative eigenvalue (Pardalos and Vavasis, 1991). Even the problem of finding a local minimizer of a nonconvex QP and of checking whether a feasible point is a local minimizer are $\mathcal{NP}$-hard (Murty and Kabadi, 1987; Pardalos and Schnitger, 1988). By contrast, convex QPs are solvable in polynomial time, e.g., by the ellipsoid method (Kozlov et al., 1980) or interior point methods (Ye and Tse, 1989; Nesterov and Nemirovskii, 1994). Computationally competitive QP solvers are typically based on an active-set strategy—similar to the simplex algorithm for linear programming—or an interior point method. For a general overview of QP algorithms we refer to Nocedal and Wright (2006).

A variety of applications are naturally modeled as quadratic programs. We refer to McCarl et al. (1977), Gupta (1995), and Gould and Toint (2012), who list examples from support vector machines, optimal control, portfolio analysis, economies of scale, and many other areas. We want to point out in particular that the linear programs over metabolic networks discussed in Chapter 4 have variations which require a convex quadratic objective. Segrè et al. (2002) use a least squares objective in order to model the difference in total flux value between two organisms, one with and one without certain genes or reactions knocked out. This methodology is also applicable to multiscale ME models and would then profit from a high-precision QP solver in order to cope with the same numerical difficulties that we have seen in Chapter 4. In addition to that, solving QPs is a crucial subroutine in sequential quadratic programming algorithms for general nonlinearly constrained programs, see Boggs and Tolle (1995) for a survey.

5.2.2 Optimality Conditions and Basic Solutions

The Karush–Kuhn–Tucker (KKT) conditions state first-order necessary conditions for local optimality (Karush, 1939; Kuhn and Tucker, 1951). The gradient vector of the objective function of (5.11) at a point x^* is $Qx^* + c$, so for a primal–dual solution x^*, y^* the KKT conditions read as follows:

1. *Primal feasibility,*

$$Ax^* = b, \tag{5.12}$$
$$x^* \geqslant \ell, \tag{5.13}$$

2. *Dual feasibility,*

$$A^T y^* \leqslant Qx^* + c, \tag{5.14}$$

3. and *Complementary slackness,*

$$(Qx^* + c - A^T y^*)^T (x^* - \ell) = 0. \tag{5.15}$$

Note that these are precisely the optimality conditions for the LP relaxation

$$\min\{(Qx^* + c)^T x \mid Ax = b, x \geqslant \ell\}, \tag{5.16}$$

which is obtained by underestimating the quadratic objective by an outer approximation "cut" at point x^* as in (5.9), i.e.,

$$\tfrac{1}{2} x^T Qx \geqslant (Qx^*)^T (x - x^*) + \tfrac{1}{2} (x^*)^T Qx^* = (Qx^*)^T x - \tfrac{1}{2} (x^*)^T Qx^*. \tag{5.17}$$

For $Q = 0$, the KKT conditions reduce to the optimality conditions for linear programming described in Section 2.1. If Q is positive-semidefinite, the KKT conditions certify global optimality.

Unlike in linear programming, an optimal solution of a QP is not always found among the set of vertices of the polyhedral feasible region. However, the following generalization of the concept of basic solutions is available.

Definition 5.1 (Basic Solutions of QPs). *A primal–dual solution* x, y *to a QP of form* (5.11) *is called* basic *if there exists a basis* $\mathcal{B} \subseteq \{1, \dots, n\}$ *such that the corresponding* basis matrix

$$B := \begin{pmatrix} -Q_{\mathcal{B}\mathcal{B}} & A_{\cdot\mathcal{B}}^\mathsf{T} \\ A_{\cdot\mathcal{B}} & 0 \end{pmatrix} \tag{5.18}$$

is regular, $x_i = \ell_i$ *for all* $i \in \mathcal{N} := \{1, \dots, n\} \setminus \mathcal{B}$, *and* $x_{\mathcal{B}}$ *and* y *are given as the unique solution to the system*

$$B \begin{pmatrix} x_{\mathcal{B}} \\ y \end{pmatrix} = \begin{pmatrix} c_{\mathcal{B}} + Q_{\mathcal{B}\mathcal{N}}\ell_{\mathcal{N}} \\ b - A_{\cdot\mathcal{N}}\ell_{\mathcal{N}} \end{pmatrix}. \tag{5.19}$$

Note that for the basis matrix B to be regular, $\mathcal{B}$ must have cardinality at least m, i.e., at most $n - m$ variables must be nonbasic, fixed at their lower bound. If exactly $n-m$ variables are nonbasic, the solution is a vertex of the feasible region. Otherwise, the solution may lie in the interior of the face defined by $x_{\mathcal{N}} = \ell_{\mathcal{N}}$.

By construction, basic solutions satisfy complementary slackness (5.15) and the equality constraints (5.12), but may violate the lower-bound constraints (5.13) or the inequalities (5.14) defining dual feasibility for basic variables. As the simplex method for linear programming, active-set QP solvers move from basic solution to basic solution until both primal and dual feasibility are satisfied.

Almost all available implementations of QP solvers are based on floating-point arithmetic. As an indication, see the list of QP implementations maintained by Gould and Toint (2012): it contains only one solver returning exact solutions for rational input data, implemented by Gärtner and Schönherr (2000) within the CGAL project.[76] Floating-point solvers return approximate solutions that may violate each of the KKT conditions by a small tolerance. However, if this violation is bounded in the way stated in Assumption 2.9 for LPs, we can apply the following refinement step to the quadratic program.

5.2.3 Iterative Refinement for Quadratic Programs

Suppose we are given an approximate reference solution x^*, y^* and a primal and dual scaling factor $\Delta_P, \Delta_D \gg 1$. As in Section 2.2, the shifting and scaling of the primal space by the affine transformation $x \mapsto x^* + \frac{1}{\Delta_P}x$ turns $Ax = b$ and $x \geqslant \ell$ into $Ax = \Delta_P(b - Ax^*)$ and $x \geqslant \Delta_P(\ell - x^*)$, respectively. The objective function transforms into

$$\begin{aligned} &\tfrac{1}{2}(x^* + \tfrac{x}{\Delta_P})^\mathsf{T}Q(x^* + \tfrac{x}{\Delta_P}) + c^\mathsf{T}(x^* + \tfrac{x}{\Delta_P}) \\ &= \tfrac{1}{2}(x^*)^\mathsf{T}Qx^* + \tfrac{1}{\Delta_P}(x^*)^\mathsf{T}Qx + \tfrac{1}{2\Delta_P^2}x^\mathsf{T}Qx + c^\mathsf{T}x^* + \tfrac{1}{\Delta_P}c^\mathsf{T}x \\ &= \tfrac{1}{2\Delta_P^2}x^\mathsf{T}Qx + \tfrac{1}{\Delta_P}(Qx^* + c)^\mathsf{T}x + \tfrac{1}{2}(x^*)^\mathsf{T}Qx^* + c^\mathsf{T}x^*, \end{aligned}$$

which through multiplication by Δ_P and dropping of the constant terms is equivalent to minimizing

$$\tfrac{1}{2\Delta_P}x^\mathsf{T}Qx + (Qx^* + c)^\mathsf{T}x.$$

---

[76] CGAL—Computational Geometry Algorithms Library. http://www.cgal.org/.

As for LPs, we may now continue to aggregate the equality constraints using the dual solution vector to obtain $(y^*)^\top Ax = b^\top y^*$ and subtract $(y^*)^\top Ax$ from the objective function, since it evaluates to the constant $b^\top y^*$ for all x. This gives the equivalent objective

$$\tfrac{1}{2\Delta_P}x^\top Qx + \underbrace{(Qx^* + c - A^\top y^*)}_{=:\hat{c}}{}^\top x.$$

Now if x^*, y^* is an exact optimal solution to (5.11) then by the KKT condition of dual feasibility (5.14), $\hat{c} \geqslant 0$. However, if x^*, y^* is only an approximate solution to the KKT conditions returned, e.g., by a floating-point solver, then we only know that $\hat{c} \geqslant -\varepsilon_D \mathbb{1}$ for some small $\varepsilon_D > 0$. The negative entries in $\hat{c}$ are dual infeasibilities that can be magnified by a factor $\Delta_D \approx 1/\varepsilon_D$ to become noticeable again by a floating-point solver. This yields the transformed objective

$$\tfrac{1}{2}x^\top(\tfrac{\Delta_D}{\Delta_P}Q)x + \Delta_D(Qx^* + c - A^\top y^*)^\top x.$$

The following theorem shows that solving the resulting transformed QP within an absolute tolerance gives a refinement of the original reference solution. The refined solution is guaranteed to satisfy the KKT conditions within a significantly smaller violation if $\Delta_P, \Delta_D \gg 1$. Formally, the theorem holds and is stated for all positive scaling factors. It is the analogue to Theorem 2.5 for LPs.

Theorem 5.2 (QP Refinement). *Suppose we are given a QP in form*

$$\min\{\tfrac{1}{2}x^\top Qx + c^\top x \mid Ax = b, x \geqslant \ell\}, \tag{P}$$

then for $x^* \in \mathbb{R}^n$, $y^* \in \mathbb{R}^m$, and scaling factors $\Delta_P, \Delta_D > 0$, consider the transformed problem

$$\min\{\tfrac{1}{2}x^\top(\tfrac{\Delta_D}{\Delta_P}Q)x + (\Delta_D\hat{c})^\top x \mid Ax = \Delta_P\hat{b}, x \geqslant \Delta_P\hat{\ell}\} \tag{$\hat{P}$}$$

where $\hat{c} = Qx^* + c - A^\top y^*$, $\hat{b} = b - Ax^*$, and $\hat{\ell} = \ell - x^*$. Then for any $\hat{x} \in \mathbb{R}^n$, $\hat{y} \in \mathbb{R}^m$ the following hold:

1. *$\hat{x}$ is primal feasible for $\hat{P}$ within an absolute tolerance $\varepsilon_P \geqslant 0$ if and only if $x^* + \tfrac{\hat{x}}{\Delta_P}$ is primal feasible for P within ε_P/Δ_P.*

2. *$\hat{y}$ is dual feasible for $\hat{P}$ within an absolute tolerance $\varepsilon_D \geqslant 0$ if and only if $y^* + \tfrac{\hat{y}}{\Delta_D}$ is dual feasible for P within ε_D/Δ_D.*

3. *$\hat{x}, \hat{y}$ satisfy complementary slackness for $\hat{P}$ within an absolute tolerance $\varepsilon_S \geqslant 0$ if and only if $x^* + \tfrac{\hat{x}}{\Delta_P}, y^* + \tfrac{\hat{y}}{\Delta_D}$ satisfy complementary slackness for P within $\varepsilon_S/(\Delta_P\Delta_D)$.*

Here, primal and dual feasibility and complementary slackness refer to the KKT conditions (5.12) to (5.15).

Proof. The constraints are linear and so the proof for primal feasibility is identical to the proof of point 1 in Theorem 2.5. According to (5.14), dual infeasibilities of $\hat{x}, \hat{y}$ in $\hat{P}$ are the negative entries of

$$(\tfrac{\Delta_D}{\Delta_P} Q)\hat{x} + \Delta_D \hat{c} - A^T \hat{y}$$
$$= \Delta_D \left(Q \tfrac{\hat{x}}{\Delta_P} + (Qx^* + c - A^T y^*) - A^T \tfrac{\hat{y}}{\Delta_D} \right)$$
$$= \Delta_D \left(Q(x^* + \tfrac{\hat{x}}{\Delta_P}) + c - A^T(y^* + \tfrac{\hat{y}}{\Delta_D}) \right).$$

Dividing this equation by Δ_D shows that the dual infeasibilities of the refined solution $x^* + \tfrac{1}{\Delta_P}\hat{x}, y^* + \tfrac{1}{\Delta_D}\hat{y}$ in the original QP are $1/\Delta_D$ times the dual infeasibilities of $\hat{x}, \hat{y}$ in $\hat{P}$. This proves point 2. The equation above also gives us that the violation of complementary slackness in $\hat{P}$ is

$$\left| \left(\tfrac{\Delta_D}{\Delta_P} Q\hat{x} + \Delta_D \hat{c} - A^T \hat{y} \right)^T (\hat{x} - \Delta_P \hat{\ell}) \right|$$
$$= \tfrac{1}{\Delta_P \Delta_D} \left| \left(Q(x^* + \tfrac{\hat{x}}{\Delta_P}) + c - A^T(y^* + \tfrac{\hat{y}}{\Delta_D}) \right)^T (x^* + \tfrac{\hat{x}}{\Delta_P} - \ell) \right|,$$

which equals $1/(\Delta_P \Delta_D)$ times the violation of complementary slackness by the refined solution $x^* + \tfrac{1}{\Delta_P}\hat{x}, y^* + \tfrac{1}{\Delta_D}\hat{y}$ in the original QP. This proves point 3. $\qquad\square$

On the basis of Theorem 5.2, only a small modification is necessary to apply the iterative refinement procedure of Algorithm 2.2 to quadratic programs: the computation of $\hat{c}$ in line 9 and the scaling of the Hessian matrix in the objective function by Δ_D/Δ_P. If the primal and dual scaling factor are chosen equal, the latter can even be skipped.

As a consequence, the results from Chapter 2 can be proven to hold also for quadratic programs if the underlying floating-point solver returns KKT solutions with violations bounded in the fashion of Assumption 2.9. The convergence rate of Corollary 2.10 and the oracle-polynomial running time of Theorem 2.12 continue to hold, and the refinement of QPs in the general form with upper bounds and inequalities works in the same way. The treatment of unbounded and infeasible QPs will be explained in the next section.

Remark 5.3 (Warm Starting Active-Set QP Solvers). If the QPs are solved with an active-set solver, the transformed problems can be warm started in the same way as when using a simplex method for LP iterative refinement. The basis information on tight inequalities carries over from P to $\hat{P}$. A slight difference compared to the LP case is that active-set QP solvers need to solve systems with the basis matrix (5.18), which involves not only the constraint matrix but also the Hessian matrix Q. The latter is scaled by Δ_D/Δ_P during the transformation.

Hence, if the active-set solver uses a factorization of (5.18) and we wish to perform a hot start and reuse this factorization, it needs to be updated inside the solver. In principle this is easy, because it amounts to a multiplication by a diagonal matrix with entries Δ_D/Δ_P and 1. If this is not possible or undesirable, it is most convenient to choose primal and dual scaling factor to be equal and only modify the bounds of the variables, the right-hand side of the constraints, and the linear part of the objective function. This can also be of advantage if Q is very dense and communicating the modification to the solver is expensive.

5.2.4 Infeasibility and Unboundedness

Because the constraints of a QP are linear, the auxiliary LP (2.28) from Section 2.4.1 can be used to test its feasibility. Applying iterative refinement to this LP yields either a feasible solution or a Farkas proof of high precision. As for LPs, we can compute a certified infeasibility box from an approximate Farkas proof by Algorithm 2.3.

For unboundedness, however, we need to take into account the quadratic objective function. If Q is positive definite, the objective function value increases quadratically in all directions, hence the QP has a finite minimum. Unboundedness can only occur in the eigenspace of Q associated with eigenvalue zero. This is accounted for in the following lemma, which shows that unboundedness of a QP is certified by a primal ray as in the case of an unbounded LP.

Lemma 5.4 (Unboundedness of QPs). *The QP (5.11),*

$$\min\{\tfrac{1}{2}x^\mathsf{T}Qx + c^\mathsf{T}x \mid Ax = b, x \geqslant \ell\},$$

is unbounded if and only if it is feasible and there exists a ray $r \in \mathbb{R}^n$ satisfying

$$\begin{aligned}
Qr &= 0, \\
c^\mathsf{T}r &= -1, \\
Ar &= 0, \; and \\
r &\geqslant 0.
\end{aligned} \qquad (5.20)$$

Proof. If x is feasible and r satisfies (5.20), then the vectors $x + kr$, $k = 1, 2, \ldots$, form a sequence of feasible solutions with objective value $c^\mathsf{T}x - k \to \infty$, hence the QP is unbounded.

Conversely, if the QP is unbounded, then so are the LP relaxations (5.16) at any point $x^* \in \mathbb{R}^n$. This implies that their dual LPs are infeasible, i.e., for all $x^* \in \mathbb{R}^n$ there exists no $y^* \in \mathbb{R}^m$ such that

$$A^\mathsf{T}y^* \leqslant Qx^* + c,$$

hence there exists no $(x^*, y^*, u^*) \in \mathbb{R}^{n+m+1}$ such that

$$A^\mathsf{T}y^* - Qx^* - cu^* \leqslant 0, u^* > 0.$$

By the Farkas lemma per Theorem 2.1, this holds if and only if there exists a primal ray $r \in \mathbb{R}^n$ satisfying (5.20). $\qquad\qquad\square$

As a consequence, unboundedness can be tested as described in Section 2.4.3 for LPs. These tests for (in)feasibility and unboundedness can be combined into an integrated refinement algorithm to handle QPs of a priori unknown status as described in Section 2.4.4.

5.2.5 Exact Solutions over the Rational Numbers

Despite the quadratic terms in the objective function, quadratic programs defined by rational data allow for rational solutions. As with linear programming, this holds because a primal and dual feasible QP always has an optimal basic solution. Basic solutions are uniquely determined by the rational linear system (5.19), hence they are rational and their size is polynomial in the size of the input A, b, ℓ, Q, and c.

Gärtner and Schönherr (2000) mention a list of problems from computational geometry that can be modeled as convex quadratic programs and for which exact solutions are desirable:

- Finding the smallest ball enclosing n points in $\mathbb{Q}^d$,

- Testing whether n points in $\mathbb{Q}^d$ lie (approximately) on a sphere by computing the smallest enclosing annulus,

- Computing the Euclidean distance between two polytopes, and

- Computing a separating hyperplane between two polytopes that maximizes the distance to each polytope.

They describe the implementation of an exact QP solver resembling the hybrid approach of the exact LP solver by Gärtner (1999). Their active-set solver uses floating-point arithmetic for pricing decisions and an exact factorization of the basis matrix (5.18).

Compared to this approach, iterative refinement for quadratic programming offers the same advantages as for linear programming. First, in a situation where high-precision solutions are sufficient and satisfactory, we can dispense of any overhead needed for exact solutions such as potentially time-consuming exact factorizations. Since the basis matrix used by an active-set QP solver is larger than in a simplex LP solver, this effect may be even more relevant than for linear programming. Second, the results of Chapter 3 for solving LPs exactly over the rational numbers carry over to quadratic programs:

- The size of basic solutions is bounded by a polynomial in the size of the input and so is the size of the smallest infeasibilities of a basic solution as proven in Lemma 3.4.

- As a consequence, if the underlying QP solver returns approximately basic solutions as in Assumption 3.7, then the sequence of bases produced by iterative refinement converges to an optimal basis in a polynomial number of refinements as guaranteed by Theorem 3.5.

- Hence, iterative refinement with an underlying active-set solver can be combined with an exact factorization of the basis matrix (5.18) to solve QPs exactly as in Algorithm 3.2.

- Alternatively, rational reconstruction can be applied to high-precision solutions obtained by iterative refinement as in Algorithm 3.3.

The proofs for the quadratic case do not feature new ideas, so we do not reiterate these results in detail. On the practical side, while rational reconstruction algorithms need no adaptation, the efficient solution of the KKT systems in active-set QP solvers usually requires different algorithms to the classical LU factorization that is typically applied by simplex-based LP solvers. Details can be found, e.g., Nocedal and Wright (2006, Sec. 16.2). Performing these methods exactly over the rational numbers may need more care than in the LP case.

5.3 Conclusion

This chapter established a link between the preceding results on exact linear programming and the algorithms for mixed-integer nonlinear programming that are the topic of the following chapters. To this end, we first introduced different classes of convex and nonconvex mixed-integer nonlinear programs and discussed basic concepts and algorithmic techniques for computing globally optimal solutions. Thereby we emphasized the role of linear programming as a workhorse in state-of-the-art solvers for nonconvex MINLPs.

As a new contribution, we showed that the iterative refinement algorithms developed in Chapters 2 and 3 can be generalized to quadratic programs. The results of the previous chapters suggest that this can provide the foundation for a practically efficient implementation of a new exact quadratic programming solver based on an active-set strategy.

Chapter 6

Three Enhancements for Optimization-Based Bound Tightening

The performance of spatial branch-and-bound algorithms for nonconvex MINLPs depends on the tightness of the convex relaxation, which in turn depends on the bounds of the variables involved in nonconvex nonlinear terms. For this reason, bound tightening algorithms have become a crucial component of global MINLP solvers. This chapter is concerned with optimization-based bound tightening (OBBT), which in its most basic form is a straightforward technique that minimizes and maximizes each variable over the relaxation at hand. While comparatively expensive, it is also one of the most effective techniques, and it seems to be common belief that OBBT is beneficial if only one is able to keep its computational cost under control. To this end, we develop methods to reduce the computational cost of OBBT and exploit dual information to learn valid inequalities that can be used to efficiently approximate iterated OBBT during the solving process.

The chapter is organized as follows. Section 6.1 gives an overview of general bound tightening algorithms for nonconvex MINLPs found in the literature and discusses the basic idea of OBBT. In Section 6.2, we show how dual solutions encountered during OBBT can be used to derive valid inequalities that approximate the effect of OBBT when propagated during the branch-and-bound search. In Section 6.3, we discuss strategies to reduce the computational cost of OBBT by saving unnecessary LP solves and exploiting the hot-starting capability of the simplex algorithm through a greedy ordering heuristic. We use an implementation based on the MINLP solver SCIP to analyze the performance impact of these methods on a set of publicly available benchmark instances in Section 6.4.

This chapter is joint work with Timo Berthold, Benjamin Müller, and Stefan Weltge and is available as preprint (Gleixner et al., 2015a). An early version of Section 6.2 is due to Gleixner and Weltge (2013).

6.1 Bound Tightening for Nonconvex MINLPs

As Belotti et al. (2012) point out, methods for tightening bounds on variables and more generally for inferring restrictions on the set of possible solutions to a (satisfaction or optimization) problem from its defining constraints are used in several areas of research, from artificial intelligence through constraint programming to mixed-integer linear and nonlinear programming. Historically, this algorithmic idea can be traced back to the first efforts in automated theorem proving, most notably by Davis and Putnam (1960), who build on the even earlier work of Herbrand (1930).

Because of its ubiquity, it is encountered under various names, which are differentiated (e.g., Achterberg, 2007) as follows. The term *constraint propagation* refers to the most general case of deriving new valid constraints from the current set of constraints, whereas *domain propagation, domain reduction,* or *domain filtering* signify the specific case of constraint propagation that only affects the domains of the variables. The terms *bound propagation* or *bound strengthening* are used when the domains of the variables are intervals of continuous or integer values without "holes". This terminology mostly stems from the constraint programming literature. In global optimization, bound propagation is called *bound tightening* or *range reduction*. In mixed-integer programming, the term *node preprocessing* denotes the application of limited presolving algorithms at nodes of a branch-and-bound tree. With the incorporation of constraint programming techniques into MIP solvers, the term *propagation* has become quite commonly used. In the following, we will discuss the particular importance of tight bounds for solving nonconvex MINLPs with spatial branch-and-bound and give an overview over existing bound tightening techniques from the literature.

6.1.1 Domain-Dependent Relaxations

The function of presolving and propagation algorithms for MINLPs is twofold: first to strengthen the associated convex relaxation and second to reduce the size of the domains of the variables over which enumerative search is performed. Though these objectives are different, they are closely related.

An important characteristic of branching on continuous variables is that the optimum x^* of the relaxation on which branching is performed typically stays inside the domains of both created subproblems. This stands in marked contrast to branching on integer variables with a fractional relaxation value, where x^* is immediately excluded from the domains of the subproblems by rounding. Besides the refinement of locally valid outer approximation cuts, bound tightening is an important feature to ensure progress of the branch-and-bound search.

Figure 6.1 illustrates how the convex envelope of the nonconvex function $x \mapsto 0.1x^3 - 1.1x$ depends on the size of the domain over which it is computed: as the lower bound on variable x increases, the convex envelope approaches the graph of the function. For $x \geq 0$, we can see that the function even becomes convex; in this case, i.e., if convexity is known, the feasibility of the constraint $x \mapsto 0.1x^3 - 1.1x \leq y$ could be enforced by separation instead of spatial branching, which is much more efficient.

This demonstrates that *tighter bounds imply tighter relaxations* beyond the trivial fact that they reduce the size of the domains. In the following we will see that, vice versa, *tighter relaxations imply tighter bounds*.

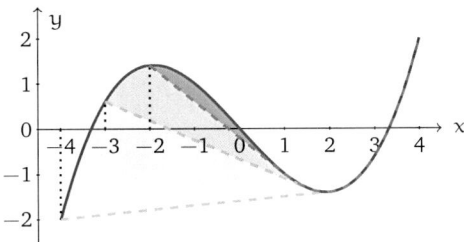

Figure 6.1: Convex envelope of the function $x \mapsto 0.1x^3 - 1.1x$ over domains $[-4, 4]$, $[-3, 4]$, and $[-2, 4]$

6.1.2 Literature Review

Presolving algorithms in mathematical programming aim at improving the performance and numerical robustness of solution algorithms by transforming the problem so that each optimal solution of the resulting problem can be transformed back to an optimal solution of the original formulation. Propagation algorithms may be considered to be a restricted form of presolving in which extensive reformulations of the problem are limited in a way that they can be applied easily during the branch-and-bound search. For mixed-integer linear programming, their study can be traced back to the seminal work of Dantzig, Fulkerson, and Johnson (1954a), who apply what is today known as *reduced cost fixing* to an instance of the traveling salesman problem with 49 cities. We refer to Savelsbergh (1994), Fügenschuh and Martin (2005), Achterberg (2007), and the recent study of Gamrath et al. (2015a) for an overview of presolving and propagation techniques for general MIPs. The following focuses on the nonlinear case.

Under *feasibility-based bound tightening* (FBBT), we subsume bound tightening algorithms that use purely primal feasibility arguments and remove parts of the domains in which no feasible solutions are contained. The most fundamental of these techniques is based on interval arithmetic; see Section 5.1.2 for a short description. It amounts to computing the activity of nonlinear expressions over the domains of the variables (forward propagation) and conversely propagating the bounds on the constraint activities back to the bounds of the variables (backward propagation). Implementations usually rely on the representation of the nonlinear terms as nodes of an expression graph, see Belotti et al. (2009) or Vigerske (2013) for details. While basic FBBT algorithms propagate the effect of constraints individually, the recent work of Belotti (2013) investigates the application of FBBT to pairs of (linear) inequalities.

As Belotti et al. (2010, 2012) point out, this is an iterative process that may produce a series of smaller and smaller domains that converge only in the limit, even if all constraints are linear. As a consequence, MIP solvers typically restrict the maximum rounds of consecutive FBBT propagations. They show that the resulting vector of intervals can be interpreted as a fixed point of an operator on a suitable lattice and—for linear constraints—can be computed in finite time by solving a large linear program. Feasibility-based reductions can be strengthened by exploiting integer information as in MIP solvers. Examples are globally available information such as implications between variables and mutually exclusive binary assignments derived from constraints with knapsack-like structure (e.g., Achterberg, 2007, Chapters 7 and 10).

Optimality-based reductions (not to be confused with OBBT) may additionally discard feasible and even optimal solutions as long as they guarantee that at least one optimal solution remains feasible for the reduced problem. They often exploit the knowledge of an *objective cut-off*, i.e., given that we are minimizing, an upper bound on the optimal objective function value of the MINLP, which can be obtained from the best known solution. A classical example is the technique of *marginals-based bound tightening* by Ryoo and Sahinidis (1996), which extends reduced cost bound tightening for MIPs by using Lagrange multipliers from convex relaxations. Suppose $g(x) \leqslant g_0$ denotes a constraint that is active in an optimal solution to a convex NLP relaxation with associated dual multiplier $\lambda > 0$, and let U be the objective cut-off and L a lower bound on the optimal objective function value (typically from the relaxation), then

$$g(x) \geqslant g_0 - (U - L)/\lambda \tag{6.1}$$

is valid for all solutions with better (lower) objective function value than U.

If the constraint $g(x) \leqslant g_0$ is not active, then λ is zero and (6.1) cannot be applied. However, if we fix $g(x) = g_0$, solve the relaxation again, and obtain a tighter bound $\tilde{L} > U$, then this proves that $g(x) < g_0$ must hold for all optimal solutions. If the associated dual multiplier $\tilde{\lambda}$ in this restricted relaxation is positive, then

$$g(x) \leqslant g_0 - (U - \tilde{L})/\tilde{\lambda} \tag{6.2}$$

is valid. This is typically applied for the special case of bound constraints, i.e., when $g(x) = x_i$ for some variable x_i. In this case, the presented inequalities immediately yield tighter domains.

Note that there are optimality-based arguments that do not rely on the explicit availability of dual multipliers. Simple examples are the FBBT-like propagation of the objective function versus the objective cut-off value (Achterberg, 2007, Sec. 7.6) or the removal of dominated columns in linear programming (Williams, 1983).

The term *shaving* or *aggressive FBBT* (see, e.g., Neumaier, 2004) refers to what is called *probing* in mixed-integer linear programming (Savelsbergh, 1994): tentatively tightening the lower or upper bound of a single variable and applying propagation in the hope of proving infeasibility and excluding this subdomain from further consideration. This is typically performed without solving relaxations, but it can include both feasibility- and optimality-based reductions. While in MIP solvers this is typically applied to binary variables only, for MINLPs this has been found beneficial also for continuous variables (Belotti et al., 2009; Nannicini et al., 2011). Iterating over all variables, it is considerably more expensive than applying FBBT to the global domains. At the same time it is potentially stronger because it exploits the appearance of variables across several constraints. Inequality (6.2) can be considered as an expensive form of shaving that solves the (restricted) relaxation.

In the related context of branching, Tawarmalani and Sahinidis (2004) consider the situation when a node is pruned because the dual bound of the relaxation exceeds the primal objective limit. They discuss the possibility of extending the dual argument to a larger box and hence remove parts of the unpruned sibling node under specific assumptions.

In general, techniques based on dual relaxation solutions as reviewed above may only be considered cheap because the vector of dual multipliers is readily available after solving the relaxation. A technique that requires the solution of *additional* relaxations is optimization-based bound tightening, which is the central topic of this chapter and will be explained in detail in the next section.

6.1.3 Optimization-Based Bound Tightening

Given an MINLP as in (5.5) with a convex relaxation $\mathcal{R}$ of its feasible region $\mathcal{X}$, $\mathcal{X} \subseteq \mathcal{R}$, classical OBBT computes the tightest bounds valid for all relaxation solutions by in turn minimizing and maximizing each variable,

$$\min/\max\{x_k \mid x \in \mathcal{R}\}. \tag{6.3}$$

Often a polyhedral relaxation is used. In this case, applying a full round of OBBT amounts to solving $2n$ linear programs. Even though this is possible efficiently, it is an expensive algorithm when compared to the average amount of work performed at a node of a spatial branch-and-bound tree.

This straightforward idea must have been folklore knowledge in the mathematical programming community for a long time. It was first mentioned in the global optimization literature by Quesada and Grossmann (1993) in the context of optimizing heat exchanger networks. Soon after, it became a regular component of generic global optimization algorithms, see Quesada and Grossmann (1995), Maranas and Floudas (1997), or Smith and Pantelides (1999) for examples. In these references it is described as a bound tightening procedure applied at the root node of a tree search, i.e., to tighten the global bounds of an MINLP.

If an upper bound on the optimal objective function value is available, OBBT can be strengthened by adding an objective cut-off to the relaxation,

$$\min/\max\{x_k \mid x \in \mathcal{R}, c^\mathsf{T}x \leqslant U\}, \tag{6.4}$$

which renders it an optimality-based procedure. Zamora and Grossmann (1999) have first used this idea in a "branch-and-contract" algorithm, which employs OBBT aggressively at every node of a spatial branch-and-bound tree.

Examples of MINLP solvers implementing OBBT are αBB (Adjiman et al., 1998, 2000), ANTIGONE (Misener and Floudas, 2012, 2014), Couenne (Belotti et al., 2009), LaGO (Nowak and Vigerske, 2008), and SCIP (Achterberg, 2007; Vigerske, 2013). It is mostly applied at the root node and within the search tree only with limited frequency or based on its success rate. ANTIGONE, for instance, measures the success of OBBT by the reduction of the box volume and disables it for all children nodes once the rate of reduction drops below a given threshold. Even then it restarts OBBT within the search tree with a probability of $2^{\lambda - \text{tree depth}}$, $\lambda = 1$ by default. Couenne implements a similar strategy.

Because OBBT uses the convex relaxation directly, it is a particularly good example for the mutual interaction between a tight relaxation and tight bounds as discussed in Section 6.1.1. If the relaxation is refined after applying OBBT, another application of OBBT may result in further bound strengthenings. In this spirit, Caprara and Locatelli (2010) present a theoretical study of an iterated version of OBBT.

The increased computational effort of OBBT is often justified because it can provide tighter bounds than the ones computed by many cheaper methods. Its strength lies in the consideration of (the relaxations of) all constraints at once and in combination with the objective cut-off. As such, it dominates any bound tightening technique that relies solely on the relaxation. This includes, e.g., FBBT on the linear relaxation and optimality-based reduced cost propagation.

Remark 6.1 (Optimization-Based and Optimality-Based Bound Tightening)**.** Note that, while OBBT is sometimes labeled *optimality*-based bound tightening, this is slightly misleading: It is only one among many bound tightening procedures that exploit optimality and is applicable also when no objective cut-off limit is available. The defining feature of OBBT is the solution of *optimization* problems, hence we use it as an abbreviation for *optimization-based bound tightening*. In this paper we consider the solution of linear programs, but the optimization problems may be of more general form such as convex NLPs (e.g., Zamora and Grossmann, 1999) or even partially solved MINLPs (e.g., Huang, 2011; Belotti et al., 2015).

6.2 One-Row Relaxations by OBBT

This section presents our main contribution: a new technique exploiting the potentially expensive solution of the OBBT LPs (6.4) beyond simply obtaining tighter bounds on variable x_k. To this end, we observe that the proof of optimality given by a dual solution of (6.4) can be used to generate globally valid inequalities that approximate OBBT locally.

6.2.1 Lagrangian Variable Bounds

Besides valid bounds for variable x_k, solving (6.4) yields dual multipliers that prove that no $x \in \mathcal{R}$ with $c^\mathsf{T} x \leqslant U$—and by that no MINLP solution with improved objective function value—can lie outside these bounds. The following theorem uses basic LP duality to derive this proof in form of a valid inequality.

Theorem 6.2 (Lagrangian Variable Bounds)**.** *Suppose we have a polyhedral relaxation of an MINLP of form* (5.5) *with feasibility set*

$$\mathcal{R} = \{x \in \mathbb{R}^n \mid Ax \geqslant b, x \in [\ell, u]\}, \tag{6.5}$$

with $A \in \mathbb{R}^{p \times n}$ and $b \in \mathbb{R}^p$, and a valid upper bound U on the optimal objective value of (5.5)*. Furthermore, suppose we have an optimal primal–dual solution $(\tilde{x}, \tilde{\lambda}, \tilde{\mu})$ to*

$$\min\{x_k \mid Ax \geqslant b, c^\mathsf{T} x \leqslant U, x \in [\ell, u]\} \tag{6.6}$$

where $\tilde{\lambda} \in \mathbb{R}^p_{\geqslant 0}$ *is the vector of dual multipliers for* $Ax \geqslant b$ *and* $\tilde{\mu} \leqslant 0$ *is the dual multiplier associated with the objective cut-off constraint.*

Let $\tilde{r} = e_k - A^T \tilde{\lambda} - c\tilde{\mu}$ *be the vector of reduced costs, where* e_k *is the k-th unit vector. Then*

$$x_k \geqslant \sum_{j=1,\dots,n} \tilde{r}_j x_j + \tilde{\mu} U + \tilde{\lambda}^T b \qquad (6.7)$$

is a valid inequality for all $x \in \mathcal{R} \cap \{x \in \mathbb{R}^n \mid c^T x \leqslant U\}$, *which is tight at* $\tilde{x}$. *If* $(\tilde{x}, \tilde{\lambda}, \tilde{\mu})$ *with* $\tilde{\lambda} \leqslant 0$ *and* $\tilde{\mu} \geqslant 0$ *is optimal for*

$$\max\{x_k \mid Ax \geqslant b, c^T x \leqslant U, x \in [\ell, u]\} \qquad (6.8)$$

then the same holds for

$$x_k \leqslant \sum_{j=1,\dots,n} \tilde{r}_j x_j + \tilde{\mu} U + \tilde{\lambda}^T b. \qquad (6.9)$$

Proof. Aggregating the rows of $Ax \geqslant b$ with $\tilde{\lambda}$ and adding $\tilde{\mu}$ times the cut-off constraint gives the valid inequality $(\tilde{\lambda}^T A + \tilde{\mu} c^T)x \geqslant \tilde{\lambda}^T b + \tilde{\mu} U$ in case of minimization and $(\tilde{\lambda}^T A + \tilde{\mu} c^T)x \leqslant \tilde{\lambda}^T b + \tilde{\mu} U$ for maximization. By definition of $\tilde{r}$, this is equivalent to (6.7) and (6.9), respectively. The tightness of (6.7) and (6.9) follows from complementary slackness. $\square$

We will refer to valid inequalities of type (6.7) and (6.9) as *Lagrangian variable bounds* (LVBs). They can be viewed as a one-row relaxation of the given MINLP. In mixed-integer linear programming one-row relaxations are a fundamental ingredient in the generation of general-purpose cutting planes such as Gomory mixed-integer cuts (Gomory, 1958, 1960), mixed-integer rounding cuts (Nemhauser and Wolsey, 1990; Marchand and Wolsey, 2001), or Chvátal-Gomory cuts (Chvátal, 1973; Caprara and Fischetti, 1996). We will use this concept for fast propagation.

Remark 6.3 (Nonexisting and Trivial LVBs). It is not guaranteed that useful LVBs exist:

1. If OBBT fails due to (6.6) or (6.8) being unbounded, then no dual feasible solution exists from which an LVB can be derived.

2. If the OBBT bound is implied only by problem constraints, i.e., $\tilde{r}$ and $\tilde{\mu}$ are zero in the dual solution, then the right-hand side of the LVB is the constant $\tilde{\lambda}^T b$ and the LVB reads $x_k \geqslant \tilde{x}_k$ or $x_k \leqslant \tilde{x}_k$.

3. If OBBT fails to tighten bounds, i.e., (6.6) equals ℓ_k or (6.8) equals u_k, then $\tilde{\lambda} = 0$, $\tilde{\mu} = 0$ are optimal dual multipliers. For these, (6.7) and (6.9) read $x_k \geqslant x_k$ and $x_k \leqslant x_k$, respectively.

In the last two cases we call the LVBs *trivial* since they do not contain useful information. We call them *nontrivial* if a variable other than x_k appears with a nonzero multiplier on the right-hand side of (6.7) or (6.9).

Note that in the presence of dual degeneracy, the optimal dual multipliers $\tilde{\lambda}$ and $\tilde{\mu}$ may be nonzero even when OBBT fails to tighten the bound as in case three. We have encountered this in our experiments. In general, LVBs are frequently nontrivial as observed empirically in Section 6.4. This is illustrated by the following example.

Example 6.4 (Nontrivial LVBs). Consider the nonconvex, polynomial NLP

$$\min\{y - x \mid y = 0.1x^3 - 1.1x, x \in [-4, 4], y \in [-2, 2]\}. \tag{6.10}$$

Figure 6.2a shows its feasible region—a one-dimensional curve—between the initial bounds -4 and 4. The shaded region over which OBBT is performed is defined by the linear relaxation

$$\begin{aligned} \mathcal{R} = \{(x, y) \in [-4, 4] \times [-2, 2] \mid & -0.1x + y \geqslant -1.6, \\ & -3.7x + y \geqslant -12.8, \\ & 0.1x - y \geqslant -1.6, \\ & 3.7x - y \geqslant -12.8\} \end{aligned} \tag{6.11}$$

and the dashed objective cut-off resulting from the zero solution. Minimizing x gives the dual solution $(\tilde{\lambda}, \tilde{\mu}) = (10/9, 0, 0, 0, -10/9)$, the lower bound $-16/9$, and the LVB

$$x \geqslant -\tfrac{10}{9}U - \tfrac{16}{9}. \tag{6.12}$$

As soon as a feasible solution with negative (better) objective function value is found, (6.12) gives an even tighter lower bound. Maximizing x does not immediately tighten its upper bound, still the nontrivial LVB

$$x \leqslant \tfrac{10}{37}y + \tfrac{128}{37} \tag{6.13}$$

can be generated from the dual solution $(\tilde{\lambda}, \tilde{\mu}) = (0, 0, 0, 10/37, 0)$. In this two-variable example, this is only the rightmost facet of $\mathcal{R}$. In general, we obtain a valid, but redundant, inequality that might not be explicitly part of the relaxation. While the inequality is redundant for the relaxation, propagating it by using integrality information on the variables might lead to bound tightenings that could not be deduced from single rows of the relaxation. Figure 6.2b shows the new bounds and tightened relaxation.

Example 6.12 above also demonstrates that applying another round of OBBT would tighten the lower bound of x even further—from $-16/9$ to almost zero. As can be seen, continuing to iterate between OBBT and the refinement of the relaxation would result in an infinite series of lower bounds on x that converge towards zero. The difficulty in iterating OBBT is its high computational cost. The propagation of LVBs as explained in the next section, however, is cheap and may still be effective.

6.2.2 OBBT via FBBT

Lagrangian variable bounds can be seen as a local approximation of the effect of OBBT. Instead of minimizing a variable x_k over the complete linear relaxation and

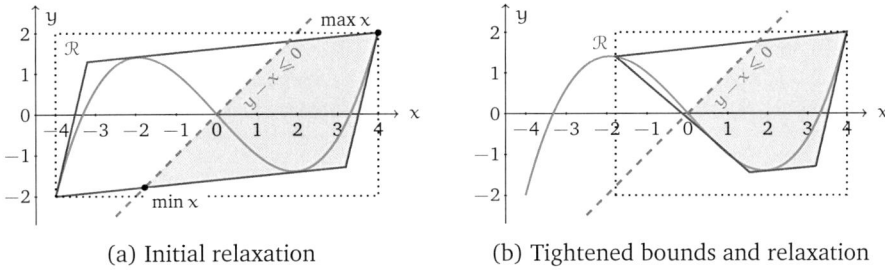

(a) Initial relaxation   (b) Tightened bounds and relaxation

Figure 6.2: Two-variable example for OBBT with objective cut-off constraint

the objective cutoff constraint as in (6.6), we could have obtained the same bound tightening by minimizing x_k over the one-row relaxation given by the optimal dual solution $(\tilde{\lambda}, \tilde{\mu})$, which is equivalent to the LVB (6.7):

$$x_k \geqslant \min \left\{ x_k \mid Ax \geqslant b, c^T x \leqslant U, x \in [\ell, u] \right\} \tag{6.14}$$

$$\geqslant \min \left\{ x_k \mid (\tilde{\lambda}^T A + \tilde{\mu} c^T) x \leqslant \tilde{\lambda}^T b + \tilde{\mu} U, x \in [\ell, u] \right\} \tag{6.15}$$

$$= \min \left\{ x_k \mid x_k \geqslant \sum_{j=1,\dots,n} \tilde{r}_j x_j + \tilde{\mu} U + \tilde{\lambda}^T b, x \in [\ell, u] \right\}. \tag{6.16}$$

The last minimization is now easily performed by replacing the right-hand side variables of the LVB with one of their bounds, giving the valid lower bound

$$\sum_{j:\tilde{r}_j > 0} \tilde{r}_j \ell_j + \sum_{j:\tilde{r}_j < 0} \tilde{r}_j u_j + \tilde{\mu} U + \tilde{\lambda}^T b. \tag{6.17}$$

For maximization, we obtain the analogous upper bound

$$\sum_{j:r_j > 0} \tilde{r}_j u_j + \sum_{j:r_j < 0} \tilde{r}_j \ell_j + \tilde{\mu} U + \tilde{\lambda}^T b. \tag{6.18}$$

This is essentially the same as applying FBBT to the LVB inequalities.

When initially created, this gives the same bound as OBBT because $(\tilde{\lambda}, \tilde{\mu})$ are optimal and hence (6.15) holds with equality. Later during the search, whenever the domain of some variable x_j appearing with $\tilde{r}_j \neq 0$ on the right-hand side is reduced, the LVB implies a tighter bound.[77] This holds both for tighter globally valid bounds as well as for bounds tightened through branching. Furthermore, if an improved primal solution is found, then the value of U may also be updated in the LVB inequalities. This is because Theorem 6.2 holds for *any* valid upper bound U. If $\tilde{\mu} < 0$, then this also leads to a tighter value in (6.17) and (6.18).

In these cases, the LVB bound may be weaker than the bound that would be obtained from reapplying OBBT for two reasons. First, the dual solution from which

---

[77] Precisely, if ℓ_j increases for $\tilde{r}_j > 0$ in (6.17) or $\tilde{r}_j < 0$ in (6.18), and if u_j decreases for $\tilde{r}_j < 0$ in (6.17) or $\tilde{r}_j > 0$ in (6.18).

the LVB was constructed is still dual feasible, but may not be optimal anymore; second, the LP relaxation may have been tightened by additional rows that were not present when performing OBBT initially. However, in stark contrast to OBBT, LVBs can be propagated very efficiently. This motivates the application of LVBs within a spatial branch-and-bound algorithm in the following scheme:

1. Generate LVBs while performing OBBT during the root node.

2. Propagate these LVBs locally at the nodes of the search tree whenever bounds appearing on the right-hand side are reduced by branching and/or propagation.

3. Propagate them globally whenever an improved primal solution is found.

The high-level observation that dual feasible solutions encountered during the solution process may be used to construct a one-row relaxation of the LP at hand and that this inequality can be used to tighten the bounds of each variable involved has been made earlier by Tawarmalani and Sahinidis (2004). Applied unconditionally, however, this idea seems overly expensive and it is rather unclear how to put it to good use. Here, we suggest selecting the dual solutions from OBBT and propagating the resulting LVBs only towards the left-hand side variable.

Remark 6.5 (Redundancy of LVBs). The main purpose of the LVBs is to identify bounds that are already implied by the relaxation and—by making them explicit—allow improvement of the convex relaxations of nonconvex nonlinear constraints. Unlike MIP cutting planes, LVBs are not designed to cut off the LP optimum. Since they are redundant inequalities, it is not beneficial to add them to the LP relaxation.

6.2.3 The LVB Graph

While propagating a single Lagrangian variable bound is a trivial task, a successful tightening may strengthen the effect of other LVBs. This raises the question in which order LVBs should be propagated and how often to repeat a round of propagation. We address this issue by considering a directed graph D defined as follows: For each variable x_j, D contains a node v_j and a node w_j representing its lower bound and upper bound, respectively. Let

$$x_k \geqslant \sum_{j=1,\dots,n} \tilde{r}_j x_j + \tilde{\mu} U + \tilde{\lambda}^\mathsf{T} b.$$

be the latest stored LVB implying a lower bound on x_k. For each variable x_j with $j \neq k$, we draw a directed arc

– From v_j to v_k if $\tilde{r}_j > 0$, or

– From w_j to v_k if $\tilde{r}_j < 0$.

If otherwise

$$x_k \leqslant \sum_{j=1,\dots,n} \tilde{r}_j x_j + \tilde{\mu} U + \tilde{\lambda}^\mathsf{T} b,$$

is the latest stored LVB implying an upper bound on x_k, for each variable x_j with $j \neq k$, we draw a directed arc

- From w_j to w_k if $\tilde{r}_j > 0$, or

- From v_j to w_k if $\tilde{r}_j < 0$.

In other words, we draw an arc (a, b) if an improvement of bound a strengthens the bound propagated by the latest stored LVB affecting bound b.

Sorting the LVB Graph. Suppose that D does not contain any directed cycle and fix some topological order of its nodes. Clearly, given some (locally) valid lower and upper variable bounds, once we have performed LVB propagation for each single LVB with respect to the order of nodes of D, no additional propagation of a single LVB can further tighten some bound.

Unfortunately, in general the graph D contains directed cycles and hence does not admit a topological order of its nodes. In this case, it is easy to construct scenarios in which every order of generated LVBs allows further improvements after one round of LVB propagation (by again propagating single LVBs). Even worse, it can be seen that the propagation of LVBs may exhibit the same non-finite convergence behavior as described by Belotti et al. (2010, 2012) for standard FBBT-like propagation of linear constraints.

In order to at least reduce this effect, we propose to compute a partial (topological) order over D. First, for each connected component of D, compute its strongly connected components, shrink each of them to a single node and topologically sort the resulting acyclic digraph. Finally, sort the LVBs according to that order by replacing each shrunk strongly connected component by some arbitrary order of its members. Note that such an order can be computed in time linear in the size of D, see, e.g., Tarjan (1972).

Efficient LVB Propagation. We are ready to propose a simple way to efficiently propagate generated LVBs. After every round of OBBT, store the latest computed LVB for each bound, update the LVB graph D and compute the partial order on D as described above. Whenever LVB propagation is called, for each (locally or globally) improved bound label its out-neighbors as *candidates*. If the global objective cut-off has been improved, add all bounds having an LVB with a nonzero coefficient $\tilde{\mu}$ on its right-hand side to the list of candidates. According to the partial order on D, for each candidate we propagate its associated LVB and, if successful, label all its out-neighbors as candidates and proceed with the next candidate with respect to the partial order.

In every LVB propagation call, this procedure is only performed once. As mentioned above, additional calls might result in further tightenings. However, this is a standard issue for all general propagation routines. For that reason, state-of-the-art solvers dynamically limit the effort spent on propagation and so we also propose to leave this decision to the solver framework itself.

6.3 Accelerating OBBT

Whereas Lagrangian variable bounds aim at leveraging the effect of OBBT, in this section we will discuss three ways to reduce the computational cost for applying OBBT in the first place. These are *selecting* promising types of variables for OBBT, *filtering* bounds a priori in which no tightening can occur, and *ordering* the LPs to be solved.

6.3.1 Selecting Variables

Because of its cost, it is a natural idea to apply OBBT not for all variables, but more selectively. Since the goal is to tighten bounds that lead to tighter convex relaxations, it has been suggested by Adjiman et al. (1998) and Zamora and Grossmann (1999) to restrict OBBT to variables appearing in nonconvex terms, i.e., to candidates for spatial branching. We will refer to variables that appear in nonlinear and nonconvex nonlinear terms as "nonlinear" and "nonconvex" variables, respectively.

Binary variables should be excluded from OBBT because the effect of OBBT can be achieved more efficiently as follows: Tentatively fix the variable to zero and one and each time solve the resulting relaxation. The relaxation is infeasible on one side if and only if OBBT would fix the variable to the other side.

It has been observed, for instance, by Adjiman et al. (2000), that this is computationally more efficient than minimizing and maximizing the binary variable. This is for two reasons: first, because it does not require the modification of the objective function and addition of the objective cut-off constraint; second, when using the simplex method to solve the LPs, then after fixing a variable the LP can be warm started by the dual simplex, while OBBT LPs are warm started by the primal simplex. Computationally, the primal simplex is often less efficient, see, e.g., Bixby (2002). Note that this tentative fixing to zero and one is equivalent to applying strong branching (Applegate et al., 1995, 2007a), which has been found to be a crucial ingredient to state-of-the-art branching rules for mixed-integer linear programming. Strong branching is also used by most state-of-the-art MINLP solvers for branching on integer variables with non-integral value in the relaxation solution, see, e.g., Belotti et al. (2009); Misener and Floudas (2013); Vigerske (2013). For more details on branching see Chapter 7.

Finally, for numerical reasons, convex nonlinearities can not always be enforced by separating outer approximation constraints; hence also variables that appear only in convex constraints can potentially be selected as branching candidates. Tighter bounds on branching candidates reduce the extent to which enumerative search must be performed. Hence, in the implementation presented in Section 6.4, we apply OBBT to all nonbinary nonlinear variables, also to those that appear only in convex nonlinear constraints.

6.3.2 Filtering Bounds

As shown above, LVBs are a dual certificate of the validity of tightened OBBT bounds. The primal counterpart is given by the simple observation that any feasible relaxation solution $x^* \in \mathcal{R} \cap \{x \in \mathbb{R}^n \mid c^T x \leqslant U\}$, certifies that the bounds that are tight at this solution cannot be improved by OBBT. More generally, if $|x_i^* - \ell_i|$ or $|x_i^* - u_i| \leqslant \varepsilon$ for some $\varepsilon \geqslant 0$, then OBBT can strengthen the corresponding bound by at most ε.

This idea is exploited in the branch-and-contract algorithm of Zamora and Grossmann (1999), which inspects the optimal solution x^* of the convex relaxation at the current node of the branch-and-bound tree. They execute OBBT only for lower and upper bounds for which $|x_i^* - \ell_i|$ and $|x_i^* - u_i|$, respectively, exceeds a specified fraction of the domain size of variable x_i. This reduces the number of expensive solutions of OBBT problems for which no or only a small improvement of the bound is possible.

We observe that it is not necessary to restrict this filtering of bounds to an optimal solution of the relaxation, but that any feasible solution of the relaxation can be used as long as it satisfies the objective cut-off. This includes, in particular, all primal solutions encountered during OBBT. These have the advantage that they are readily available at no additional cost. Beyond that, we can ask the question of how to generate solutions or a sequence of solutions that filter a large amount of bounds at a lower cost than applying OBBT to each of these bounds. To rephrase the task: compute points in $\mathcal{R} \cap \{x \in \mathbb{R}^n \mid c^\mathsf{T} x \leqslant U\}$ that are tight at or close to a large number of yet unfiltered bounds.

A natural method to achieve this is given by Algorithm 6.1, called *aggressive bound filtering*. It takes a general direction $v \in \mathbb{R}^n$ with positive objective coefficient for the variables that we want to become tight at their upper bound and negative ones for variables that we want to push towards their lower bound. The algorithm maximizes $v^\mathsf{T} x$ over the feasible region over which OBBT operates and filters all bounds at which the resulting solution is tight. Subsequently, positive objective coefficients are set to zero if the corresponding upper bound was filtered and negative objective coefficients are set to zero if the lower bound was filtered. Reoptimizing the filtering problem with reduced v may now give a point at which further bounds become tight and can be filtered. This procedure is iterated until either all bounds have been filtered, until v is reduced to a unit direction—in which case the filtering problem would be identical to applying OBBT in this direction—, or until the success rate, i.e., the number of bounds filtered in the last iteration, drops below a specified threshold δ.

In order to ensure that the filtering problem is bounded, we force objective coefficients to be nonpositive for variables without upper bound and nonnegative for variables without lower bound. We assume that the relaxation is not empty because this procedure is applied after solving this relaxation for computing the dual bound. The resulting optimal solution can be used to compute the initial sets of unfiltered bounds that are required as input. The repeated solution of the filtering problem is particularly efficient if the relaxation is polyhedral and can be solved using the dual simplex algorithm with warm starts.

Remark 6.6 (Filtering Directions). A natural candidate for an initial direction v in order to filter upper bounds is $v_i = 1$ for all $i \in \mathcal{U}_0$ and zero otherwise; and $v_i = -1$ for all $i \in \mathcal{L}_0$, zero otherwise, in order to filter lower bounds. Note that even if $v_i > 0$, the resulting filtering solution x^k may be tight at—and hence filter—the lower bound ℓ_i. One frequently encounters MINLP models that feature nonnegative variables only and allow the zero solution as feasible, at least for the relaxation. In this case, if the

---

Algorithm 6.1: Aggressive Bound Filtering

---

input:　　variable bounds $[\ell, u]$, convex relaxation $\mathcal{R} \subseteq [\ell, u]$, objective function
　　　　　　vector $c \in \mathbb{R}^n$, objective limit U, unfiltered bound indices $\mathcal{L}_0, \mathcal{U}_0 \subseteq \{1, \dots, n\}$,
　　　　　　initial direction $v \in \mathbb{R}^n$

params: termination threshold δ

output:　reduced index sets $\tilde{\mathcal{L}} \subseteq \mathcal{L}_0$, $\tilde{\mathcal{U}} \subseteq \mathcal{U}_0$

```
 1  begin
 2  │   𝓛₁ ← {i ∈ 𝓛₀ | ℓᵢ ≠ −∞}                       /* 0. exclude unbounded directions */
 3  │   𝒰₁ ← {i ∈ 𝒰₀ | uᵢ ≠ ∞}
 4  │   for k ← 1, 2, … do                                               /* filtering loop */
                                                      /* 1. reduce v to unfilt. directions */
 5  │   │    foreach i ∈ 𝓛ₖ₋₁ \ 𝓛ₖ, vᵢ < 0 do vᵢ ← 0
 6  │   │    foreach i ∈ 𝒰ₖ₋₁ \ 𝒰ₖ, vᵢ > 0 do vᵢ ← 0
 7  │   │    if 𝓛ₖ ∪ 𝒰ₖ = ∅ or v = 0 then                         /* 2. check termination */
 8  │   │    │   break
 9  │   │    else if k ⩾ 2 and |𝓛ₖ₋₁ \ 𝓛ₖ| + |𝒰ₖ₋₁ \ 𝒰ₖ| ⩽ δ then
10  │   │    │   break
11  │   │    xᵏ = arg max{vᵀx | x ∈ 𝓡, cᵀx ⩽ U}                        /* 3. minimize v */
12  │   │    𝓛ₖ ← {i ∈ 𝓛ₖ₋₁ | xᵢᵏ ≠ ℓᵢ}                       /* 4a. filter lower bounds */
13  │   │    𝒰ₖ ← {i ∈ 𝒰ₖ₋₁ | xᵢᵏ ≠ uᵢ}                       /* 4a. filter upper bounds */
14  │   return 𝓛̃ ← 𝓛ₖ, 𝒰̃ ← 𝒰ₖ
```

---

objective cut-off bound U is positive, Algorithm 6.1 started with the "-1" direction requires only one iteration to detect that no lower bounds can be strengthened.

To summarize, we have identified three different types of solutions that can be used to filter bounds for which applying OBBT is fruitless. In order of increasing computational cost, these are

- Optimal solutions of the convex relaxation computed for bounding,

- The solutions encountered during OBBT, and

- The solutions produced by aggressive bound filtering.

We will refer to the first two as *trivial* filtering. If tightness is tested exactly as in lines 12 and 13 of Algorithm 6.1, then the filtering is exact in the sense that only bounds are excluded for which OBBT is guaranteed to find no tightening. This is easily replaced by a test using an absolute or relative threshold. A practical implementation based on floating-point arithmetic needs to use a small nonzero tolerance, even when testing for exact equality.

Remark 6.7 (Filtering vs. LVB Generation). As noted in Section 6.2, nontrivial LVBs can be learned also from dual OBBT solutions that do not tighten the current bound. In this sense, the filtering of such bounds might reduce the number of generated LVBs.

In order to counteract this effect, our implementation uses the following technique that is specific to the simplex method.

Suppose x^* is a solution that we want to use for filtering tight bounds, which might be the initial relaxation solution, a solution of an OBBT LP, or the solution computed during aggressive filter in line 11 of Algorithm 6.1. Suppose further that $x_i^* = \ell_i$, but due to primal degeneracy variable x_i is basic in the simplex solution. Then we know that its lower bound is not part of the proof of optimality in the current simplex basis. In this case, there is the chance of generating a nontrivial LVB by using the primal simplex to minimize x_i starting from the current basis. We perform this step because it should take only few simplex iterations; afterwards we can mark the bound as processed. Note that the LP solution may change in the course of these calls to the simplex solver if pivots occur due to floating-point numerics.

By contrast, if x_i^* is nonbasic in the simplex solution then we know that the simplex solver will (in all likelihood) return the trivial dual solution that only involves this bound, leading to no useful LVB. Hence we filter nonbasic variables immediately without solving the above LP.

6.3.3 Ordering LP Solves

By construction, the optimal solutions computed during OBBT are typically widely scattered across the feasible region. When using a polyhedral relaxation, this fact challenges the warm-starting capabilities of the simplex algorithm, which is the algorithm of choice for solving LP relaxations in most spatial branch-and-bound implementations. In the following, we devise schemes for the order in which OBBT explores the bounds such as to heuristically minimize the time and iterations spent by a simplex-based LP solver.

Since a practical implementation will employ working limits on the number of LP iterations spent by OBBT, ideally, a good ordering strategy does not only minimize the total number of LP iterations, but also keeps low the number of LP iterations that are spent between any two successive candidates. An intuitive strategy is the following greedy heuristic: select the next bound candidate by considering the distance between the value of a variable in the current LP solution and the bound of the variable that we want to tighten. This is formalized in Algorithm 6.2, which executes a complete round of OBBT in this greedy order, interleaved with filtering steps. Note that if OBBT tightens the bound of an integer variable to a non-integral value, the bound may be tightened further by rounding it, see Lines 13 and 18 of Algorithm 6.2. As argued in Remark 6.7, LVBs are not only generated in Line 19, but also in Lines 4 and 20.

When OBBT is applied with working limits, e.g., on the number of simplex iterations, the order in which OBBT LPs are processed becomes important. There are two criteria for an efficient ordering: finding as many reductions as possible within the given limits and finding "important" reductions, which are often those with a large ratio between the original and the reduced domain. These criteria might be mutually exclusive.

In Lines 7 and 8, we use a greedy strategy for ordering the OBBT LPs. This is motivated by the expectation that we need few simplex iterations when a variable's

---

Algorithm 6.2: OBBT with Filtering and Greedy Ordering

input: variable bounds $[\ell, u]$, convex relaxation $\mathcal{R} \subseteq [\ell, u]$, objective function
vector $c \in \mathbb{R}^n$, objective limit U, bound indices $\mathcal{L}, \mathcal{U} \subseteq \{1, \dots, n\}$, index set of
integer variables $\mathcal{I}$, initial relaxation solution x^*

output: tightened variable bounds $[\ell', u'] \subseteq [\ell, u]$

1 **begin**
2 $\quad [\ell', u'] \leftarrow [\ell, u]$
3 $\quad$ reduce $(\mathcal{L}, \mathcal{U})$ by trivial filtering on x^*
4 $\quad$ reduce $(\mathcal{L}, \mathcal{U})$ by aggressive bound filtering (optional)
5 $\quad \tilde{x} \leftarrow$ last LP solution
6 $\quad$ **while** $\mathcal{L} \cup \mathcal{U} \neq \emptyset$ **do**   /* 1. select bound */
7 $\quad\quad val_{\mathcal{L}} \leftarrow \min\{\tilde{x}_k - \ell_k \mid k \in \mathcal{L}\}$ /* ∞ if $\mathcal{L} = \emptyset$ */
8 $\quad\quad val_{\mathcal{U}} \leftarrow \min\{u_k - \tilde{x}_k \mid k \in \mathcal{U}\}$ /* ∞ if $\mathcal{U} = \emptyset$ */
9 $\quad\quad$ **if** $val_{\mathcal{L}} \leqslant val_{\mathcal{U}}$ **then**
10 $\quad\quad\quad i \leftarrow \arg\min\{\tilde{x}_k - \ell_k \mid k \in \mathcal{L}\}$
11 $\quad\quad\quad \mathcal{L} \leftarrow \mathcal{L} \setminus \{i\}$
12 $\quad\quad\quad \tilde{x} \leftarrow \arg\min\{\tilde{x}_i \mid \tilde{x} \in \mathcal{R}, c^\mathsf{T}x \leqslant U\}$ /* 2a. minimize */
13 $\quad\quad\quad$ **if** $i \in \mathcal{I}$ **then** $\ell'_i \leftarrow \lceil \tilde{x}_i \rceil$ **else** $\ell'_i \leftarrow \tilde{x}_i$
14 $\quad\quad$ **else**
15 $\quad\quad\quad i \leftarrow \arg\min\{u_k - \tilde{x}_k \mid k \in \mathcal{U}\}$
16 $\quad\quad\quad \mathcal{U} \leftarrow \mathcal{U} \setminus \{i\}$
17 $\quad\quad\quad \tilde{x} \leftarrow \arg\max\{\tilde{x}_i \mid \tilde{x} \in \mathcal{R}, c^\mathsf{T}x \leqslant U\}$ /* 2b. maximize */
18 $\quad\quad\quad$ **if** $i \in \mathcal{I}$ **then** $u'_i \leftarrow \lfloor \tilde{x}_i \rfloor$ **else** $u'_i \leftarrow \tilde{x}_i$
19 $\quad\quad$ generate LVB, store if nontrivial
20 $\quad\quad$ reduce $(\mathcal{L}, \mathcal{U})$ by trivial filtering on $\tilde{x}$ /* 3. filter */
21 $\quad$ **return** $[\ell', u']$

---

relaxation value is close to the bound that should be tightened. One disadvantage
may be, though, that it explores those variables first which are closest to their bound
and therefore show the smallest potential for a tightening. Therefore just the opposite
strategy could be equally beneficial: a *reverse greedy* ordering that always chooses
variables with largest distance between their value in the current LP solution and the
bound of interest. This might lead to more bound tightenings among the early LPs
solved; at the same time, these LPs may be more expensive to solve.

A different heuristic, which we called *min–max*, first solves all minimization LPs,
then all maximization LPs. This is motivated by the conjecture that the solutions to
the minimization LPs are close to each other and that processing them sequentially
may indirectly exploit the simplex algorithm's warm-starting capabilities; likewise for
the solutions to the maximization LPs. Finally, as a control setting, the experiments
in Section 6.4 also used a trivial *random* choice strategy as a baseline that any more
sophisticated strategy should outperform.

Remark 6.8 (Minimum Spanning Tree Orders). During our experiments we obtained
the following "negative" result. If the simplex basis at an optimal OBBT solution can

be stored and restored, this allows for the design of more sophisticated ordering algorithms since one is not bound to warm start the LP solves from the last solution. We have experimented with a heuristic similar to the algorithm of Prim (1957) for computing a minimum spanning tree in a graph. To this end, consider the bounds of the variables as the nodes of a complete graph. The number of LP iterations spent between the LP solves associated with two bounds gives weights on the edges; these weights are, in general, not well-defined and a priori unknown, but the distances used in the above greedy ordering can serve as a simple proxy. If we assume symmetric weights then, at least in theory, a minimum-weight spanning tree of the undirected graph provides an order of LP solves that minimizes the total number of LP iterations. In more detail, if during the algorithm of Prim we add an edge $\{v, w\}$ to the current tree T, where $v \in T$ and $w \notin T$, we load the LP basis stored at v and solve the corresponding OBBT LP for w. In our experiments, however, this was not efficient due to the large overhead for storing and restoring the basis information, which additionally invalidates the underlying factorization of the simplex solver.

6.4 Computational Study

In this section, we investigate the computational impact of OBBT with the enhancements described in Section 6.3 and the effectiveness of the LVB propagation introduced in Section 6.2. We implemented these techniques as part of the MINLP solver SCIP.[78] For an overview of SCIP's general solving algorithm and MINLP features we refer to Section 5.1.3.

6.4.1 Implementation

We extended SCIP by two propagator plug-ins for OBBT and LVB propagation. The OBBT propagator is based on the diving mode of SCIP, which allows direct access and modification of the LP relaxation. Since the validity of OBBT is based on dual feasibility, we chose a dual feasibility tolerance of 10^{-9} for the OBBT LP solves. (The SCIP default is 10^{-7}.) When using the simplex method to solve the LPs, an upper limit on the condition number of the basis matrix at the optimal LP solution can be specified, beyond which the resulting bound is not trusted and discarded; this feature is optional and disabled by default.

Because of its potentially high computational cost, we gave OBBT the lowest priority and executed it only when no other propagators found problem reductions. By default, it is only called at the root node and the number of LP iterations spent by OBBT during one call is limited to ten times the LP iterations spent so far at the root node. Note that SCIP may call OBBT repeatedly at one node. In this case we continue with the variable directions not yet processed at this node; each variable direction is tested at most once per node. The LP optimum is used for the initial filtering, followed by aggressive bound filtering, and the filtering via OBBT solutions.

---

[78] Zuse Institute Berlin. SCIP—Solving Constraint Integer Programs. Available for download under `http://scip.zib.de/`.

From each successfully solved OBBT LP at the root node, globally valid Lagrangian variable bounds are transferred to the LVB propagator and propagated at each node as described in Section 6.2.3. In order to avoid frequent resorting triggered by the addition of new LVBs, we sort only after the root node processing. Before that, we propagate without making use of the graph structure. When a restart is performed by SCIP, the LVB information is preserved and propagated during the subsequent presolving loop.

6.4.2 Experimental Setup

We conducted two main experiments. For the first experiment, we processed only the root node of SCIP. For the second one, we analyzed the overall performance impact on the complete tree search.

The goal of the root node experiment was to answer the following questions: How many nontrivial LVBs can be generated during OBBT? As explained in Remark 6.3, LVBs may be trivial and not contain any information useful for bound tightening. Furthermore, what is the effect of filtering? And what is the effect of different ordering strategies on the performance of OBBT?

In our root node experiment, we chose a "pure" OBBT setting: perform OBBT once at the root node, solving one OBBT LP for the bounds of each unfixed, nonbinary, nonlinear variable as explained in Section 6.3.1. In order to ensure comparability, we did not apply any working limits on the number of LPs or LP iterations spent in the root node experiments. We compared the ordering strategies described in Section 6.3.3 (greedy, reverse greedy, min–max, and random) without aggressive bound filtering. We analyzed the impact of filtering on the number of bound tightenings and LVBs found by OBBT using the following settings:

– No filtering,

– Trivial filtering after each solved OBBT LP, and

– Trivial plus aggressive filtering.

The goal of the tree experiment was to analyze the impact of OBBT alone and of OBBT enhanced by LVB propagation on the overall performance of SCIP. To this end, we measured the running time and the number of branch-and-bound nodes taken by SCIP when solving to ε-global optimality[79] for the gap limit $\varepsilon = 10^{-4}$. In this second experiment, we compared

– SCIP without OBBT,

– SCIP with OBBT at the root node, with working limits as stated in Section 6.4.1, using greedy ordering and aggressive filtering, and

– SCIP with OBBT as above and LVB propagation during the tree search.

---

[79] Let U and L be the global upper and lower bounds, respectively. Then SCIP terminates if either $U - L \leqslant \varepsilon$ (absolute gap limit reached) or U and L have same sign, $|U|, |L| > 10^{-9}$, and $|U - L| / \min\{|U|, |L|\} \leqslant \varepsilon$ (relative gap limit reached).

We imposed a time limit of one hour per instance for both the root node and the tree experiment.

We also conducted preliminary experiments applying OBBT within the tree, as opposed to calling it only once at the root and exploiting its deductions during tree search. While this can be beneficial on some instances, a simple execution of OBBT at each node of the tree easily incurs a large overhead in running time on average. The main research question that we address in this paper is not how to tune OBBT for application at local nodes of the tree, but how to efficiently use it for the deduction of global bounds and how to generate and harness LVBs as an approximation of iterated OBBT during the entire solving process. It turns out that this already requires a careful balancing of the effort that OBBT spends a priori versus the savings gained later through smaller search trees. Achieving similar results for local OBBT constitutes a far more difficult challenge. This being said, the ordering and filtering strategies found superior in our root node experiments may be a good starting point for investigating how to perform OBBT locally in the tree.

Test Sets. We used the publicly available instances of the MINLPlib2 described in Section 1.4.3, from which 1,279 instances were available in OSiL format and contained nonlinear expressions that could be handled by SCIP. Of these, 36 instances were solved by SCIP's presolving or were linear after presolving; we removed them because they are not relevant for our experiments. For 287 instances OBBT was never called because SCIP solved them to proven optimality or hit the time limit before; we excluded those instances because they are not relevant to our experiments. For a more detailed analysis, we subdivided the remaining 952 instances into the 605 MINLPs which contained integer variables after presolving (test set INT) and the 347 instances that were NLPs after presolving (test set GO, for *global optimization*).

Note that OBBT terminates if the LP solver fails to solve an OBBT LP, e.g., due to numerical difficulties. Such an early termination for one of the settings would affect the observed success of this setting and compromise the comparability of the results. Hence, only for the root node experiments, we removed instances for which an LP error occurred for at least one of the settings. This was the case for 28 of the INT instances and for 5 of the GO instances. For the tree experiments we did not exclude instances because of LP errors in order to measure the performance of the actual OBBT implementation used by the solver SCIP as accurately as possible.

Hardware and Software. The experiments were conducted on a cluster of 64-bit Intel Xeon X5672 CPUs at 3.2 GHz with 12 MB cache and 48 GB main memory. In order to safeguard against a potential mutual slowdown of parallel processes, we ran only one job per node at a time. We used SCIP developer version 3.1.0.1 with SoPlex[80] 2.0.1 as LP solver, CppAD[81] 20140000.1, and Ipopt[82] 3.12.4 (Wächter and Biegler, 2006).

---

[80] Zuse Institute Berlin. SoPlex—the Sequential object-oriented simPlex. Available for download under `http://soplex.zib.de/`.

[81] COIN-OR. CppAD, a Package for Differentiation of C++ Algorithms. `http://www.coin-or.org/CppAD`.

[82] COIN-OR. Ipopt, Interior point optimizer. `http://www.coin-or.org/Ipopt`.

6.4.3 Computational Results for Root Node Experiments

For each of the root node experiments, detailed instance-wise results can be found in Tables A.6, A.7, A.8, and A.9 in the appendix. In the following we discuss these results in an aggregated fashion. First, we analyze how frequently LVBs are generated and how LVB generation interacts with different filtering strategies. Second, we compare various ordering heuristics w.r.t. the number of bound reductions and LVBs they find when given working limits on the overall number of LP iterations. For computing shifted geometric means as explained in Section 1.3, we use shift values of ten for averaging over the number of LP solves, 100 for averaging over the number of LP iterations, ten seconds for averaging over the running time, and 100 nodes for averaging over the size of the search trees.

Generated LVBs. We counted the number of nontrivial LVBs that were generated per number of OBBT LPs solved. This gives a *success rate* for the generation of nonlinear LVBs, which lies between 0% and 100% for each instance. Figure 6.3 plots a histogram of these success rates on test set INT when using greedy ordering and the three different filtering strategies; Figure 6.4 plots the same for the NLP test set GO. In Figure 6.3a, for example, we can see that for eleven instances it was possible to generate an LVB from 95–100% of the LPs solved.

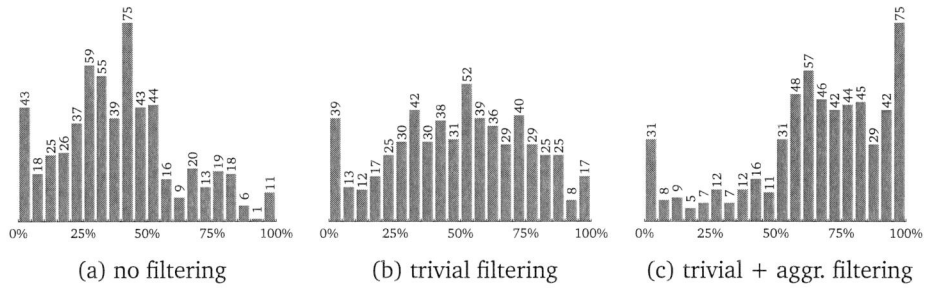

| (a) no filtering | (b) trivial filtering | (c) trivial + aggr. filtering |

Figure 6.3: Histograms of LVB success rates for test set INT

Our first observation is that both trivial and aggressive filtering each increase the success rates significantly: visually, the "mass" of the histogram moves towards higher success rates as we go from Figure 6.3a through 6.3b to 6.3c; likewise for Figure 6.4. This shows that we filter primarily those OBBT LPs from which no LVB could have been generated, so filtering seems to affect the generation of LVBs only slightly. This is a positive result: the increase in efficiency of OBBT does not compromise the effective generation of LVBs. To highlight specific numbers, with aggressive filtering, for more than 75% of the instances the LVB success rate was 50% or higher, i.e., on these instances, more than half of the OBBT LPs led to a nontrivial LVB.

For the GO instances, there is a much larger group on which almost no LVBs could be generated. Even with aggressive filtering, 59 out of 342 instances are in the 0–5% bin; for 48 of these, no LVB could be generated at all. Except for this difference, the distribution of success rates and the effect of filtering is very similar.

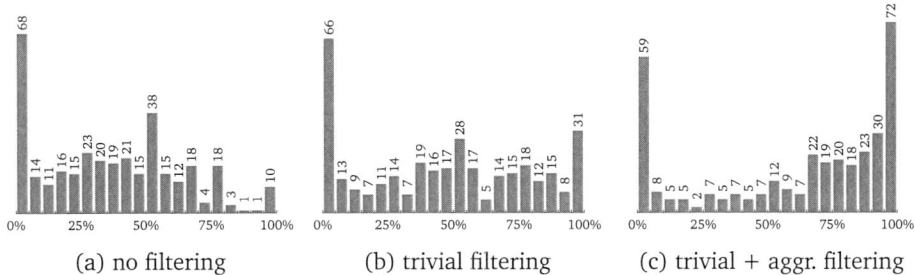

Figure 6.4: Histograms of LVB success rates for test set GO

Overall Performance of Ordering and Filtering Strategies. Table 6.1 shows average performance measures on six different OBBT strategies. The base setting is the greedy ordering with trivial filtering. In the first three strategies, we changed the ordering heuristic to one of greedy reverse, min–max, and random. Additionally, we tested one setting with no filtering and one setting with aggressive bound filtering activated. We compare the number of LPs solved, the number of LP iterations spent, the number of bounds tightened, and the number of LVBs generated.

As a first observation, the average number of bounds that could be tightened (see column "b_{obbt}") is almost identical for all strategies. This is expected since we processed the same variables for all strategies. However, the limited precision of floating-point operations can potentially lead to different outcomes, as can be seen, e.g., at the average value 13.2 for "greedy (base)" versus the value 13.1 for the other strategies on the INT test set. The number of LVBs generated is similar over different orders. However, in column "lvb" it can be noted that turning filtering off causes 3.1% (39.4 vs. 38.2) and 4.0% (18.1 vs. 17.4) more LVBs to be generated on the INT and GO test set, respectively. This can be explained by the fact that we filter bounds for which, at the current simplex basis, no nontrivial LVB can be generated because the corresponding variable is nonbasic, see also Remark 6.7. Without filtering, these bounds may become basic at a later stage and admit a nontrivial LVB.

In general, the number of bound tightenings found is rather low: only about 15.6% on the INT (13.2 from 84.5) and 13.0% on the GO test set (7.6 from 58.4) of the OBBT LPs solved by the base setting led to a tighter bound. However, we can see that many more LVBs could be generated: 2.9 times as many LVBs as tightened bounds on the INT test set, and 2.3 times as many on the GO test set. This shows that in many cases valuable information in the form of nontrivial LVBs can be learned even if OBBT does not yield tighter bounds.

To evaluate the efficiency of the different strategies, we first look at the number of LP iterations. Although the greedy strategy had to solve more OBBT LPs than the other ordering heuristics, it did so in significantly fewer iterations: between 14.7% and 26.6% fewer than for reverse greedy, random, and min–max. Not surprisingly, reverse greedy even took more iterations than the random order. The min–max order was better than the random order, but could not outperform the greedy order. Also

Table 6.1: Aggregated results comparing six OBBT strategies at the root node

lp — number of LPs solved by OBBT (shifted geo. mean)
lp_{filt} — number of LPs solved by filtering (not incl. in lp, shifted geo. mean)
iter — number of LP iterations by OBBT (shifted geo. mean)
$iter_{filt}$ — number of LP iterations by filtering (not incl. in iter, shifted geo. mean)
b_{obbt} — number of bounds tightened by OBBT (shifted geo. mean)
t_{obbt} — time used by OBBT (shifted geo. mean)
lvb — number of LVBs found by OBBT (shifted geo. mean)

Test set	strategy	lp	lp_{filt}	iter	$iter_{filt}$	b_{obbt}	t_{obbt}	lvb
INT	random	77.3	15.6	1830.0	45.1	13.1	0.93	37.9
	min–max	77.7	15.5	1765.2	44.6	13.1	0.89	37.8
	reverse greedy	76.0	16.8	1847.5	47.6	13.1	0.92	37.8
	greedy (base)	84.5	10.6	1478.1	32.5	13.2	0.79	38.2
	greedy+filt. off	109.9	0.0	1505.1	0.0	13.1	0.79	39.4
	greedy+filt. aggr.	63.1	27.5	982.9	701.8	13.1	0.79	38.7
GO	random	55.1	9.5	429.6	22.5	7.6	0.42	17.7
	min–max	55.5	9.7	401.6	24.2	7.6	0.39	17.7
	reverse greedy	54.3	10.2	443.7	22.6	7.6	0.42	17.7
	greedy (base)	58.4	6.8	351.7	17.8	7.6	0.37	17.4
	greedy+filt. off	74.8	0.0	358.7	0.0	7.6	0.40	18.1
	greedy+filt. aggr.	40.2	18.2	212.4	204.6	7.6	0.31	18.3

w.r.t. time, the greedy strategy is superior to the other ordering strategies.

Aggressive filtering increases the number of LP iterations in total, but decreases the number of LPs solved. Because each LP solve incurs a fixed setup time, OBBT with aggressive filtering is overall faster than OBBT with trivial filtering only: in combination with the greedy strategy it leads to a 16% speedup of OBBT on the GO test set; for the INT test it is performance-neutral. For single instances we observe tremendous improvements. For elec200 of the GO test set, aggressive filtering reduces the time for OBBT from 714.3s to 62.6s and the LP iterations from 143,948 to 38,811. For space960 from the INT test set, it reduces the time spent for OBBT from 502.1s to 89.7s and the LP iterations from 804,391 to 120,398. Therefore, we applied aggressive filtering in the tree experiments of Section 6.4.4.

Moreover, we compared the relative domain reductions found or implied by the OBBT or LVB propagator in the root node between the different ordering strategies. The relative reduction for all different strategies are almost the same. For the INT test set all ordering strategies reduced the volume of the domain by 19% compared to 15% reduction without using OBBT. Also, for the GO test set there are almost no differences in the volume reduction between the different ordering strategies. All strategies reduce the volume by 25% compared to 13% if we are not using OBBT.

We conclude that the greedy strategy is most promising when OBBT shall be performed for all variables. The next paragraph considers the case that we conduct only a partial execution of OBBT.

Bound Tightening Profiles. The following analysis is relevant to the tree experiments presented later. An efficient implementation as part of an MINLP solver will impose working limits, e.g., on the number of LP iterations, in order to control the effort spent on OBBT. Thus, the above comparison based on a full execution of OBBT does not show the full picture: we do not so much want to find the strategy that is fastest at processing all bounds, but rather the strategy that succeeds most often with limited computational effort. This of course depends on the chosen limits; ideally, a solver uses dynamic limits. Therefore, we performed a slightly involved graphical analysis that is inspired by performance profiles as used for benchmarking optimization software (Dolan and Moré, 2002). The result, comparing the ordering strategies from Table 6.1, can be seen in Figure 6.5.

These profiles were generated as follows: For one run of OBBT on one instance, we recorded the number of bound tightenings found within the first n iterations vs. n, resulting in an increasing step function. This gives one profile per instance per strategy. For each instance, we normalized its profiles on both axes, dividing the number of iterations by the maximum number of iterations over all strategies compared, and likewise dividing the number of bound tightenings by the maximum number of bound tightenings over all strategies compared. The scaled profiles are between zero and one on both axes and can be averaged pointwise over all instances. This gives one profile per strategy as displayed in Figure 6.5. Loosely speaking, a point (x, y) in this profile means that after $x\%$ of the total LP iterations, $y\%$ of the total bound tightenings have been found (on average). Profiles that are higher and further left indicate better performance: more bound tightenings and more LVBs are found earlier. If the profile of strategy A is consistently above the profile of strategy B, we know that A outperforms B *independently* of how we choose working limits on the number of LP iterations: at whichever iteration we stop OBBT, we know that strategy A will (on average) have found more bound tightenings and LVBs than strategy B.

Only for plotting these profiles, we excluded instances where all strategies took less than 1,000 LP iterations or achieved less than ten bound tightenings. The former criterion handles instances for which OBBT is trivial and different orders have hardly any effect. The latter criterion excludes instances for which OBBT, and hence different strategies for OBBT, have limited impact on the overall performance.

We generated profiles both for the number of bounds tightened and the number of LVBs generated; for the latter, the same procedure can be applied. We did not distinguish between the two test sets INT and GO, because their profiles look almost identical.

As can be seen in Figure 6.5, all ordering heuristics reach the same level of success eventually. The greedy order, however, significantly outperforms the others during the intermediate stages of OBBT. The min–max strategy performs slightly better than the random order.

To summarize the results from the root node experiments, we saw that the greedy ordering heuristic outperformed the other orderings in our experiments. Furthermore, the overhead of aggressive bound filtering in terms of LP iterations was compensated by the fewer numbers of LPs being solved, leading to faster overall running times

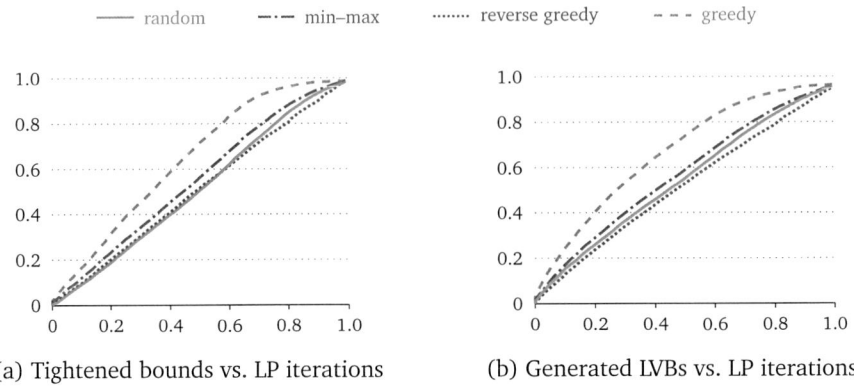

(a) Tightened bounds vs. LP iterations   (b) Generated LVBs vs. LP iterations

Figure 6.5: Profiles for different ordering strategies

of OBBT. Finally, a greedy ordering strategy is expected to have a higher success rate when operated with limits on LP iterations. Thus, we made greedy ordering and aggressive filtering the default strategies for our tree experiment. We close this section with a remark on the relationship between greedy ordering and trivial filtering.

Remark 6.9 (Greedy Filtering). If some of the variables in the current LP solution are tight at a yet unprocessed bound (and filtering is turned off), then the greedy strategy would select one of these tight bounds for the next OBBT LP. Hence, it "filters" this bound automatically by explicitly solving the LP, usually without a single pivot when warm starting the primal simplex method. As explained in Remark 6.7, we already partially solve these LPs even when filtering is turned on in order to generate LVBs. In this sense, a plain greedy ordering makes trivial filtering almost superfluous, with the exception of variables with nonbasic status that should be skipped.

6.4.4 Computational Results for Tree Experiments

Having arrived at efficient default settings for OBBT, we now turn to the results for comparing the overall performance of SCIP without OBBT, SCIP with OBBT only, and SCIP with OBBT and LVB propagation. Tables A.10 and A.11 in the appendix give detailed results for each instance of the INT and GO test set, respectively. We do not compare instances that could not be solved by any of the settings within the time limit (226 on INT and 160 on GO). Furthermore, we exclude 38 INT instances and 21 GO instances where one of the settings aborted due to insufficient memory, returned a solution that violated feasibility by more than 10^{-6}, or returned primal or dual bounds that were inconsistent with the values stated on the MINLPlib2 webpage. (None of the settings ran into these cases significantly more often than the others.)

This leaves 341 instances of the INT test set and 166 instances of the GO test set. For each of these test sets, Table 6.2 gives aggregated results over several subgroups. The group "all solved" contains the instances that could be solved by all settings within the time limit. Additionally, we analyze the instances which could be solved by all, but for which at least one setting took at least 1, 10, and 1000 seconds, respectively ("$\geqslant 1, 10, 100s$"). These groups are contained in "all solved" and form a hierarchy of harder and harder instances which are defined in an unbiased way by excluding instances that are easy to *all* settings; see, e.g., Achterberg and Wunderling (2013) for this benchmarking methodology. To be able to condense the presentation into one table, we state running time and number of branch-and-bound nodes on top of each other in one column. The key observations are:

– On all subgroups, SCIP+OBBT and SCIP+OBBT+LVB reduce the average number of branch-and bound nodes: by at least 6.2% (SCIP+OBBT on test set GO, group "all solved"), up to 41.9% (SCIP+OBBT+LVB on test set INT, group "$\geqslant 100s$").

– For the two easier INT subgroups "$\geqslant 1s$" and "$\geqslant 10s$", SCIP+OBBT alone gives a small slowdown, but the combination SCIP+OBBT+LVB is slightly faster than plain SCIP.

– On the harder INT subgroup "$\geqslant 100s$", SCIP+OBBT and SCIP+OBBT+LVB reduce the average running time by 5.7% and 17.1%, respectively.

– On the GO test set, SCIP+OBBT did not change the average running time significantly (at most 1.4% slower or faster than plain SCIP).

– For the GO subgroups $\geqslant 1, 10s$ and $\geqslant 100s$, SCIP+OBBT+LVB results in a speedup of 9.8%, 14.1%, and 18.8%: the harder the instances are, the larger are the reductions in running time achieved.

– Since LVBs are applied during the entire solving process, LVBs perform significantly more (one to three orders of magnitude more) bound tightenings than OBBT at the root node.

– On all subgroups LVBs lead to an additional reduction of running time and tree size as compared to SCIP+OBBT.

Note that the above groups do not contain instances that could be solved with one of the settings, but not with all. There were 11 such "timeout" instances in the INT test set and 9 in the GO test set. Out of the 11 "timeout" INT instances, plain SCIP solved only 1, while SCIP+OBBT solved 6, and SCIP+OBBT+LVB solved even 10 instances. Out of the 9 "timeout" GO instances, plain SCIP solved only 3, while SCIP+OBBT and SCIP+OBBT+LVB each solved 7 instances. Comparing plain SCIP with SCIP+OBBT+LVB on both test sets, SCIP+OBBT+LVB solved all but 3 instances which were solved by plain SCIP, and most importantly solved 16 additional instances for which plain SCIP timed out.

On the INT subgroups "all solved", "$\geqslant 1s$", and "$\geqslant 10s$" the reduction in the number of branch-and-bound nodes is consumed by the overhead in time spent on the solution of the auxiliary LPs. On the harder instances in the subgroup "$\geqslant 100s$", OBBT reduces running time by 5.7%. We believe that the slowdown on easier instances is

Table 6.2: Aggregated results comparing SCIP without OBBT, SCIP with OBBT only, and SCIP with OBBT and LVB propagation

N — number of instances
$t \lceil n$ — running time of SCIP in seconds (first line) and number of branch-and-bound nodes (second line, both shifted geo. mean)
% — relative time/nodes w.r.t. SCIP plain (in %)
b_{obbt} — number of bounds tightened by OBBT (shifted geo. mean)
b_{lvb} — number of bounds tightened by LVB propagation (shifted geo. mean)

Test set	subgroup	N	SCIP plain $t \lceil n$	SCIP+OBBT $t \lceil n$	%	b_{obbt}	SCIP+OBBT+LVB $t \lceil n$	%	b_{lvb}
INT	all solved	330	20.7	21.1	101.6	12.0	20.3	97.7	130.4
			2084.4	1894.3	90.9		1742.3	83.6	
	$\geqslant 1$s	228	40.1	40.9	101.9	14.3	39.0	97.2	292.4
			7196.2	6226.8	86.5		5605.0	77.9	
	$\geqslant 10$s	142	102.1	104.8	102.6	17.1	98.4	96.4	659.4
			24245.4	20140.6	83.1		18363.7	75.7	
	$\geqslant 100$s	69	353.5	333.5	94.3	11.9	293.0	82.9	811.0
			119914.6	86908.9	72.5		69714.2	58.1	
GO	all solved	157	4.6	4.6	99.1	5.2	4.3	93.5	54.6
			451.4	423.5	93.8		412.0	91.3	
	$\geqslant 1$s	43	27.7	27.3	98.5	7.2	25.0	90.2	2087.5
			16819.0	14147.3	84.1		13560.1	80.6	
	$\geqslant 10$s	20	124.8	123.0	98.6	8.2	107.2	85.9	9012.2
			140227.9	123725.6	88.2		116555.9	83.1	
	$\geqslant 100$s	11	362.7	367.7	101.4	7.8	294.6	81.2	11004.5
			357429.9	306267.6	85.7		294799.8	82.5	

an acceptable price to pay given the drastic improvements on harder instances and, even more importantly, the additional instances that can be solved with OBBT, but not with plain SCIP.

The number of LVB propagations grows with tree size on the subgroups of increasingly harder instances. This shows that LVB propagation is not only effective early, but throughout the solving process. However, it does not increase as drastically as tree size; this indicates that for most instances LVBs are most effective at upper and medium levels of the tree. LVBs are an approximation of OBBT generated from the LP relaxation at the root node; hence, it is intuitive that their effectiveness would decrease as the LP relaxation at local nodes of the search tree differs more and more from the root node relaxation.

Finally, we the Wilcoxon signed rank test allows us to analyze how consistently the improvements seen in the shifted geometric means are distributed across the test set. First, we analyze the changes due to OBBT alone, i.e., the change in average running time between plain SCIP and SCIP+OBBT. The small slowdowns on the INT test set are all judged significant, while the 5.7% speedup on the INT subgroup "$\geqslant 100s$" is not. Though this may be surprising at first, it shows the importance of significance testing: the statistical test here reveals the fact that this average speedup mainly comes from 5 instances with high speedup factors of 90, 163, 310, 334, and 357.[83] None of the small changes on the GO test set are significant. Second, the improvements due to LVBs, i.e., the reductions in average running time between SCIP+OBBT and SCIP+OBBT+LVB, are all confirmed as significant except for the two GO subgroups "$\geqslant 10s$" and "$\geqslant 100s$". The failing of the Wilcoxon test on these two subgroups is most likely due to their small sample size of 20 and 11 instances, respectively.

Finally, we discuss some distinguished results for individual instances. To ensure that the observed effects are indeed due to OBBT and LVBs and not just random noise, we exploit the idea of performance variability, see, e.g., Koch et al. (2011). To this end, we have solved each of the instances ten times with different permutations of their constraints. In the following, we will present average running times, taken over the ten permuted runs.

First, consider instance `fin2bb`. OBBT causes a slowdown from 224.3 to 291.7 seconds in shifted geometric mean. However, by using LVBs we could reduce the run time to 132.9 seconds. This is an example where the immediate bound changes of OBBT at the root do not weigh out its computational overhead, but the implied bound changes by LVBs during tree search lead to a significant improvement. For the instance `rsyn0805h`, OBBT leads to a mean running time of only 0.8 seconds, while for all permutations SCIP times out with an average gap of 36.3% when not using OBBT. Applying LVB in addition does not have an impact on the running time. This is one of the instances for which OBBT makes the difference between extremely hard (for SCIP) and almost trivial to solve. Finally, instance `smallinvDAXr3b150-165` is one of the biggest success stories for both, plain OBBT and LVBs. By applying OBBT we could solve it in 106.4s instead of 1367.3s. This is a speedup factor of 12.8. By using LVBs we can increase the speedup factor to 213.6 and solve the instance in 6.4s in shifted geomatric mean.

Our computational results show that the impact of OBBT on the INT and GO instances lies not in delivering a consistent speedup over all instances, but in being a "game changer": using OBBT, significantly more instances can be solved and some hard instances are solved much faster.

6.5 Conclusion

In this chapter, we have presented three enhancements to optimization-based bound tightening for MINLP and an extensive computational study of their performance impact over a large and heterogeneous test set. In reverse order of presentation these are: ordering strategies to reduce the number of LP iterations for a simplex-based implementation, filtering techniques to recude the number of LPs solved, and an

---

[83] `smallinvDAXr2b100-110`, `smallinvDAXr4b150-165`, `smallinvDAXr1b100-110`, `rsyn0805h`, `nvs09`

approximation of OBBT in form of a one-row relaxation, which can be propagated repeatedly and efficiently via FBBT, which we termed LVB propagation.

Our experiments on benchmark instances from MINLPlib2, using an implementation that is part of the MINLP solver SCIP, show that both OBBT and LVB propagation each improve the computational performance.We analyzed the results separately for instances with and without integer variables after SCIP's presolving. On those instances that could be solved with and without our OBBT algorithms within a time limit of one hour, our implementation reduced the average running time by 6.5% on the continuous test set and by 2.3% on the integer test set. On the subset of harder instances that took at least 100 seconds with one of the settings, the speedup was more pronounced: 17.1% and 18.8% on the continuous and integer test set, respectively. We noticed that OBBT alone does not give a significant speedup on average, because the average reduction in the number of branch-and-bound nodes is mostly compensated by the overhead in time for solving LPs at the root node. LVB propagation, however, leverages the effect of OBBT and leads to a statistically significant reduction in average running time.

Most importantly, OBBT and LVB propagation lead to significantly more instances being solved on each test set. With OBBT and LVB propagation SCIP could solve 16 additional instances, while it exceeded the time limit only on three instances which plain SCIP could solve without OBBT.

Chapter 7

Branching on Nonlinear Integer Variables

The resolution of infeasibilities through branching on integer variables and through spatial branching is a key component of global MINLP solvers. Predominantly, however, both types of branching have been studied independently from each other in the fields of mixed-integer linear programming and global optimization. In this chapter, we take a close look at mixed-integer nonlinear programs that truly combine these two sources of complexity in MINLP, in the sense that they feature variables that are both required to be integral and are contained in nonlinear terms.

In Section 7.1, we give an overview of the literature on branching strategies from mixed-integer linear programming and global optimization. In Section 7.2, we introduce the new concept of a *minimum cover* of a mixed-integer nonlinear program that leads to a non-standard, discrete measure for nonlinearity. We present a computational study in Section 7.3 that sheds some light on the structure of mixed-integer nonlinear programs that are typically used for benchmarking MINLP algorithms. To this end, we analyze the distribution of nonlinearities in relation to integer variables, minimum covers, and typical branching candidates. We use this as a starting point for investigating three new branching strategies in Section 7.4. Section 7.5 contains our conclusions.

Parts of this chapter are joint work with Timo Berthold. This includes in particular the notion of minimum covers presented in Section 7.2, see also their use in the Undercover heuristic (Berthold and Gleixner, 2014), and the branching rule presented in Section 7.4.2, an early version of which is available as proceedings article (Berthold and Gleixner, 2013).

7.1 Branching in Mixed-Integer Nonlinear Programming

In Section 5.1.2, we discussed spatial branch-and-bound as the basis for the algorithms implemented by global MINLP solvers. Branching is used in order to resolve the infeasibility of a solution, x^*, of a convex relaxation: The problem is subdivided into several subproblems such that x^* can be cut off, i.e., excluded from the feasible region, in each of the subproblems. The two types of infeasibilities in MINLP that are resolved by branching are violations of the integrality requirements on some of the variables and of nonconvex nonlinear constraints. We will refer to subdivisions for the first type as *integer branching* and subdivisions for the second type as *spatial branching*.

There is a large degree of freedom in choosing branchings so that the theoretically necessary conditions for convergence to a global optimum are satisfied, see, e.g. Horst and Tuy (1990). Algorithms that make this choice are called *branching rules* or *branching heuristics*; we refer to Berthold (2014, Sec. 10.2) for a discussion on their heuristic nature. Their performance is—usually empirically—judged by the size of the resulting search trees and the overall running time to ε-global optimality. In the following, we will survey common rules for integer and spatial branching found in the literature.

7.1.1 Integer Branching

The most common methodology applied to resolve a violation of the integrality constraints through branching is variable-based branching. Given an optimal relaxation solution $x^* \in \mathbb{R}^n$ of an MINLP of form (5.5),

$$\min\{c^\mathsf{T}x \mid g_k(x) \leqslant 0, k = 1, \ldots, m,$$
$$x \in [\ell, u],$$
$$x_i \in \mathbb{Z} \text{ for } i \in \mathcal{I}\},$$

the candidate variables for integer branching are the integer variables x_i, $i \in \mathcal{I}$, with $x_i^* \notin \mathbb{Z}$. A variable-based branching rule selects one of these and creates the disjunction $x_i \leqslant \lfloor x_i^* \rfloor \vee x_i \geqslant \lceil x_i^* \rceil$. State-of-the-art branching rules for mixed-integer linear programming estimate the impact that splitting a variable's domain will have on the dual bound of the created subproblems, i.e., the optimal objective value of their LP relaxations.

A prominent approach is the use of so-called *pseudocosts* introduced by Benichou et al. (1971). Pseudocosts estimate the increase in the dual bound relative to the reduction of integer infeasibility on one variable. For each variable, this is averaged over the search tree and used to predict the increase in the dual bound caused by branching on this variable. Variables with large predicted increase are preferred. Linderoth and Savelsbergh (1999) show that it is beneficial to initialize pseudocosts by *strong branching*. Strong branching, first used by Applegate et al. (1995, 2007a) for solving the traveling salesman problem, tentatively applies a branching disjunction and computes the increase in the dual bound exactly by solving the resulting LP relaxations. Achterberg et al. (2005) refine this approach further to *reliability branching*, which applies multiple strong branchings per variable before it deems its pseudocost is deemed reliable. *Nonchimerical branching* by Fischetti and Monaci (2012) and *cloud branching* by Berthold and Salvagnin (2013) are two recent strategies proposed to speed up strong branching by reducing the set of candidate variables in an exact manner.

Achterberg (2007) and Achterberg and Berthold (2009) suggest combining multiple selection criteria into one decision via a weighted sum of scores

$$\sum_{k=1}^{K} \omega_k \nu(s_k(i)),$$ (7.1)

where $s_1(i), \ldots, s_K(i) \geq 0$ is a list of K scores available for each variable x_i. These are normalized by the monotone, i.e., rank-preserving, function

$$\nu : \mathbb{R}_{\geq 0} \to [0, 1), x \mapsto \frac{x}{x+1},$$ (7.2)

which transforms each nonnegative score value to a value below one, and summed up using variable-independent weights $\omega_1, \ldots, \omega_K \geq 0$. Finally, a variable with the highest score is selected for branching. In this way, their *hybrid branching* rule combines the following criteria: reliability branching (highest weight), so-called VSIDS branching of Moskewicz et al. (2001) from satisfiability solving (medium weight), and inference values and cut-off scores of Li and Anbulagan (1997) from constraint programming (lowest weight).

Except for strong branching, the above approaches are all based on historical information: from statistics on how branching on a variable influenced a certain performance measure collected during the search, an attempt is made to predict which branched variable will have the most impact on the resulting subproblems. Methods that combine the above criteria can be considered to be state-of-the-art for general-purpose MIP solvers.

In recent years, several publications have investigated new paradigms for variable-based branching schemes that show superior performance on important classes of hard MIPs. The *active constraint* method by Patel and Chinneck (2007) uses the impact of variables on the set of active constraints in order to find integer feasible solutions early during the search. The paper of Kılınç-Karzan et al. (2009) suggests using conflict learning information for branching on 0-1 integer programs. To this end, they run a sampling phase of 500 branch-and-bound nodes during which they collect conflict constraints, restart the solution process, and prefer branching on variables that appear in short conflict constraints during the second phase. *Backdoor branching* by Fischetti and Monaci (2013) goes one step further. It applies multiple restarts, attempting to find a good approximation of a so-called backdoor: a set of variables, preferably small, such that, whenever these variables get assigned integer values, solving an LP on the remaining variables gives a proof of feasibility or infeasibility. After each restart, the approximated backdoor is computed by solving a set-covering problem. Branching is exclusively performed on backdoor variables until all of them are fixed. Though different in nature, backdoors and minimum covers are general structures of MIPs and MINLPs, respectively, both encoding knowledge about easier subproblems that can be obtained by fixing variables.

Finally, we want to mention that there is a stream of research that investigates the use of general disjunctions for branching. This goes back to the work of Ryan and Foster (1981) on solving scheduling problems with integer programming. We refer to the recent overview of Berthold (2014, Sec. 10.2) for more details on this topic.

A new technique to integrate variable-based branching and branching on general disjunctions is branching on multi-aggregated variables by Gamrath et al. (2015b).

7.1.2 Spatial Branching

For resolving violations of nonconvex nonlinear constraints, variable-based branching is also the most common methodology used in general-purpose solvers. There are two main differences to integer branching, however. First, branching on a continuous variable, x_i, that has solution value, x_i^*, in the relaxation leads to the inclusive disjunction $x_i \leqslant x_i^* \vee x_i \geqslant x_i^*$. Unlike for integer branching, this alone does not cut off x^* in any of the parts; a subsequent tightening of bounds and relaxation is necessary.

Second, there is no one-to-one relation between violations and variables: a violated constraint usually contains more than one variable and a variable may appear in several violated constraints. One strategy suggested by Tawarmalani and Sahinidis (2004) is a so-called *violation transfer*. This estimates the impact of each variable on the problem by minimizing and maximizing a Lagrangian function over a neighborhood of the current relaxation solution when holding all other variables fixed.

Belotti et al. (2009) have transferred the concept of *pseudocost branching* from integer to spatial branching by investigating suitable counterparts for integer infeasibility. Their computational analysis suggests that pseudocost-based branching is superior for hard MINLPs, while for easy instances and nonconvex NLPs it is outperformed by violation transfer or simpler violation-based rules. As in integer branching, it is a straightforward idea to combine pseudocosts with *strong branching* on candidate variables. This is implemented in the general-purpose solvers GloMIQO[84] and ANTIGONE[85] by Misener and Floudas (2012) in the fashion of *reliability branching* for mixed-integer linear programs.

Finally, variable-based branching can be seen as a branching of the domain into hyperrectangles. The term *geometric branching* is commonly used if general subdivisions are used. We refer to Locatelli and Schoen (2013, Chap. 5) for a recent overview on this topic.

7.1.3 Interleaving Integer and Spatial Branching

At the nodes of a branch-and-bound search tree, the solution, x^*, of the relaxation may violate integrality requirements and nonlinear constraints simultaneously. In this case, the MINLP solver must decide whether to perform an integer or a spatial branching.

The violation transfer scheme of Tawarmalani and Sahinidis (2004) implemented

---

[84] Ruth Misener and Christodoulos A. Floudas. GloMIQO: Global Mixed-Integer Quadratic Optimizer. `http://helios.princeton.edu/GloMIQO/`.

[85] Ruth Misener and Christodoulos A. Floudas. ANTIGONE: Algorithms for coNTinuous / Integer Global Optimization of Nonlinear Equations. `http://helios.princeton.edu/ANTIGONE/`.

in the solver BARON[86] is one unifying selection algorithm for all types of variables; the special treatment of integer variables is restricted to an additional rounding step. Belotti et al. (2009) instead suggest a prioritized approach in which integer branching is always performed first if an integer variable with non-integral value in x^* is available. This is the approach taken by the solvers Couenne[87] and SCIP[88], see also Vigerske (2013).

A special case is the solver ANTIGONE since it solves MIP relaxations at the nodes of its branch-and-bound tree (Misener and Floudas, 2012, 2014). Because of this, it does not need to perform integer branching itself but delegates this task to the underlying MIP solver. This leads to a nested branching scheme with spatial branching on the outer level and integer branching on the inner level.

7.2 Covers of Mixed-Integer Nonlinear Programs

The concepts described in this section were originally motivated by the design of a primal heuristic, the *Undercover heuristic* (Berthold and Gleixner, 2010, 2014), which proved useful not only as a standalone procedure for computing good primal solutions but also as component of a global MINLP solver for improving the overall performance. It utilizes structural properties of an MINLP that are of more general interest, also to branching heuristics, as we will show in Section 7.4.

7.2.1 A Discrete Measure of Nonlinearity

The following formalization is useful in answering the questions: How "linear" is a given MINLP? Does it contain large linear subproblems? Concretely, how many variables must be fixed in order to arrive at a mixed-integer linear program?

Definition 7.1 (Cover of a Function, Cover of an MINLP). *Let* P *be an MINLP of form* (5.5) *and let* $C \subseteq \{1, \ldots, n\}$ *be a set of variable indices of* P. *Then we call* C *a* cover *of function* g_k, $k \in \{1, \ldots, m\}$, *if and only if for all* $x^* \in [\ell, u]$ *the set*

$$\left\{ (x, g_k(x)) : x \in [\ell, u], x_i = x_i^* \text{ for all } i \in C \right\} \tag{7.3}$$

is an affine set intersected with $[\ell, u] \times \mathbb{R}$. *We call* C *a* cover *of* P *if and only if* C *is a cover of all constraint functions* $g_1, \ldots, g_m$.

To rephrase the definition, a cover of an MINLP corresponds to a set of variables for which, when they are fixed to any value in their domain, all constraints have a linear representation and the resulting subproblem is a mixed-integer *linear* program. The set of all variables trivially "covers" any MINLP, and the empty set is a cover if

[86] Mohit Tawarmalani and Nikolaos V. Sahinidis. BARON. Branch-And-Reduce Optimization Navigator. `http://archimedes.cheme.cmu.edu/?q=baron`.

[87] COIN-OR. Couenne, an exact solver for nonconvex MINLPs. `https://projects.coin-or.org/Couenne`.

[88] Zuse Institute Berlin. SCIP—Solving Constraint Integer Programs. Available for download under `http://scip.zib.de/`.

and only if the problem has no nonlinearities at all. More generally, the *minimum size of a cover* constitutes a quantity to measure the nonlinearity of an MINLP in a discrete, non-standard way that goes beyond just counting potentially nonzero entries of the Hessian matrix of the constraint functions. The following example shows that, even for well-structured problems with obvious choices for small covers, computing a minimum cover may give additional insight.

Example 7.2 (Bilinearly Constrained Programming). A QCP with a bipartition of its variables, $\{1, \ldots, n\} = \mathcal{U} \cup \mathcal{V}$, $\mathcal{U} \cap \mathcal{V} = \emptyset$, such that each nonlinear term has the form $x_i x_j$, $i \in \mathcal{U}$, $j \in \mathcal{V}$, is called a *bilinearly constrained program*. By definition, holding the variables of either $\mathcal{U}$ or $\mathcal{V}$ fixed results in a linear program. This basic property has been used in solution approaches such as the cutting-plane algorithm of Konno (1976).

In global optimization, the variables in the smaller of the two set are sometimes called *complicating variables*, and by construction they constitute a cover. However, this cover is not guaranteed to be minimal and, in general, its size can be larger than the size of a minimum cover by an arbitrary factor. This can be seen on the following family of examples on the two sets of variables x_i, $i \in \mathcal{U} = \{1, \ldots, n_{\mathcal{U}}\}$, and y_j, $j \in \mathcal{V} = \{1, \ldots, n_{\mathcal{V}}\}$, $n_{\mathcal{V}} \geqslant n_{\mathcal{U}} \geqslant 3$,

$$\min\{x_1 + y_1 \mid x_1 y_j \leqslant 1, j \in \mathcal{V}, \\ x_i y_1 \leqslant 1, i \in \mathcal{U}\}. \tag{7.4}$$

The set $\mathcal{U}$ can grow arbitrarily large whereas the actual minimum cover given by x_1 and y_1 has constant size two.

In the above example, each of the two partitions $\mathcal{U}$ and $\mathcal{V}$ is even a *minimal* cover in the sense that one cannot obtain a smaller cover by simply discarding an element. This shows that a simple greedy heuristic for generating small covers may fail, which leads us to the question of how to compute minimum cover sizes.

7.2.2 Computing Minimum Covers

In this section, for clarity of presentation, we make the standard assumptions that the nonlinear functions involved are twice continuously differentiable and that the domain $[\ell, u]$ has a nonempty interior such that partial derivatives are well-defined. For nonsmooth constraint functions, the methodology can be extended on an ad hoc basis. The following definition generalizes the notion of the *co-occurrence graph* of a quadratically constrained quadratic programs used by Hansen and Jaumard (1992).

Definition 7.3 (Co-occurrence Graph). *Let P be an MINLP of form (5.5) and let $g_1, \ldots, g_m$ be twice continuously differentiable on the interior of $[\ell, u]$. We call the undirected graph $G_P = (V_P, E_P)$ the co-occurrence graph of P if the node set is given by the variable indices of P, $V_P = \{1, \ldots, n\}$, and the edge set*

$$E_P = \left\{ \{i, j\} \mid i, j \in V_P, \exists k \in \{1, \ldots, m\} : \frac{\partial^2}{\partial x_i \partial x_j} g_k(x) \not\equiv 0 \right\},$$

i.e., we draw an edge between i *and* j *if and only if the Hessian matrix of some constraint has a structurally nonzero entry* $\{i, j\}$.

Since the Hessian of a twice continuously differentiable function is symmetric, G_P is a well-defined, undirected graph. It may contain loops, e.g., if square terms are present. Trivially, the co-occurrence graph of a MIP is edge-free. The co-occurrence graph of a bilinear program is bipartite, as is illustrated by Figure 7.1, which shows the co-occurence graph of the bilinearly constrained program (7.4) from Example 7.2.

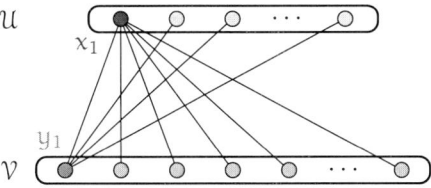

Figure 7.1: Co-occurrence graph of the bilinearly constrained program (7.4), for which the cover $\mathcal{U}$ of so-called complicating variables may be arbitrarily larger than the minimum cover given by x_1 and y_1.

Our algorithm for computing minimum covers is based on the following observation.

Theorem 7.4 (Covers as Vertex Covers). *Let* P *be an MINLP of form* (5.5) *with* $g_1, \ldots, g_m$ *twice continuously differentiable on the interior of* $[\ell, u]$. *Then* $\mathcal{C} \subseteq \{1, \ldots, n\}$ *is a cover of* P *if and only if it is a vertex cover of the co-occurrence graph* G_P.

Proof. If $h : \mathbb{R}^n \to \mathbb{R}$ is twice continuously differentiable, $x^* \in \mathbb{R}^n$, $\mathcal{C} \subseteq \{1, \ldots, n\}$, then fixing variables $x_i = x_i^*$, $i \in \mathcal{C}$, and projecting to the nonfixed variables yields another twice continuously differentiable function $\bar{h} : \mathbb{R}^{n-|\mathcal{C}|} \to \mathbb{R}$. Let $\pi : \mathbb{R}^n \to \mathbb{R}^{n-|\mathcal{C}|}$ be the projection $x \mapsto (x_i)_{i \notin \mathcal{C}}$.

Now the Hessian matrix of $\bar{h}$ is obtained from the Hessian of h simply by taking the columns and rows of nonfixed variables:

$$\nabla^2 \bar{h}_{\pi(x)} = \left(\frac{\partial^2}{\partial x_i \partial x_j} h(x) \right)_{i, j \notin \mathcal{C}}$$

for any $x \in \mathbb{R}^n$ with $x_i = x_i^*$, $i \in \mathcal{C}$. A twice continuously differentiable function is affine if and only if its Hessian vanishes on its domain. Hence,

$$\mathcal{C} \text{ is a cover of } h \iff \forall i, j \notin \mathcal{C} : \frac{\partial^2}{\partial x_i \partial x_j} h(x) \equiv 0.$$

For the MINLP P, this gives that

$$\mathcal{C} \text{ is a cover of } P \Leftrightarrow \forall i, j \notin \mathcal{C}, k \in \{1, \ldots, m\} : \frac{\partial^2}{\partial x_i \partial x_j} g_k(x) \equiv 0$$

$$\Leftrightarrow \forall i, j \notin \mathcal{C} : \{i, j\} \notin E_P$$

$$\Leftrightarrow \forall \{i, j\} \in E_P : i \in \mathcal{C} \lor j \in \mathcal{C},$$

i.e., the Hessian vanishes if and only if $\mathcal{C}$ is a vertex cover of the co-occurrence graph G_P. $\qquad\qquad\square$

Note that any undirected graph $G = (V, E)$ is the co-occurrence graph of the QCP $\min\{c^T x \mid x_i x_j \leqslant 0, \{i, j\} \in E\}$ for all $c \in \mathbb{R}^n$. Hence, minimum vertex cover can be transformed to computing a minimum cover of an MINLP. Since minimum vertex cover is $\mathcal{NP}$-hard, see Garey and Johnson (1979), the following holds.

Corollary 7.5 (Hardness of Computing Minimum Covers). *Computing a minimum cover of an MINLP is $\mathcal{NP}$-hard. This holds even when restricted to quadratic constraints.*

There are, however, many polynomial-time algorithms that approximate a minimum vertex cover within a factor of 2, such as taking the vertices of a maximal matching. It is conjectured that 2 is also the optimal approximation factor (Khot and Regev, 2008) and it is proven that vertex cover is $\mathcal{NP}$-hard to approximate within a factor of less than $10\sqrt{5} - 21 \approx 1.3606$ (Dinur and Safra, 2005), hence no polynomial-time approximation scheme exists. Approximation ratios $2 - \varepsilon(G)$ can be found with $\varepsilon(G) > 0$ depending on particular properties of the graph such as number of nodes (Karakostas, 2009) or bounded degree (Halperin, 2002).

Minimum covers can be computed exactly using the following binary programming formulation. For an MINLP of form (5.5), define auxiliary binary variables α_i, $i = 1, \ldots, n$, equal to 1 if and only if the original variable x_i is fixed. Then

$$\mathcal{C}(\alpha) := \{i \in \{1, \ldots, n\} \mid \alpha_i = 1\}$$

forms a cover of P if and only if $\alpha_i + \alpha_j \geqslant 1$ for all $\{i, j\} \in E_P$. For a QCP, for example, this requires all square terms and at least one variable in each bilinear term to be fixed. For a given general MINLP we solve the binary program

$$\min\left\{ \sum_{i=1}^{n} \alpha_i \,\Big|\, \alpha_i + \alpha_j \geqslant 1 \text{ for all } \{i, j\} \in E_P, \alpha \in \{0, 1\}^n \right\} \qquad (7.5)$$

to minimize the sum of auxiliary variables.

Note that for particular classes of MINLPs, it is possible to exploit special features of the co-occurrence graph in order to exactly compute a minimum cover in polynomial time—a simple example is the class of bilinear programs mentioned above—or to approximate it within a factor sufficiently close to one. However, for the vast majority of instances in our experiments, the binary program (7.5) could be solved in under a fraction of a second by a standard MIP solver. In addition, the computational effort spent can easily be controlled by imposing working limits, e.g., on the number of branch-and-bound nodes explored.

Remark 7.6 (Formulation-Dependency of Minimum Covers). Note that our treatment relies on the functions g_k given by one formulation of the problem and not on the feasible region of the MINLP directly. It does not exploit, for example, the fact that some of the functions might be redundant or the fact that some of the variables may implicitly be equal. As an example, if the formulation contains the term x^2 for a binary variable x, then x is contained in any cover, although $x^2 = x$ for $x \in \{0, 1\}$.

Considering the formulation is motivated by taking the perspective of an MINLP solver, which also obtains a problem in a specific formulation as a list of individual constraints. In this respect, it is an advantage that for improving the performance of the solver we are interested in the presolved problem, rather than the original problem: standard presolving routines in MINLP solvers aim at removing these kinds of redundancies in the formulation. For the example above, they would certainly replace the square of a binary variable by the variable itself. See also the discussion of our implementation based on algorithmic differentiation in Section 7.3.1.

7.2.3 Minimum Splitting Covers

While the minimum cover size of an MINLP is a well-defined quantity, minimum covers are not necessarily unique. This degree of freedom can be used to try to identify minimum covers that are particularly suitable for certain algorithmic purposes. For the purpose of branching, if we want to use a minimum cover to make a distinction among several candidate variables, the following concept is useful. Given a set of variables, it classifies minimum covers by the degree to which they split this set into cover and non-cover variables.

Definition 7.7 (Minimum Splitting Cover). *Let P be an MINLP of form (5.5) and let $\mathcal{S} \subseteq \{1, \ldots, n\}$ be a nonempty set of variable indices of P. Then we call $\mathcal{C} \subseteq \{1, \ldots, n\}$ a minimum splitting cover for $\mathcal{S}$ if $\mathcal{C}$ is a minimum cover of P and $\mathcal{C} \cap \mathcal{S}$ is neither empty nor equal to $\mathcal{S}$. In this case, the relative cardinality of $\mathcal{C} \cap \mathcal{S}$ with respect to $\mathcal{S}$,*

$$\sigma(\mathcal{C}) := \frac{\#(\mathcal{C} \cap \mathcal{S})}{\#\mathcal{S}}$$

is called the splitting ratio *of $\mathcal{C}$ and lies strictly between zero and one.*

In general, the existence of a minimum splitting covers for a given set $\mathcal{S}$ is not guaranteed. It can be tested using a binary programming formulation similar to (7.5) as follows. Suppose C is the minimum cover size of an MINLP. Adding the constraint $\sum_{i=1}^{n} \alpha_i \leq C$ to (7.5) restricts the feasible region to the set of minimum covers. The sum $\sum_{i \in \mathcal{S}} \alpha_i$ expresses the cardinality of $\mathcal{C}(\alpha) \cap \mathcal{S}$. Hence, the existence of a minimum splitting cover for S can be tested by solving the feasibility problem

$$\alpha_i + \alpha_j \geqslant 1 \text{ for all } \{i, j\} \in E_P, \tag{7.6}$$

$$\sum_{i=1}^{n} \alpha_i \leqslant C, \tag{7.7}$$

$$1 \leqslant \sum_{i \in S} \alpha_i \leqslant \#S - 1, \tag{7.8}$$

$$\alpha \in \{0, 1\}^n. \tag{7.9}$$

The existence of a minimum splitting cover with a specified splitting ratio $\tilde{\sigma}$ can be tested by replacing (7.8) by the constraint $\sum_{i \in S} \alpha_i = \tilde{\sigma} \cdot \#S$.

In many cases, this may be too strict a requirement. A minimum splitting covers with splitting ratio as close as possible to $\tilde{\sigma}$ can be computed by solving an optimization problem: minimize $\left| \sum_{i \in S} \alpha_i - \tilde{\sigma} \cdot \#S \right|$ over (7.6), (7.7), and (7.9). A standard reformulation of the absolute value gives the mixed-binary program

$$\min \Big\{ \beta \mid \beta - \sum_{i \in S} \alpha_i \geqslant -\tilde{\sigma} \cdot \#S,$$

$$\beta + \sum_{i \in S} \alpha_i \geqslant \tilde{\sigma} \cdot \#S, \tag{7.10}$$

$$(7.6), (7.7), (7.9), \beta \in \mathbb{R}_{\geqslant 0} \Big\}.$$

As before, this can be solved by a standard MIP solver. For one of the branching rules presented in Section 7.4, we will use a minimum splitting cover obtained by solving (7.10) with $\tilde{\sigma} = \frac{1}{2}$.

7.3 Nonlinearity, Integrality, Covers

In this section, we present an empirical analysis of benchmark instances from the library MINLPlib2, which are widely used for the computational evaluation of MINLP algorithms. We study these instances with respect to their degree of nonlinearity and integrality, and compute minimum covers and minimum splitting covers. Details on MINLPlib2 are given in Section 1.4.3.

7.3.1 Implementation and Experimental Setup

As part of the MINLP solver SCIP[89] we have implemented the functionality to generate the co-occurrence graph from Definition 7.3 and compute minimum covers and minimum splitting covers via separate SCIP instantiations that solve the mixed-binary programming formulations (7.5) and (7.10), respectively. We collect the edges of the co-occurrence graph constraint-wise. We skip second-order cone constraints,

---

[89] Zuse Institute Berlin. SCIP—Solving Constraint Integer Programs. Available for download under `http://scip.zib.de/`.

since they are convex and can be enforced very efficiently by separation. For general quadratic constraints, the sparsity pattern of the Hessian matrix is readily available. For general nonlinear constraints, we detect which entries of the Hessian matrix are structurally, i.e., potentially, nonzero by using *algorithmic differentiation* (AD), see, e.g., Griewank and Walther (2008) for details. In order to apply AD, the constraints must be given explicitly, i.e., in the form of oracle methods that are accessible in source code by the AD tool and compute the value and derivative of the constraint functions.

The results of algorithmic differentiation may depend on the formulation of the function. For example, the linear term $3x + 1$ may equivalently be written in the nonlinear form $(x+1)^3 - x^3 + 3x^2$; in this case an AD package may unnecessarily return a structural nonzero entry for x on the diagonal of the Hessian. This is one reason why we performed the analysis not on the original problem, but after the presolving phase of SCIP, at which point simple redundancies should have been eliminated. This is related to Remark 7.6 on the formulation-dependency of minimum covers.

With the later experiments on branching rules in mind, we performed the analysis at the end of the root node, at the first occasion when SCIP calls the branching rule plug-ins. Note that this setup is solver-centric by intention, since our prime interest is in improving computational performance. The following details apply equally to the experiments in Section 7.4.

Hardware and Software. The experiments were conducted on a cluster of 64-bit Intel Xeon X5672 CPUs at 3.2 GHz with 12 MB cache and 48 GB main memory, running only one job per node at a time. We used an extended version of SCIP 3.1.1 with SoPlex[90] 2.0.1 as LP solver, CppAD[91] 20140000.1, and Ipopt[92] 3.11.7 (Wächter and Biegler, 2006).

7.3.2 Nonlinear Integer Variables

We first describe the selection of instances that are relevant for the analysis. Out of the 1,279 instances of MINLPlib2 that could be parsed by SCIP, eleven instances were solved already during presolving, 95 instances had only linear constraints after presolving, in two of the instances SCIP ran out of memory during root node processing, and 130 instances were solved at the root node before branching. Of the remaining set, 355 instances had only continuous variables after presolving and 686 instances contained integer variables after presolving. On 318 instances, the integer variables appear only in linear constraints. Finally, 368 instances feature what we call *nonlinear integer variables*, i.e., integer variables that appear in nonlinear constraints. We denote this subset by NLINT.

For the NLINT instances, Figure 7.2 plots the fractions of nonlinear variables among all variables and the fractions of nonlinear integer variables among nonlin-

---

[90] Zuse Institute Berlin. SoPlex—the Sequential object-oriented simPlex. Available for download under `http://soplex.zib.de/`.

[91] COIN-OR. CppAD, a Package for Differentiation of C++ Algorithms. `http://www.coin-or.org/CppAD`.

[92] COIN-OR. Ipopt, Interior point optimizer. `http://www.coin-or.org/Ipopt`.

ear variables. Here, by *nonlinear variables* we refer to the variables that appear in nonlinear terms, but may also be continuous. Due to several problem classes that are overrepresented in MINLPlib2, the histograms show three large bins, each containing more than 130 instances: classes autocorr_bern*, graphpart_*, and smallinv* dominate the 140 instances for which more than 95% of the variables are nonlinear and 138 instances for which more than 95% of the nonlinear variables are integers; classes nvs*, smallinvSNP*, and squfl*persp dominate the 130 instances for which no nonlinear variable is binary.

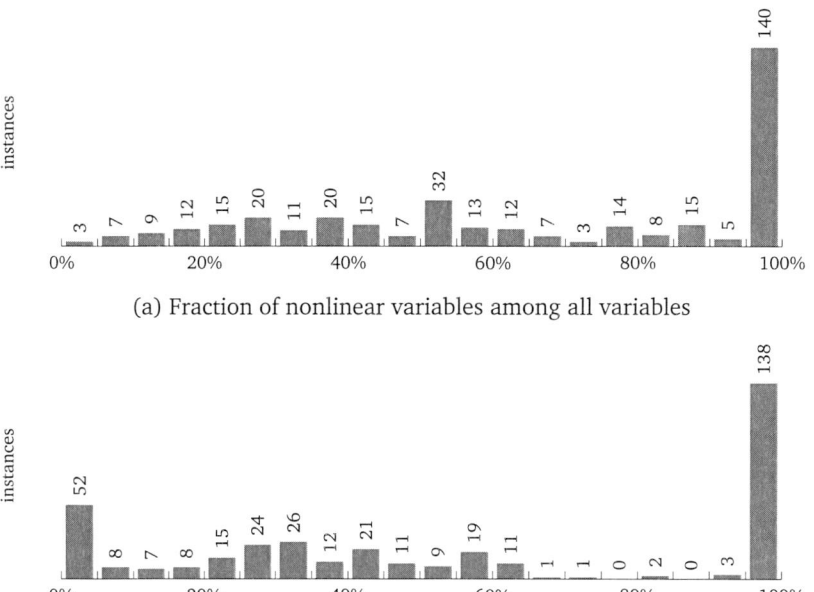

(a) Fraction of nonlinear variables among all variables

(b) Fraction of nonlinear integer variables (including binaries) among all nonlinear variables

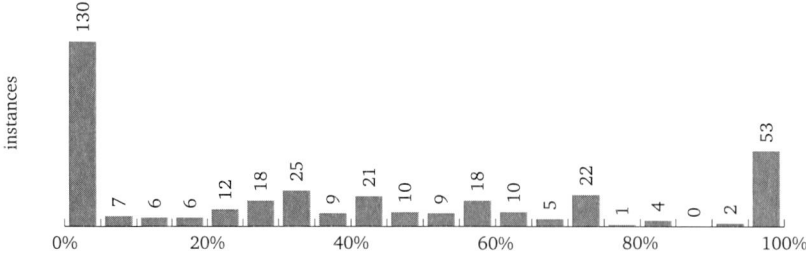

(c) Fraction of nonlinear binary variables among all nonlinear variables (74 instances in the 0–5% bin have no nonlinear binary variables)

Figure 7.2: Proportions of nonlinear variables and nonlinear integer variables over NLINT test set (368 nontrivial MINLPlib2 instances with nonlinear integer variables)

As can be seen by comparing Figure 7.2b with 7.2c, there is a group of instances with almost all nonlinear variables being integer that feature almost no or few binary variables. Of these instances, 60 are of type `smallinv*`. For the majority of instances, however, most nonlinear integer variables are binary.

7.3.3 Minimum Covers

Next, we computed the size of a minimum cover for the instances in test set NLINT. Figure 7.3 shows the distribution of minimum cover sizes relative to the total number of variables and relative to the number of nonlinear variables in the problem. The latter is relevant, since the set of nonlinear variables trivially constitutes a cover, and hence their number bounds the size of a minimum cover from above.

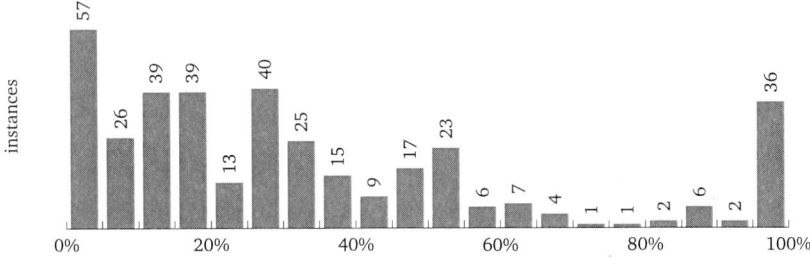

(a) Size of minimum covers relative to the total number of variables

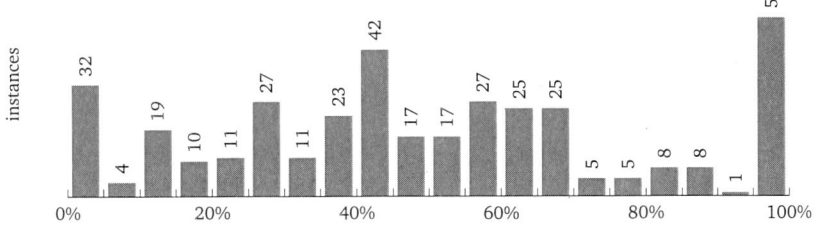

(b) Size of minimum covers relative to the number of nonlinear variables

Figure 7.3: Relative sizes of minimum covers as distributed over test set NLINT (368 nontrivial MINLPlib2 instances with nonlinear integer variables)

First it can be seen that most instances admit small covers in total: in 57 instances minimum covers have at most 5% of the variables, and 65% of the instances have minimum covers with under 35% of the variables. Second, except on 48 instances[93], all minimum covers where nontrivial in the sense that they were a strict subset of the set of nonlinear variables. On 196 instances, the minimum cover consisted of under half the nonlinear variables.

---

[93] 33 `smallinv*` instances, 14 instances with less thanunder 30 variables, and instance ibs2

In addition to the size of the minimum covers, we tried to quantify the degree of freedom in choosing a minimum cover. To this end, we used the "countsols" constraint handler of SCIP, in order to count distinct solutions to the feasibility problem (7.6–7.9), see Achterberg et al. (2008) for details. We imposed a time limit of 1,800 seconds, which was hit in 131 cases; for these instances, the computed numbers give a lower bound on the number of distinct minimum covers. Figure 7.4a shows the results in a logarithmic histogram: For 169 instances, i.e., for more than half of the test set, the minimum cover is unique; for 82 instances there are up to 100 distinct minimum covers and twelve instances have one million or more distinct covers.

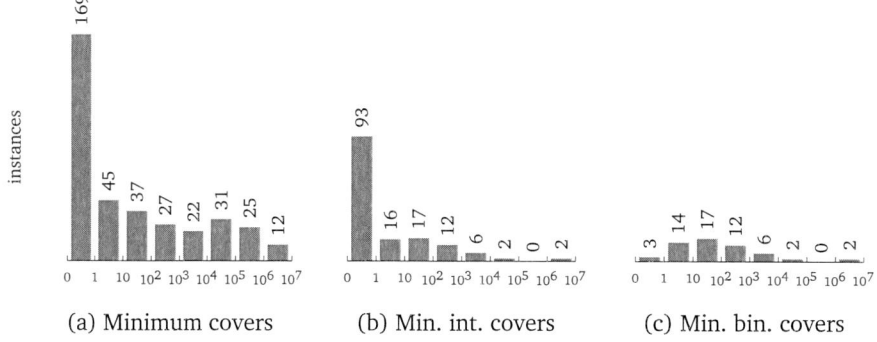

(a) Minimum covers   (b) Min. int. covers   (c) Min. bin. covers

Figure 7.4: Number of distinct minimum covers as distributed over test set NLINT (368 nontrivial MINLPlib2 instances with nonlinear integer variables)

We extended this experiment in order to evaluate how this degree of freedom is reduced if we restrict ourselves to minimum covers that contain only integer variables or only binary variables. We refer to these as *integer minimum covers* and *binary minimum covers*, respectively. The number of distinct integer minimum covers can be counted by fixing all α_i for which x_i is not an integer variable to zero; likewise we can count binary minimum covers by fixing all α_i for which x_i is not a binary variable. The results are shown in Figures 7.4b and 7.4c. First, we can see that many instances do not even admit these restricted covers: only 148 instances have a integer minimum cover, and only 56 instances have a binary minimum cover. In the same way, if they exist then integer minimum covers are often unique (on 93 instances); for binary minimum covers, however, the majority of instances (43 instances) exhibit between two and 1,000 distinct binary minimum covers.

7.3.4 Minimum Splitting Covers

In Section 7.4, we will be interested in splitting the set of candidate variables for integer branching in order to apply a preselection of the branching candidates. To analyze whether this is viable, we solved the mixed-binary program (7.10) with S defined as the first set of integer branching candidates, i.e., at the point when SCIP called the branching rule plug-ins for the first time at the root node. Note that, in general, the first branching can also be a spatial branching step. On the NLINT test set, however, the solution of the LP relaxation at the root node always violated the integrality requirements, with SCIP calling the plug-ins for integer branching. Heuristically, we chose the target splitting ratio $\tilde{\sigma} = \frac{1}{2}$. This will be motivated further in Section 7.4.2.

We recorded the resulting splitting ratios and plot their distribution as a histogram in Figure 7.5. If the splitting ratio is zero or 100%, this means that no minimum splitting cover actually exists because either all minimum covers lie outside of $\mathcal{S}$ or they contain $\mathcal{S}$. This occurred 164 times and is not included in the histogram. On the remaining 204 instances, the splitting ratio was between 0.1% and 89.7%. Furthermore, we see a peak of 32 instances in the 45–50% bin. Instances with smaller splitting ratios are more frequent: only 41 out of the 204 instances have splitting ratios above 50%. For the experiments on branching rules in Section 7.4.4, we define the test set SPLIT of the 204 instances that admit a minimum splitting cover with respect to the integer branching candidates at the root node.

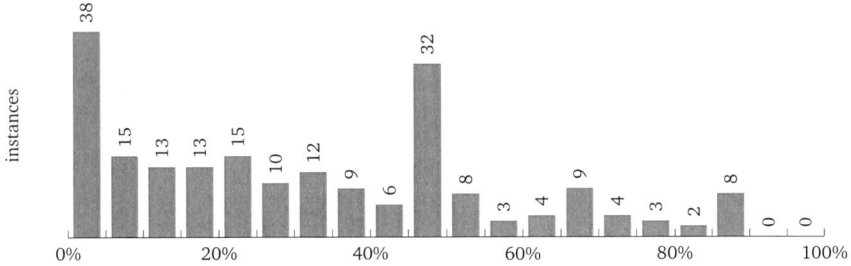

Figure 7.5: Histogram of splitting ratios as distributed over test set NLINT (368 nontrivial MINLPlib2 instances with nonlinear integer variables)

For visualization, Figure 7.6 displays the co-occurrence graphs of four selected instances `gasprod_sarawak01`, `clay0303h`, `sepasequ_convent`, and `kport20`. These were produced using the Gephi[94] graph drawing tool by Bastian et al. (2009). The nodes contained in the minimum splitting cover are marked orange. As can be seen, the graphs often consist of several connected components. The most prevalent structures are stars, single edges, and isolated vertices with loops. The latter correspond to variables that are included in any cover because, for one of the constraint functions, they have a structurally nonzero entry on the diagonal of the Hessian matrix.

7.3.5 Summary

We analyzed 368 instances of the MINLPlib2 which are nontrivial in the sense that SCIP is currently not able to solve them at the root node and contain nonlinear integer variables after presolving. Although in general minimum covers may comprise all nonlinear variables, on NLINT we have seen that minimum covers are frequently

---

[94] Gephi—The Open Graph Viz Platform. `http://gephi.github.io/`

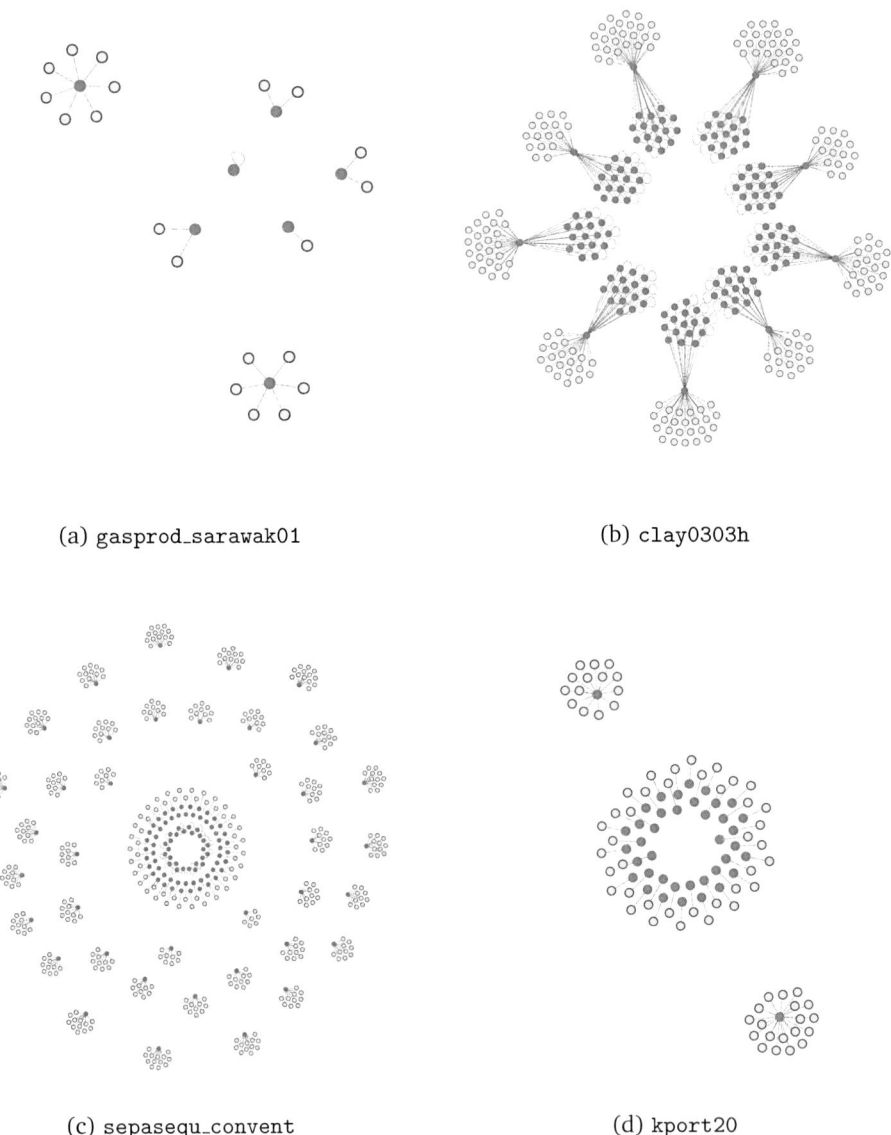

(a) `gasprod_sarawak01`

(b) `clay0303h`

(c) `sepasequ_convent`

(d) `kport20`

Figure 7.6: Four examples of co-occurence graphs. Variables corresponding to a minimum splitting cover are marked orange.

nontrivial, i.e., contain fewer than all nonlinear variables, and small: both with respect to the total number of variables and the number of nonlinear variables. We refer to this set of instances by NLINT.

For more than half of the instances minimum covers were unique. On some of the remaining instances we counted over one million distinct minimum covers. We saw that more than half of the instances do not allow for a purely integer minimum cover. Finally, out of the 368 NLINT instances, 204 instances admit a minimum splitting cover that partitions the set of integer branching candidates at the root node. We refer to this set of instances by SPLIT.

7.4 Three New Branching Rules for Mixed-Integer Nonlinear Programming

As mentioned earlier, the available literature on the interaction of integer and spatial branching in MINLP solvers is very limited. Branching, however, is one of the key components of state-of-the-art MINLP solvers and, as shown by Berthold et al. (2012a), has a large impact on their computational performance. In the following, we develop three new branching rules that use integrality information during spatial branching (in Section 7.4.1) and information on the nonlinearity of the problem during integer branching (in Section 7.4.2 and 7.4.3).

7.4.1 Variable Type Information in Spatial Branching

One possible way to use information about integrality during spatial branching is by means of a hybrid branching rule in the fashion of Achterberg (2007) and Achterberg and Berthold (2009); see Section 7.1. In the context of MIP solvers, this technique of combining multiple branching criteria has proven successful and often outperforms branching rules based on a single criterion alone. In the context of spatial branching in MINLP solvers, we propose combining a score computed from the type of the variable with nonlinear pseudocost scores of Belotti et al. (2009), which are used by state-of-the-art solvers such as ANTIGONE, Couenne, and SCIP.

Suppose we are at a node of the branch-and-bound tree and want to apply spatial branching, then let $S \subseteq \{1, \ldots, n\}$ be the index set of candidate variables contained in violated nonconvex constraints. Let $s_{ps}(i)$ be the nonlinear pseudocost score for variable x_i, $i \in S$, that has been computed by the solver.[95] Additionally define the following simple variable type score

$$s_{type}(i) := \begin{cases} 3 & \text{if } x_i \text{ is binary,} \\ 2 & \text{if } x_i \text{ is integer,} \\ 1 & \text{if } x_i \text{ is implicit integer, and} \\ 0 & \text{if } x_i \text{ is continuous.} \end{cases} \tag{7.11}$$

---

[95] Belotti et al. (2009) develop several methods for this computation. The solver SCIP used in our experiments implements their strategy "rb-int-br", see Vigerske (2013).

This score system prefers binary over integer variables, integer over implicit integer variables, and implicit integer over continuous variables. Here an *implicit integer variable* is a variable that is required to be integer in any feasible solution, but for which integrality does not need to be enforced explicitly: it has been identified, e.g., during the presolving phase of the solver, that in any optimal solution the variable will necessarily take an integer value, see, e.g., Achterberg (2007).

The above preferences have several motivations. First, in contrast to branching on continuous variables, branching on integer variables always leads to a disjunction into two disjoint parts; branching on binary variables even leads to the fixing of the variable in both child nodes. Second, even if all integer variables take integral values in the solution of the current relaxation, this may change again at lower levels of the tree if they are not fixed. In this case, preferring to branch on integer variables even if currently unnecessary may save work in later stages of the search. Note that for an integer variable x_i, $i \in \mathcal{I}$, with integral value in the relaxation solution, $x_i^* \in \mathbb{Z}$, SCIP creates up to three subproblems: $x_i \leqslant x_i^* - 1$, $x_i = x_i^*$, and $x_i \geqslant x_i^* + 1$.

Both scores can be combined to a weighted score

$$\omega_{ps} \nu(s_{ps}(i)) + \omega_{type} \frac{s_{type}(i)}{4} \tag{7.12}$$

for each branching candidate x_i, $i \in \mathcal{S}$. The pseudocost values are normalized by function ν as in (7.2); $\frac{s_{type}(i)}{4}$ is already between zero and one. Finally, the variable with highest weighted score is selected. We call this rule *hybrid pseudocost/variable type branching* (VARTYPE).

In Section 7.4.4, we test the new branching rule taken from SCIP using the weights $\omega_{ps} = 1$ and $\omega_{type} = 0.1$. This puts most emphasis on the pseudocost score, which predicts improvement in the dual bound and is important for pruning, but it uses the variable type score in a tie-breaking fashion for cases when the pseudocost scores are small or very similar across all candidates. The latter situation is encountered, for instance, when solving a pure feasibility problem or when the relaxation is very weak and branching has no or little impact on the dual bound.

The VARTYPE branching rule as presented in this section uses integrality information during spatial branching. In the next two sections we focus on the complementary question: how can information on the nonlinearity of the problem be used when performing integer branching.

7.4.2 Minimum Splitting Covers in Integer Branching

Although MIP and MINLP are both $\mathcal{NP}$-hard, arguably MIPs are computationally easier than MINLPs. For MIP, it is possible to compute a relaxation solution in polynomial time that only drops the integrality requirements, but respects all constraint functions. For MINLP, solving a (nonconvex) NLP relaxation is already $\mathcal{NP}$-hard. Also generic cutting plane algorithms, which contribute a lot to the practical success of MIP solvers, do not have a direct equivalent in MINLP. Although they can be used to strengthen a MIP relaxation, they do not yield a finite algorithm. Last, but not least, in considering today's state-of-the-art in optimization software, MIP solvers have reached an impressive maturity and have become a standard industry tool, whereas the range of MINLPs that can be solved reliably is still comparatively small.

From this point of view, it is an important observation that a cover of an MINLP presents a structure that turns an MINLP into a MIP for any assignment of the variables in the cover. Branching shrinks variables' domains, ideally fixes them, and therefore, branching on cover variables offers itself as a promising strategy to drive an MINLP towards linearity. In particular, the following observation holds:

Lemma 7.8. *Let* P *be an MINLP of form* (5.5) *and* $\mathcal{C} \subseteq \mathcal{I}$ *be a cover of* P *with* $\ell_i, u_i \in \mathbb{Z}$ *for all* $i \in \mathcal{C}$*. Then,* P *can be solved up to optimality by solving a sequence of at most* $\prod_{i \in \mathcal{C}} (u_i - \ell_i + 1)$ *MIPs.*

In the case of variables with unbounded integer variables, branching on cover variables does not necessarily enforce linearity in a bounded number of steps. Nevertheless, branching on such a variable, and thereby tightening its domain, is likely to produce better underestimates than, e.g., branching on an integer variable which is not even part of a nonlinear expression. Better underestimates lead to better relaxation bounds which lead to earlier pruning of the created subproblems.

We therefore suggest a branching strategy that aims at *driving the subproblems towards linearity* and which does so by preferring variables in a minimum cover over other branching candidates. As with SMALL VARTYPE branching, we are interested in a strategy that can be combined with other efficient, variable-based branching rules. And because computing a minimum cover from scratch at each node seems prohibitively expensive, it is strongly desirable to use one fixed cover $\mathcal{C}$ throughout the solution process.

To this end, we propose a filtering method. Suppose we are at a node of the branch-and-bound tree and want to apply integer branching. If $\mathcal{S} \subseteq \{1, \ldots, n\}$ is the index set of candidate variables for integer branching and $\mathcal{C}$ is the given minimum cover, then two cases may arise:

- If $\mathcal{C} \cap \mathcal{S}$ is not empty, then we restrict the set of branching candidates to $\mathcal{C} \cap \mathcal{S}$ and apply the branching rule of choice to this restricted set of candidates.

- If $\mathcal{C} \cap \mathcal{S}$ is empty, then we do not force branching on a variable in $\mathcal{C}$, but apply the branching rule of choice directly to the unfiltered set $\mathcal{S}$.

This scheme benefits from the use of a minimum splitting cover $\mathcal{C}$ rather than any minimum cover: it ensures that, at least at the root node, the filtering step is applied. Heuristically, the closer the splitting ratio is to $\frac{1}{2}$, the more we may hope that the filtering step is also applied at many of the integer branchings performed throughout the tree. Hence, we propose to use a minimum splitting cover computed via (7.10) with $\tilde{\sigma} = \frac{1}{2}$.

We call this branching rule *cover branching* (COVER) and in Section 7.4.4 we analyze its effect computationally in combination with the state-of-the-art MIP branching rule *hybrid reliability/inference branching*, see Achterberg (2007) and Achterberg and Berthold (2009). For the filtering step, it does not use history-based information, but exploits structural information about the nonlinearity of the problem encoded in a minimum cover. In this sense of using structure, it is related to the schemes of Kılınç-Karzan et al. (2009) or Fischetti and Monaci (2013). As a distinguishing feature, however, the structure used for cover branching is computed exactly, at negligible cost, and no sampling phases are required.

7.4.3 Nonlinear Scores in Integer Branching

Last, but not least, this section makes use of a straightforward measure for the "non-linearity" of a variable in an MINLP: the number of nonlinear terms of the constraint functions in which it appears. This measure can be cheaply computed for those fac-torable MINLPs whose constraint functions are given explicitly via expression graphs. It promises to be a good proxy for the impact that branching on this variable has on the nonlinear part of the problem. If η_i denotes the number of nonlinear terms in which variable x_i appears, $i \in \{1, \ldots, n\}$, then we can compute a normalized score value for x_i as

$$s_{nl}(i) := \frac{\eta_i}{\max\{1, \max_{j=1,\ldots,n} \eta_j\}}.$$

We propose to combine this score with hybrid reliability/inference branching of Achter-berg and Berthold (2009), which is a state-of-the-art branching rule for MIP and more generally for branching on integer variables in MINLP. Note that this score is depen-dent on the formulation and hence computed on the presolved form of the problem.

Let $\mathcal{S}$ again denote the index set of candidates for integer branching and let $s_{hybrel}(i)$ denote the hybrid reliability/inference score for a candidate variable x_i, $i \in \mathcal{S}$. Then, we suggest selecting the candidate variable for branching which has the highest combined score

$$\omega_{hybrel} s_{hybrel}(i) + \omega_{nl} s_{nl}(i) \tag{7.13}$$

among all candidates.

We refer to this branching rule as *nlscore branching* (NLSCORE). In the next section we test the rule in SCIP using the weights $\omega_{hybrel} = 1$ and $\omega_{nl} = 0.1$. This emphasizes the hybrid reliability/inference score and uses the nonlinearity of the variable as a tie breaker in cases when the hybrid reliability/inference scores are small or very similar across all candidates.

7.4.4 Computational Results

We implemented the newly introduced branching rules as part of the MINLP solver SCIP and compared them to SCIP's default branching rule, using the computational environment described in the previous section. We used a time limit of one hour for each run and a primal–dual gap limit of zero, solving to global optimality. VARTYPE and NLSCORE branching can lead to different branching decisions as soon as nonlinear integer variables are present, hence we compared them on the full NLINT test set of 368 instances. COVER branching can only make a difference if the splitting ratio of the cover is neither zero nor one, hence we tested it on the subset SPLIT of 208 instances, where this is guaranteed to hold for the root node. We want to note that SCIP's default rule for spatial branching, pseudocost branching, can be interpreted as degenerate VARTYPE branching with weights $\omega_{ps} = 1$ and $\omega_{type} = 0$ in (7.12). Likewise, SCIP's default rule for integer branching, hybrid reliability/inference branching, is obtained from NLSCORE branching when we set $\omega_{hybrel} = 1$ and $\omega_{nl} = 0$ in see (7.13) or from COVER branching when we use the trivial cover containing all nonlinear variables.

Detailed results can be found in Table A.12 in the appendix. In the following, we discuss aggregated results, which are shown in Tables 7.1, 7.2 and 7.3. We report shifted geometric means for the number of branch-and-bound nodes and the overall running time, using a shift of 100 nodes and ten seconds, respectively, see also Section 1.3 for our benchmarking methodology. In each table, we state the results for the subgroups of instances solved by both branching rules ("all solved") and the instances that could be solved by only one branching rule ("timeout"). We disregard the instances that both branching rules could not solve within the time limit. Additionally, we consider the sets of instances which could be solved by both, but for which one of the settings took at least 1, 10, 100, and 1,000 seconds, respectively ("$\geqslant 1, 10, 100, 1000$s"). The latter subgroups form a hierarchy of harder and harder instances which are defined in an unbiased way by excluding instances that are easy to *all* settings; see, e.g., Achterberg and Wunderling (2013) for this benchmarking methodology.

In the aggregated tables, columns three to ten list the number of instances solved, the average running time, the average number of branch-and-bound nodes, and the relative reduction in the average performance measures. By definition, these numbers for groups "all solved" and "$\geqslant 1, 10, 100, 1000$s" do not include the instances from the "timeout" category, which are solved with only one of the branching rules. Otherwise, the average number of branch-and-bound nodes could not be compared easily, since they would include incomplete search trees.

This comparison favors SCIP's default branching, since it always solved fewer instances than any of the new branching rules. Note that, if one instance was solved after time t using one setting and reached the time limit T with another setting, then the ratio $\frac{T}{t}$ gives a lower bound on the speedup factor achieved by the better setting. In this sense, instances from the "timeout" category can be included into the other subgroups to compute an underestimation of the speedups. These values are reported in the last column of the aggregated tables. For the same reason, the Wilcoxon signed-rank test described in Section 1.3 should also be performed on subgroups including instances that could only be solved with one setting.

Results for VARTYPE Branching. First and foremost, VARTYPE branching solved three more instances[96] than the default, taking on average only 226 seconds. The default did not solve any instance for which VARTYPE branching hit the time limit. By contrast, the average running time on the instances that can be solved by both increased slightly. This can be attributed to a parallel increase in the size of the search trees. If we included the three "timeout" instances into the other subgroups, however, we would see a reduction of the average running time of between 3.6% on "all solved"

---

[96] `clay0304h` in 953.8 seconds, `supplychainp1_020306` in 11.1 seconds, and `tspn05` in 636.7 seconds

Table 7.1: Computational performance of spatial pseudocost branching with variable type information

N — total number of instances solved
n — number of branch-and-bound nodes (shifted geom. mean)
t — running time (shifted geom. mean in seconds)
% — relative number of nodes/time w.r.t. SCIP default (in %)
t_{time} — relative time w.r.t. SCIP default including instances from category "timeout" (in %)

Test set	subgroup	DEFAULT			VARTYPE			%		
		N	n	t	N	n	t	n	t	t_{time}
NLINT	all solved	201	1069.1	14.3	201	1081.7	14.7	101.2	103.1	96.4
	⩾1s	115	4654.7	35.9	115	4727.1	37.4	101.6	104.1	95.3
	⩾10s	61	23141.5	128.4	61	24877.7	137.3	107.5	106.9	92.9
	⩾100s	33	104352.9	390.0	33	119326.9	439.0	114.3	112.6	88.3
	⩾1000s	10	272896.5	1157.4	10	421579.2	1696.5	154.5	146.6	71.2
	timeout (3 inst.)	0	—	3600.0	3	—	226.0	—	6.3	6.3

and 28.8% on "⩾1000s".

The Wilcoxon signed-rank test judges the slowdown on the subgroups "all solved", "⩾1s". "⩾10s", and "⩾1000s" significant at the 5% level. When including the three "timeout" instances into the other subgroups, the speedup on all categories is judged highly significant at the 2.5% level.

Results for COVER Branching.   Table 7.2 shows a reduction in running time for all subgroups: between 3% for "all solved" and 9% on the 22 instances for "⩾ 100s". SCIP with COVER branching solved two more instances[97] than the default, while the default could not solve any instance that COVER branching did not. However, none of the changes is judged significant by the Wilcoxon signed-rank test, not even if we included the two "timeout" instances.

Results for NLSCORE Branching.   The largest improvements were obtained by the branching rule NLSCORE. It reduced running time for all subgroups: by 6.1% on "all solved", by 12.3% on "⩾ 10s", and by as much as 31.6% on "⩾ 1000s" (which, however, only includes eight instances and must be read with caution). Even more importantly, NLSCORE branching could solve seven instances[98] that were not solved by SCIP default, while SCIP default only solved one additional instance[99] that NLSCORE branching did not. If we included the "timeout" category into to other subgroups, we would observe a decrease in running time by 17.1% already on "all solved".

The Wilcoxon signed-rank test judges the improvement in the "timeout" category highly significant on the 1% level. Without the "timeout" instances, the speedups

---

[97] clay0304h in 108.5 seconds and tspn05 in 1139.4 seconds
[98] clay0304h in 108.5 seconds, rsyn0830m04h in 208.7 seconds, rsyn0840m03h in 313.1 seconds, rsyn0840m04h in 2,244.1 seconds, sepasequ_convent in 1,046 seconds, supplychainp1_020306 in 0.9 seconds, and tspn05 in 1,242.4 seconds
[99] rsyn0815m04m in 3,366.5 seconds

Table 7.2: Computational performance of integer branching with cover information

N — total number of instances solved
n — number of branch-and-bound nodes (shifted geom. mean)
t — running time (shifted geom. mean in seconds)
% — relative number of nodes/time w.r.t. SCIP default (in %)
t_{time} — relative time w.r.t. SCIP default including instances from category "timeout" (in %)

Test set	subgroup	DEFAULT			COVER			%		
		N	n	t	N	n	t	n	t	t_{time}
SPLIT	all solved	103	1784.2	21.3	103	1773.6	20.7	99.4	97.0	91.3
	$\geqslant 1s$	70	6689.3	42.8	70	6534.6	41.2	97.7	96.1	89.2
	$\geqslant 10s$	37	45981.2	169.4	37	41785.9	157.2	90.9	92.8	82.4
	$\geqslant 100s$	22	141507.0	492.2	22	117407.0	447.7	83.0	91.0	75.5
	$\geqslant 1000s$	8	500738.5	1581.4	8	423427.5	1519.3	84.6	96.1	61.2
	timeout (2 inst.)	0	—	3600.0	2	—	359.1	—	10.0	10.0

in the other subgroups in Table 7.3 are not statistically significant. However, if we include the eight "timeout" instances (also the one which NLSCORE branching could not solve) then the Wilcoxon signed-rank test confirms the improvement of NLSCORE branching to be highly significant at the 1% level for subgroups "$\geqslant 10s$" and "$\geqslant 1000s$"; at the 2.5% level for subgroups "all solved", and "$\geqslant 1s$"; and at the 5% level for subgroup "$\geqslant 100s$".

NLSCORE VS. COVER. While NLSCORE branching is conceptually much simpler than COVER branching, it results in significantly better computational performance. We see two possible explanations for this: First, COVER enforces a hard restriction on the branching candidates that cannot be overruled by the underlying hybrid reliability/inference branching rule, even if there was a variable outside the cover with an extremely high score value. By contrast, NLSCORE branching allows this, because it combines both scores in a weighted sum.

As a second explanation, observe that co-occurence graphs often seem to exhibit single edges as components as in the examples shown in Figure 7.6c. While NLSCORE distributes the nonlinearity of these single edges equally over the two corresponding variables, a minimum cover may contain only one of them, and by that has to decide which variable to prefer. This decision may sometimes be arbitrary and poor.

7.5 Conclusion

In this chapter, we have presented three new branching rules for mixed-integer nonlinear programming. They are targeted specifically at integer variables that appear in nonlinear constraints. VARTYPE branching prefers branching on integer variables with small domains when performing spatial branching; COVER branching filters the set of candidate variables for integer branching using the new notion of a minimum splitting cover; and NLSCORE branching favors branching on variables that appear in many nonlinear terms during integer branching.

Table 7.3: Computational performance of integer branching with nonlinearity information

N — total number of instances solved
n — number of branch-and-bound nodes (shifted geom. mean)
t — running time (shifted geom. mean in seconds)
% — relative number of nodes/time w.r.t. SCIP default (in %)
t_{time} — relative time w.r.t. SCIP default including instances from category "timeout" (in %)

Test set	subgroup	DEFAULT			NLSCORE			%		
		N	n	t	N	n	t	n	t	t_{time}
NLINT	all solved	199	994.0	13.2	199	918.2	12.4	92.4	93.9	82.9
	⩾1s	113	4236.2	32.9	113	3678.5	30.4	86.8	92.2	78.1
	⩾10s	60	19997.3	112.2	60	15197.8	98.4	76.0	87.7	68.1
	⩾100s	32	89660.9	332.6	32	71182.0	289.6	79.4	87.1	57.7
	⩾1000s	8	227603.7	1466.5	8	116707.7	1002.6	51.3	68.4	28.0
	timeout (8 inst.)	1	—	3570.0	7	—	408.8	—	11.5	11.5

The presented branching rules are hybrid rules in the sense that they can be combined with other selection criteria. As part of the MINLP solver SCIP, we tested the rules in combination with the hybrid reliability/inference rule for integer branching and the pseudocost rule for spatial branching. However, they could equally be integrated with other strategies such as strong branching or violation transfer.

Our experiments on benchmark instances from MINLPlib2 show that VARTYPE branching solves more instances, but leads to a slowdown on instances that can be solved also without VARTYPE branching. With the inclusion of the additionally solved instances, we observed an average reduction in running time by 3.6%; on the more challenging instances that took at least ten seconds with at least one setting, the reduction was as large as 7.1%. These improvements are statistically significant.

COVER branching also leads to a reduction in running time and solves two additional instances, but these results failed to be statistically significant. Furthermore, COVER branching is clearly outperformed by NLSCORE branching: NLSCORE branching solves six instances more than the default. On the instances that could be solved by at least one of default or NLSCORE branching, NLSCORE branching reduces the average running time by 17.1%; on the more challenging instances that took at least ten seconds with at least one setting, the reduction was as large as 31.9%. Both improvements are highly significant.

Even more importantly, both VARTYPE and NLSCORE branching increase the number of instances solved. With VARTYPE branching SCIP could solve three additional instances. With NLSCORE branching, SCIP could solve all but one instance that could be solved by default SCIP and, in addition seven instances, for which default SCIP exceeded the time limit.

Chapter 8

Conclusion

With a focus on linear programming and mixed-integer nonlinear programming, this thesis has presented new methods that improve general-purpose optimization algorithms in two respects: exactness and performance. Here, the term *exact* is meant in the same sense as *rigorous* in Neumaier (2004): guaranteeing proven results also in the presence of rounding errors that may occur under limited-precision arithmetic. In this sense, the first part of the thesis was devoted to the development of new algorithms for high-precision and exact linear programming over the rational numbers.

Our guiding motivation was to improve upon the computational efficiency of the state-of-the-art approach in order to extend the scope of high-precision linear programming as an out-of-the-box technique. We achieved this by generalizing the method of iterative refinement for linear systems of equations (Wilkinson, 1963) to an algorithm for linear programs that corrects primal and dual residual errors simultaneously. On the theoretical side we proved that the resulting oracle algorithms have polynomial running time in the encoding length of the input. On the practical side, we demonstrated that they lead to large reductions in average running time because they reduce the amount of computation that must be performed in extended-precision or rational arithmetic (see Chapters 2 and 3). As only one example, the new methods have applications in systems biology for the study of numerically difficult, multiscale LP formulations for metabolic networks. We applied them successfully to so-called ME models that integrate both metabolism and gene expression to better predict bacterial growth (see Chapter 4). We showed that our general iterative refinement scheme can be extended to the case of quadratic programs (see Chapter 5).

Considering problems with rational input data includes most of the problems encountered in practical applications. We want to mention, however, that high-precision solutions for irrationally defined linear programs can be obtained by first computing a sufficiently accurate rational approximation of the input data and subsequently applying the iterative refinement algorithm to the resulting rational linear program.

The second part of the thesis has been concerned with improving the computational performance of asymptotically complete algorithms for MINLP. We showed that the proofs of optimality obtained from auxiliary linear programs that are solved during optimization-based bound tightening (OBBT) give useful one-row relaxations, which we called Lagrangian variable bounds. Their propagation leverages the effect

173

of OBBT and helps to reduce the average size of the search trees and the running time to optimality (see Chapter 6).

Furthermore, we developed new branching strategies that make use of nonlinear integer variables, i.e., decision variables that are both required to be integral *and* contained in nonlinear terms. In this context, we introduced the notion of a minimum cover of an MINLP, which gives a new measure of nonlinearity that can be detected automatically and exploited algorithmically. The most effective of the new branching rules, however, were hybrid rules that refine the branching scores used by state-of-the-art methods and are easy to implement in an MINLP solver (see Chapter 7).

In this thesis we have studied the question of numerical accuracy on the fundamental level of solving linear programs. In the context of MINLP, this is only a first step and further research is necessary to investigate how to best apply the technique of LP iterative refinement as part of MINLP solvers. We want to emphasize that this is more difficult than only solving LP relaxations within a branch-and-bound search to high levels of accuracy. If the definitions of feasibility within numerical tolerances are not coordinated between the MINLP solver and the underlying LP solver, this may result in situations with undefined behavior. To achieve this level of coordination for the complex code base of modern, state-of-the-art MINLP solvers is a nontrivial task.

In the introduction, we have mentioned the impressive progress in MIP solving over the last decades as one of the motivations to focus on the computational performance of MINLP solvers. We have approached this task constructively. We developed an exact extension of the LP solver SoPlex, which proved on average 1.85 to 3 times faster than current state-of-the-art software. As part of the MINLP solver SCIP, our implementation of OBBT reduces the average running time by 17–19% on hard mixed-integer nonlinear programs. The new branching rules reduce the average running time of SCIP by 17% on affected instances. But even more importantly, these methods extend the scope of LP and MINLP as tools in practice. SoPlex can now solve LPs to proven feasibility and optimality under exact arithmetic on which its standard floating-point routines had failed before due to particularly difficult numerics. As one example, the numerical challenges posed by the recently developed ME models from the field of systems biology can now be addressed reliably. From the heterogeneous test set MINLPlib2, which is the most extensive benchmark set for mixed-integer nonlinear programs available today, SCIP can now solve significantly more instances, instances for which it previously exceeded customary limits on computing resources.

List of Abbreviations and Names

AD . . . . Algorithmic Differentiation

CPLEX . . . . . Optimization software for LPs, MIPs, and (MI)QPs

FBBT . . . . . Feasibility-Based Bound Tightening

LP . . . . Linear Program/Programming

LVB . . . . . . Lagrangian Variable Bound

ME model . . . *In silico* model of a cell as linear program that integrates metabolism and gene expression

MINLP . . . . Mixed-Integer Nonlinear Program/Programming

MIP . . . . . . Mixed-Integer (Linear) Program/Programming

MIQCQP . . . Mixed-Integer Quadratically Constrained Quadratic Program/ Programming

NLP . . . . . . Nonlinear Program/Programming

OBBT . . . . . Optimization-Based Bound Tightening

QCP . . . . . . Quadratically Constrained Program/Programming

QP . . . . Quadratic Program/Programming

QSopt_ex . . . Optimization software for exact linear programming

SCIP . . . . . . Optimization software "Solving Constraint Integer Programs" for MIPs, MINLPs, and more general constraint integer programs

SoPlex . . . . . Optimization software "Sequential object-oriented simPlex" for LPs

List of Algorithms

List of Figures

179

List of Tables

Bibliography

The page numbers in brackets at the end of each citation refer to the text.

J. Abbott and T. Mulders. How tight is Hadamard's bound? *Experimental Mathematics*, 10(3): 331–336, 2001. http://projecteuclid.org/euclid.em/1069786341. [77]

K. Abhishek, S. Leyffer, and J. Linderoth. FilMINT: An outer approximation-based solver for convex mixed-integer nonlinear programs. *INFORMS Journal on Computing*, 22(4):555–567, 2010. DOI:10.1287/ijoc.1090.0373. [106]

T. Achterberg. *Constraint Integer Programming*. PhD thesis, Technische Universität Berlin, 2007. URN:nbn:de:kobv:83-opus-16117. [11, 111, 122, 123, 124, 125, 150, 165, 166, and 167]

T. Achterberg. SCIP: Solving Constraint Integer Programs. *Mathematical Programming Computation*, 1(1):1–41, 2009. DOI:10.1007/s12532-008-0001-1. [111]

T. Achterberg and T. Berthold. Hybrid branching. In W.-J. van Hoeve and J. N. Hooker, editors, *Integration of AI and OR Techniques in Constraint Programming for Combinatorial Optimization Problems*, volume 5547 of *Lecture Notes in Computer Science*, pages 309–311. 2009. DOI:10.1007/978-3-642-01929-6_23. [150, 165, 167, and 168]

T. Achterberg and R. Wunderling. Mixed integer programming: Analyzing 12 years of progress. In M. Jünger and G. Reinelt, editors, *Facets of Combinatorial Optimization*, pages 449–481. Springer Berlin Heidelberg, 2013. DOI:10.1007/978-3-642-38189-8_18. [11, 83, 110, 145, and 169]

T. Achterberg, T. Koch, and A. Martin. Branching rules revisited. *Operations Research Letters*, 33(1):42–54, Jan. 2005. DOI:10.1016/j.orl.2004.04.002. [150]

T. Achterberg, S. Heinz, and T. Koch. Counting solutions of integer programs using unrestricted subtree detection. In L. Perron and M. A. Trick, editors, *Integration of AI and OR Techniques in Constraint Programming for Combinatorial Optimization Problems*, volume 5015 of *Lecture Notes in Computer Science*, pages 278–282. Springer, 2008. DOI: 10.1007/978-3-540-68155-7_22. [162]

C. S. Adjiman, I. P. Androulakis, and C. A. Floudas. A global optimization method, αBB, for general twice-differentiable constrained NLPs—II. Implementation and computational results. *Computers & Chemical Engineering*, 22(9):1159–1179, 1998. DOI:10.1016/S0098-1354(98) 00218-X. [125, 132]

C. S. Adjiman, I. P. Androulakis, and C. A. Floudas. Global optimization of mixed-integer nonlinear problems. *AIChE Journal*, 46(9):1769–1797, 2000. DOI:10.1002/aic.690460908. [125, 132]

B. Alberts, A. Johnson, J. Lewis, M. Raff, K. Roberts, and P. Walter. *Molecular Biology of the Cell*. Garland Science, New York, 4th edition, 2002. [93]

E. Althaus and D. Dumitriu. Fast and accurate bounds on linear programs. In J. Vahrenhold, editor, *Proc. 8th International Symposium on Experimental Algorithms*, volume 5526 of *LNCS*, pages 40–50. Springer, June 2009. DOI:10.1007/978-3-642-02011-7_6. [63]

D. L. Applegate, R. E. Bixby, V. Chvátal, and W. J. Cook. Finding cuts in the TSP (A preliminary report). Technical Report 95-05, Center for Discrete Mathematics & Theoretical Computer Science (DIMACS), March 1995. [132, 150]

D. L. Applegate, R. E. Bixby, V. Chvátal, and W. J. Cook. *The Traveling Salesman Problem: A Computational Study*. Princeton Series in Applied Mathematics. Princeton University Press, Princeton, NJ, USA, 2007a. [132, 150]

D. L. Applegate, W. Cook, S. Dash, and D. G. Espinoza. Exact solutions to linear programming problems. *Operations Research Letters*, 35(6):693–699, 2007b. DOI:10.1016/j.orl.2006.12.010. [3, 31, 46, 55, 59, and 94]

A. Atamtürk and V. Narayanan. Conic mixed-integer rounding cuts. *Mathematical Programming*, 122(1):1–20, 2010. DOI:10.1007/s10107-008-0239-4. [106]

D. Avis and K. Fukuda. A pivoting algorithm for convex hulls and vertex enumeration of arrangements and polyhedra. *Discrete & Computational Geometry*, 8(1):295–313, 1992. DOI:10.1007/BF02293050. [57]

D.-O. Azulay and J.-F. Pique. Optimized Q-pivot for exact linear solvers. In M. Maher and J.-F. Puget, editors, *Principles and Practice of Constraint Programming — CP98*, volume 1520 of *Lecture Notes in Computer Science*, pages 55–71. Springer, 1998. DOI:10.1007/3-540-49481-2_6. [58]

D.-O. Azulay and J.-F. Pique. A revised simplex method with integer Q-matrices. *ACM Transactions on Mathematical Software*, 27(3):350–360, 2001. DOI:10.1145/502800.502804. [58]

M. Bastian, S. Heymann, and M. Jacomy. Gephi: An open source software for exploring and manipulating networks. In *International AAAI Conference on Weblogs and Social Media*, 2009. http://www.aaai.org/ocs/index.php/ICWSM/09/paper/view/154. [163]

E. M. L. Beale. An alternative method for linear programming. *Mathematical Proceedings of the Cambridge Philosophical Society*, 50(04):513–523, 1954. DOI:10.1017/S0305004100029650. [18]

P. Belotti. Bound reduction using pairs of linear inequalities. *Journal of Global Optimization*, 56(3):787–819, 2013. DOI:10.1007/s10898-012-9848-9. [123]

P. Belotti, J. Lee, L. Liberti, F. Margot, and A. Wächter. Branching and bounds tightening techniques for non-convex MINLP. *Optimization Methods & Software*, 24:597–634, 2009. DOI:10.1080/10556780903087124. [111, 123, 124, 125, 132, 152, 153, and 165]

P. Belotti, S. Cafieri, J. Lee, and L. Liberti. Feasibility-based bounds tightening via fixed points. In W. Wu and O. Daescu, editors, *Combinatorial Optimization and Applications*, volume 6508 of *Lecture Notes in Computer Science*, pages 65–76. Springer Berlin Heidelberg, 2010. DOI:10.1007/978-3-642-17458-2_7. [123, 131]

P. Belotti, S. Cafieri, J. Lee, and L. Liberti. On feasibility based bounds tightening. Technical Report 3325, Optimization Online, January 2012. http://www.optimization-online.org/DB_HTML/2012/01/3325.html. [121, 123, and 131]

P. Belotti, C. Kirches, S. Leyffer, J. Linderoth, J. Luedtke, and A. Mahajan. Mixed-integer non-linear optimization. *Acta Numerica*, 22:1–131, 5 2013. DOI:10.1017/S0962492913000032. [104, 106, and 109]

P. Belotti, P. Bonami, M. Fischetti, A. Lodi, M. Monaci, A. Nogales-Gómez, and D. Salvagnin. On handling indicator constraints in mixed integer programming. Technical report, Università degli Studi di Padova, April 2015. http://www.dei.unipd.it/~salvagni/pdf/indicator.pdf. [126]

M. Benichou, J. M. Gauthier, P. Girodet, G. Hentges, G. Ribiere, and O. Vincent. Experiments in mixed-integer linear programming. *Mathematical Programming*, 1(1):76–94, 1971. DOI:10.1007/BF01584074. [150]

T. Berthold. *Heuristic algorithms in global MINLP solvers*. PhD thesis, Technische Universität Berlin, 2014. [111, 112, 150, and 151]

T. Berthold and A. M. Gleixner. Undercover – a primal heuristic for MINLP based on sub-MIPs generated by set covering. In P. Bonami, L. Liberti, A. J. Miller, and A. Sartenaer, editors, *Proceedings of the European Workshop on Mixed Integer Nonlinear Programming (EWMINLP)*, pages 103–112, CIRM Marseille, France, April 2010. http://www.lix.polytechnique.fr/~liberti/ewminlp/ewminlp-proceedings.pdf. [6, 153]

T. Berthold and A. M. Gleixner. Undercover branching. In V. Bonifaci, C. Demetrescu, and A. Marchetti-Spaccamela, editors, *Experimental Algorithms, 12th International Symposium, SEA 2013, Rome, Italy, June 5-7, 2013, Proceedings*, volume 7933 of *LNCS*, pages 212–223. Springer Berlin Heidelberg, 2013. DOI:10.1007/978-3-642-38527-8_20. [6, 149]

T. Berthold and A. M. Gleixner. Undercover: a primal MINLP heuristic exploring a largest sub-MIP. *Mathematical Programming*, 144(1-2):315–346, 2014. DOI:10.1007/s10107-013-0635-2. [6, 149, and 153]

T. Berthold and D. Salvagnin. Cloud branching. In C. Gomes and M. Sellmann, editors, *Integration of AI and OR Techniques in Constraint Programming for Combinatorial Optimization Problems*, volume 7874 of *Lecture Notes in Computer Science*, pages 28–43. Springer Berlin Heidelberg, 2013. DOI:10.1007/978-3-642-38171-3_3. [150]

T. Berthold, S. Heinz, and M. E. Pfetsch. Nonlinear pseudo-boolean optimization: Relaxation or propagation? In O. Kullmann, editor, *Theory and Applications of Satisfiability Testing – SAT 2009*, volume 5584 of *Lecture Notes in Computer Science*, pages 441–446. Springer Berlin Heidelberg, 2009. DOI:10.1007/978-3-642-02777-2_40. [113]

T. Berthold, S. Heinz, M. E. Lübbecke, R. H. Möhring, and J. Schulz. A constraint integer programming approach for resource-constrained project scheduling. In A. Lodi, M. Milano, and P. Toth, editors, *Integration of AI and OR Techniques in Constraint Programming for Combinatorial Optimization Problems*, volume 6140 of *Lecture Notes in Computer Science*, pages 313–317. Springer Berlin Heidelberg, 2010. DOI:10.1007/978-3-642-13520-0_34. [113]

T. Berthold, A. M. Gleixner, S. Heinz, and S. Vigerske. Analyzing the computational impact of MIQCP solver components. *Numerical Algebra, Control and Optimization (NACO)*, 2(4):739–748, 2012a. DOI:10.3934/naco.2012.2.739. [165]

T. Berthold, S. Heinz, and S. Vigerske. Extending a CIP framework to solve MIQCPs. In J. Lee and S. Leyffer, editors, *Mixed Integer Nonlinear Programming*, volume 154 of *The IMA Volumes in Mathematics and its Applications*, pages 427–444. Springer New York, 2012b. DOI:10.1007/978-1-4614-1927-3_15. [111, 112]

R. Bixby and E. Rothberg. Progress in computational mixed integer programming—a look back from the other side of the tipping point. *Annals of Operations Research*, 149(1):37–41, 2007. DOI:10.1007/s10479-006-0091-y. [1]

R. E. Bixby. Solving real-world linear programs: A decade and more of progress. *Operations Research*, 50(1):3–15, 2002. DOI:10.1287/opre.50.1.3.17780. [132]

R. E. Bixby and M. J. Saltzman. Recovering an optimal LP basis from an interior point solution. *Operations Research Letters*, 15(4):169–178, 1994. DOI:10.1016/0167-6377(94)90074-4. [19]

R. E. Bixby, M. Fenelon, Z. Gu, E. Rothberg, and R. Wunderling. MIP: Theory and practice – closing the gap. In *Proceedings of the 19th IFIP TC7 Conference on System Modelling and Optimization: Methods, Theory and Applications*, pages 19–50, Deventer, The Netherlands, 2000. Kluwer, B.V. [1, 110]

A. Bley, A. M. Gleixner, T. Koch, and S. Vigerske. Comparing MIQCP solvers to a specialised algorithm for mine production scheduling. In H. G. Bock, H. X. Phu, R. Rannacher, and J. P. Schlöder, editors, *Modeling, Simulation and Optimization of Complex Processes: Proceedings of the Fourth International Conference on High Performance Scientific Computing, March 2-6, 2009, Hanoi, Vietnam*, pages 25–39. Springer, 2012. DOI:10.1007/978-3-642-25707-0_3. [103]

C. Bliek and P. Bonami. IBM CPLEX global non-convex MIQP. Talk at MINLP 2014, Carnegie Mellon University, Pittsburgh, PA, USA, June 2014. http://minlp.cheme.cmu.edu/2014/papers/bliek.ppt. [111]

P. T. Boggs and J. W. Tolle. Sequential quadratic programming. *Acta Numerica*, 4:1–51, January 1995. DOI:10.1017/S0962492900002518. [114]

I. M. Bomze, M. Dür, E. de Klerk, C. Roos, A. J. Quist, and T. Terlaky. On copositive programming and standard quadratic optimization problems. *Journal of Global Optimization*, 18(4): 301–320, 2000. DOI:10.1023/A:1026583532263. [104]

P. Bonami, L. T. Biegler, A. R. Conn, G. Cornuéjols, I. E. Grossmann, C. D. Laird, J. Lee, A. Lodi, F. Margot, N. Sawaya, and A. Wächter. An algorithmic framework for convex mixed integer nonlinear programs. *Discrete Optimization*, 5(2):186–204, 2008. DOI:10.1016/j.disopt.2006.10.011. [106]

P. Bonami, M. Kılınç, and J. Linderoth. Algorithms and software for convex mixed integer nonlinear programs. In J. Lee and S. Leyffer, editors, *Mixed Integer Nonlinear Programming*, volume 154 of *The IMA Volumes in Mathematics and its Applications*, pages 1–39. Springer New York, 2012. DOI:10.1007/978-1-4614-1927-3_1. [104, 106]

P. Bonami, J. Lee, S. Leyffer, and A. Wächter. On branching rules for convex mixed-integer nonlinear optimization. *Journal of Experimental Algorithmics*, 18(2):6:1–6:31, 2013. DOI:10.1145/2532568. [106]

J. Bonwick. The slab allocator: An object-caching kernel memory allocator. In *USENIX Summer 1994 Technical Conference, Boston, Massachusetts, USA, June 6-10, 1994, Conference Proceeding*, pages 87–98, 1994. http://www.usenix.org/publications/library/proceedings/bos94/full_papers/bonwick.ps. [48]

C. Bragalli, C. D'Ambrosio, J. Lee, A. Lodi, and P. Toth. On the optimal design of water distribution networks: a practical MINLP approach. *Optimization and Engineering*, 13(2): 219–246, 2012. DOI:10.1007/s11081-011-9141-7. [106]

C. Buchheim, M. Chimani, D. Ebner, C. Gutwenger, M. Jünger, G. W. Klau, P. Mutzel, and R. Weiskircher. A branch-and-cut approach to the crossing number problem. *Discrete Optimization*, 5(2):373–388, 2008. In Memory of George B. Dantzig. DOI:10.1016/j.disopt. 2007.05.006. [15]

D. A. Bulutoglu and D. M. Kaziska. Improved WLP and GWP lower bounds based on exact integer programming. *Journal of Statistical Planning and Inference*, 140(5):1154–1161, 2010. DOI:10.1016/j.jspi.2009.10.013. [15]

S. Burer and A. N. Letchford. Non-convex mixed-integer nonlinear programming: A survey. *Surveys in Operations Research and Management Science*, 17(2):97–106, 2012. DOI:10.1016/j.sorms.2012.08.001. [103]

B. A. Burton and M. Ozlen. Computing the crosscap number of a knot using integer programming and normal surfaces. *ACM Transactions on Mathematical Software*, 39(1):4:1–4:18, 2012. DOI:10.1145/2382585.2382589. [15]

M. R. Bussieck and S. Vigerske. MINLP solver software. In J. J. Cochran, L. A. C. Jr., P. Keskinocak, J. P. Kharoufeh, and J. C. Smith, editors, *Wiley Encyclopedia of Operations Research and Management Science*. Wiley & Sons, Inc., 2010. DOI:10.1002/9780470400531.eorms0527. [110]

M. R. Bussieck, A. S. Drud, and A. Meeraus. MINLPLib—a collection of test models for mixed-integer nonlinear programming. *INFORMS Journal on Computing*, 15(1):114–119, 2003. DOI:10.1287/ijoc.15.1.114.15159. [14]

A. Caprara and M. Fischetti. $\{0, \frac{1}{2}\}$-Chvátal-Gomory cuts. *Mathematical Programming*, 74(3): 221–235, 1996. DOI:10.1007/BF02592196. [127]

A. Caprara and M. Locatelli. Global optimization problems and domain reduction strategies. *Mathematical Programming*, 125:123–137, 2010. DOI:10.1007/s10107-008-0263-4. [126]

S. Ceria and J. Soares. Convex programming for disjunctive convex optimization. *Mathematical Programming*, 86(3):595–614, 1999. DOI:10.1007/s101070050106. [106]

Z. Chen and A. Storjohann. A BLAS based C library for exact linear algebra on integer matrices. In *Proceedings of the 2005 International Symposium on Symbolic and Algebraic Computation*, ISSAC '05, pages 92–99, 2005. DOI:10.1145/1073884.1073899. [61, 77]

V. Chvátal. Edmonds polytopes and a hierarchy of combinatorial problems. *Discrete Mathematics*, 4(4):305–337, 1973. DOI:10.1016/0012-365X(73)90167-2. [127]

V. Chvátal. *Linear Programming*. W. H. Freeman and Company, New York, 1983. [16]

M. Collins, L. Cooper, R. Helgason, J. Kennington, and L. LeBlanc. Solving the pipe network analysis problem using optimization techniques. *Management Science*, 24(7):747–760, 1978. DOI:10.1287/mnsc.24.7.747. [106]

W. Cook and D. E. Steffy. Solving very sparse rational systems of equations. *ACM Transactions on Mathematical Software*, 37(4):39:1–39:21, 2011. DOI:10.1145/1916461.1916463. [61, 79]

W. Cook, T. Koch, D. E. Steffy, and K. Wolter. A hybrid branch-and-bound approach for exact rational mixed-integer programming. *Mathematical Programming Computation*, 5(3):305–344, 2013. DOI:10.1007/s12532-013-0055-6. [15]

R. J. Dakin. A tree-search algorithm for mixed integer programming problems. *The Computer Journal*, 8(3):250–255, 1965. DOI:10.1093/comjnl/8.3.250. [105]

E. Danna. Performance variability in mixed integer programming. Talk at Workshop on Mixed Integer Programming 2008, Columbia University, New York, NY, USA, June 2014. [11]

G. B. Dantzig. *Linear programming and extensions*. Princeton University Press, Princeton, NJ, 1963. [16, 18]

G. B. Dantzig, D. R. Fulkerson, and S. M. Johnson. Solution of a large-scale traveling-salesman problem. *Operations Research*, 2:393–410, 1954a. DOI:10.1287/opre.2.4.393. [2, 110, and 123]

G. B. Dantzig, D. R. Fulkerson, and S. M. Johnson. Solution of a large-scale traveling-salesman problem. Technical Report P-510, The RAND Corporation, April 1954b. [2, 110]

M. Davis and H. Putnam. A computing procedure for quantification theory. *Journal of the ACM*, 7(3):201–215, 1960. DOI:10.1145/321033.321034. [122]

F. M. de Oliveira Filho and F. Vallentin. Fourier analysis, linear programming, and densities of distance avoiding sets in R^n. *Journal of the European Mathematical Society*, 12(6):1417–1428, 2010. DOI:10.4171/JEMS/236. [15]

M. Dhiflaoui, S. Funke, C. Kwappik, K. Mehlhorn, M. Seel, E. Schömer, R. Schulte, and D. Weber. Certifying and repairing solutions to large LPs: How good are LP-solvers? In *Proceedings of the 14th annual ACM-SIAM symposium on Discrete algorithms*, SODA '03, pages 255–256. SIAM, 2003. [21, 53, and 59]

I. Dinur and S. Safra. On the hardness of approximating vertex cover. *Annals of Mathematics*, 162:439–485, 2005. DOI:10.4007/annals.2005.162.439. [156]

E. D. Dolan and J. J. Moré. Benchmarking optimization software with performance profiles. *Mathematical Programming*, 91(2):201–213, 2002. DOI:10.1007/s101070100263. [143]

F. Domes and A. Neumaier. A scaling algorithm for polynomial constraint satisfaction problems. *Journal of Global Optimization*, 42(3):327–345, 2008. DOI:10.1007/s10898-008-9317-7. [2]

S. Drewes. *Mixed Integer Second Order Cone Programming*. PhD thesis, Technische Universität Darmstadt, 2009. [106]

M. A. Duran and I. E. Grossmann. An outer-approximation algorithm for a class of mixed-integer nonlinear programs. *Mathematical Programming*, 36(3):307–339, 1986. DOI:10.1007/BF02592064. [106]

C. D'Ambrosio and A. Lodi. Mixed integer nonlinear programming tools: an updated practical overview. *Annals of Operations Research*, 204(1):301–320, 2013. DOI:10.1007/s10479-012-1272-5. [110]

A. Ebrahim, J. A. Lerman, B. Ø. Palsson, and D. R. Hyduke. COBRApy: COnstraints-Based Reconstruction and Analysis for Python. *BMC Systems Biology*, 7:1–6, 2013. DOI:10.1186/1752-0509-7-74. [6, 91]

J. Edmonds. Systems of distinct representatives and linear algebra. *Journal of Research of the National Bureau of Standards*, 71B(4):241–245, 1967. [63, 73]

J. Edmonds. Exact pivoting. Talk at ECCO VII, Milan, Italy, February 1994. [58]

J. Edmonds and J.-F. Maurras. Note sur les Q-matrices d'Edmonds (in French). *RAIRO. Recherche Opérationnelle*, 31(2):203–209, 1997. http://www.numdam.org/item?id=RO_1997__31_2_203_0. [58]

D. G. Espinoza. *On Linear Programming, Integer Programming and Cutting Planes*. PhD thesis, Georgia Institute of Technology, 2006. http://hdl.handle.net/1853/10482. [58, 59]

J. E. Falk and R. M. Soland. An algorithm for separable nonconvex programming problems. *Management Science*, 15(9):550–569, 1969. DOI:10.1287/mnsc.15.9.550. [107]

A. V. Fiacco and G. P. McCormick. *Nonlinear Programming: Sequential Unconstrained Minimization Techniques*. John Wiley and Sons, New York, 1968. [19]

M. Fischetti and M. Monaci. Branching on nonchimerical fractionalities. *Operations Research Letters*, 40(3):159–164, 2012. DOI:10.1016/j.orl.2012.01.008. [150]

M. Fischetti and M. Monaci. Backdoor branching. *INFORMS Journal on Computing*, 25(4): 693–700, 2013. DOI:10.1287/ijoc.1120.0531. [151, 168]

C. A. Floudas and C. E. Gounaris. A review of recent advances in global optimization. *Journal of Global Optimization*, 45(1):3–38, 2009. DOI:10.1007/s10898-008-9332-8. [102]

A. Frangioni and C. Gentile. Perspective cuts for a class of convex 0–1 mixed integer programs. *Mathematical Programming*, 106(2):225–236, 2006. DOI:10.1007/s10107-005-0594-3. [106]

R. M. Freund and J. R. Vera. Some characterizations and properties of the "distance to ill-posedness" and the condition measure of a conic linear system. *Mathematical Programming*, 86(2):225–260, 1999. DOI:10.1007/s10107990063a. [21, 98]

A. Fügenschuh and J. Humpola. A unified view on relaxations for a nonlinear network flow problem. ZIB-Report 13-31, Zuse Institute Berlin, July 2013. URN:nbn:de:0297-zib-18857. [106]

A. Fügenschuh and A. Martin. Computational integer programming and cutting planes. In K. Aardal, G. L. Nemhauser, and R. Weismantel, editors, *Discrete Optimization*, volume 12 of *Handbooks in Operations Research and Management Science*, pages 69–121. Elsevier, 2005. DOI:10.1016/S0927-0507(05)12002-7. [123]

K. Fukuda and A. Prodon. Double description method revisited. In M. Deza, R. Euler, and I. Manoussakis, editors, *Combinatorics and Computer Science*, volume 1120 of *Lecture Notes in Computer Science*, pages 91–111. Springer, 1996. DOI:10.1007/3-540-61576-8_77. [57]

G. Gamrath, T. Koch, A. Martin, M. Miltenberger, and D. Weninger. Progress in presolving for mixed integer programming. *Mathematical Programming Computation*, 7(4):367–398, 2015a. DOI:10.1007/s12532-015-0083-5. [123]

G. Gamrath, A. Melchiori, T. Berthold, A. M. Gleixner, and D. Salvagnin. Branching on multi-aggregated variables. In L. Michel, editor, *Integration of AI and OR Techniques in Constraint Programming*, volume 9075 of *Lecture Notes in Computer Science*, pages 141–156. 2015b. DOI:10.1007/978-3-319-18008-3_10. [152]

M. R. Garey and D. S. Johnson. *Computers and Intractability: A Guide to the Theory of NP-Completeness*. W. H. Freeman & Co., New York, NY, USA, 1979. [156]

B. Gärtner. Exact arithmetic at low cost – a case study in linear programming. *Computational Geometry*, 13(2):121–139, 1999. DOI:10.1016/S0925-7721(99)00012-7. [58, 119]

B. Gärtner and S. Schönherr. An efficient, exact, and generic quadratic programming solver for geometric optimization. In *Proceedings of the Sixteenth Annual Symposium on Computational Geometry*, SCG '00, pages 110–118, 2000. DOI:10.1145/336154.336191. [115, 119]

B. Geißler, A. Martin, A. Morsi, and L. Schewe. Using piecewise linear functions for solving MINLPs. In J. Lee and S. Leyffer, editors, *Mixed Integer Nonlinear Programming*, volume 154 of *The IMA Volumes in Mathematics and its Applications*, pages 287–314. Springer New York, 2012. DOI:10.1007/978-1-4614-1927-3_10. [107]

A. M. Geoffrion. Generalized Benders decomposition. *Journal of Optimization Theory and Applications*, 10(4):237–260, 1972. DOI:10.1007/BF00934810. [105]

A. M. Gleixner. Factorization and update of a reduced basis matrix for the revised simplex method. ZIB-Report 12-36, Zuse Institute Berlin, October 2012. URN:nbn:de: 0297-zib-16349. [81]

A. M. Gleixner and S. Weltge. Learning and propagating Lagrangian variable bounds for mixed-integer nonlinear programming. In C. Gomes and M. Sellmann, editors, *Integration of AI and OR Techniques in Constraint Programming for Combinatorial Optimization Problems, 10th International Conference, CPAIOR 2013, Yorktown Heights, NY, USA, May 18-22, 2013*, volume 7874 of *LNCS*, pages 355–361. Springer, 2013. DOI:10.1007/978-3-642-38171-3_26. [5, 121]

A. M. Gleixner, H. Held, W. Huang, and S. Vigerske. Towards globally optimal operation of water supply networks. *Numerical Algebra, Control and Optimization (NACO)*, 2(4):695–711, 2012a. DOI:10.3934/naco.2012.2.695. [103]

A. M. Gleixner, D. E. Steffy, and K. Wolter. Improving the accuracy of linear programming solvers with iterative refinement. In *ISSAC '12. Proceedings of the 37th International Symposium on Symbolic and Algebraic Computation*, pages 187–194. ACM, July 2012b. DOI:10.1145/2442829.2442858. [5, 16, and 64]

A. M. Gleixner, T. Berthold, B. Müller, and S. Weltge. Three enhancements for optimization-based bound tightening. ZIB-Report 15-16, Zuse Institute Berlin, 2015a. Submitted to the Journal of Global Optimization. [5, 121]

A. M. Gleixner, D. E. Steffy, and K. Wolter. Iterative refinement for linear programming. ZIB-Report 15-15, Zuse Institute Berlin, 2015b. Accepted for publication in the INFORMS Journal on Computing. [5, 16, and 64]

G. H. Golub and C. F. van Loan. *Matrix Computations*. Johns Hopkins University Press, Baltimore, Maryland, USA, 1983. [20, 23]

R. E. Gomory. Outline of an algorithm for integer solutions to linear programs. *Bulletin of the American Society*, 64:275–278, 1958. http://projecteuclid.org/euclid.bams/1183522679. [127]

R. E. Gomory. An algorithm for the mixed integer problem. Technical Report P-1885, The RAND Corporation, June 1960. [127]

N. I. M. Gould and P. L. Toint. A quadratic programming page, 2012. http://www.numerical.rl.ac.uk/people/nimg/qp/qp.html. [113, 115]

A. Griewank and A. Walther. *Evaluating derivatives: principles and techniques of algorithmic differentiation*. Society for Industrial and Applied Mathematics, 2008. [159]

I. E. Grossmann. Review of nonlinear mixed-integer and disjunctive programming techniques. *Optimization and Engineering*, 3(3):227–252, 2002. DOI:10.1023/A:1021039126272. [104, 106]

M. Grötschel, L. Lovász, and A. Schrijver. The ellipsoid method and its consequences in combinatorial optimization. *Combinatorica*, 1(2):169–197, 1981. DOI:10.1007/BF02579273. [79]

M. Grötschel, L. Lovász, and A. Schrijver. *Geometric Algorithms and Combinatorial Optimization*, volume 2 of *Algorithms and Combinatorics*. Springer, Berlin / Heidelberg, 1988. [7, 9, 19, 31, 62, 63, 76, and 104]

O. Günlük and J. Linderoth. Perspective relaxation of mixed integer nonlinear programs with indicator variables. In A. Lodi, A. Panconesi, and G. Rinaldi, editors, *Integer Programming and Combinatorial Optimization*, volume 5035 of *Lecture Notes in Computer Science*, pages 1–16. Springer Berlin Heidelberg, 2008. DOI:10.1007/978-3-540-68891-4_1. [106]

O. K. Gupta. *Branch and Bound Experiments in Nonlinear Integer Programming*. PhD thesis, Purdue University, 1980. http://search.proquest.com/docview/303090473. [105]

O. K. Gupta. Applications of quadratic programming. *Journal of Information and Optimization Sciences*, 16(1):177–194, 1995. DOI:10.1080/02522667.1995.10699213. [113]

O. K. Gupta and A. Ravindran. Branch and bound experiments in convex nonlinear integer programming. *Management Science*, 31(12):1533–1546, 1985. DOI:10.1287/mnsc.31.12.1533. [105]

T. C. Hales. A proof of the Kepler conjecture. *Annals of Mathematics*, 162(3):1065–1185, 2005. http://www.jstor.org/stable/20159940. [15, 64]

J. A. J. Hall and Q. Huangfu. A high performance dual revised simplex solver. In R. Wyrzykowski, J. Dongarra, K. Karczewski, and J. Waśniewski, editors, *Parallel Processing and Applied Mathematics*, volume 7203 of *Lecture Notes in Computer Science*, pages 143–151. Springer Berlin Heidelberg, 2012. DOI:10.1007/978-3-642-31464-3_15. [19]

E. Halperin. Improved approximation algorithms for the vertex cover problem in graphs and hypergraphs. *SIAM Journal on Computing*, 31:1608–1623, 2002. DOI:10.1137/S0097539700381097. [156]

E. R. Hansen. Global optimization using interval analysis: The one-dimensional case. *Journal of Optimization Theory and Applications*, 29(3):331–344, 1979. DOI:10.1007/BF00933139. [109]

E. R. Hansen. Global optimization using interval analysis–the multi-dimensional case. *Numerische Mathematik*, 34(3):247–270, 1980. DOI:10.1007/BF01396702. [109]

P. Hansen and B. Jaumard. Reduction of indefinite quadratic programs to bilinear programs. *Journal of Global Optimization*, 2(1):41–60, 1992. DOI:10.1007/BF00121301. [154]

S. Held, W. Cook, and E. C. Sewell. Maximum-weight stable sets and safe lower bounds for graph coloring. *Mathematical Programming Computation*, 4(4):363–381, 2012. DOI:10.1007/s12532-012-0042-3. [15]

R. Hemmecke, M. Köppe, J. Lee, and R. Weismantel. Nonlinear integer programming. In M. Jünger, T. M. Liebling, D. Naddef, G. L. Nemhauser, W. R. Pulleyblank, G. Reinelt, G. Rinaldi, and L. A. Wolsey, editors, *50 Years of Integer Programming 1958-2008*, pages 561–618. 2010. DOI:10.1007/978-3-540-68279-0_15. [103]

G. Hendel. Empirical analysis of solving phases in mixed integer programming. Master's thesis, Technische Universität Berlin, August 2014. URN:nbn:de:0297-zib-54270. [11]

J. Herbrand. *Recherches sur la théorie de la démonstration (in French)*. PhD thesis, Faculté des Sciences de Paris, 1930. http://eudml.org/doc/192791. [122]

I. V. Hicks and N. B. McMurray Jr. The branchwidth of graphs and their cycle matroids. *Journal of Combinatorial Theory, Series B*, 97(5):681–692, 2007. DOI:10.1016/j.jctb.2006.12.007. [15]

N. J. Higham. A survey of condition number estimation for triangular matrices. *SIAM Review*, 29(4):575–596, 1987. DOI:10.1137/1029112. [63]

D. S. Hochbaum. Complexity and algorithms for nonlinear optimization problems. *Annals of Operations Research*, 153(1):257–296, 2007. DOI:10.1007/s10479-007-0172-6. [103]

J. N. Hooker. Testing heuristics: We have it all wrong. *Journal of Heuristics*, 1(1):33–42, 1995. DOI:10.1007/BF02430364. [10]

R. Horst and H. Tuy. *Global Optimization. Deterministic Approaches*. Springer-Verlag, Berlin Heidelberg, 1990. [108, 150]

W. Huang. Operative planning of water supply networks by mixed integer nonlinear programming. Master's thesis, Freie Universität Berlin, April 2011. [126]

IEEE Std 754-2008, Standard for Floating-Point Arithmetic. IEEE, New York, NY, USA, 2008. DOI:10.1109/IEEESTD.2008.4610935. [20]

C. Jansson. Rigorous lower and upper bounds in linear programming. *SIAM Journal on Optimization*, 14(3):914–935, 2004. DOI:10.1137/S1052623402416839. [63]

E. Kaltofen and B. D. Saunders. On Wiedemann's method of solving sparse linear systems. In H. F. Mattson, T. Mora, and T. R. N. Rao, editors, *Applied Algebra, Algebraic Algorithms and Error-Correcting Codes*, volume 539 of *Lecture Notes in Computer Science*, pages 29–38. 1991. DOI:10.1007/3-540-54522-0_93. [61]

L. V. Kantorovich. Mathematical methods of organizing and planning production. *Management Science*, 6(4):366–422, 1960. http://www.jstor.org/stable/2627082. [18]

G. Karakostas. A better approximation ratio for the vertex cover problem. *ACM Transactions on Algorithms*, 5:41:1–41:8, 2009. DOI:10.1145/1597036.1597045. [156]

N. Karmarkar. A new polynomial-time algorithm for linear programming. *Combinatorica*, 4 (4):373–395, 1984. DOI:10.1007/BF02579150. [19, 62]

W. Karush. Minima of functions of several variables with inequalities as side constraints. Master's thesis, Department of Mathematics, University of Chicago, 1939. [114]

L. G. Khachiyan. A polynomial algorithm in linear programming (in Russian). *Doklady Akademii Nauk SSSR*, 244:1093–1096, 1979. English translation: *Soviet Mathematics Doklady*, 20(1):191–194, 1979. [19]

L. G. Khachiyan. Polynomial algorithms in linear programming (in Russian). *Zhurnal Vychislitel'noi Matematiki i Matematicheskoi Fiziki*, 20(1):51–68, 1980. English translation: *USSR Computational Mathematics and Mathematical Physics*, 20(1):53–72, 1980. DOI:10.1016/0041-5553(80)90061-0. [19]

S. Khot and O. Regev. Vertex cover might be hard to approximate to within $2 - \varepsilon$. *Journal of Computer and System Sciences*, 74(3):335–349, 2008. DOI:10.1016/j.jcss.2007.06.019. [156]

M. Kılınç and N. V. Sahinidis. Solving MINLPs with BARON. Talk at MINLP 2014, Carnegie Mellon University, Pittsburgh, PA, USA, June 2014. http://minlp.cheme.cmu.edu/2014/papers/kilinc.pdf. [110]

M. Kılınç, J. Linderoth, J. Luedtke, and A. Miller. Strong-branching inequalities for convex mixed integer nonlinear programs. *Computational Optimization and Applications*, 59(3): 639–665, 2014. DOI:10.1007/s10589-014-9690-8. [106]

F. Kılınç-Karzan, G. L. Nemhauser, and M. W. P. Savelsbergh. Information-based branching schemes for binary linear mixed integer problems. *Mathematical Programming Computation*, 1:249–293, 2009. DOI:10.1007/s12532-009-0009-1. [151, 168]

E. Klotz. Identification, assessment, and correction of ill-conditioning and numerical instability in linear and integer programs. In A. Newman and J. Leung, editors, *Bridging Data and Decisions*, TutORials in Operations Research, pages 54–108. 2014. DOI:10.1287/educ.2014.0130. [19, 30]

T. Koch. The final NETLIB-LP results. *Operations Research Letters*, 32(2):138–142, 2004. DOI:10.1016/S0167-6377(03)00094-4. [21, 53, and 59]

T. Koch, T. Achterberg, E. Andersen, O. Bastert, T. Berthold, R. E. Bixby, E. Danna, G. Gamrath, A. M. Gleixner, S. Heinz, A. Lodi, H. Mittelmann, T. Ralphs, D. Salvagnin, D. E. Steffy, and K. Wolter. MIPLIB 2010. *Mathematical Programming Computation*, 3(2):103–163, 2011. DOI:10.1007/s12532-011-0025-9. [107, 110, and 147]

T. Koch, A. Martin, and M. E. Pfetsch. Progress in academic computational integer programming. In M. Jünger and G. Reinelt, editors, *Facets of Combinatorial Optimization*, pages 483–506. 2013. DOI:10.1007/978-3-642-38189-8_19. [1]

H. Konno. A cutting plane algorithm for solving bilinear programs. *Mathematical Programming*, 11:14–27, 1976. DOI:10.1007/BF01580367. [154]

M. Köppe. On the complexity of nonlinear mixed-integer optimization. In J. Lee and S. Leyffer, editors, *Mixed Integer Nonlinear Programming*, volume 154 of *The IMA Volumes in Mathematics and its Applications*, pages 533–557. Springer New York, 2012. DOI:10.1007/978-1-4614-1927-3_19. [103]

M. K. Kozlov, S. P. Tarasov, and L. G. Khachiyan. The polynomial solvability of convex quadratic programming (in Russian). *Zhurnal Vychislitel'noi Matematiki i Matematicheskoi Fiziki*, 20(5):1319–1323, 1980. English translation: *USSR Computational Mathematics and Mathematical Physics*, 20(5):223–228, 1980. DOI:10.1016/0041-5553(80)90098-1. [113]

H. W. Kuhn and A. W. Tucker. Nonlinear programming. In *Proceedings of the Second Berkeley Symposium on Mathematical Statistics and Probability*, pages 481–492. University of California Press, 1951. http://projecteuclid.org/euclid.bsmsp/1200500249. [114]

C. Kwappik. Exact linear programming. Master's thesis, Universität des Saarlandes, May 1998. [59]

L. Ladányi. IBM USA. Personal communication, November 26, 2011. [27]

J. Lee and S. Leyffer, editors. *Mixed Integer Nonlinear Programming*, volume 154 of *The IMA Volumes in Mathematics and its Applications*. Springer New York, 2012. DOI:10.1007/978-1-4614-1927-3. [2]

C. E. Lemke. The dual method of solving the linear programming problem. *Naval Research Logistics Quarterly*, 1(1):36–47, 1954. DOI:10.1002/nav.3800010107. [18]

A. K. Lenstra, H. W. Lenstra, and L. Lovász. Factoring polynomials with rational coefficients. *Mathematische Annalen*, 261(4):515–534, 1982. DOI:10.1007/BF01457454. [61, 62]

J. A. Lerman, D. R. Hyduke, H. Latif, V. A. Portnoy, N. E. Lewis, J. D. Orth, A. C. Schrimpe-Rutledge, R. D. Smith, J. N. Adkins, K. Zengler, and B. Ø. Palsson. In silico method for modelling metabolism and gene product expression at genome scale. *Nature Communications*, 3, 2012. DOI:10.1038/ncomms1928. [93, 94]

N. E. Lewis, H. Nagarajan, and B. Ø. Palsson. Constraining the metabolic genotype–phenotype relationship using a phylogeny of in silico methods. *Nature Reviews Microbiology*, 10:291–305, 2012. DOI:10.1038/nrmicro2737. [92]

S. Leyffer, A. Sartenaer, and E. Wanufelle. Branch-and-refine for mixed-integer nonconvex global optimization. Preprint ANL/MCS-P1547-0908, Argonne National Laboratory, October 2008. http://wiki.mcs.anl.gov/leyffer/images/1/15/SOS-OA-ANL.pdf. [107]

C. M. Li and Anbulagan. Look-ahead versus look-back for satisfiability problems. In *Principles and Practice of Constraint Programming - CP97, Third International Conference, Linz, Austria, October 29 - November 1, 1997, Proceedings*, pages 341–355, 1997. DOI: 10.1007/BFb0017450. [151]

L. Liberti. Introduction to global optimization. Lecture Notes, Ecole Polytechnique, 2013. [1, 110]

L. Liberti and N. Maculan, editors. *Global Optimization. From Theory to Implementation*, volume 84 of *Nonconvex Optimization and Its Applications*. Springer US, 2006. [110]

J. T. Linderoth and M. W. P. Savelsbergh. A computational study of search strategies for mixed integer programming. *INFORMS Journal on Computing*, 11(2):173–187, 1999. DOI: 10.1287/ijoc.11.2.173. [150]

M. Locatelli and F. Schoen. *Global Optimization: Theory, Algorithms, and Applications*. MOS-SIAM Series on Optimization, Philadelphia, PA, USA, 2013. [103, 104, 109, and 152]

M. Lubin, J. A. J. Hall, C. G. Petra, and M. Anitescu. Parallel distributed-memory simplex for large-scale stochastic LP problems. *Computational Optimization and Applications*, 55(3): 571–596, 2013. DOI:10.1007/s10589-013-9542-y. [19]

D. Ma and M. A. Saunders. Solving multiscale linear programs using the simplex method in quadruple precision. In M. Al-Baali, L. Grandinetti, and A. Purnama, editors, *Numerical Analysis and Optimization*, volume 134 of *Springer Proceedings in Mathematics & Statistics*, pages 223–235. Springer International Publishing, 2015. DOI:10.1007/978-3-319-17689-5_9. [94]

C. Maes. Gurobi Optimization, Inc. Personal communication, September 6, 2013. [22]

R. Mahadevan and C. H. Schilling. The effects of alternate optimal solutions in constraint-based genome-scale metabolic models. *Metabolic Engineering*, 5(4):264–276, 2003. DOI: 10.1016/j.ymben.2003.09.002. [98]

A. Mahajan. Solving convex MINLPs with MINOTAUR: Presolving, cuts and more. Talk at MINLP 2014, Carnegie Mellon University, Pittsburgh, PA, USA, June 2014. http://minlp.cheme.cmu.edu/2014/papers/mahajan.pdf. [111]

A. Mahajan, S. Leyffer, and C. Kirches. Solving mixed-integer nonlinear programs by QP-diving. Preprint ANL/MCS-P2071-0312, Argonne National Laboratory, March 2012. http://www.mcs.anl.gov/papers/P2071-0312.pdf. [106]

C. D. Maranas and C. A. Floudas. Global optimization in generalized geometric programming. *Computers & Chemical Engineering*, 21(4):351–369, 1997. DOI:10.1016/S0098-1354(96)00282-7. [125]

H. Marchand and L. A. Wolsey. Aggregation and mixed integer rounding to solve MIPs. *Operations Research*, 49(3):363–371, 2001. DOI:10.1287/opre.49.3.363.11211. [127]

H. M. Markowitz. The elimination form of the inverse and its application to linear programming. *Management Science*, 3(3):255–269, 1957. DOI:10.1287/mnsc.3.3.255. [82]

H. M. Markowitz and A. S. Manne. On the solution of discrete programming problems. *Econometrica*, 25(1):84–110, 1957. http://www.jstor.org/stable/1907744. [107]

Y. V. Matiyasevich. Enumerable sets are diophantine (in Russian). *Doklady Akademii Nauk SSSR*, 191:279–282, 1970. English translation: *Soviet Mathematics Doklady*, 11(2):354–357, 1970. [103]

J. J. Maugis. Étude de réseaux de transport et de distribution de fluide (in French). *RAIRO - Operations Research - Recherche Opérationnelle*, 11(2):243–248, 1977. http://eudml.org/doc/104665. [106]

B. A. McCarl, H. Moskowitz, and H. Furtan. Quadratic programming applications. *Omega*, 5 (1):43–55, 1977. DOI:10.1016/0305-0483(77)90020-2. [113]

G. P. McCormick. Converting general nonlinear programming problems to separable nonlinear programming problems. Technical Report T-267, The George Washington University, December 1972. [107]

G. P. McCormick. Computability of global solutions to factorable nonconvex programs: Part I — Convex underestimating problems. *Mathematical Programming B*, 10(1):147–175, 1976. DOI:10.1007/BF01580665. [102, 107]

G. Michal and D. Schomburg, editors. *Biochemical Pathways: An Atlas of Biochemistry and Molecular Biology*. 2nd edition, 2012. [92]

K. J. Millman and M. Aivazis. Python for scientists and engineers. *Computing in Science & Engineering*, 13(2):9–12, 2011. DOI:10.1109/MCSE.2011.36. [12]

R. Misener. *Novel Global Optimization Methods: Theoretical and Computational Studies on Pooling Problems with Environmental Constraints*. PhD thesis, Princeton University, Dept. of Chemical and Biological Engineering, 2012. http://arks.princeton.edu/ark:/88435/dsp015q47rn787. [103]

R. Misener and C. A. Floudas. Global optimization of mixed-integer quadratically-constrained quadratic programs (MIQCQP) through piecewise-linear and edge-concave relaxations. *Mathematical Programming*, 136(1):155–182, 2012. DOI:10.1007/s10107-012-0555-6. [125, 152, and 153]

R. Misener and C. A. Floudas. GloMIQO: Global mixed-integer quadratic optimizer. *Journal of Global Optimization*, 57:3–50, 2013. DOI:10.1007/s10898-012-9874-7. [110, 132]

R. Misener and C. A. Floudas. ANTIGONE: Algorithms for coNTinuous / Integer Global Optimization of Nonlinear Equations. *Journal of Global Optimization*, 59(2-3):503–526, 2014. DOI:10.1007/s10898-014-0166-2. [110, 125, and 153]

R. E. Moore, R. B. Kearfott, and M. J. Cloud. *Introduction to Interval Analysis*. Society for Industrial and Applied Mathematics, 2009. DOI:10.1137/1.9780898717716. [43, 63, and 109]

M. W. Moskewicz, C. F. Madigan, Y. Zhao, L. Zhang, and S. Malik. Chaff: Engineering an efficient SAT solver. In *Proceedings of the 38th Annual Design Automation Conference*, DAC '01, pages 530–535, 2001. DOI:10.1145/378239.379017. [151]

K. G. Murty and S. N. Kabadi. Some NP-complete problems in quadratic and nonlinear programming. *Mathematical Programming*, 39(2):117–129, 1987. DOI:10.1007/BF02592948. [113]

G. Nannicini, P. Belotti, J. Lee, J. Linderoth, F. Margot, and A. Wächter. A probing algorithm for MINLP with failure prediction by SVM. In T. Achterberg and J. C. Beck, editors, *Integration of AI and OR Techniques in Constraint Programming for Combinatorial Optimization Problems*, volume 6697 of *Lecture Notes in Computer Science*, pages 154–169. Springer Berlin Heidelberg, 2011. DOI:10.1007/978-3-642-21311-3_15. [124]

J. L. Nazareth. *Computer solutions of linear programs.* Monographs on Numerical Analysis. Oxford Univ. Press, New York, Oxford, 1987. [48]

G. L. Nemhauser and L. A. Wolsey. A recursive procedure to generate all cuts for 0–1 mixed integer programs. *Mathematical Programming*, 46(1-3):379–390, 1990. DOI:10.1007/BF01585752. [127]

Y. Nesterov and A. Nemirovskii. *Interior-Point Polynomial Algorithms in Convex Programming.* Society for Industrial and Applied Mathematics, 1994. DOI:10.1137/1.9781611970791. [104, 113]

A. Neumaier. Complete search in continuous global optimization and constraint satisfaction. *Acta Numerica*, 13:271–369, 5 2004. DOI:10.1017/S0962492904000194. [124, 173]

A. Neumaier and O. Shcherbina. Safe bounds in linear and mixed-integer linear programming. *Mathematical Programming*, 99(2):283–296, 2004. DOI:10.1007/s10107-003-0433-3. [40, 63]

P. Q. Nguyen and D. Stehlé. An LLL algorithm with quadratic complexity. *SIAM Journal on Computing*, 39(3):874–903, 2009. DOI:10.1137/070705702. [61]

J. Nocedal and S. J. Wright. *Numerical Optimization.* Springer Series in Operations Research and Financial Engineering. Springer Verlag, 2006. Second edition. [113, 120]

I. Nowak and S. Vigerske. LaGO: a (heuristic) branch and cut algorithm for nonconvex MINLPs. *Central European Journal of Operations Research*, 16(2):127–138, 2008. DOI:10.1007/s10100-007-0051-x. [125]

E. J. O'Brien, J. A. Lerman, R. L. Chang, D. R. Hyduke, and B. Ø. Palsson. Genome-scale models of metabolism and gene expression extend and refine growth phenotype prediction. *Molecular Systems Biology*, 9, 2013. DOI:10.1038/msb.2013.52. [93, 94, and 95]

S. Obua and T. Nipkow. Flyspeck II: The basic linear programs. *Annals of Mathematics and Artificial Intelligence*, 56(3-4):245–272, 2009. DOI:10.1007/s10472-009-9168-z. [15, 64]

T. E. Oliphant. Python for scientific computing. *Computing in Science & Engineering*, 9(3): 10–20, 2007. DOI:10.1109/MCSE.2007.58. [12]

F. Ordóñez and R. M. Freund. Computational experience and the explanatory value of condition measures for linear optimization. *SIAM Journal on Optimization*, 14(2):307–333, 2003. DOI:10.1137/S1052623402401804. [21]

J. D. Orth, I. Thiele, and B. Ø. Palsson. What is flux balance analysis? *Nature Biotechnology*, 28:245–248, 2010. DOI:10.1038/nbt.1614. [92]

V. Y. Pan. Nearly optimal solution of rational linear systems of equations with symbolic lifting and numerical initialization. *Computers & Mathematics with Applications*, 62(4):1685–1706, 2011. DOI:10.1016/j.camwa.2011.06.006. [23, 55]

P. M. Pardalos and G. Schnitger. Checking local optimality in constrained quadratic programming is NP-hard. *Operations Research Letters*, 7(1):33–35, 1988. DOI:10.1016/0167-6377(88)90049-1. [113]

P. M. Pardalos and S. A. Vavasis. Quadratic programming with one negative eigenvalue is NP-hard. *Journal of Global Optimization*, 1(1):15–22, 1991. DOI:10.1007/BF00120662. [113]

J. Patel and J. W. Chinneck. Active-constraint variable ordering for faster feasibility of mixed integer linear programs. *Mathematical Programming*, 110(3):445–474, 2007. DOI:10.1007/s10107-006-0009-0. [151]

R. Pörn and T. Westerlund. A cutting plane method for minimizing pseudo-convex functions in the mixed integer case. *Computers & Chemical Engineering*, 24(12):2655–2665, 2000. DOI:10.1016/S0098-1354(00)00622-0. [106]

S. Powell. A development of the product form algorithm for the simplex method using reduced transformation vectors. In M. L. Balinski and E. Hellerman, editors, *Computational Practice in Mathematical Programming*, volume 4 of *Mathematical Programming Studies*, pages 93–107. North-Holland, Amsterdam, 1975. DOI:10.1007/BFb0120713. [81]

J. W. Pratt. Remarks on zeros and ties in the Wilcoxon signed rank procedures. *Journal of the American Statistical Association*, 54(287):655–667, 1959. http://www.jstor.org/stable/2282543. [12]

R. C. Prim. Shortest connection networks and some generalizations. *Bell System Technical Journal*, 36(6):1389–1401, 1957. DOI:10.1002/j.1538-7305.1957.tb01515.x. [137]

Y. Puranik and N. Sahinidis. A systematic approach for infeasibility diagnosis in NLP and MINLP. Talk at AIChE 2014, Atlanta, GA, USA, November 2014. https://aiche.confex.com/aiche/2014/webprogram/Paper371248.html. [111]

A. Qualizza, P. Belotti, and F. Margot. Linear programming relaxations of quadratically constrained quadratic programs. In J. Lee and S. Leyffer, editors, *Mixed Integer Nonlinear Programming*, volume 154 of *The IMA Volumes in Mathematics and its Applications*, pages 407–426. Springer New York, 2012. DOI:10.1007/978-1-4614-1927-3_14. [111]

I. Quesada and I. E. Grossmann. An LP/NLP based branch and bound algorithm for convex MINLP optimization problems. *Computers & Chemical Engineering*, 16(10-11):937–947, 1992. DOI:10.1016/0098-1354(92)80028-8. [106]

I. Quesada and I. E. Grossmann. Global optimization algorithm for heat exchanger networks. *Industrial & engineering chemistry research*, 32(3):487–499, 1993. DOI:10.1021/ie00015a012. [125]

I. Quesada and I. E. Grossmann. A global optimization algorithm for linear fractional and bilinear programs. *Journal of Global Optimization*, 6:39–76, 1995. DOI:10.1007/BF01106605. [125]

J. Renegar. A polynomial-time algorithm based on Newton's method, for linear programming. *Mathematical Programming*, 40(1–3):59–93, 1988. DOI:10.1007/BF01580724. [63, 72, and 76]

J. Renegar. Some perturbation theory for linear programming. *Mathematical Programming*, 65 (1-3):73–91, 1994. DOI:10.1007/BF01581690. [21, 98]

R. Rovatti, C. D'Ambrosio, A. Lodi, and S. Martello. Optimistic MILP modeling of non-linear optimization problems. *European Journal of Operational Research*, 239(1):32–45, 2014. DOI:10.1016/j.ejor.2014.03.020. [107]

D. M. Ryan and B. A. Foster. An integer programming approach to scheduling. In A. Wren, editor, *Computer Scheduling of Public Transport Urban Passenger Vehicle and Crew Scheduling*, pages 269–280. North Holland, Amsterdam, 1981. [151]

H. S. Ryoo and N. V. Sahinidis. A branch-and-reduce approach to global optimization. *Journal of Global Optimization*, 8(2):107–138, 1996. DOI:10.1007/BF00138689. [124]

B. D. Saunders, D. H. Wood, and B. S. Youse. Numeric-symbolic exact rational linear system solver. In *ISSAC '11. Proceedings of the 36th International Symposium on Symbolic and Algebraic Computation*, pages 305–312. ACM, June 2011. DOI:10.1145/1993886.1993932. [23, 55]

M. A. Saunders and L. Tenenblat. The zoom strategy for accelerating and warm-starting interior methods. Talk at INFORMS Annual Meeting, Pittsburgh, PA, USA, November 2006. `http://www.stanford.edu/group/SOL/talks/saunders-tenenblat-INFORMS2006.pdf`. [27]

M. W. P. Savelsbergh. Preprocessing and probing techniques for mixed integer programming problems. *ORSA Journal on Computing*, 6(4):445–454, 1994. `DOI:10.1287/ijoc.6.4.445`. [123, 124]

A. Schrijver. *Theory of Linear and Integer Programming*. John Wiley & Sons, 1986. [16, 60, and 76]

D. Segrè, D. Vitkup, and G. M. Church. Analysis of optimality in natural and perturbed metabolic networks. *Proceedings of the National Academy of Sciences*, 99(23):15112–15117, 2002. `DOI:10.1073/pnas.232349399`. [101, 114]

D. Shanno. Who invented the interior-point method? In M. Grötschel, editor, *Optimization Stories*, volume Extra Volume: Optimization Stories, pages 55–64. Documenta Mathematica, Bielefeld, 2012. [19]

E. M. B. Smith. *On the Optimal Design of Continuous Processes*. PhD thesis, Imperial College of Science, Technology and Medicine, University of London, 1996. [109]

E. M. B. Smith and C. C. Pantelides. A symbolic reformulation/spatial branch-and-bound algorithm for the global optimisation of nonconvex MINLPs. *Computers & Chemical Engineering*, 23:457–478, 1999. `DOI:10.1016/S0098-1354(98)00286-5`. [109, 125]

R. M. Soland. An algorithm for separable nonconvex programming problems II: Nonconvex constraints. *Management Science*, 17(11):759–773, 1971. `DOI:10.1287/mnsc.17.11.759`. [107]

D. E. Steffy. *Topics in exact precision mathematical programming*. PhD thesis, Georgia Institute of Technology, 2011. `http://hdl.handle.net/1853/39639`. [21, 77, and 80]

D. E. Steffy and K. Wolter. Valid linear programming bounds for exact mixed-integer programming. *INFORMS Journal on Computing*, 25(2):271–284, 2013. `DOI:10.1287/ijoc.1120.0501`. [63]

R. A. Stubbs and S. Mehrotra. A branch-and-cut method for 0-1 mixed convex programming. *Mathematical Programming*, 86(3):515–532, 1999. `DOI:10.1007/s101070050103`. [106]

Y. Sun, R. M. T. Fleming, I. Thiele, and M. A. Saunders. Robust flux balance analysis of multiscale biochemical reaction networks. *BMC Bioinformatics*, 14(1), 2013. `DOI:10.1186/1471-2105-14-240`. [94, 96]

R. Tarjan. Depth-first search and linear graph algorithms. *SIAM Journal on Computing*, 1(2):146–160, 1972. `DOI:10.1137/0201010`. [131]

M. Tawarmalani and N. V. Sahinidis. *Convexification and Global Optimization in Continuous and Mixed-Integer Nonlinear Programming: Theory, Algorithms, Software, and Applications*. Kluwer Academic Publishers, Dordrecht Boston London, 2002. [103, 104]

M. Tawarmalani and N. V. Sahinidis. Global optimization of mixed-integer nonlinear programs: A theoretical and computational study. *Mathematical Programming*, 99:563–591, 2004. `DOI:10.1007/s10107-003-0467-6`. [124, 130, and 152]

M. Tawarmalani and N. V. Sahinidis. A polyhedral branch-and-cut approach to global optimization. *Mathematical Programming*, 103(2):225–249, 2005. `DOI:10.1007/s10107-005-0581-8`. [109, 110]

I. Thiele, R. M. T. Fleming, A. Bordbar, J. Schellenberger, and B. Ø. Palsson. Functional characterization of alternate optimal solutions of Escherichia coli's transcriptional and translational machinery. *Biophysical Journal*, 98:2072–2081, 2010. DOI:10.1016/j.bpj.2010.01.060. [96]

I. Thiele, R. M. T. Fleming, R. Que, A. Bordbar, D. Diep, and B. Ø. Palsson. Multiscale modeling of metabolism and macromolecular synthesis in E. coli and its application to the evolution of codon usage. *PLoS ONE*, page e45635, 2012. DOI:10.1371/journal.pone.0045635. [1, 93]

S. Ursic and C. Patarra. Exact solution of systems of linear equations with iterative methods. *SIAM Journal on Matrix Analysis and Applications*, 4(1):111–115, 1983. DOI:10.1137/0604014. [23]

R. J. Vanderbei. *Linear Programming: Foundations and Extensions*. Springer US, 1996. DOI:10.1007/978-1-4614-7630-6. [16]

S. A. Vavasis. Quadratic programming is in NP. *Information Processing Letters*, 36(2):73–77, 1990. DOI:10.1016/0020-0190(90)90100-C. [113]

S. Vigerske. *Decomposition in multistage stochastic programming and a constraint integer programming approach to mixed-integer nonlinear programming*. PhD thesis, Humboldt-Universität zu Berlin, Mathematisch-Naturwissenschaftliche Fakultät II, 2013. URN:nbn:de:kobv:11-100208240. [103, 104, 111, 123, 125, 132, 153, and 165]

J. Viswanathan and I. E. Grossmann. A combined penalty function and outer-approximation method for MINLP optimization. *Computers & Chemical Engineering*, 14(7):769–782, 1990. DOI:10.1016/0098-1354(90)87085-4. [106]

J. von zur Gathen and J. Gerhard. *Modern Computer Algebra*. Cambridge University Press, Cambridge, UK, 1999. [62]

A. Wächter and L. T. Biegler. On the implementation of an interior-point filter line-search algorithm for large-scale nonlinear programming. *Mathematical Programming*, 106(1):25–57, 2006. DOI:10.1007/s10107-004-0559-y. [139, 159]

Z. Wan. An algorithm to solve integer linear systems exactly using numerical methods. *Journal of Symbolic Computation*, 41(6):621–632, 2006. DOI:10.1016/j.jsc.2005.11.001. [23, 28, 33, 55, 60, and 61]

T. Westerlund and F. Pettersson. An extended cutting plane method for solving convex MINLP problems. *Computers & Chemical Engineering*, 19, Supplement 1:131–136, 1995. DOI:10.1016/0098-1354(95)87027-X. [106]

F. Wilcoxon. Individual comparisons by ranking methods. *Biometrics Bulletin*, 1(6):80–83, 1945. http://www.jstor.org/stable/3001968. [12]

J. H. Wilkinson. *Rounding Errors in Algebraic Processes*. Prentice Hall, Englewood Cliffs, NJ, 1963. [3, 23, 27, 55, and 173]

H. P. Williams. A reduction procedure for linear and integer programming models. In *Redundancy in Mathematical Programming*, volume 206 of *Lecture Notes in Economics and Mathematical Systems*, pages 87–107. Springer Berlin Heidelberg, 1983. DOI:10.1007/978-3-642-45535-3_9. [124]

R. Wunderling. *Paralleler und objektorientierter Simplex-Algorithmus*. PhD thesis, Technische Universität Berlin, 1996. URN:nbn:de:0297-zib-5386. [47, 49]

R. Wunderling. The kernel simplex method. Talk at the 21st International Symposium on Mathematical Programming, Berlin, Germany, August 2012. [81]

Y. Ye. On affine scaling algorithms for nonconvex quadratic programming. *Mathematical Programming*, 56(1-3):285–300, 1992. DOI:10.1007/BF01580903. [104]

Y. Ye and E. Tse. An extension of Karmarkar's projective algorithm for convex quadratic programming. *Mathematical Programming*, 44(1-3):157–179, 1989. DOI:10.1007/BF01587086. [113]

J. M. Zamora and I. E. Grossmann. A branch and contract algorithm for problems with concave univariate, bilinear and linear fractional terms. *Journal of Global Optimization*, 14:217–249, 1999. DOI:10.1023/A:1008312714792. [125, 126, 132, and 133]

A. R. Zomorrodi, P. F. Suthers, S. Ranganathan, and C. D. Maranas. Mathematical optimization applications in metabolic networks. *Metabolic Engineering*, 14(6):672–686, 2012. DOI:10.1016/j.ymben.2012.09.005. [1, 91]

G. Zoutendijk. A product-form algorithm using contracted transformation vectors. In J. Abadie, editor, *Integer and Nonlinear Programming*, pages 511–523. North-Holland, Amsterdam, 1970. [81]

Appendix A

Experimental Data and Results

This appendix offers instance-wise problem statistics and computational results concerning the experiments presented in the thesis.

A.1 Linear Programming

A.1.1 Problem Statistics

Table A.1: Statistics on LP test instances. Column "sparsity" reports the number of columns times the number of rows divided by the number of nonzeros in the constraint matrix. The last two columns refer to the absolute values of the nonzeros in the constraint matrix.

Instance	columns	rows	nonzeros	sparsity	min. abs.	max. abs.
10teams	230	2025	12150	38	1	1
16_n14	16384	262144	524288	8.2e+03	1	1
22433	198	429	3408	28	1	1.1e+03
23588	137	368	3701	14	1	1.1e+03
25fv47	821	1571	10400	1.4e+02	0.0002	2.4e+02
30_70_45_095_100	12526	10976	46640	3.1e+03	1	1
30n20b8	576	18380	109706	1.2e+02	1	2.2e+02
50v-10	233	2013	2745	2.3e+02	1	2.2e+02
80bau3b	2262	9799	21002	1.1e+03	0.00022	1e+02
Test3	50680	72215	617906	6.3e+03	0.0001	1.8e+07
a1c1s1	3312	3648	10178	1.7e+03	1	4.8e+02
aa01	823	8904	72965	1e+02	1	1
aa03	825	8627	70806	1e+02	1	1
aa3	825	8627	70806	1e+02	1	1
aa4	426	7195	52121	61	1	1
aa5	801	8308	65953	1.1e+02	1	1
aa6	646	7292	51728	92	1	1
acc-tight4	3285	1620	17073	3.3e+02	1	1
acc-tight5	3052	1339	16134	2.5e+02	1	1
acc-tight6	3047	1335	16108	2.5e+02	1	1
adlittle	56	97	383	19	0.0012	64
afiro	27	32	83	14	0.11	2.4
aflow30a	479	842	2091	2.4e+02	1	1e+02
aflow40b	1442	2728	6783	7.2e+02	1	1e+02
agg	488	163	2410	35	2e-05	4.2e+02
agg2	516	302	4284	37	2e-05	4.2e+02
agg3	516	302	4300	37	2e-05	4.2e+02
air02	50	6774	61555	5.6	1	1
air03	124	10757	91028	16	1	1
air04	823	8904	72965	1e+02	1	1
air05	426	7195	52121	61	1	1
air06	825	8627	70806	1e+02	1	1

Table A.1 continued

Instance	columns	rows	nonzeros	sparsity	min. abs.	max. abs.
aircraft	3754	7517	20267	1.9e+03	1	81
aligninq	340	1831	15734	42	1	3.0e+02
app1-2	53467	26871	199175	7.6e+03	1e-05	1
arki001	1048	1388	20439	75	0.0002	2.3e+07
ash608gpia-3col	24748	3651	74244	1.2e+03	1	1
atlanta-ip	21732	48738	257532	4.3e+03	0.028	64
atm20-100	4380	6480	58878	4.9e+02	0.1	1.3e+04
b2c1s1	3904	3872	11408	2e+03	0.2	4.8e+02
bab1	60680	61152	854392	4.7e+03	0.05	4
bab3	23069	393800	3301838	2.9e+03	0.09	8
bab5	4964	21600	155520	7.1e+02	0.09	8
bal8x12	116	192	384	58	1	35
bandm	305	472	2494	61	0.001	2e+02
bas1lp	5411	4461	582411	42	1	14
baxter	27441	15128	95971	4.6e+03	0.001	3.2e+05
bc	1913	1751	276842	12	1.1e-13	10
bc1	1913	1751	276842	12	1.1e-13	10
beaconfd	173	262	3375	14	0.0012	5e+02
beasleyC3	1750	2500	5000	8.7e+02	1	82
bell3a	123	133	347	61	8.3e-05	1.3e+03
bell5	91	104	266	46	8.3e-05	1.3e+03
berlin_5_8_0	1532	1083	4507	3.8e+02	1	2.4e+02
bg512142	1307	792	3953	3.3e+02	1	5.6e+03
biella1	1203	7328	71489	1.3e+02	1	1e+08
bienst1	576	505	2184	1.4e+02	1	81
bienst2	576	505	2184	1.4e+02	1	81
binkar10_1	1026	2298	4496	1e+03	1	29
bk4x3	19	24	48	9.5	1	40
blend	74	83	491	15	0.003	66
blend2	274	353	1409	91	1	7.2e+03
blp-ar98	1128	16021	200601	94	1	1e+03
blp-ic97	923	9845	118149	77	1	9.6e+02
bnatt350	4923	3150	19061	8.2e+02	0.12	1
bnatt400	5614	3600	21698	9.4e+02	0.12	1
bnl1	643	1175	5121	1.6e+02	0.0011	78
bnl2	2324	3489	13999	5.8e+02	0.0006	78
boeing1	351	384	3485	39	0.011	3.1e+03
boeing2	166	143	1196	21	0.01	3e+03
bore3d	233	315	1429	58	0.0001	1.4e+03
brandy	220	249	2148	28	0.0008	2e+02
buildingenergy	277594	154978	788969	5.6e+04	0.05	64
cap6000	2176	6000	48243	2.7e+02	1	9.9e+04
capri	271	353	1767	54	9e-05	2.2e+02
car4	16384	33052	63724	1.6e+04	0.00098	1
cari	400	1200	152800	3.1	7.8e-05	1
cep1	1521	3248	6712	7.6e+02	0.01	1
ch	3700	5062	20873	9.2e+02	4.3e-05	2.6e+02
circ10-3	42620	2700	307320	3.8e+02	1	16
co-100	2187	48417	1995817	53	1	2e+04
co5	5774	7993	53661	9.6e+02	1e-05	2.7e+03
co9	10789	14851	101578	1.8e+03	1e-05	2.7e+03
complex	1023	1408	46463	32	1	1
cont1	160792	40398	399990	1.8e+04	0.005	4
cont4	160792	40398	398398	1.8e+04	0.005	4
core2536-691	2539	15293	177739	2.3e+02	0.1	1e+02
core4872-1529	4875	24656	218762	6.1e+02	0.1	1e+02
cov1075	637	120	14280	5.4	1	1
cq5	5048	7530	47353	8.4e+02	1.6e-05	1e+03
cq9	9278	13778	88897	1.5e+03	1.6e-05	1e+03
cr42	905	1513	6614	2.3e+02	0.012	40
cre-a	3516	4067	14987	1.2e+03	0.6	71
cre-b	9648	72447	256095	3.2e+03	0.6	71
cre-c	3068	3678	13244	1e+03	0.5	71
cre-d	8926	69980	242604	3e+03	0.5	71
crew1	135	6469	46950	19	1	1
csched007	351	1758	6379	1.2e+02	1	1.9e+02
csched008	351	1536	5687	1.2e+02	1	1.8e+02
csched010	351	1758	6376	1.2e+02	1	1.8e+02
cycle	1903	2857	20720	2.7e+02	1e-05	9.1e+02
czprob	929	3523	10669	3.1e+02	0.0016	1.4e+02
d10200	947	2000	57637	34	1	1.4e+02
d20200	1502	4000	189389	32	1	2.8e+02
d2q06c	2171	5167	32417	3.6e+02	0.0002	2.3e+03
d6cube	415	6184	37704	69	1	3.6e+02

Table A.1 continued

Instance	columns	rows	nonzeros	sparsity	min. abs.	max. abs.
dano3_3	3202	13873	79655	6.4e+02	0.5	1e+03
dano3_4	3202	13873	79655	6.4e+02	0.5	1e+03
dano3_5	3202	13873	79655	6.4e+02	0.5	1e+03
dano3mip	3202	13873	79655	6.4e+02	0.5	1e+03
danoint	664	521	3232	1.1e+02	0.5	66
dbir1	18804	27355	1058605	4.9e+02	1	1.5e+05
dbir2	18906	27355	1139637	4.6e+02	1	1.1e+05
dc1c	1649	10039	121158	1.4e+02	1	1e+07
dc1l	1653	37297	448754	1.4e+02	1	1e+02
dcmulti	290	548	1315	1.4e+02	1	6e+02
de063155	852	1488	4553	2.8e+02	2.1e-07	8.4e+11
de063157	936	1488	4699	3.1e+02	1.3e-09	2.3e+18
de080285	936	1488	4662	3.1e+02	1.6e-17	9.7e+02
degen2	444	534	3978	63	1	1
degen3	1503	1818	24646	1.2e+02	1	1
delf000	3128	5464	12606	1.6e+03	1e-06	1.9e+03
delf001	3098	5462	13214	1.5e+03	1e-06	1.9e+03
delf002	3135	5460	13287	1.6e+03	1e-06	1.9e+03
delf003	3065	5460	13269	1.5e+03	1e-06	1.9e+03
delf004	3142	5464	13546	1.6e+03	1e-06	1.9e+03
delf005	3103	5464	13494	1.6e+03	1e-06	1.9e+03
delf006	3147	5469	13604	1.6e+03	1e-06	1.9e+03
delf007	3137	5471	13758	1.6e+03	1e-06	1.9e+03
delf008	3148	5472	13821	1.6e+03	1e-06	1.9e+03
delf009	3135	5472	13750	1.6e+03	1e-06	1.9e+03
delf010	3147	5472	13802	1.6e+03	1e-06	1.9e+03
delf011	3134	5471	13777	1.6e+03	1e-06	1.9e+03
delf012	3151	5471	13793	1.6e+03	1e-06	1.9e+03
delf013	3116	5472	13809	1.6e+03	1e-06	1.9e+03
delf014	3170	5472	13866	1.6e+03	1e-06	1.9e+03
delf015	3161	5471	13793	1.6e+03	1e-06	1.9e+03
delf017	3176	5471	13732	1.6e+03	1e-06	1.9e+03
delf018	3196	5471	13774	1.6e+03	1e-06	1.9e+03
delf019	3185	5471	13762	1.6e+03	1e-06	1.9e+03
delf020	3213	5472	14070	1.6e+03	1e-06	1.9e+03
delf021	3208	5471	14068	1.6e+03	1e-06	1.9e+03
delf022	3214	5472	14060	1.6e+03	1e-06	1.9e+03
delf023	3214	5472	14098	1.6e+03	1e-06	1.9e+03
delf024	3207	5466	14456	1.6e+03	1e-06	1.9e+03
delf025	3197	5464	14447	1.6e+03	1e-06	1.9e+03
delf026	3190	5462	14220	1.6e+03	1e-06	1.9e+03
delf027	3187	5457	14200	1.6e+03	1e-06	1.9e+03
delf028	3177	5452	14402	1.6e+03	1e-06	1.9e+03
delf029	3179	5454	14402	1.6e+03	1e-06	1.9e+03
delf030	3199	5469	14262	1.6e+03	1e-06	1.9e+03
delf031	3176	5455	14205	1.6e+03	1e-06	1.9e+03
delf032	3196	5467	14251	1.6e+03	1e-06	1.9e+03
delf033	3173	5456	14205	1.6e+03	1e-06	1.9e+03
delf034	3175	5455	14208	1.6e+03	1e-06	1.9e+03
delf035	3193	5468	14284	1.6e+03	1e-06	1.9e+03
delf036	3170	5459	14202	1.6e+03	1e-06	1.9e+03
deter0	1923	5468	11173	9.6e+02	0.72	1.4
deter1	5527	15737	32187	2.8e+03	0.55	1.5
deter2	6095	17313	35731	3e+03	0.62	1.5
deter3	7647	21777	44854	3.8e+03	0.55	1.5
deter4	3235	9133	19231	1.6e+03	0.61	1.5
deter5	5103	14529	29715	2.6e+03	0.41	1.4
deter6	4255	12113	24771	2.1e+03	0.6	1.5
deter7	6375	18153	37131	3.2e+03	0.58	1.4
deter8	3831	10905	22299	1.9e+03	0.67	1.4
df2177	630	9728	21706	3.2e+02	1	1
dfl001	6071	12230	35632	3e+03	0.083	2
dfn-gwin-UUM	158	938	2632	79	1	6.2e+02
dg012142	6310	2080	14795	9e+02	1	7.2e+03
disctom	399	10000	30000	1.3e+02	1	1
disp3	2182	1856	6407	7.3e+02	0.016	4.5
dolom1	1803	11612	190413	1.1e+02	1	1e+08
ds	656	67732	1024059	44	1	1
ds-big	1042	174997	4623442	40	1	1
dsbmip	1182	1886	7366	3.9e+02	0.062	3.6e+04
e18	24617	14231	132095	2.7e+03	1	2e+03
e226	223	282	2578	25	0.00026	1.5e+03
egout	98	141	282	49	1	1.2e+02
eil33-2	32	4516	44243	3.6	1	1

Table A.1 continued

Instance	columns	rows	nonzeros	sparsity	min. abs.	max. abs.
eilA101-2	100	65832	959373	7.1	1	1
eilB101	100	2818	24120	12	1	1
enigma	21	100	289	10	1	9e+05
enlight13	169	338	962	84	1	2
enlight14	196	392	1120	98	1	2
enlight15	225	450	1290	1.1e+02	1	2
enlight16	256	512	1472	1.3e+02	1	2
enlight9	81	162	450	40	1	2
etamacro	400	688	2409	1.3e+02	0.019	2e+03
ex10	69608	17680	1162000	1.1e+03	1	1
ex1010-pi	1468	25200	102114	3.7e+02	1	1
ex3sta1	17443	8156	59419	2.5e+02	0.29	1e+02
ex9	40962	10404	517112	8.4e+02	1	1
f2000	10500	4000	29500	1.5e+03	1	1
farm	7	12	36	2.3	1	2.5e+02
fast0507	507	63009	409349	84	1	1
fffff800	524	854	6227	75	0.008	1.1e+05
fiball	3707	34219	104792	1.2e+03	0.6	94
fiber	363	1298	2944	1.8e+02	1	2.2e+02
finnis	497	614	2310	1.7e+02	0.00046	32
fit1d	24	1026	13404	1.8	0.01	1.9e+03
fit1p	627	1677	9868	1.3e+02	0.01	1.9e+03
fit2d	25	10500	129018	2.1	0.05	2.6e+03
fit2p	3000	13525	50284	1e+03	0.05	2.6e+03
fixnet6	478	878	1756	2.4e+02	1	5e+02
flugpl	18	18	46	9	0.9	1.5e+02
fome11	12142	24460	71264	6.1e+03	0.083	2
fome12	24284	48920	142528	1.2e+04	0.083	2
fome13	48568	97840	285056	2.4e+04	0.083	2
fome20	33874	105728	230200	1.7e+04	1	1
fome21	67748	211456	460400	3.4e+04	1	1
forplan	161	421	4563	16	0.0074	2.8e+03
fxm2-16	3900	5602	31239	7.8e+02	0.0005	1.3e+02
fxm2-6	1520	2172	12139	3e+02	0.0005	1.3e+02
fxm3_16	41340	64162	370839	8.3e+03	0.0005	1.3e+02
fxm3_6	6200	9492	54589	1.2e+03	0.0005	1.3e+02
fxm4_6	22400	30732	248989	2.8e+03	0.0005	1.3e+02
g200x740i	940	1480	2960	4.7e+02	1	2e+02
gams10a	114	61	297	28	0.38	1
gams30a	354	181	937	71	0.14	1
ganges	1309	1681	6912	3.3e+02	0.0014	1
ge	10099	11098	39554	3.4e+03	4.5e-06	1.4e+04
gen	780	870	2592	3.9e+02	1	6.7e+02
gen1	769	2560	63085	32	2e-07	1
gen2	1121	3264	81855	45	2e-09	1
gen4	1537	4297	107102	64	5.1e-08	1
ger50_17_trans	499	22414	172035	71	1	3.2e+04
germanrr	10779	10813	175547	6.7e+02	1	8.2e+05
germany50-DBM	2526	8189	24479	1.3e+03	1	40
gesa2	1392	1224	5064	3.5e+02	0.069	1.2e+02
gesa2-o	1248	1224	3672	4.2e+02	0.069	1.2e+02
gesa2_o	1248	1224	3672	4.2e+02	0.069	1.2e+02
gesa3	1368	1152	4944	3.4e+02	0.069	1.2e+02
gesa3_o	1224	1152	3624	4.1e+02	0.069	1.2e+02
gfrd-pnc	616	1092	2377	3.1e+02	1	1.1e+02
glass4	396	322	1815	79	1	8.4e+06
gmu-35-40	424	1205	4843	1.1e+02	0.8	2.6e+03
gmu-35-50	435	1919	8643	1.1e+02	0.8	2.6e+03
gmut-75-50	2565	68865	571475	3.2e+02	0.95	7.7e+03
gmut-77-40	2554	24338	159902	4.3e+02	0.95	6.1e+03
go19	441	441	1885	1.1e+02	1	1
gr4x6	34	48	96	17	1	35
greenbea	2392	5405	30877	4.8e+02	6e-05	1e+02
greenbeb	2392	5405	30877	4.8e+02	6e-05	1e+02
grow15	300	645	5620	38	6e-06	1
grow22	440	946	8252	55	6e-06	1
grow7	140	301	2612	18	6e-06	1
gt2	29	188	376	14	1	2.5e+03
hanoi5	16399	3862	39718	1.6e+03	1	1
haprp	1048	1828	3628	1e+03	1	1.8e+04
harp2	112	2993	5840	1.1e+02	1	4.2e+09
i_n13	8192	741455	1482910	4.1e+03	1	1
ic97_potential	1046	728	3138	2.6e+02	1	60
iiasa	669	2970	6648	3.3e+02	0.51	8.8e+03

Table A.1 continued

Instance	columns	rows	nonzeros	sparsity	min. abs.	max. abs.
iis-100-0-cov	3831	100	22986	17	1	1
iis-bupa-cov	4803	345	38392	43	1	1
iis-pima-cov	7201	768	71941	77	1	1
israel	174	142	2269	12	0.001	1.6e+03
ivu06-big	1177	2277736	23125770	1.2e+02	1	1
ivu52	2116	157591	2179476	1.6e+02	0.0027	9
janos-us-DDM	760	2184	6384	3.8e+02	1	64
jendrec1	2109	4228	89608	1e+02	0.00018	1.5e+02
k16x240	256	480	960	1.3e+02	1	1e+03
kb2	43	41	286	7.2	0.17	1.1e+02
ken-07	2426	3602	8404	1.2e+03	1	1
ken-11	14694	21349	49058	7.3e+03	1	1
ken-13	28632	42659	97246	1.4e+04	1	1
ken-18	105127	154699	358171	5.3e+04	1	1
kent	31300	16620	184710	2.8e+03	0.03	1.4e+03
khb05250	101	1350	2700	50	1	5e+03
kl02	71	36699	212536	14	1	1
kleemin3	3	3	6	1.5	1	2e+02
kleemin4	4	4	10	2	1	2e+03
kleemin5	5	5	15	1.7	1	2e+04
kleemin6	6	6	21	2	1	2e+05
kleemin7	7	7	28	1.8	1	2e+06
kleemin8	8	8	36	2	1	2e+07
l152lav	97	1989	9922	24	1	43
l30	2701	15380	51169	9e+02	0.017	1.8
l9	244	1401	4577	81	0.056	1.8
large000	4239	6833	16573	2.1e+03	1e-06	1.9e+03
large001	4162	6834	17225	2.1e+03	1e-06	1.9e+03
large002	4249	6835	18330	2.1e+03	1e-06	1.9e+03
large003	4200	6835	18016	2.1e+03	1e-06	1.9e+03
large004	4250	6836	17739	2.1e+03	1e-06	1.9e+03
large005	4237	6837	17575	2.1e+03	1e-06	1.9e+03
large006	4249	6837	17887	2.1e+03	1e-06	1.9e+03
large007	4236	6836	17856	2.1e+03	1e-06	1.9e+03
large008	4248	6837	17898	2.1e+03	1e-06	1.9e+03
large009	4237	6837	17878	2.1e+03	1e-06	1.9e+03
large010	4247	6837	17887	2.1e+03	1e-06	1.9e+03
large011	4236	6837	17878	2.1e+03	1e-06	1.9e+03
large012	4253	6838	17919	2.1e+03	1e-06	1.9e+03
large013	4248	6838	17941	2.1e+03	1e-06	1.9e+03
large014	4271	6838	17979	2.1e+03	1e-06	1.9e+03
large015	4265	6838	17957	2.1e+03	1e-06	1.9e+03
large016	4287	6838	18029	2.1e+03	1e-06	1.9e+03
large017	4277	6837	17983	2.1e+03	1e-06	1.9e+03
large018	4297	6837	17791	2.1e+03	1e-06	1.9e+03
large019	4300	6836	17786	2.2e+03	1e-06	1.9e+03
large020	4315	6837	18136	2.2e+03	1e-06	1.9e+03
large021	4311	6838	18157	2.2e+03	1e-06	1.9e+03
large022	4312	6834	18104	2.2e+03	1e-06	1.9e+03
large023	4302	6835	18123	2.2e+03	1e-06	1.9e+03
large024	4292	6831	18599	2.1e+03	1e-06	1.9e+03
large025	4297	6832	18743	2.1e+03	1e-06	1.9e+03
large026	4284	6824	18631	2.1e+03	1e-06	1.9e+03
large027	4275	6821	18562	2.1e+03	1e-06	1.9e+03
large028	4302	6833	18886	2.2e+03	1e-06	1.9e+03
large029	4301	6832	18952	2.2e+03	1e-06	1.9e+03
large030	4285	6823	18843	2.1e+03	1e-06	1.9e+03
large031	4294	6826	18867	2.1e+03	1e-06	1.9e+03
large032	4292	6827	18850	2.1e+03	1e-06	1.9e+03
large033	4273	6817	18791	2.1e+03	1e-06	1.9e+03
large034	4294	6831	18855	2.1e+03	1e-06	1.9e+03
large035	4293	6829	18881	2.1e+03	1e-06	1.9e+03
large036	4282	6822	18840	2.1e+03	1e-06	1.9e+03
lectsched-1	50108	28718	310792	5e+03	1	1.3e+03
lectsched-1-obj	50108	28718	310792	5e+03	1	1.3e+03
lectsched-2	30738	17656	186520	3.1e+03	1	1.3e+03
lectsched-3	45262	25776	279967	4.5e+03	1	1.3e+03
lectsched-4-obj	14163	7901	82428	1.4e+03	1	1.3e+03
leo1	593	6731	131218	31	1	9e+07
leo2	593	11100	219959	31	1	1.7e+08
liu	2178	1156	10626	2.4e+02	1	8.4e+03
lo10	46341	406225	812450	2.3e+04	1	1
long15	32769	753687	1507374	1.6e+04	1	1
lotfi	153	308	1078	51	0.019	1e+03

Table A.1 continued

Instance	columns	rows	nonzeros	sparsity	min. abs.	max. abs.
lotsize	1920	2985	6565	9.6e+02	1	2e+04
lp22	2958	13434	65560	7.4e+02	1	1
lpl1	39951	125000	381259	1.3e+04	1	5.4e+02
lpl2	3294	10755	32106	1.6e+03	1	5e+02
lpl3	10828	33538	100377	5.4e+03	1	5e+02
lrn	8491	7253	46123	1.4e+03	0.00098	6.2e+07
lrsa120	14521	3839	39956	1.5e+03	1	2
lseu	28	89	309	9.3	1	5.2e+02
m100n500k4r1	100	500	2000	25	1	1
macrophage	3164	2260	9492	7.9e+02	1	1
manna81	6480	3321	12960	2.2e+03	1	1
map06	328818	164547	549920	1.1e+05	1	1.1e+07
map10	328818	164547	549920	1.1e+05	1	1.1e+07
map14	328818	164547	549920	1.1e+05	1	1.1e+07
map18	328818	164547	549920	1.1e+05	1	1.1e+07
map20	328818	164547	549920	1.1e+05	1	1.1e+07
markshare1	6	62	312	1.2	1	99
markshare2	7	74	434	1.4	1	99
markshare_5_0	5	45	203	1.2	1	1e+02
maros	846	1443	9614	1.4e+02	0.0001	1.7e+04
maros-r7	3136	9408	144848	2.1e+02	0.0017	1
mas74	13	151	1706	1.2	1	9.9e+03
mas76	12	151	1640	1.2	1	9.9e+03
maxgasflow	7160	7437	19717	3.6e+03	1	1e+04
mc11	1920	3040	6080	9.6e+02	1	2.1e+02
mcf2	664	521	3232	1.1e+02	0.5	66
mcsched	2107	1747	8088	5.3e+02	1	1e+03
methanosarcina	14604	7930	43812	2.9e+03	1	1
mik-250-1-100-1	151	251	5351	7.2	1	2e+03
mine-166-5	8429	830	19412	3.7e+02	1	2.5e+04
mine-90-10	6270	900	15407	3.7e+02	1	4.7e+04
misc03	96	160	2053	8	1	9.6e+02
misc06	820	1808	5859	2.7e+02	0.12	2.6e+02
misc07	212	260	8619	6.4	1	7e+02
mitre	2054	10724	39704	6.8e+02	1	1.1e+03
mkc	3411	5325	17038	1.1e+03	1	38
mkc1	3411	5325	17038	1.1e+03	1	38
mod008	6	319	1243	2	0.5	7.6
mod010	146	2655	11203	36	1	20
mod011	4480	10958	22254	2.2e+03	0.14	4.2e+04
mod2	34774	31728	165129	7e+03	0.0019	5.6e+03
model1	362	798	3028	1.2e+02	0.001	3
model10	4400	15447	149000	4.9e+02	2.2e-05	1.9e+04
model11	7056	18288	55859	2.4e+03	2e-05	2e+02
model2	379	1212	7498	63	0.0001	4e+02
model3	1609	3840	23236	2.7e+02	1e-05	6.9e+03
model4	1337	4549	45340	1.5e+02	2.2e-05	3.1e+03
model5	1888	11360	89483	2.7e+02	1e-05	1.7e+02
model6	2096	5001	27340	4.2e+02	8e-05	1.1e+03
model7	3358	8007	49452	5.6e+02	1e-05	7.3e+03
model8	2896	6464	25277	9.7e+02	0.001	3
model9	2879	10257	55274	5.8e+02	0.0001	1e+03
modglob	291	422	968	1.5e+02	0.9	1.7e+04
modszk1	687	1620	3168	6.9e+02	0.00074	1.2
momentum1	42680	5174	103198	2.2e+03	2.5e-10	10
momentum2	24237	3732	349695	2.6e+02	3.3e-21	20
momentum3	56822	13532	949495	8.1e+02	2.1e-19	20
msc98-ip	15850	21143	92918	4e+03	0.0078	2.3e+06
mspp16	561657	29280	27678735	5.9e+02	1	1.5e+03
multi	61	102	961	6.8	0.0001	10
mzzv11	9499	10240	134603	7.3e+02	1	2e+03
mzzv42z	10460	11717	151261	8.7e+02	1	1.1e+03
n15-3	29494	153140	611000	9.8e+03	1	2.6e+02
n3-3	2425	9028	35380	8.1e+02	1	2.6e+02
n3700	5150	10000	20000	2.6e+03	1	3.8e+03
n3701	5150	10000	20000	2.6e+03	1	2.1e+03
n3702	5150	10000	20000	2.6e+03	1	2.5e+03
n3703	5150	10000	20000	2.6e+03	1	2.2e+03
n3704	5150	10000	20000	2.6e+03	1	2.7e+03
n3705	5150	10000	20000	2.6e+03	1	3.1e+03
n3706	5150	10000	20000	2.6e+03	1	2.1e+03
n3707	5150	10000	20000	2.6e+03	1	2.3e+03
n3708	5150	10000	20000	2.6e+03	1	2.9e+03
n3709	5150	10000	20000	2.6e+03	1	2.5e+03

Table A.1 continued

Instance	columns	rows	nonzeros	sparsity	min. abs.	max. abs.
n370a	5150	10000	20000	2.6e+03	1	2.2e+03
n370b	5150	10000	20000	2.6e+03	1	2.4e+03
n370c	5150	10000	20000	2.6e+03	1	2.3e+03
n370d	5150	10000	20000	2.6e+03	1	2.3e+03
n370e	5150	10000	20000	2.6e+03	1	2.5e+03
n3div36	4484	22120	340740	3e+02	1	24
n3seq24	6044	119856	3232340	2.3e+02	1	24
n4-3	1236	3596	14036	4.1e+02	1	2.6e+02
n9-3	2364	7644	30072	7.9e+02	1	2.6e+02
nag	5840	2884	26499	6.5e+02	1	1e+04
nemsafm	334	2252	2730	3.3e+02	0.99	1
nemscem	651	1570	3698	3.3e+02	0.18	1
nemsemm1	3945	71413	1050047	2.8e+02	1e-05	4.5e+03
nemsemm2	6943	42133	175267	1.7e+03	1e-07	5e+03
nemspmm1	2372	8622	55586	4e+02	0.001	8.8e+03
nemspmm2	2301	8413	67904	2.9e+02	0.00097	8.8e+03
nemswrld	7138	27174	190907	1e+03	0.00055	1.6e+02
neos	479119	36786	1047675	1.7e+04	0.5	1
neos-1053234	2596	5621	14920	1.3e+03	0.0099	9.8e+05
neos-1053591	1263	1386	3543	6.3e+02	1e-05	1e+05
neos-1056905	900	463	3510	1.3e+02	1	1e+02
neos-1058477	1529	2805	9376	5.1e+02	0.0099	1.2e+06
neos-1061020	10618	14010	114508	1.3e+03	1	1
neos-1062641	1677	1748	4544	8.4e+02	0.0028	1e+05
neos-1067731	3423	8779	30998	1.1e+03	1	1
neos-1096528	550339	1520	2171928	3.9e+02	1	2
neos-1109824	28979	1520	89528	5e+02	1	1
neos-1112782	2115	4140	8145	2.1e+03	1	2.9e+07
neos-1112787	1680	3280	6440	1.7e+03	1	2.5e+07
neos-1120495	21739	1140	67146	3.7e+02	1	1
neos-1121679	6	62	312	1.2	1	99
neos-1122047	57791	5100	163640	1.8e+03	1	1e+04
neos-1126860	36709	2565	105219	9e+02	1	1e+04
neos-1140050	3795	40320	808080	1.9e+02	5.6e-07	2.9e+04
neos-1151496	982	1549	27817	58	1	1
neos-1171448	13206	4914	131859	5.1e+02	1	1.2e+03
neos-1171692	4239	1638	42945	1.6e+02	1	3.2e+03
neos-1171737	4179	2340	58620	1.7e+02	1	1.6e+03
neos-1173026	893	1314	6933	1.8e+02	0.0099	2.7e+06
neos-1200887	633	234	6084	24	1	6.4e+03
neos-1208069	1150	2322	27242	1e+02	0.2	1
neos-1208135	1040	2322	24034	1e+02	0.25	1
neos-1211578	356	260	1540	71	1	8.2e+03
neos-1215259	1236	1601	38435	52	0.2	1
neos-1215891	6068	5035	44590	7.6e+02	0.25	7
neos-1223462	5890	5495	47040	7.4e+02	0.33	7
neos-1224597	3276	3395	25090	4.7e+02	0.5	7
neos-1225589	675	1300	2525	6.8e+02	1	1.8e+06
neos-1228986	356	260	1540	71	1	1.2e+04
neos-1281048	522	739	8808	47	1	2
neos-1311124	1643	1092	7140	2.7e+02	1	4.1e+03
neos-1324574	5904	5256	20880	2e+03	1	1
neos-1330346	4248	2664	13032	1.1e+03	1	1
neos-1330635	2717	1736	8260	6.8e+02	1	1e+06
neos-1337307	5687	2840	30799	5.7e+02	1	1.6e+04
neos-1337489	356	260	1540	71	1	8.2e+03
neos-1346382	796	520	3400	1.3e+02	1	8.2e+03
neos-1354092	3135	13702	187187	2.4e+02	1	1
neos-1367061	102750	36600	260250	1.5e+04	1	8e+03
neos-1396125	1494	1161	5511	3.7e+02	1	2.5e+02
neos-1407044	6908	16604	206653	5.8e+02	1	1
neos-1413153	2500	2451	9653	8.3e+02	1	4.2e+02
neos-1415183	2809	2757	10868	9.4e+02	1	5e+02
neos-1417043	3284	573315	1146630	1.6e+03	1	1
neos-1420205	383	231	1050	96	1	1e+06
neos-1420546	12671	26055	67959	6.3e+03	0.00072	7.9
neos-1420790	2310	4926	12720	1.2e+03	0.00067	8
neos-1423785	25721	21506	64082	1.3e+04	1	4.1
neos-1425699	89	105	430	22	1	2e+05
neos-1426635	796	520	3400	1.3e+02	1	8.2e+03
neos-1426662	1914	832	8048	2.1e+02	1	1e+03
neos-1427181	1786	832	7792	2e+02	1	2e+03
neos-1427261	2226	1040	9740	2.5e+02	1	2e+03
neos-1429185	1346	624	5844	1.5e+02	1	2e+03

Table A.1 continued

Instance	columns	rows	nonzeros	sparsity	min. abs.	max. abs.
neos-1429212	58726	416040	1855220	1.5e+04	0.067	27
neos-1429461	1096	520	4780	1.2e+02	1	3.1e+03
neos-1430701	668	312	2868	74	1	4.1e+03
neos-1430811	73661	519704	2474280	1.8e+04	0.01	27
neos-1436709	1417	676	6214	1.6e+02	1	3.1e+03
neos-1436713	2666	1248	11688	3e+02	1	2e+03
neos-1437164	187	2256	9016	62	1	1e+04
neos-1439395	775	364	3346	86	1	8.2e+03
neos-1440225	330	1285	14168	30	1	1
neos-1440447	561	260	2390	62	1	6.1e+03
neos-1440457	1952	936	8604	2.2e+02	1	3.1e+03
neos-1440460	989	468	4302	1.1e+02	1	6.1e+03
neos-1441553	316	960	11138	29	1	1e+04
neos-1442119	1524	728	6692	1.7e+02	1	4.1e+03
neos-1442657	1310	624	5736	1.5e+02	1	4.1e+03
neos-1445532	1924	14406	27736	1.9e+03	1	1.3e+02
neos-1445738	2145	20631	40256	2.1e+03	1	44
neos-1445743	2148	20344	39685	2.1e+03	1	43
neos-1445755	2139	20516	40020	2.1e+03	1	47
neos-1445765	2147	20617	40230	2.1e+03	1	48
neos-1451294	1238	1626	21036	1e+02	1	1
neos-1456979	6770	4605	36440	9.7e+02	1	5e+02
neos-1460246	306	285	2303	38	1	10
neos-1460265	1656	1728	11902	2.8e+02	1	10
neos-1460543	2012	1700	15121	2.5e+02	1	10
neos-1460641	1532	1641	11697	2.2e+02	1	10
neos-1461051	4370	528	14220	1.7e+02	1	2
neos-1464762	1632	1721	12313	2.3e+02	1	9.2
neos-1467067	1084	1196	4692	3.6e+02	1	3
neos-1467371	1628	1693	12084	2.3e+02	1	9.2
neos-1467467	1644	1693	12116	2.3e+02	1	9.2
neos-1480121	363	222	1060	91	1	1e+03
neos-1489999	1046	534	2186	2.6e+02	0.015	1
neos-1516309	489	4500	30400	82	1	5
neos-1582420	10180	10100	24814	5.1e+03	0.1	3
neos-1593097	798	18460	113308	1.3e+02	1	7.3e+02
neos-1595230	1750	490	3885	2.5e+02	1	1
neos-1597104	109833	714	331373	2.4e+02	1	1
neos-1599274	1237	4500	46800	1.2e+02	1	5
neos-1601936	3131	4446	72500	2e+02	1	1
neos-1603512	555	730	13541	31	1	3
neos-1603518	880	1272	25716	44	1	3
neos-1603965	28984	15003	86947	5.8e+03	0.05	1e+10
neos-1605061	3474	4111	93483	1.6e+02	1	1e+04
neos-1605075	3467	4173	91377	1.7e+02	1	1
neos-1616732	1999	200	3998	1.1e+02	1	1
neos-1620770	9296	792	19292	3.9e+02	1	1
neos-1620807	1340	231	2860	1.1e+02	1	1
neos-1622252	9695	828	20125	4e+02	1	1
neos-430149	990	395	2895	1.4e+02	0.004	2.5e+02
neos-476283	10015	11915	3945693	30	0.0002	1e+04
neos-480878	1321	534	44370	16	0.0002	1e+04
neos-494568	2215	6889	115463	1.4e+02	1	1e+02
neos-495307	3	9423	27831	1.5	1	1.8e+03
neos-498623	2047	9861	148434	1.4e+02	1	1e+02
neos-501453	40	165	535	13	0.62	40
neos-501474	265	206	2228	26	0.62	40
neos-503737	500	2850	16850	1e+02	1	1
neos-504674	1344	844	3450	3.4e+02	0.1	78
neos-504815	1067	674	2736	2.7e+02	0.1	63
neos-506422	6811	2527	31815	5.7e+02	1	40
neos-506428	129925	42981	343466	1.9e+04	4.5e-08	33
neos-512201	1337	838	3418	3.3e+02	0.1	78
neos-520729	31178	91149	322203	1e+02	1	5e+03
neos-522351	1705	1524	5436	5.7e+02	0.1	5e+04
neos-525149	144120	3640	1519200	3.5e+02	1	2
neos-530627	113	103	324	38	1	1.3e+02
neos-538867	1170	792	3888	2.9e+02	1	1
neos-538916	1314	864	4272	3.3e+02	1	1
neos-544324	732	10080	1757280	4.2	1	1
neos-547911	693	3528	615048	4	1	1
neos-548047	3970	2020	26405	3.1e+02	1	1
neos-548251	2386	1922	5791	8e+02	1	31
neos-551991	3332	1730	31631	1.9e+02	1	1

Table A.1 continued

Instance	columns	rows	nonzeros	sparsity	min. abs.	max. abs.
neos-555001	3474	3855	16649	8.7e+02	0.2	40
neos-555298	2755	4827	20145	6.9e+02	0.2	40
neos-555343	3326	3815	16967	8.3e+02	0.2	40
neos-555424	2676	3815	15667	6.7e+02	0.2	40
neos-555694	1948	4139	39543	2.2e+02	0.01	1e+02
neos-555771	1978	4170	40349	2.2e+02	0.01	1e+02
neos-555884	4331	3815	19067	1.1e+03	0.2	40
neos-555927	1403	1945	7965	3.5e+02	0.2	40
neos-565672	318334	190589	809816	8e+04	2.9e-07	44
neos-565815	15413	1276	124071	1.6e+02	1	1
neos-570431	931	511	12041	40	1	1
neos-574665	3790	740	16792	1.7e+02	0.93	1.8e+05
neos-578379	21703	17010	101560	4.3e+03	1	1
neos-582605	1240	1265	3735	6.2e+02	1	2
neos-583731	1491	1350	5220	5e+02	0.5	1
neos-584146	936	811	3035	3.1e+02	0.5	2
neos-584851	661	445	1709	2.2e+02	1	1
neos-584866	9009	3674	21338	1.8e+03	1	1
neos-585192	2628	2597	72396	97	0.81	1e+06
neos-585467	2166	2116	50058	94	0.89	1e+06
neos-593853	1606	2400	6000	8e+02	1	1.1e+07
neos-595904	2452	4508	22364	6.1e+02	1	1e+03
neos-595905	704	1200	5788	1.8e+02	1	1e+03
neos-595925	956	1276	5960	2.4e+02	1	1e+03
neos-598183	992	1696	8388	2.5e+02	1	1e+03
neos-603073	992	1696	8388	2.5e+02	1	1e+03
neos-611135	5277	6400	769300	44	0.25	2e+02
neos-611838	1876	9954	37027	6.3e+02	1	3e+04
neos-612125	1795	9554	35791	6e+02	1	3e+04
neos-612143	1842	9832	36643	6.1e+02	1	3e+04
neos-612162	1859	9893	36835	6.2e+02	1	3e+04
neos-619167	6800	3452	20020	1.4e+03	0.053	1e+06
neos-631164	406	1282	3156	2e+02	1	1.6e+05
neos-631517	351	1090	2743	1.8e+02	1	1.6e+05
neos-631694	3996	3725	18523	1e+03	1	4
neos-631709	46496	45150	225148	1.2e+04	1	4
neos-631710	169576	167056	834166	4.2e+04	1	4
neos-631784	23996	22725	113023	6e+03	1	4
neos-632335	24864	12719	73025	5e+03	1	1
neos-633273	21781	11154	63910	4.4e+03	1	1
neos-641591	1085	18235	200055	1.1e+02	1	5
neos-655508	13573	13572	40484	6.8e+02	1	1
neos-662469	1085	18235	200055	1.1e+02	1	5
neos-686190	3664	3660	18085	9.2e+02	1	69
neos-691058	2667	3006	30837	2.7e+02	1	7
neos-691073	2667	1935	29766	1.8e+02	1	7
neos-693347	3192	1576	113472	44	1	1
neos-702280	1600	7199	2421882	4.8	1	1
neos-709469	469	224	4432	25	1	10
neos-717614	891	3049	10477	3e+02	0.0024	1.6e+04
neos-738098	25849	9093	101360	2.3e+03	1	12
neos-775946	6602	4710	107876	3e+02	0.01	1e+02
neos-777800	479	6400	32000	96	1	1
neos-780889	73910	182700	497210	3.7e+04	1	3
neos-785899	1653	1320	17180	1.3e+02	1	1
neos-785912	1714	1380	16610	1.4e+02	1	1
neos-785914	1590	1260	15290	1.3e+02	1	1
neos-787933	1897	236376	298320	1.9e+03	1	1.3e+02
neos-791021	3694	9448	29708	1.2e+03	1	12
neos-796608	286	311	778	1.4e+02	1	4
neos-799711	59218	41998	147164	2e+04	0.00058	1e+08
neos-799838	5976	20844	57888	3e+03	1	50
neos-801834	3300	3220	55200	1.9e+02	1	1
neos-803219	901	640	3020	2.3e+02	0.0079	1e+02
neos-803220	891	630	2980	2.2e+02	0.0079	1e+02
neos-806323	1541	1060	5650	3.1e+02	0.0079	6.9e+02
neos-807454	1622	1638	35272	77	1	1
neos-807456	840	1635	4905	2.8e+02	1	1
neos-807639	1541	1030	5520	3.1e+02	0.0079	6e+02
neos-807705	1541	1030	5520	3.1e+02	0.0079	6e+02
neos-808072	1713	1702	38054	78	1	1
neos-808214	640	1308	22530	38	1	1
neos-810286	2675	2915	69952	1.2e+02	1	1
neos-810326	1749	1702	38810	80	1	1

Appendix A. Experimental Data and Results

Table A.1 continued

Instance	columns	rows	nonzeros	sparsity	min. abs.	max. abs.
neos-820146	830	600	3225	1.7e+02	1	1
neos-820157	1015	1200	4875	2.5e+02	1	1
neos-820879	361	9522	72356	52	1	8e+03
neos-824661	18804	45390	138890	6.3e+03	1	1e+03
neos-824695	9576	23970	72590	3.2e+03	1	1e+03
neos-825075	328	800	5480	55	1	1
neos-826224	17266	41820	127840	5.8e+03	1	1e+03
neos-826250	5250	12250	37520	1.7e+03	1	1e+03
neos-826650	2414	5912	20440	8e+02	1	1e+03
neos-826694	6904	16410	59268	2.3e+03	1	1e+03
neos-826812	6844	15864	53808	2.3e+03	1	1e+03
neos-826841	2354	5516	18460	7.8e+02	1	1e+03
neos-827015	7688	79347	166239	3.8e+03	1	1
neos-827175	14187	32504	110790	4.7e+03	1	1e+03
neos-829552	5153	40971	86952	2.6e+03	1	1
neos-830439	1375	1468	4804	4.6e+02	1	1e+02
neos-831188	2185	4612	11256	1.1e+03	1	1
neos-839838	12751	7700	47800	2.1e+03	1	1e+08
neos-839859	3251	1975	12025	5.4e+02	1	1e+08
neos-839894	33201	16325	98825	5.5e+03	1	1e+06
neos-841664	3135	2925	10920	1e+03	1	1e+04
neos-847051	4731	5417	19372	1.6e+03	0.00011	1e+06
neos-847302	609	737	9566	51	1	1
neos-848150	731	949	12300	61	1	1
neos-848198	924	10164	29106	4.6e+02	1	1e+03
neos-848589	1484	550539	1101078	7.4e+02	1	1e+06
neos-848845	1050	1737	19470	95	1	1
neos-849702	1041	1737	19308	95	1	1
neos-850681	2067	2594	37113	1.5e+02	1	6.7e+02
neos-856059	17827	450	35654	2.3e+02	1	1
neos-859770	2065	2504	880736	5.9	1	1
neos-860244	675	3105	413305	5.1	1	1
neos-860300	850	1385	384329	3.1	1	21
neos-862348	5801	3835	81027	2.8e+02	0.01	1e+02
neos-863472	523	588	5440	58	0.43	45
neos-872648	93291	175219	350438	4.7e+04	1	1
neos-873061	93360	175288	350576	4.7e+04	1	1
neos-876808	85808	87268	682376	1.2e+04	1	22
neos-880324	348	261	1484	70	1	1e+03
neos-881765	278	712	7208	28	1	1
neos-885086	11574	4860	248310	2.3e+02	1	1.6e+03
neos-885524	65	91670	258309	32	3	1.3e+03
neos-886822	1089	1057	4128	3.6e+02	1	1.5e+04
neos-892255	2137	1800	10005	4.3e+02	1	25
neos-905856	403	686	6601	45	1	1
neos-906865	1634	1184	5728	4.1e+02	1	1
neos-911880	83	888	2568	42	1	1.4e+02
neos-911970	107	888	3408	36	1	1.4e+02
neos-912015	617	686	14742	29	1	1
neos-912023	623	686	14728	30	1	1
neos-913984	1076	76000	152000	5.4e+02	1	6
neos-914441	15129	15007	59658	5e+03	1	1.8e+08
neos-916173	1413	1084	72701	21	0.015	1e+03
neos-916792	1909	1474	134442	21	0.017	1e+03
neos-930752	6549	9674	27864	3.3e+03	1	1
neos-931517	5529	7920	29565	1.8e+03	1	2
neos-931538	5964	7920	33480	1.5e+03	1	2
neos-932721	18085	22266	107908	4.5e+03	1	1
neos-932816	30823	21007	484926	1.3e+03	1	1
neos-933364	1006	1728	6768	3.4e+02	1	1.4e+02
neos-933550	2288	3032	13776	5.7e+02	1	1
neos-933562	3200	3032	28800	3.6e+02	1	1
neos-933638	13658	32417	187173	2.7e+03	1	1
neos-933815	947	1728	5088	4.7e+02	1	1.4e+02
neos-933966	12047	31762	180618	2.4e+03	1	1
neos-934184	1006	1728	6768	3.4e+02	1	1.4e+02
neos-934278	11495	23123	125577	2.3e+03	1	1
neos-934441	11691	23362	127383	2.3e+03	1	1
neos-934531	47078	1082	136119	3.8e+02	1	1e+05
neos-935234	9568	10309	55271	1.9e+03	1	1
neos-935348	7859	10301	40476	2.6e+03	1	1
neos-935496	2890	2820	27984	3.2e+02	1	1
neos-935627	7859	10301	40476	2.6e+03	1	1
neos-935674	2890	3108	28560	3.2e+02	1	1

Table A.1 continued

Instance	columns	rows	nonzeros	sparsity	min. abs.	max. abs.
neos-935769	6741	9799	36447	2.2e+03	1	1
neos-936660	7311	10019	39546	2.4e+03	1	1
neos-937446	8176	11341	44697	2.7e+03	1	1
neos-937511	8158	11332	44237	2.7e+03	1	1
neos-937815	9251	11646	48013	2.3e+03	1	1
neos-941262	6703	9480	35659	2.2e+03	1	1
neos-941313	13189	167910	484080	6.6e+03	1	1.3e+02
neos-941698	844	946	13002	65	1	1
neos-941717	1092	1350	20214	78	1	1
neos-941782	968	1094	17086	65	1	1
neos-942323	754	732	10884	54	1	1
neos-942830	803	882	13290	54	1	1
neos-942886	359	464	7109	24	1	1
neos-948126	7271	9551	38219	1.8e+03	1	1
neos-948268	4773	7550	26410	1.6e+03	1	1
neos-948346	1570	57855	540443	1.7e+02	1	1e+04
neos-950242	34224	5760	104160	1.9e+03	1	1
neos-952987	354	31329	90384	1.8e+02	1	1.0e+03
neos-953928	12498	23305	169861	1.8e+03	1	53
neos-954925	2989	84718	844983	3.3e+02	1	53
neos-955215	723	1302	3822	3.6e+02	1	1.1e+02
neos-955800	6516	1848	19536	6.5e+02	1	1
neos-956971	2527	57756	483560	3.2e+02	1	53
neos-957143	2767	57756	497676	3.5e+02	1	53
neos-957270	3282	5929	417968	47	1	43
neos-957323	3757	57756	499656	4.7e+02	1	53
neos-957389	5115	6036	355172	88	1	30
neos-960392	4744	59376	189503	1.6e+03	1	53
neos-983171	6711	8965	36691	1.7e+03	1	1
neos-984165	6962	8883	36742	1.7e+03	1	1
neos1	131581	1892	468009	5.3e+02	1	1
neos13	20852	1827	253842	1.5e+02	1.5e-05	2.3e+02
neos15	552	792	1766	2.8e+02	1	1e+03
neos16	1018	377	2801	1.5e+02	1	7
neos18	11402	3312	24614	1.6e+03	1	1
neos2	132568	1560	552519	3.7e+02	1	1
neos6	1036	8786	251946	37	1	18
neos788725	433	352	4912	33	1	5
neos808444	18329	19846	120512	3.1e+03	1	79
neos858960	132	160	2770	7.8	1	1
nesm	662	2923	13288	1.7e+02	0.001	33
net12	14021	14115	80384	2.8e+03	1	12
netdiversion	119589	129180	615282	3e+04	1	1
netlarge2	40000	1160000	2320000	2e+04	1	1
newdano	576	505	2184	1.4e+02	1	81
nl	7039	9718	41428	1.8e+03	0.0004	2.2e+02
nobel-eu-DBE	879	3771	11313	2.9e+02	1	7.7e+04
noswot	182	128	735	36	0.25	21
npmv07	76342	220686	859614	2.5e+04	0.003	7.1e+05
ns1111636	13895	360822	568444	1.4e+04	1	1.9e+02
ns1116954	131991	12648	410582	4.1e+03	1	10
ns1208400	4289	2883	81746	1.5e+02	0.2	1
ns1456591	1997	8399	199862	87	1	3e+04
ns1606230	3503	4173	92133	1.6e+02	1	1
ns1631475	24496	22696	116733	4.9e+03	1	4.8e+03
ns1644855	40698	30200	2110696	5.9e+02	1	20
ns1663818	172017	124626	20433649	1.1e+03	1	1e+03
ns1685374	44121	10000	220859	2e+05	0.01	1
ns1686196	4055	2738	68529	1.6e+02	1	1e+03
ns1688347	4191	2685	66908	1.7e+02	1	1e+03
ns1696083	11063	7982	384129	2.3e+02	1	1e+03
ns1702808	1474	804	5856	2.1e+02	1	1e+04
ns1745726	4687	3208	90278	1.7e+02	1	1e+03
ns1758913	624166	17956	1283444	8.8e+03	1	2.7e+03
ns1766074	182	100	666	30	1	10
ns1769397	5527	3772	117383	1.8e+02	1	1e+03
ns1778858	10666	4720	32673	1.8e+03	1	7.3e+05
ns1830653	2932	1629	100933	48	1	2.9e+02
ns1856153	35407	11998	105882	4.4e+03	0.5	1e+03
ns1904248	149437	38458	378770	1.7e+04	6.1e-17	98
ns1905797	51884	18192	239700	4e+03	0.2	50
ns1905800	8289	3228	38100	7.5e+02	0.75	50
ns1952667	41	13264	335643	1.6	1	1.9e+02
ns2017839	54510	55224	317840	1.1e+04	0.33	3.5e+07

Table A.1 continued

Instance	columns	rows	nonzeros	sparsity	min. abs.	max. abs.
ns2081729	1190	661	5680	1.5e+02	0.5	1e+02
ns2118727	163354	167440	646864	5.4e+04	0.042	8.8
ns2122603	24754	19300	77044	8.3e+03	0.042	1e+08
ns2124243	139280	156083	429032	7e+04	0.5	1
ns2137859	206726	103361	923682	2.6e+04	1	2e+03
ns4-pr3	2210	8601	25986	7.4e+02	1	60
ns4-pr9	2220	7350	22176	7.4e+02	1	35
ns894236	8218	9666	41067	2.1e+03	1	1
ns894244	12129	21856	90864	3e+03	1	1
ns894786	16794	27278	113575	4.2e+03	1	1
ns894788	2279	3463	14381	5.7e+02	1	1
ns903616	18052	21582	91641	4.5e+03	1	1
ns930473	23240	11328	121764	2.3e+03	1	1.2e+05
nsa	1297	388	4204	1.3e+02	1	1
nsct1	22901	14981	656259	5.3e+02	1	1.4e+05
nsct2	23003	14981	675156	5.1e+02	1	1.4e+05
nsic1	451	463	2853	75	1	5e+05
nsic2	465	463	3015	78	1	5e+05
nsir1	4407	5717	138955	1.8e+02	1	2e+05
nsir2	4453	5717	150599	1.7e+02	1	2e+05
nsr8k	6284	38356	371608	7e+02	1	1e+08
nsrand-ipx	735	6621	223261	22	1	1.8e+04
nu120-pr3	2210	8601	25986	7.4e+02	1	1.2e+02
nu60-pr9	2220	7350	22176	7.4e+02	1	60
nug05	210	225	1050	52	1	1
nug06	372	486	2232	93	1	1
nug07	602	931	4214	1.5e+02	1	1
nug08	912	1632	7296	2.3e+02	1	1
nug08-3rd	19728	20448	139008	3.3e+03	1	1
nug12	3192	8856	38304	8e+02	1	1
nw04	36	87482	636666	5.1	1	1
nw14	73	123409	904910	10	1	1
ofi	422587	420434	1778754	1.1e+05	3e-22	1e+11
opm2-z10-s2	160633	6250	371243	2.7e+03	1	4.1e+03
opm2-z11-s8	223082	8019	510283	3.5e+03	1	4.1e+03
opm2-z12-s14	319508	10800	725376	4.8e+03	1	4.1e+03
opm2-z12-s7	319508	10800	725385	4.8e+03	1	4.1e+03
opm2-z7-s2	31798	2023	79762	8.2e+02	1	4.1e+03
opt1217	64	769	1542	32	1	8
orna1	882	882	3108	2.9e+02	1.5	1.4e+04
orna2	882	882	3108	2.9e+02	1.5	1.4e+04
orna3	882	882	3108	2.9e+02	1.5	1.4e+04
orna4	882	882	3108	2.9e+02	1.5	1.4e+04
orna7	882	882	3108	2.9e+02	1.5	1.4e+04
orswq2	80	80	264	27	0.023	42
osa-07	1118	23949	143694	1.9e+02	0.29	13
osa-14	2337	52460	314760	3.9e+02	0.29	13
osa-30	4350	100024	600138	8.7e+02	0.16	13
osa-60	10280	232966	1397793	2.1e+03	0.35	13
p0033	16	33	98	8	1	4e+02
p0040	23	40	110	12	1	2.2e+03
p010	10090	19000	117910	1.7e+03	0.1	3
p0201	133	201	1923	15	1	64
p0282	241	282	1966	40	1	2e+02
p0291	252	291	2031	42	0.4	52
p05	5090	9500	58955	8.5e+02	0.1	3
p0548	176	548	1711	59	1	1e+04
p100x588b	688	1176	2352	3.4e+02	1	9e+02
p19	284	586	5305	32	0.0001	5.4
p2756	755	2756	8937	2.5e+02	1	1e+04
p2m2p1m1p0n100	1	100	100	1	6.6e+03	1.4e+04
p6000	2095	5872	17731	7e+02	1	9.9e+04
p6b	5852	462	11704	2.3e+02	1	1
p80x400b	480	800	1600	2.4e+02	1	8e+02
pcb1000	1565	2428	20071	2e+02	1	2
pcb3000	3960	6810	56557	5e+02	1	2
pds-02	2953	7535	16390	1.5e+03	1	1
pds-06	9881	28655	62524	4.9e+03	1	1
pds-10	16558	48763	106436	8.3e+03	1	1
pds-100	156243	505360	1086785	7.8e+04	1	1
pds-20	33874	105728	230200	1.7e+04	1	1
pds-30	49944	154998	337144	2.5e+04	1	1
pds-40	66844	212859	462128	3.3e+04	1	1
pds-50	83060	270095	585114	4.2e+04	1	1

Table A.1 continued

Instance	columns	rows	nonzeros	sparsity	min. abs.	max. abs.
pds-60	99431	329643	712779	5e+04	1	1
pds-70	114944	382311	825771	5.7e+04	1	1
pds-80	129181	426278	919524	6.5e+04	1	1
pds-90	142823	466671	1005359	7.1e+04	1	1
perold	625	1376	6018	1.6e+02	5.3e-05	2.4e+04
pf2177	9728	900	21706	4.1e+02	1	1
pg	125	2700	5200	1.2e+02	1	1.6e+03
pg5_34	225	2600	7700	1.1e+02	1	1.5e+03
pgp2	4034	9220	18440	2e+03	1	16
pigeon-10	931	490	8150	58	1	1e+03
pigeon-11	1123	572	9889	66	1	1e+03
pigeon-12	1333	660	11796	78	1	1e+03
pigeon-13	1561	754	13871	87	1	1e+03
pigeon-19	3307	1444	29849	1.7e+02	1	1e+03
pilot	1441	3652	43167	1.3e+02	1e-06	1.5e+02
pilot-ja	940	1988	14698	1.3e+02	2e-06	5.9e+06
pilot-we	722	2789	9126	2.4e+02	0.00014	4.8e+04
pilot4	410	1000	5141	82	3.7e-05	2.8e+04
pilot87	2030	4883	73152	1.4e+02	1e-06	1e+03
pilotnov	975	2172	13057	1.6e+02	2e-06	5.9e+06
pk1	45	86	915	4.5	1	55
pldd000b	3069	3267	8980	1.5e+03	1e-06	1.1e+02
pldd001b	3069	3267	8981	1.5e+03	1e-06	1.1e+02
pldd002b	3069	3267	8982	1.5e+03	1e-06	1.1e+02
pldd003b	3069	3267	8983	1.5e+03	1e-06	1.1e+02
pldd004b	3069	3267	8984	1.5e+03	1e-06	1.1e+02
pldd005b	3069	3267	8985	1.5e+03	1e-06	1.1e+02
pldd006b	3069	3267	8986	1.5e+03	1e-06	1.1e+02
pldd007b	3069	3267	8987	1.5e+03	1e-06	1.1e+02
pldd008b	3069	3267	9047	1.5e+03	1e-06	1.1e+02
pldd009b	3069	3267	9050	1.5e+03	1e-06	1.1e+02
pldd010b	3069	3267	9053	1.5e+03	1e-06	1.1e+02
pldd011b	3069	3267	9055	1.5e+03	1e-06	1.1e+02
pldd012b	3069	3267	9057	1.5e+03	1e-06	1.1e+02
pltexpa2-16	1726	4540	9233	8.6e+02	1	1e+03
pltexpa2-6	686	1820	3703	3.4e+02	1	1e+03
pltexpa3_16	28350	74172	150801	1.4e+04	1	1e+03
pltexpa3_6	4430	11612	23611	2.2e+03	1	1e+03
pltexpa4_6	26894	70364	143059	1.3e+04	1	1e+03
pp08a	136	240	480	68	1	5e+02
pp08aCUTS	246	240	839	82	1	5e+02
primagaz	1554	10836	21665	1.6e+03	1	1
problem	12	46	86	12	1	1
probportfolio	302	320	6620	15	0.8	1.5
prod1	208	250	5350	9.9	6.2e-05	7
prod2	211	301	10501	6.2	6.2e-05	10
progas	1650	1425	8422	3.3e+02	3.9e-05	1e+04
protfold	2112	1835	23491	1.8e+02	1	1
pw-myciel4	8164	1059	17779	5.1e+02	1	1
qap10	1820	4150	18200	4.6e+02	1	1
qiu	1192	840	3432	3e+02	0.26	22
qiulp	1192	840	3432	3e+02	0.26	22
qnet1	503	1541	4622	2.5e+02	1	4.1e+03
qnet1_o	456	1541	4214	2.3e+02	1	4.1e+03
queens-30	960	900	93440	9.3	1	7
r05	5190	9500	103955	5.2e+02	0.1	3
r80x800	880	1600	3200	4.4e+02	1	1e+03
rail01	46843	117527	392086	1.6e+04	1	2.3e+02
rail2586	2586	920683	8008776	3.2e+02	1	1
rail4284	4284	1092610	11279748	4.3e+02	1	1
rail507	509	63019	468878	73	1	1
rail516	516	47311	314896	86	1	1
rail582	582	55515	401708	83	1	1
ramos3	2187	2187	32805	1.5e+02	1	1
ran10x10a	120	200	400	60	1	16
ran10x10b	120	200	400	60	1	19
ran10x10c	120	200	400	60	1	17
ran10x12	142	240	480	71	1	30
ran10x26	296	520	1040	1.5e+02	1	56
ran12x12	168	288	576	84	1	26
ran12x21	285	504	1008	1.4e+02	1	50
ran13x13	195	338	676	98	1	28
ran14x18	284	504	1008	1.4e+02	1	38
ran14x18-disj-8	447	504	10277	22	3.2e-09	38

Table A.1 continued

Instance	columns	rows	nonzeros	sparsity	min. abs.	max. abs.
ran14x18.disj-8	447	504	10277	22	3.2e-09	38
ran14x18.1	284	504	1008	1.4e+02	1	38
ran16x16	288	512	1024	1.4e+02	1	54
ran17x17	323	578	1156	1.6e+02	1	35
ran4x64	324	512	1024	1.6e+02	1	44
ran6x43	307	516	1032	1.5e+02	1	49
ran8x32	296	512	1024	1.5e+02	1	44
rat1	3136	9408	88267	3.5e+02	0.00023	1
rat5	3136	9408	137413	2.2e+02	0.0012	1
rat7a	3136	9408	268908	1.1e+02	0.0004	1
rd-rplusc-21	125899	622	852384	92	0.2	1e+07
reblock166	17024	1660	39442	7.4e+02	1	2.5e+04
reblock354	19906	3540	52901	1.4e+03	1	7.2e+02
reblock420	62800	4200	138670	1.9e+03	1	9.1e+03
reblock67	2523	670	7495	2.3e+02	1	6.5e+03
recipe	91	180	663	30	0.12	1.4e+02
refine	29	33	124	9.7	0.5	66
rentacar	6803	9557	41842	1.7e+03	0.01	1e+05
rgn	24	180	460	12	1	4.6
rlfddd	4050	57471	260577	1e+03	1	1
rlfdual	8052	66918	273079	2e+03	1	1
rlfprim	58866	8052	265927	1.8e+03	1	1
rlp1	68	461	836	68	1	14
rmatr100-p10	7260	7359	21877	3.6e+02	1	1
rmatr100-p5	8685	8784	26152	4.3e+02	1	1
rmatr200-p10	35055	35254	105362	1.8e+04	1	1
rmatr200-p20	29406	29605	88415	1.5e+04	1	1
rmatr200-p5	37617	37816	113048	1.9e+04	1	1
rmine10	65274	8439	162264	3.4e+03	1	1e+02
rmine14	268535	32205	660346	1.3e+04	1	1e+02
rmine6	7078	1096	18084	4.4e+02	1	1e+02
rocII-4-11	21738	9234	243106	8.4e+02	0.38	12
rocII-7-11	37215	16101	423661	1.4e+03	0.38	12
rocII-9-11	47533	20679	544031	1.8e+03	0.38	12
rococoB10-011000	1667	4456	16517	5.6e+02	1	1.6e+03
rococoC10-001000	1293	3117	11751	4.3e+02	1	3.4e+04
rococoC11-011100	2367	6491	30472	5.9e+02	1	4.1e+03
rococoC12-111000	10776	8619	48920	2.2e+03	1	3.2e+04
roll3000	2295	1166	29386	92	0.25	3.2e+02
rosen1	520	1024	23274	24	1	9
rosen10	2056	4096	62136	1.4e+02	1	9
rosen2	1032	2048	46504	47	1	9
rosen7	264	512	7770	18	1	9
rosen8	520	1024	15538	35	1	9
rout	291	556	2431	73	0.22	1.4e+02
route	20894	23923	187686	3e+03	1	1.5e+03
roy	162	149	411	81	1	30
rvb-sub	225	33765	984143	7.8	0.00035	1
satellites1-25	5996	9013	59023	1e+03	0.36	2.3e+05
satellites2-60	20916	35378	283668	2.6e+03	0.36	2.4e+05
satellites2-60-fs	16516	35378	125048	5.5e+03	0.36	2.4e+05
satellites3-40	44804	81681	698176	5.6e+03	0.36	2.4e+05
satellites3-40-fs	35553	81681	291161	1.2e+04	0.36	2.4e+05
sc105	105	103	280	52	0.1	2
sc205	205	203	551	1e+02	0.1	2
sc205-2r-100	2213	2214	6030	1.1e+03	1	2
sc205-2r-16	365	366	990	1.8e+02	1	2
sc205-2r-1600	35213	35214	96030	1.8e+04	1	2
sc205-2r-200	4413	4414	12030	2.2e+03	1	2
sc205-2r-27	607	608	1650	3e+02	1	2
sc205-2r-32	717	718	1950	3.6e+02	1	2
sc205-2r-4	101	102	270	50	1	2
sc205-2r-400	8813	8814	24030	4.4e+03	1	2
sc205-2r-50	1113	1114	3030	5.6e+02	1	2
sc205-2r-64	1421	1422	3870	7.1e+02	1	2
sc205-2r-8	189	190	510	94	1	2
sc205-2r-800	17613	17614	48030	8.8e+03	1	2
sc50a	50	48	130	25	0.1	2
sc50b	50	48	118	25	0.3	3
scagr25	471	500	1554	1.6e+02	0.2	9.3
scagr7	129	140	420	43	0.2	9.3
scagr7-2b-16	623	660	2058	2.1e+02	0.2	9.3
scagr7-2b-4	167	180	546	56	0.2	9.3
scagr7-2b-64	9743	10260	32298	3.2e+03	0.2	9.3

Table A.1 continued

Instance	columns	rows	nonzeros	sparsity	min. abs.	max. abs.
scagr7-2c-16	623	660	2058	2.1e+02	0.2	9.3
scagr7-2c-4	167	180	546	56	0.2	9.3
scagr7-2c-64	2447	2580	8106	8.2e+02	0.2	9.3
scagr7-2r-108	4119	4340	13542	1.4e+03	0.2	9.3
scagr7-2r-16	623	660	2058	2.1e+02	0.2	9.3
scagr7-2r-216	8223	8660	27042	2.7e+03	0.2	9.3
scagr7-2r-27	1041	1100	3444	3.5e+02	0.2	9.3
scagr7-2r-32	1231	1300	4074	4.1e+02	0.2	9.3
scagr7-2r-4	167	180	546	56	0.2	9.3
scagr7-2r-432	16431	17300	54042	5.5e+03	0.2	9.3
scagr7-2r-54	2067	2180	6846	6.9e+02	0.2	9.3
scagr7-2r-64	2447	2580	8106	8.2e+02	0.2	9.3
scagr7-2r-8	319	340	1050	1.1e+02	0.2	9.3
scagr7-2r-864	32847	34580	108042	1.1e+04	0.2	9.3
scfxm1	330	457	2589	66	0.0005	1.3e+02
scfxm1-2b-16	2460	3714	13959	8.2e+02	0.001	62
scfxm1-2b-4	684	1014	3999	2.3e+02	0.001	62
scfxm1-2b-64	19036	28914	106919	6.3e+03	0.001	62
scfxm1-2c-4	684	1014	3999	2.3e+02	0.001	62
scfxm1-2r-128	19036	28914	106919	6.3e+03	0.001	62
scfxm1-2r-16	2460	3714	13959	8.2e+02	0.001	62
scfxm1-2r-256	37980	57714	213159	1.3e+04	0.001	62
scfxm1-2r-27	4088	6189	23089	1.4e+03	0.001	62
scfxm1-2r-32	4828	7314	27239	1.6e+03	0.001	62
scfxm1-2r-4	684	1014	3999	2.3e+02	0.001	62
scfxm1-2r-64	9564	14514	53799	3.2e+03	0.001	62
scfxm1-2r-8	1276	1914	7319	4.3e+02	0.001	62
scfxm1-2r-96	14300	21714	80359	4.8e+03	0.001	62
scfxm2	660	914	5183	1.3e+02	0.0005	1.3e+02
scfxm3	990	1371	7777	2e+02	0.0005	1.3e+02
scorpion	388	358	1426	1.3e+02	0.01	1
scrs8	490	1169	3182	2.4e+02	0.001	3.9e+02
scrs8-2b-16	476	645	1633	2.4e+02	0.001	36
scrs8-2b-4	140	189	457	70	0.001	36
scrs8-2b-64	1820	2469	6337	9.1e+02	0.001	36
scrs8-2c-16	476	645	1633	2.4e+02	0.001	36
scrs8-2c-32	924	1253	3201	4.6e+02	0.001	36
scrs8-2c-4	140	189	457	70	0.001	36
scrs8-2c-64	1820	2469	6337	9.1e+02	0.001	36
scrs8-2c-8	252	341	849	1.3e+02	0.001	36
scrs8-2r-128	3612	4901	12609	1.8e+03	0.001	36
scrs8-2r-16	476	645	1633	2.4e+02	0.001	36
scrs8-2r-256	7196	9765	25153	3.6e+03	0.001	36
scrs8-2r-27	784	1063	2711	3.9e+02	0.001	36
scrs8-2r-32	924	1253	3201	4.6e+02	0.001	36
scrs8-2r-4	140	189	457	70	0.001	36
scrs8-2r-512	14364	19493	50241	7.2e+03	0.001	36
scrs8-2r-64	1820	2469	6337	9.1e+02	0.001	36
scrs8-2r-64b	1820	2469	6337	9.1e+02	0.001	36
scrs8-2r-8	252	341	849	1.3e+02	0.001	36
scsd1	77	760	2388	26	0.24	1
scsd6	147	1350	4316	49	0.24	1
scsd8	397	2750	8584	1.3e+02	0.24	1
scsd8-2b-16	330	2310	7170	1.1e+02	0.24	1
scsd8-2b-4	90	630	1890	30	0.24	1
scsd8-2b-64	5130	35910	112770	1.7e+03	0.24	1
scsd8-2c-16	330	2310	7170	1.1e+02	0.24	1
scsd8-2c-4	90	630	1890	30	0.24	1
scsd8-2c-64	5130	35910	112770	1.7e+03	0.24	1
scsd8-2r-108	2170	15190	47650	7.2e+02	0.24	1
scsd8-2r-16	330	2310	7170	1.1e+02	0.24	1
scsd8-2r-216	4330	30310	95170	1.4e+03	0.24	1
scsd8-2r-27	550	3850	12010	1.8e+02	0.24	1
scsd8-2r-32	650	4550	14210	2.2e+02	0.24	1
scsd8-2r-4	90	630	1890	30	0.24	1
scsd8-2r-432	8650	60550	190210	2.9e+03	0.24	1
scsd8-2r-54	1090	7630	23890	3.6e+02	0.24	1
scsd8-2r-64	1290	9030	28290	4.3e+02	0.24	1
scsd8-2r-8	170	1190	3650	57	0.24	1
scsd8-2r-8b	170	1190	3650	57	0.24	1
sct1	12154	22886	105571	3e+03	2.1e-06	6e+04
sct32	5440	9767	109654	4.9e+02	0.0012	1.5e+05
sct5	13304	37265	147037	4.4e+03	0.0024	1.9e+05
sctap1	300	480	1692	1e+02	1	80

Table A.1 continued

Instance	columns	rows	nonzeros	sparsity	min. abs.	max. abs.
sctap1-2b-16	990	1584	5740	3.3e+02	1	80
sctap1-2b-4	270	432	1516	90	1	80
sctap1-2b-64	15390	24624	90220	5.1e+03	1	80
sctap1-2c-16	990	1584	5740	3.3e+02	1	80
sctap1-2c-4	270	432	1516	90	1	80
sctap1-2c-64	3390	5424	19820	1.1e+03	1	80
sctap1-2r-108	6510	10416	38124	2.2e+03	1	80
sctap1-2r-16	990	1584	5740	3.3e+02	1	80
sctap1-2r-216	12990	20784	76140	4.3e+03	1	80
sctap1-2r-27	1650	2640	9612	5.5e+02	1	80
sctap1-2r-32	1950	3120	11372	6.5e+02	1	80
sctap1-2r-4	270	432	1516	90	1	80
sctap1-2r-480	28830	46128	169068	9.6e+03	1	80
sctap1-2r-54	3270	5232	19116	1.1e+03	1	80
sctap1-2r-64	3870	6192	22636	1.3e+03	1	80
sctap1-2r-8	510	816	2924	1.7e+02	1	80
sctap1-2r-8b	510	816	2924	1.7e+02	1	80
sctap2	1090	1880	6714	3.6e+02	1	80
sctap3	1480	2480	8874	4.9e+02	1	80
seba	515	1028	4352	1.3e+02	1	1.6e+02
self	960	7364	1148845	6.2	2.3e-07	1
set1ch	492	712	1412	4.9e+02	1	1.1e+03
set3-10	3747	4019	13747	1.2e+03	0.022	1.9e+04
set3-15	3747	4019	13747	1.2e+03	0.022	1.8e+04
set3-20	3747	4019	13747	1.2e+03	0.022	1.8e+04
seymour	4944	1372	33549	2.1e+02	1	1
seymour-disj-10	5108	1209	64704	96	1e-08	9
seymour.disj-10	5108	1209	64704	96	1e-08	9
seymourl	4944	1372	33549	2.1e+02	1	1
sgpf5y6	246077	308634	828070	1.2e+05	1	1
share1b	117	225	1151	23	0.1	1.3e+03
share2b	96	79	694	12	0.01	1e+02
shell	536	1775	3556	2.7e+02	1	1
ship04l	402	2118	6332	2e+02	0.014	4.7
ship04s	402	1458	4352	2e+02	0.014	4.7
ship08l	778	4283	12802	3.9e+02	0.011	5
ship08s	778	2387	7114	3.9e+02	0.011	5
ship12l	1151	5427	16170	5.8e+02	0.0062	1.6
ship12s	1151	2763	8178	5.8e+02	0.0062	1.6
shipsched	45554	13594	121571	5.7e+03	1	7.3e+04
shs1023	133944	444625	1044725	6.7e+04	9.2e-06	42
siena1	2220	13741	258915	1.2e+02	1	1e+08
sierra	1227	2036	7302	4.1e+02	1	1e+05
sing2	28891	31630	149712	7.2e+03	0.36	4e+02
sing245	143161	235146	652817	7.2e+04	0.044	4.3e+02
sing359	437116	713762	1975605	2.2e+05	0.044	4.3e+02
slptsk	2861	3347	72465	1.4e+04	0.0044	16
small000	709	1140	2749	3.5e+02	6e-06	1e+03
small001	687	1140	2871	3.4e+02	6e-06	1e+03
small002	713	1140	2946	3.6e+02	6e-06	1e+03
small003	711	1140	2945	3.6e+02	6e-06	1e+03
small004	717	1140	2983	3.6e+02	1e-06	1e+03
small005	717	1140	3017	3.6e+02	1e-06	1e+03
small006	710	1138	3024	3.6e+02	6e-06	1e+03
small007	711	1137	3079	3.6e+02	6e-06	1e+03
small008	712	1134	3042	3.6e+02	6e-06	1e+03
small009	710	1135	3030	3.6e+02	6e-06	1e+03
small010	711	1138	3027	3.6e+02	6e-06	1e+03
small011	705	1133	3005	3.5e+02	6e-06	1e+03
small012	706	1134	3014	3.5e+02	6e-06	1e+03
small013	701	1131	2989	3.5e+02	2e-06	1e+03
small014	687	1130	2927	3.4e+02	6e-06	1e+03
small015	683	1130	2967	3.4e+02	6e-06	1e+03
small016	677	1130	2937	3.4e+02	1e-06	1e+03
south31	18425	35421	111498	6.1e+03	0.021	1.1e+04
sp97ar	1761	14101	290968	88	1	21
sp97ic	2086	1662	66632	52	1	2.2e+02
sp98ar	4680	5478	231756	1.1e+02	1	8.4e+02
sp98ic	2311	2508	138053	42	1	4.5e+02
sp98ir	1531	1680	71704	36	1	5.7e+02
square15	32762	753526	1507052	1.6e+04	1	1
stair	356	467	3856	44	1e-05	9.9
standata	359	1075	3031	1.8e+02	1	3e+02
standmps	467	1075	3679	1.6e+02	1	3e+02

Table A.1 continued

Instance	columns	rows	nonzeros	sparsity	min. abs.	max. abs.
stat96v1	5995	197472	588798	3e+03	0.065	2
stat96v4	3173	62212	490472	4.5e+02	0.00091	20
stat96v5	2307	75779	233921	7.7e+02	1.6e-06	20
stein27	118	27	378	8.4	1	1
stein45	331	45	1034	15	1	1
stocfor1	117	111	447	29	0.063	3.4e+02
stocfor2	2157	2031	8343	5.4e+02	0.2	3.4e+02
stocfor3	16675	15695	64875	4.2e+03	0.063	3.4e+02
stockholm	57346	20644	171076	7.2e+03	1	1e+01
stormG2_1000	528185	1259121	3341696	2.6e+05	1	71
stormg2-125	66185	157496	418321	3.3e+04	1	71
stormg2-27	14441	34114	90903	7.2e+03	1	71
stormg2-8	4409	10193	27424	2.2e+03	1	71
stormg2_1000	528185	1259121	3341696	2.6e+05	1	71
stp3d	159488	204880	662128	5.3e+04	1	1
sts405	27270	405	81810	1.4e+02	1	1
sts729	88452	729	265356	2.4e+02	1	1
swath	884	6805	34965	1.8e+02	1	1.1e+03
sws	14310	12465	93015	2e+03	0.03	1
t0331-4l	664	46915	430982	74	1	1
t1717	551	73885	325689	1.4e+02	1	1
t1722	338	36630	133096	1.1e+02	1	1
tanglegram1	68342	34759	205026	1.4e+04	1	1
tanglegram2	8980	4714	26940	1.8e+03	1	1
testbig	17613	31223	61639	1.8e+04	1	2
timtab1	171	397	829	86	1	60
timtab2	294	675	1482	1.5e+02	1	60
toll-like	4408	2883	13224	1.1e+03	1	1
tr12-30	750	1080	2508	3.8e+02	1	1.5e+03
transportmoment	9616	9685	29541	3.2e+03	1	1e+04
triptim1	15706	30055	515436	9.2e+02	0.0001	6.3e+01
triptim2	14427	27326	521898	7.6e+02	0.0001	1e+03
triptim3	14939	28440	524124	8.3e+02	0.0001	4.1e+03
truss	1000	8806	27836	3.3e+02	0.45	1
tuff	333	587	4520	48	1e-05	1e+04
tw-myciel4	8146	760	27961	2.3e+02	1	1
uc-case11	51438	34134	202042	1e+04	1	1.1e+03
uc-case3	52003	37749	273618	7.4e+03	0.044	3.7e+02
uct-subprob	1973	2256	10147	4.9e+02	1	1
ulevimin	6590	44605	162206	2.2e+03	4.4e-05	2.1e+07
umts	4465	2947	23016	6.4e+02	1	1e+09
unitcal_7	48939	25755	127595	1.2e+04	1	1e+03
us04	163	28016	297538	16	1	1
usAbbrv-8-25_70	3291	2312	9628	8.2e+02	1	36
van	27331	12481	487296	7e+02	1	1.6e+02
vpm1	234	378	749	2.3e+02	1	6e+02
vpm2	234	378	917	1.2e+02	0.025	6e+02
vpphard	47280	51471	372305	6.8e+03	1	1
vpphard2	198450	199999	648540	6.6e+04	1	1
vtp-base	198	203	908	49	0.13	4e+03
wachplan	1553	3361	89361	60	1	1
watson_1	201155	383927	1052028	1e+05	0.018	8.8
watson_2	352013	671861	1841028	1.8e+05	0.015	8.3
wide15	32769	753687	1507374	1.6e+04	1	1
wnq-n100-mw99-14	656900	10000	1333400	4.9e+03	1	1
wood1p	244	2594	70215	9	3e-05	1e+03
woodw	1098	8405	37474	2.7e+02	0.01	1e+03
world	34506	32734	164470	6.9e+03	0.0028	5.6e+03
zed	116	43	567	8.9	0.05	3.5e+03
zib54-UUE	1809	5150	15288	9e+02	1	2e+03

A.1.2 Iterative Refinement

Table A.2: Detailed results comparing iterative refinement to floating-point performance. Entries corresponding to unsolved instances are printed in italics. See Table 2.2 for aggregated results.

iter — number of simplex iterations
R — number of refinements
R_0 — number of refinements to final basis
t — total running time (in seconds)

Instance	SoPlex$_9$ iter	SoPlex$_9$ t	SoPlex$_{50}$ iter	SoPlex$_{50}$ R	SoPlex$_{50}$ R_0	SoPlex$_{50}$ t	SoPlex$_{250}$ iter	SoPlex$_{250}$ R	SoPlex$_{250}$ R_0	SoPlex$_{250}$ t
10teams	1611	0.2	1611	3	0	0.3	1611	17	0	0.3
16_n14	329933	376.0	329933	0	0	371.8	329933	0	0	377.1
22433	1041	0.1	1041	3	0	0.1	1041	17	0	0.1
23588	548	0.0	548	3	0	0.0	548	17	0	0.1
25fv47	4359	0.8	4359	3	0	0.7	4359	17	0	0.8
30_70_45_095_100	16103	7.4	16103	3	0	7.6	16103	17	0	8.3
30n20b8	269	0.3	269	3	0	0.3	269	18	0	0.5
50v-10	220	0.0	220	3	0	0.0	220	15	0	0.1
80bau3b	7797	1.0	7797	3	0	1.1	7797	16	0	1.3
Test3	6948	3.3	6948	3	0	3.7	6948	16	0	5.0
a1c1s1	1742	0.1	1742	3	0	0.1	1742	15	0	0.1
aa01	11898	3.5	11898	3	0	3.6	11898	17	0	3.9
aa03	7584	2.0	7584	3	0	1.9	7584	16	0	2.5
aa3	7584	2.1	7584	3	0	2.0	7584	16	0	2.2
aa4	4486	1.1	4486	3	0	1.2	4486	17	0	1.2
aa5	9706	2.4	9706	3	0	2.6	9706	16	0	2.8
aa6	4870	1.1	4870	3	0	1.2	4870	16	0	1.4
acc-tight4	11142	2.6	11142	3	0	2.2	11142	17	0	2.3
acc-tight5	10469	2.0	10469	3	0	2.2	10469	17	0	2.2
acc-tight6	10672	2.2	10672	3	0	2.3	10672	17	0	2.0
adlittle	88	0.0	88	3	0	0.0	88	16	0	0.0
afiro	16	0.0	16	3	0	0.0	16	15	0	0.0
aflow30a	396	0.0	396	3	0	0.0	396	15	0	0.0
aflow40b	1826	0.2	1826	3	0	0.1	1826	15	0	0.3
agg	80	0.0	80	3	0	0.0	80	16	0	0.1
agg2	152	0.0	152	3	0	0.0	152	16	0	0.1
agg3	165	0.0	165	3	0	0.0	165	16	0	0.1
air02	95	0.1	95	3	0	0.1	95	15	0	0.2
air03	626	0.3	626	2	0	0.4	626	2	0	0.4
air04	11898	3.4	11898	3	0	3.4	11898	17	0	4.0
air05	4486	1.2	4486	3	0	1.3	4486	17	0	1.4
air06	7584	2.0	7584	3	0	2.0	7584	16	0	2.2
aircraft	1912	0.4	1912	3	0	0.2	1912	16	0	0.6
aligninq	1492	0.1	1492	3	0	0.2	1492	16	0	0.3
app1-2	14617	7.2	14622	4	1	8.0	14622	18	1	10.0
arki001	1570	0.2	1571	4	1	0.3	1571	18	1	0.2
ash608gpia-3col	3123	0.2	3123	3	0	0.3	3123	15	0	0.4
atlanta-ip	38285	34.8	38286	3	1	35.2	38286	18	1	37.3
atm20-100	4586	0.4	4699	4	1	0.5	4699	17	1	0.8
b2c1s1	2621	0.1	2621	3	0	0.1	2621	16	0	0.2
bab1	5711	4.0	5711	3	0	4.7	5711	16	0	7.6
bab3	697791	5263.4	697791	3	0	5266.5	697791	16	0	5292.6
bab5	29788	10.3	29788	3	0	10.5	29788	16	0	11.3
bal8x12	102	0.0	102	3	0	0.0	102	15	0	0.0
bandm	474	0.0	474	3	0	0.0	474	17	0	0.2
bas1lp	2582	1.0	2582	3	0	1.1	2582	16	0	1.4
baxter	12347	2.6	12347	3	0	2.8	12347	16	0	3.6
bc	3759	1.3	3759	3	0	1.4	3759	18	0	1.6
bc1	3759	1.3	3759	3	0	1.4	3759	18	0	1.6
beaconfd	88	0.0	88	3	0	0.0	88	15	0	0.0
beasleyC3	1140	0.1	1140	3	0	0.1	1140	15	0	0.3
bell3a	81	0.0	81	3	0	0.0	81	16	0	0.0
bell5	66	0.0	66	3	0	0.0	66	16	0	0.0
berlin_5_8_0	1017	0.0	1017	3	0	0.1	1017	15	0	0.1
bg512142	2238	0.3	2238	3	0	0.1	2238	17	0	0.4
biella1	16158	4.2	16158	3	0	4.2	16158	18	0	4.7
bienst1	455	0.0	455	3	0	0.0	455	15	0	0.0
bienst2	455	0.0	455	3	0	0.0	455	15	0	0.0
binkar10_1	1267	0.1	1267	3	0	0.1	1267	16	0	0.1

Table A.2 continued

| Instance | SoPlex$_9$ | | SoPlex$_{50}$ | | | | SoPlex$_{250}$ | | | |
	iter	t	iter	R	R$_0$	t	iter	R	R$_0$	t
bk4x3	16	0.0	16	3	0	0.0	16	15	0	0.0
blend	97	0.0	97	3	0	0.0	97	16	0	0.0
blend2	175	0.0	175	3	0	0.0	175	17	0	0.0
blp-ar98	368	0.2	368	3	0	0.5	368	16	0	1.4
blp-ic97	446	0.3	446	3	0	0.4	446	16	0	1.3
bnatt350	685	0.1	685	3	0	0.1	685	15	0	0.2
bnatt400	831	0.1	831	3	0	0.1	831	15	0	0.2
bnl1	1474	0.1	1474	3	0	0.2	1474	17	0	0.5
bnl2	2718	0.4	2718	3	0	0.5	2718	16	0	0.4
boeing1	458	0.0	458	3	0	0.0	458	16	0	0.1
boeing2	145	0.0	145	3	0	0.0	145	16	0	0.0
bore3d	100	0.0	100	3	0	0.0	100	16	0	0.0
brandy	482	0.0	482	3	0	0.0	482	16	0	0.1
buildingenergy	144595	827.1	144595	4	0	835.0	144595	17	0	837.0
cap6000	815	0.2	815	3	0	0.3	815	15	0	0.5
capri	338	0.0	338	3	0	0.0	338	16	0	0.0
car4	10442	1.9	10442	4	0	2.0	10442	21	0	2.6
cari	681	0.1	681	3	0	0.3	681	18	0	0.5
cep1	1399	0.1	1399	3	0	0.1	1399	16	0	0.4
ch	10747	1.4	10747	3	0	1.5	10747	16	0	1.7
circ10-3	10910	10.2	10910	3	0	10.4	10910	17	0	11.3
co-100	783	1.2	783	3	0	2.3	783	16	0	6.8
co5	12067	3.2	12067	3	0	3.5	12067	17	0	4.0
co9	19820	11.6	19820	3	0	11.7	19820	17	0	13.2
complex	9592	2.4	9646	4	1	2.6	9646	20	1	2.8
cont1	40707	985.6	40707	3	0	1226.7	40707	18	0	1234.3
cont4	40802	2698.5	40802	3	0	2839.8	40802	18	0	2842.6
core2536-691	41182	16.0	41182	3	0	16.8	41182	16	0	16.7
core4872-1529	69516	69.1	69516	4	0	70.6	69516	19	0	70.9
cov1075	3270	0.5	3270	3	0	0.4	3270	18	0	0.6
cq5	12350	3.1	12350	3	0	3.1	12350	17	0	3.7
cq9	17826	7.1	17826	3	0	7.2	17826	17	0	8.2
cr42	999	0.1	999	3	0	0.1	999	16	0	0.4
cre-a	3555	0.3	3555	3	0	0.3	3555	16	0	0.6
cre-b	12712	5.0	12712	3	0	5.4	12712	16	0	6.6
cre-c	2842	0.3	2842	3	0	0.3	2842	16	0	0.5
cre-d	9271	3.5	9271	3	0	3.9	9271	16	0	5.0
crew1	1849	0.6	1849	3	0	0.4	1849	16	0	0.8
csched007	5522	0.7	5522	3	0	0.4	5522	17	0	0.6
csched008	3102	0.4	3102	3	0	0.4	3102	17	0	0.2
csched010	6832	0.8	6832	3	0	0.8	6832	17	0	0.7
cycle	920	0.1	920	3	0	0.1	920	17	0	0.1
czprob	1572	0.2	1572	3	0	0.1	1572	16	0	0.4
d10200	2852	0.6	2852	4	0	0.5	2852	19	0	0.8
d20200	9723	1.8	9723	4	0	1.9	9723	22	0	2.5
d2q06c	13920	3.9	13920	4	0	3.7	13920	18	0	4.2
d6cube	1179	0.4	1179	3	0	0.4	1179	16	0	0.4
dano3_3	46251	28.8	46251	3	0	29.0	46251	18	0	29.4
dano3_4	46251	28.9	46251	3	0	28.7	46251	18	0	29.5
dano3_5	46251	30.5	46251	3	0	29.0	46251	18	0	29.3
dano3mip	46251	28.8	46251	3	0	29.1	46251	18	0	29.4
danoint	2763	0.3	2763	3	0	0.3	2763	17	0	0.2
dbir1	14306	11.2	14306	3	0	11.4	14306	16	0	12.2
dbir2	11641	4.9	11641	3	0	5.3	11641	16	0	6.7
dc1c	19795	6.2	19795	4	0	6.2	19795	19	0	6.9
dc1l	24825	21.1	24825	3	0	21.4	24825	18	0	23.3
dcmulti	479	0.0	479	3	0	0.0	479	16	0	0.0
de063155	2546	0.3	2623	7	3	0.4	2623	20	3	0.4
de063157	31021	1.8	*145532883*	*5*	*0*	*7200.0*	*145916183*	*20*	*0*	*7200.0*
de080285	888	0.1	888	3	0	0.1	888	16	0	0.2
degen2	1325	0.1	1325	3	0	0.1	1325	16	0	0.2
degen3	5832	0.9	5832	3	0	0.8	5832	16	0	0.9
delf000	1675	0.4	1675	3	0	0.5	1675	16	0	0.8
delf001	1703	0.4	1703	3	0	0.5	1703	16	0	0.7
delf002	2009	0.4	2009	3	0	0.2	2009	16	0	0.8
delf003	3118	0.6	3180	3	1	0.8	3180	16	1	0.7
delf004	2624	0.6	2704	3	1	0.3	2704	16	1	0.6
delf005	3226	0.7	3294	3	1	0.8	3294	16	1	1.1
delf006	3066	0.6	3125	3	1	0.3	3125	16	1	0.8
delf007	2757	0.5	2826	3	1	0.3	2826	16	1	0.8
delf008	3393	0.7	3493	3	2	0.6	3493	17	2	0.8
delf009	3307	0.6	3383	3	1	0.8	3383	16	1	0.7

Table A.2 continued

Instance	SoPlex$_9$		SoPlex$_{50}$				SoPlex$_{250}$			
	iter	t	iter	R	R_0	t	iter	R	R_0	t
delf010	3132	0.7	3209	3	2	0.4	3209	16	2	0.9
delf011	3030	0.6	3104	3	2	0.4	3104	16	2	0.7
delf012	2890	0.6	2963	3	1	0.5	2963	16	1	0.6
delf013	3103	0.6	3184	3	1	0.5	3184	16	1	1.0
delf014	4257	0.8	4322	3	1	0.9	4322	16	1	0.8
delf015	3176	0.7	3242	3	1	0.4	3242	16	1	0.9
delf017	3075	0.5	3131	3	1	0.4	3131	17	1	0.8
delf018	3281	0.5	3335	3	1	0.4	3335	16	1	0.6
delf019	3160	0.6	3160	3	0	0.6	3160	16	0	0.7
delf020	3701	0.6	3784	3	2	0.6	3784	17	2	0.9
delf021	3247	0.5	3331	3	1	0.3	3331	16	1	0.7
delf022	3673	0.6	3757	3	2	0.6	3757	17	2	0.9
delf023	4603	0.8	4747	3	2	0.6	4747	17	2	0.8
delf024	3973	0.8	4150	3	1	0.7	4150	16	1	0.8
delf025	4032	0.7	4152	3	2	0.8	4152	16	2	1.0
delf026	3378	0.8	3495	3	1	0.5	3495	16	1	1.0
delf027	3402	0.8	3519	3	2	0.5	3519	16	2	0.7
delf028	3269	0.6	3425	3	2	0.4	3425	17	2	1.1
delf029	2949	0.6	3085	3	1	0.5	3085	16	1	0.7
delf030	3002	0.6	3135	3	1	0.6	3135	16	1	0.9
delf031	2835	0.6	2969	3	1	0.5	2969	16	1	0.7
delf032	2958	0.5	3091	3	1	0.6	3091	16	1	0.7
delf033	2210	0.4	2344	3	1	0.3	2344	16	1	0.5
delf034	2699	0.6	2832	3	1	0.4	2832	16	1	0.7
delf035	2407	0.5	2545	3	2	0.4	2545	16	2	0.6
delf036	2522	0.5	2655	3	2	0.4	2655	16	2	1.0
deter0	3725	0.2	3725	3	0	0.2	3725	16	0	0.5
deter1	9022	0.6	9022	3	0	0.8	9022	16	0	1.2
deter2	11017	0.6	11017	3	0	0.8	11017	16	0	1.3
deter3	12348	0.8	12348	3	0	1.0	12348	16	0	1.6
deter4	5633	0.3	5633	3	0	0.2	5633	16	0	0.6
deter5	9277	0.6	9277	3	0	0.8	9277	16	0	1.3
deter6	7482	0.4	7482	3	0	0.3	7482	16	0	1.0
deter7	11149	0.8	11149	3	0	0.8	11149	16	0	1.6
deter8	6977	0.4	6977	3	0	0.3	6977	16	0	0.9
df2177	1391	0.7	1391	3	0	0.7	1391	18	0	1.0
dfl001	30478	18.5	30478	3	0	18.6	30478	18	0	18.9
dfn-gwin-UUM	373	0.0	373	3	0	0.0	373	15	0	0.0
dg012142	11646	2.0	11646	3	0	1.9	11646	18	0	2.0
disctom	13965	1.7	13965	3	0	1.7	13965	17	0	1.8
disp3	490	0.0	490	3	0	0.0	490	16	0	0.1
dolom1	19382	7.3	19382	4	0	7.5	19382	19	0	8.3
ds	13089	17.8	13089	3	0	19.0	13089	18	0	25.0
ds-big	44732	604.9	44732	4	0	610.6	44732	21	0	638.7
dsbmip	2179	0.1	2179	3	0	0.2	2179	17	0	0.3
e18	12395	4.6	12395	3	0	4.8	12395	16	0	5.3
e226	402	0.0	402	3	0	0.0	402	16	0	0.1
egout	96	0.0	96	3	0	0.0	96	15	0	0.0
eil33-2	308	0.1	308	3	0	0.2	308	16	0	0.7
eilA101-2	5338	20.6	5338	3	0	21.9	5338	17	0	28.7
eilB101	1459	0.3	1459	3	0	0.4	1459	16	0	0.6
enigma	44	0.0	44	3	0	0.0	44	15	0	0.0
enlight13	0	0.0	0	0	0	0.0	0	0	0	0.0
enlight14	0	0.0	0	0	0	0.0	0	0	0	0.0
enlight15	0	0.0	0	0	0	0.0	0	0	0	0.0
enlight16	0	0.0	0	0	0	0.0	0	0	0	0.0
enlight9	0	0.0	0	0	0	0.0	0	0	0	0.0
etamacro	766	0.0	771	4	1	0.1	771	16	1	0.1
ex10	115687	1791.6	115687	3	0	1799.0	115687	18	0	1797.9
ex1010-pi	19385	10.0	19385	3	0	10.1	19385	18	0	10.8
ex3sta1	7713	4.8	7713	4	0	5.0	7713	19	0	6.2
ex9	57559	349.2	57559	3	0	348.8	57559	18	0	350.3
f2000	40611	60.6	40611	3	0	60.2	40611	19	0	61.4
farm	0	0.0	0	0	0	0.0	0	0	0	0.0
fast0507	13213	11.4	13213	3	0	11.9	13213	17	0	13.5
fffff800	855	0.1	855	3	0	0.0	855	16	0	0.1
fiball	2818	0.4	2818	3	0	0.7	2818	16	0	1.5
fiber	278	0.0	278	3	0	0.0	278	16	0	0.1
finnis	524	0.0	524	3	0	0.0	524	16	0	0.1
fit1d	1006	0.1	1006	3	0	0.0	1006	16	0	0.2
fit1p	3573	0.4	3573	3	0	0.4	3573	18	0	0.5
fit2d	10781	1.9	10781	3	0	2.1	10781	16	0	2.8

Table A.2 continued

Instance	SoPlex$_9$ iter	t	SoPlex$_{50}$ iter	R	R$_0$	t	SoPlex$_{250}$ iter	R	R$_0$	t
fit2p	16070	3.8	16070	3	0	4.0	16070	16	0	4.3
fixnet6	184	0.0	184	3	0	0.0	184	15	0	0.0
flugpl	11	0.0	11	3	0	0.0	11	16	0	0.0
fome11	45759	38.5	46151	8	1	39.9	46151	38	1	41.0
fome12	93828	108.4	95445	8	1	115.1	95445	38	1	118.8
fome13	175861	309.8	177979	8	1	321.1	177979	39	1	327.6
fome20	37294	13.1	37294	3	0	13.3	37294	15	0	14.4
fome21	81051	77.4	81051	3	0	76.9	81051	16	0	78.9
forplan	369	0.0	369	3	0	0.0	369	18	0	0.1
fxm2-16	6817	0.6	6817	3	0	0.8	6817	17	0	1.0
fxm2-6	2277	0.1	2277	3	0	0.1	2277	17	0	0.4
fxm3_16	48534	22.2	48534	3	0	22.7	48534	17	0	27.1
fxm3_6	9986	0.8	9986	3	0	1.1	9986	17	0	1.5
fxm4_6	25805	4.9	25805	3	0	5.7	25805	17	0	8.5
g200x740i	721	0.0	721	3	0	0.0	721	15	0	0.1
gams10a	38	0.0	44	4	1	0.0	44	16	1	0.0
gams30a	146	0.0	184	4	1	0.0	184	17	1	0.0
ganges	1291	0.1	1291	3	0	0.1	1291	18	0	0.3
ge	11171	2.4	11172	4	1	2.5	11172	19	1	3.3
gen	384	0.0	384	3	0	0.0	384	16	0	0.0
gen1	11914	12.1	12282	4	1	12.4	12282	20	1	13.2
gen2	12239	25.4	12239	4	0	26.4	12239	19	0	27.9
gen4	14397	56.3	14823	4	1	57.2	14823	20	1	59.1
ger50_17_trans	4819	1.3	4819	3	0	1.6	4819	15	0	2.6
germanrr	8775	2.7	8775	3	0	2.8	8775	16	0	3.6
germany50-DBM	8399	1.2	8399	3	0	0.9	8399	15	0	0.9
gesa2	1118	0.0	1118	3	0	0.0	1118	16	0	0.1
gesa2-o	653	0.0	653	3	0	0.0	653	16	0	0.1
gesa2_o	653	0.0	653	3	0	0.0	653	16	0	0.1
gesa3	974	0.1	974	3	0	0.1	974	16	0	0.2
gesa3_o	530	0.0	530	3	0	0.0	530	16	0	0.1
gfrd-pnc	664	0.0	664	3	0	0.1	664	16	0	0.1
glass4	73	0.0	73	3	0	0.0	73	15	0	0.0
gmu-35-40	316	0.0	316	3	0	0.0	316	16	0	0.1
gmu-35-50	359	0.0	359	3	0	0.1	359	16	0	0.1
gmut-75-50	6042	8.1	6042	3	0	8.5	6042	16	0	10.7
gmut-77-40	4047	1.7	4047	3	0	2.0	4047	16	0	2.6
go19	2606	0.2	2606	3	0	0.1	2606	17	0	0.3
gr4x6	36	0.0	36	3	0	0.0	36	15	0	0.0
greenbea	18382	4.7	18382	4	0	4.8	18382	23	0	5.0
greenbeb	11957	2.5	11957	4	0	2.5	11957	21	0	2.7
grow15	2102	0.2	2102	3	0	0.1	2102	16	0	0.2
grow22	3334	0.4	3334	3	0	0.2	3334	17	0	0.6
grow7	1071	0.1	1071	3	0	0.0	1071	16	0	0.1
gt2	19	0.0	19	3	0	0.0	19	15	0	0.0
hanoi5	7389	1.6	7389	3	0	1.7	7389	15	0	1.6
haprp	1303	0.0	1303	3	0	0.0	1303	16	0	0.1
harp2	323	0.0	323	3	0	0.0	323	16	0	0.1
i_n13	948541	2586.7	948541	1	0	2571.1	948541	1	0	2573.6
ic97_potential	339	0.0	339	3	0	0.0	339	15	0	0.0
iiasa	1562	0.1	1562	3	0	0.1	1562	16	0	0.2
iis-100-0-cov	1279	0.5	1279	3	0	0.5	1279	18	0	0.7
iis-bupa-cov	4196	1.1	4196	3	0	1.1	4196	17	0	1.1
iis-pima-cov	3532	1.2	3532	3	0	1.1	3532	16	0	1.5
israel	149	0.0	149	3	0	0.0	149	16	0	0.1
ivu06-big	37753	3149.6	37753	3	0	3170.1	37753	19	0	3335.5
ivu52	18983	107.6	18983	3	0	110.3	18983	17	0	124.0
janos-us-DDM	1042	0.0	1042	0	0	0.0	1042	0	0	0.0
jendrec1	11116	2.4	11116	3	0	2.8	11116	17	0	4.3
k16x240	39	0.0	39	3	0	0.0	39	15	0	0.0
kb2	58	0.0	58	3	0	0.0	58	17	0	0.0
ken-07	2777	0.1	2777	3	0	0.2	2777	16	0	0.5
ken-11	17739	2.7	17739	3	0	3.0	17739	16	0	3.8
ken-13	41646	20.1	41646	3	0	20.5	41646	16	0	22.0
ken-18	192333	500.8	192333	3	0	501.6	192333	16	0	505.9
kent	1680	0.3	1680	3	0	0.5	1680	16	0	1.2
khb05250	119	0.0	119	3	0	0.0	119	15	0	0.0
kl02	263	0.2	263	3	0	0.4	263	15	0	0.6
kleemin3	0	0.0	0	0	0	0.0	0	0	0	0.0
kleemin4	0	0.0	0	0	0	0.0	0	0	0	0.0
kleemin5	0	0.0	0	0	0	0.0	0	0	0	0.0
kleemin6	0	0.0	0	0	0	0.0	0	0	0	0.0

Table A.2 continued

Instance	SoPlex$_9$		SoPlex$_{50}$				SoPlex$_{250}$			
	iter	t	iter	R	R$_0$	t	iter	R	R$_0$	t
kleemin7	0	0.0	0	0	0	0.0	0	0	0	0.0
kleemin8	0	0.0	0	0	0	0.0	0	0	0	0.0
l152lav	700	0.1	700	3	0	0.1	700	16	0	0.2
l30	*1379080*	*3042.6*	*3304695*	*6*	*0*	*7200.0*	*3304140*	*34*	*0*	*7200.0*
l9	746	0.1	746	3	0	0.1	746	17	0	0.3
large000	3748	0.6	3748	3	0	0.7	3748	16	0	1.2
large001	7844	1.4	7844	3	0	1.4	7844	17	0	1.6
large002	3866	0.8	4065	3	2	0.6	4065	16	2	1.5
large003	4179	0.8	4254	3	1	0.7	4254	17	1	1.2
large004	4647	1.0	4681	4	2	0.8	4681	17	2	1.4
large005	4501	0.8	4567	3	1	0.8	4567	16	1	0.9
large006	4864	0.8	4942	3	2	0.7	4942	17	2	1.2
large007	5010	0.8	5094	3	1	0.9	5094	17	1	1.3
large008	5203	0.8	5291	3	1	0.9	5291	16	1	1.2
large009	4988	0.9	5074	3	1	1.1	5074	16	1	1.2
large010	4639	0.7	4725	3	1	0.8	4725	16	1	1.2
large011	5135	0.9	5221	3	1	0.9	5221	16	1	1.0
large012	4924	0.8	5009	3	1	0.9	5009	16	1	1.0
large013	4975	0.8	5062	3	1	0.7	5062	16	1	1.2
large014	5082	0.9	5148	3	1	0.9	5148	16	1	1.1
large015	4318	0.7	4380	3	1	0.6	4380	16	1	0.9
large016	4571	0.6	4633	3	1	0.8	4633	16	1	0.9
large017	3980	0.8	3980	3	0	0.8	3980	16	0	0.9
large018	4459	0.8	4459	3	0	0.8	4459	16	0	1.0
large019	4909	0.8	4909	3	0	0.8	4909	16	0	1.1
large020	6984	1.1	7059	3	2	0.9	7059	17	2	1.5
large021	6201	0.9	6288	3	2	0.9	6288	16	2	1.5
large022	6907	0.9	6993	3	2	0.8	6993	16	2	1.4
large023	4224	0.9	4398	3	2	0.7	4398	17	2	1.1
large024	5788	1.0	5988	4	2	1.2	5988	17	2	1.6
large025	4811	0.8	5000	3	2	1.0	5000	16	2	1.2
large026	4196	0.8	4367	3	2	0.7	4367	17	2	1.1
large027	4172	0.8	4353	3	2	0.8	4353	16	2	1.0
large028	4691	1.1	4906	3	1	0.8	4906	16	1	1.4
large029	4158	0.8	4372	4	2	0.6	4372	17	2	1.2
large030	3732	0.6	3930	3	2	0.7	3930	16	2	1.2
large031	3729	0.6	3931	3	1	0.6	3931	16	1	1.2
large032	4851	0.9	5052	3	1	0.9	5052	16	1	1.1
large033	3675	0.7	3877	3	2	0.5	3877	16	2	1.2
large034	4009	0.7	4201	3	2	0.7	4201	16	2	0.9
large035	3450	0.8	3655	3	1	0.7	3655	17	1	1.1
large036	3111	0.6	3314	3	2	0.6	3314	17	2	1.0
lectsched-1	7	1.1	7	3	0	1.3	7	15	0	1.9
lectsched-1-obj	963	1.2	963	3	0	1.5	963	15	0	2.1
lectsched-2	3	0.5	3	3	0	0.7	3	15	0	1.1
lectsched-3	7	0.9	7	3	0	1.1	7	15	0	1.6
lectsched-4-obj	174	0.3	174	3	0	0.2	174	15	0	0.7
leo1	862	0.3	862	3	0	0.5	862	16	0	0.9
leo2	1637	0.6	1637	4	0	0.9	1637	17	0	1.8
liu	543	0.1	543	3	0	0.1	543	15	0	0.1
lo10	953870	5941.4	953870	0	0	5931.8	953870	0	0	5940.8
long15	229488	2753.1	229488	0	0	2738.1	229488	0	0	2736.6
lotfi	226	0.0	226	3	0	0.0	226	15	0	0.0
lotsize	1460	0.0	1460	3	0	0.1	1460	15	0	0.2
lp22	38451	35.6	38451	3	0	35.8	38451	18	0	36.3
lpl1	36759	59.4	36759	3	0	60.0	36759	17	0	64.2
lpl2	1465	0.2	1465	3	0	0.1	1465	15	0	0.4
lpl3	5040	1.1	5040	3	0	1.2	5040	15	0	1.4
lrn	11450	2.6	11452	4	1	2.5	11452	17	1	3.0
lrsa120	9787	2.5	9789	3	1	2.4	9789	17	1	2.8
lseu	25	0.0	25	3	0	0.0	25	15	0	0.0
m100n500k4r1	174	0.0	174	3	0	0.0	174	17	0	0.0
macrophage	706	0.0	706	0	0	0.0	706	0	0	0.0
manna81	3018	0.1	3018	0	0	0.1	3018	0	0	0.1
map06	23840	39.4	23840	3	0	40.6	23840	16	0	44.4
map10	23747	40.6	23747	3	0	41.5	23747	16	0	45.2
map14	23178	38.3	23178	3	0	39.4	23178	16	0	42.9
map18	20964	33.6	20964	3	0	34.7	20964	16	0	37.8
map20	19686	33.1	19686	3	0	30.9	19686	15	0	33.8
markshare1	35	0.0	35	3	0	0.0	35	16	0	0.0
markshare2	43	0.0	43	3	0	0.0	43	16	0	0.0
markshare_5_0	24	0.0	24	3	0	0.0	24	16	0	0.0

Table A.2 continued

Instance	SoPlex$_9$ iter	t	SoPlex$_{50}$ iter	R	R$_0$	t	SoPlex$_{250}$ iter	R	R$_0$	t
maros	1255	0.1	1255	4	0	0.2	1255	19	0	0.4
maros-r7	7953	2.9	7953	3	0	3.5	7953	16	0	4.6
mas74	224	0.0	224	3	0	0.0	224	17	0	0.0
mas76	132	0.0	132	3	0	0.0	132	16	0	0.0
maxgasflow	6737	0.6	6737	3	0	0.8	6737	16	0	1.1
mc11	1239	0.1	1239	3	0	0.1	1239	15	0	0.3
mcf2	2763	0.3	2763	3	0	0.1	2763	17	0	0.4
mcsched	2546	0.3	2546	3	0	0.3	2546	17	0	0.5
methanosarcina	655	0.1	655	0	0	0.2	655	0	0	0.2
mik-250-1-100-1	100	0.0	100	3	0	0.0	100	15	0	0.1
mine-166-5	1642	0.5	1642	3	0	0.5	1642	16	0	0.8
mine-90-10	1948	0.5	1948	3	0	0.4	1948	16	0	0.4
misc03	45	0.0	45	0	0	0.0	45	0	0	0.0
misc06	816	0.1	816	3	0	0.1	816	16	0	0.2
misc07	157	0.0	157	3	0	0.0	157	15	0	0.1
mitre	2451	0.3	2451	3	0	0.4	2451	17	0	0.7
mkc	538	0.1	538	3	0	0.1	538	16	0	0.1
mkc1	538	0.1	538	3	0	0.0	538	16	0	0.1
mod008	27	0.0	27	3	0	0.0	27	15	0	0.0
mod010	1062	0.1	1062	3	0	0.2	1062	15	0	0.4
mod011	6153	0.9	6153	3	0	0.8	6153	17	0	1.0
mod2	58340	100.4	58534	14	11	103.8	58534	26	11	104.2
model1	180	0.0	182	3	1	0.0	182	16	1	0.0
model10	44687	22.8	44687	3	0	23.2	44687	17	0	24.5
model11	5273	1.0	5273	3	0	1.0	5273	17	0	1.2
model2	3466	0.4	3466	3	0	0.2	3466	18	0	0.5
model3	9208	1.3	9208	3	0	1.3	9208	18	0	1.7
model4	15098	3.0	15098	3	0	2.8	15098	18	0	3.1
model5	19877	3.8	19877	3	0	4.1	19877	17	0	4.7
model6	14770	3.3	14770	4	0	3.2	14770	19	0	3.5
model7	16070	4.0	16070	3	0	4.0	16070	18	0	4.6
model8	2522	0.3	2538	3	1	0.4	2538	16	1	0.7
model9	12397	2.3	12397	4	0	2.5	12397	20	0	3.2
modglob	359	0.0	359	3	0	0.0	359	15	0	0.0
modszk1	653	0.0	653	3	0	0.0	653	17	0	0.1
momentum1	3305	0.8	3306	4	1	1.1	3306	17	1	1.9
momentum2	45882	18.8	45950	5	2	19.7	45950	18	2	21.9
momentum3	46184	189.5	46185	4	1	192.0	46185	18	1	198.4
msc98-ip	9496	1.3	9549	4	1	1.7	9549	16	1	2.2
mspp16	52	19.3	52	3	0	20.0	52	15	0	25.1
multi	59	0.0	59	3	0	0.0	59	16	0	0.0
mzzv11	37921	26.4	37921	3	0	26.2	37921	18	0	26.6
mzzv42z	34373	17.6	34373	3	0	17.2	34373	17	0	17.5
n15-3	43662	114.6	43662	3	0	115.3	43662	16	0	118.7
n3-3	2965	0.6	2965	3	0	0.5	2965	15	0	0.9
n3700	8698	1.6	8698	3	0	1.6	8698	15	0	1.6
n3701	8106	1.3	8106	3	0	1.3	8106	15	0	1.5
n3702	7987	1.2	7987	3	0	1.1	7987	15	0	1.5
n3703	7397	1.1	7397	3	0	1.0	7397	15	0	1.4
n3704	7325	1.1	7325	3	0	0.9	7325	15	0	1.4
n3705	6974	1.3	6974	3	0	1.2	6974	16	0	1.4
n3706	7305	1.1	7305	3	0	1.1	7305	15	0	1.4
n3707	8681	1.2	8681	3	0	1.4	8681	15	0	1.6
n3708	7133	1.6	7133	3	0	1.2	7133	15	0	1.7
n3709	8030	1.5	8030	3	0	1.4	8030	15	0	1.8
n370a	8608	1.2	8608	3	0	1.3	8608	15	0	1.3
n370b	8553	1.4	8553	3	0	1.2	8553	15	0	1.6
n370c	7273	1.1	7273	3	0	0.9	7273	15	0	1.4
n370d	7273	1.1	7273	3	0	0.9	7273	15	0	1.4
n370e	7852	1.2	7852	3	0	1.0	7852	15	0	1.3
n3div36	306	0.4	306	3	0	0.7	306	16	0	1.9
n3seq24	4646	14.1	4720	4	1	15.9	4720	18	1	20.4
n4-3	1341	0.2	1341	3	0	0.2	1341	15	0	0.1
n9-3	3531	0.6	3531	3	0	0.5	3531	16	0	0.9
nag	2476	0.2	2476	3	0	0.1	2476	15	0	0.3
nemsafm	755	0.0	755	3	0	0.0	755	16	0	0.1
nemscem	932	0.1	932	3	0	0.1	932	17	0	0.2
nemsemm1	10608	3.6	10608	3	0	5.4	10608	16	0	12.7
nemsemm2	10445	1.5	10445	3	0	1.9	10445	16	0	3.8
nemspmm1	15049	4.3	15049	4	0	4.5	15049	21	0	5.4
nemspmm2	15056	4.2	15058	4	1	4.4	15058	19	1	5.4
nemswrld	35922	32.6	35922	4	0	33.7	35922	20	0	35.3

Table A.2 continued

Instance	SoPlex$_9$		SoPlex$_{50}$				SoPlex$_{250}$			
	iter	t	iter	R	R$_0$	t	iter	R	R$_0$	t
neos	89942	497.3	90253	4	1	497.8	90253	17	1	500.9
neos-1053234	257	0.1	257	3	0	0.0	257	16	0	0.1
neos-1053591	821	0.0	821	3	0	0.0	821	16	0	0.1
neos-1056905	140	0.0	140	3	0	0.0	140	15	0	0.0
neos-1058477	48	0.0	48	3	0	0.0	48	15	0	0.1
neos-1061020	16447	7.3	16447	3	0	7.5	16447	17	0	8.1
neos-1062641	873	0.0	902	3	1	0.0	902	16	1	0.1
neos-1067731	12132	2.4	12132	3	0	2.5	12132	16	0	2.8
neos-1096528	115	47.5	115	3	0	54.0	115	15	0	56.4
neos-1109824	115	0.5	115	3	0	0.6	115	15	0	0.5
neos-1112782	622	0.1	622	3	0	0.0	622	16	0	0.1
neos-1112787	557	0.1	557	3	0	0.0	557	16	0	0.1
neos-1120495	96	0.2	96	0	0	0.3	96	0	0	0.2
neos-1121679	35	0.0	35	3	0	0.0	35	16	0	0.0
neos-1122047	5499	1.2	5499	3	0	1.5	5499	16	0	3.0
neos-1126860	5216	0.9	5216	3	0	1.0	5216	16	0	1.9
neos-1140050	13327	30.1	13327	5	0	32.6	13327	29	0	38.6
neos-1151496	4930	0.8	4930	3	0	0.8	4930	17	0	0.6
neos-1171448	4075	0.6	4075	1	0	0.5	4075	1	0	0.5
neos-1171692	1290	0.1	1290	3	0	0.1	1290	16	0	0.1
neos-1171737	2060	0.3	2060	3	0	0.2	2060	15	0	0.3
neos-1173026	33	0.0	33	3	0	0.0	33	15	0	0.0
neos-1200887	294	0.0	294	3	0	0.0	294	15	0	0.0
neos-1208069	3008	0.6	3008	3	0	0.6	3008	16	0	0.6
neos-1208135	2364	0.4	2364	3	0	0.4	2364	17	0	0.2
neos-1211578	187	0.0	187	3	0	0.0	187	15	0	0.0
neos-1215259	6319	1.0	6319	3	0	1.0	6319	16	0	1.2
neos-1215891	3541	0.9	3541	3	0	0.7	3541	17	0	1.1
neos-1223462	13116	3.7	13116	3	0	3.6	13116	16	0	4.0
neos-1224597	10673	1.8	10673	3	0	1.5	10673	15	0	1.7
neos-1225589	438	0.0	438	3	0	0.0	438	15	0	0.1
neos-1228986	197	0.0	197	3	0	0.0	197	15	0	0.0
neos-1281048	2276	0.2	2276	3	0	0.3	2276	15	0	0.3
neos-1311124	639	0.0	639	3	0	0.0	639	15	0	0.1
neos-1324574	6564	0.8	6564	3	0	0.9	6564	16	0	0.9
neos-1330346	1983	0.3	1983	3	0	0.1	1983	16	0	0.2
neos-1330635	141	0.0	141	3	0	0.0	141	16	0	0.1
neos-1337307	10964	2.1	10964	3	0	2.2	10964	16	0	2.5
neos-1337489	187	0.0	187	3	0	0.0	187	15	0	0.0
neos-1346382	350	0.0	350	3	0	0.0	350	15	0	0.0
neos-1354092	12518	7.0	12518	3	0	7.1	12518	18	0	8.2
neos-1367061	18400	23.5	18400	4	0	24.1	18400	18	0	25.9
neos-1396125	4834	0.6	4834	3	0	0.7	4834	16	0	0.7
neos-1407044	27160	31.7	27160	3	0	32.1	27160	19	0	33.5
neos-1413153	1027	0.2	1027	3	0	0.2	1027	16	0	0.3
neos-1415183	1763	0.3	1763	3	0	0.4	1763	16	0	0.2
neos-1417043	13642	57.3	13642	0	0	57.2	13642	0	0	57.2
neos-1420205	944	0.0	944	3	0	0.0	944	15	0	0.0
neos-1420546	102727	92.3	102727	3	0	92.2	102727	17	0	93.4
neos-1420790	12930	2.5	12930	3	0	2.4	12930	17	0	2.3
neos-1423785	19268	3.1	19268	3	0	3.6	19268	15	0	4.2
neos-1425699	26	0.0	26	3	0	0.0	26	15	0	0.0
neos-1426635	350	0.0	350	3	0	0.0	350	15	0	0.0
neos-1426662	665	0.0	665	3	0	0.0	665	16	0	0.0
neos-1427181	591	0.0	591	0	0	0.0	591	0	0	0.0
neos-1427261	883	0.1	883	3	0	0.1	883	15	0	0.1
neos-1429185	466	0.0	466	3	0	0.0	466	15	0	0.1
neos-1429212	28835	220.5	28835	3	0	218.8	28835	16	0	223.0
neos-1429461	399	0.0	399	3	0	0.0	399	16	0	0.1
neos-1430701	257	0.0	257	3	0	0.0	257	15	0	0.0
neos-1430811	32707	340.4	32707	3	0	338.3	32707	17	0	343.2
neos-1436709	505	0.0	505	3	0	0.0	505	15	0	0.1
neos-1436713	835	0.0	835	3	0	0.1	835	15	0	0.1
neos-1437164	296	0.0	296	3	0	0.1	296	16	0	0.2
neos-1439395	312	0.0	312	3	0	0.0	312	16	0	0.1
neos-1440225	486	0.1	486	3	0	0.1	486	17	0	0.1
neos-1440447	239	0.0	239	3	0	0.0	239	15	0	0.0
neos-1440457	820	0.0	820	3	0	0.1	820	15	0	0.1
neos-1440460	359	0.0	359	3	0	0.0	359	16	0	0.1
neos-1441553	288	0.0	288	3	0	0.1	288	16	0	0.1
neos-1442119	607	0.0	607	3	0	0.1	607	15	0	0.1
neos-1442657	498	0.0	498	3	0	0.0	498	15	0	0.1

Table A.2 continued

Instance	SoPlex$_9$		SoPlex$_{50}$				SoPlex$_{250}$			
	iter	t	iter	R	R$_0$	t	iter	R	R$_0$	t
neos-1445532	14103	1.7	14103	3	0	1.8	14103	16	0	1.9
neos-1445738	15164	2.8	15164	3	0	2.9	15164	16	0	3.0
neos-1445743	16103	3.2	16103	3	0	3.1	16103	16	0	3.4
neos-1445755	15634	3.1	15634	3	0	3.2	15634	16	0	3.4
neos-1445765	15960	3.3	15960	3	0	3.4	15960	16	0	4.0
neos-1451294	10087	1.8	10087	3	0	1.6	10087	18	0	1.7
neos-1456979	957	0.4	957	3	0	0.2	957	15	0	0.5
neos-1460246	276	0.0	276	3	0	0.0	276	16	0	0.1
neos-1460265	808	0.1	808	3	0	0.1	808	16	0	0.1
neos-1460543	9199	1.3	9199	3	0	1.2	9199	18	0	1.6
neos-1460641	10168	1.1	10168	3	0	1.1	10168	16	0	1.1
neos-1461051	418	0.0	418	3	0	0.1	418	15	0	0.1
neos-1464762	10563	1.2	10563	3	0	1.2	10563	16	0	1.0
neos-1467067	673	0.0	673	3	0	0.0	673	15	0	0.1
neos-1467371	8604	1.1	8604	3	0	0.9	8604	16	0	1.0
neos-1467467	8764	0.9	8764	3	0	0.9	8764	16	0	1.0
neos-1480121	86	0.0	86	3	0	0.0	86	15	0	0.0
neos-1489999	835	0.1	835	3	0	0.1	835	16	0	0.1
neos-1516309	134	0.0	134	3	0	0.1	134	15	0	0.1
neos-1582420	2433	0.5	2433	3	0	0.7	2433	16	0	0.7
neos-1593097	921	0.6	921	3	0	0.7	921	17	0	1.3
neos-1595230	518	0.0	518	3	0	0.0	518	16	0	0.1
neos-1597104	150	0.5	150	3	0	0.7	150	15	0	1.7
neos-1599274	357	0.1	357	3	0	0.1	357	15	0	0.3
neos-1601936	14758	5.3	14758	3	0	5.4	14758	18	0	5.7
neos-1603512	814	0.1	814	3	0	0.1	814	15	0	0.2
neos-1603518	2362	0.4	2362	3	0	0.4	2362	16	0	0.6
neos-1603965	30552	4.3	*1109741*	*4*	*4*	*1033.8*	*1109741*	*16*	*16*	*1032.2*
neos-1605061	23399	11.5	23399	3	0	11.5	23399	19	0	11.7
neos-1605075	16980	8.8	16980	3	0	8.8	16980	18	0	9.1
neos-1616732	314	0.0	314	0	0	0.0	314	0	0	0.0
neos-1620770	772	0.1	772	0	0	0.1	772	0	0	0.0
neos-1620807	222	0.0	222	0	0	0.0	222	0	0	0.0
neos-1622252	807	0.1	807	0	0	0.1	807	0	0	0.1
neos-430149	240	0.0	240	3	0	0.0	240	15	0	0.1
neos-476283	5536	3.8	5536	3	0	4.3	5536	17	0	5.8
neos-480878	466	0.1	466	3	0	0.1	466	16	0	0.2
neos-494568	553	0.2	553	3	0	0.2	553	15	0	0.4
neos-495307	963	0.3	963	3	0	0.2	963	15	0	0.5
neos-498623	1034	0.3	1037	3	1	0.6	1037	16	1	0.7
neos-501453	37	0.0	37	3	0	0.0	37	16	0	0.0
neos-501474	288	0.0	288	3	0	0.0	288	16	0	0.0
neos-503737	4883	0.8	4883	3	0	0.8	4883	18	0	0.9
neos-504674	548	0.0	548	3	0	0.0	548	15	0	0.1
neos-504815	468	0.0	468	3	0	0.0	468	15	0	0.0
neos-506422	98	0.1	98	3	0	0.1	98	15	0	0.3
neos-506428	2092	1.0	2092	0	0	1.1	2092	0	0	1.0
neos-512201	543	0.0	543	3	0	0.0	543	15	0	0.0
neos-520729	32688	19.9	32688	3	0	20.2	32688	16	0	20.8
neos-522351	205	0.0	205	3	0	0.0	205	15	0	0.1
neos-525149	1124	0.7	1124	0	0	0.8	1124	0	0	0.8
neos-530627	68	0.0	68	3	0	0.0	68	15	0	0.0
neos-538867	169	0.0	171	3	1	0.0	171	15	1	0.1
neos-538916	168	0.0	168	3	0	0.0	168	15	0	0.1
neos-544324	2629	1.2	2629	3	0	1.5	2629	19	0	3.3
neos-547911	1895	0.3	1895	3	0	0.5	1895	17	0	0.9
neos-548047	11782	1.6	11782	3	0	1.8	11782	16	0	1.6
neos-548251	2037	0.1	2037	3	0	0.1	2037	15	0	0.1
neos-551991	7378	1.1	7378	3	0	1.1	7378	16	0	1.1
neos-555001	10198	1.0	10198	3	0	0.6	10198	16	0	0.9
neos-555298	3995	0.2	3995	3	0	0.1	3995	16	0	0.2
neos-555343	11043	1.1	11043	3	0	0.9	11043	16	0	1.3
neos-555424	5895	0.4	5895	3	0	0.2	5895	16	0	0.3
neos-555694	565	0.1	565	3	0	0.1	565	16	0	0.3
neos-555771	575	0.1	575	3	0	0.1	575	16	0	0.3
neos-555884	7183	0.6	7183	3	0	0.4	7183	16	0	0.8
neos-555927	2202	0.1	2202	3	0	0.1	2202	16	0	0.1
neos-565672	87785	259.8	87785	3	0	260.9	87785	16	0	264.9
neos-565815	6971	2.3	6971	3	0	2.4	6971	17	0	2.9
neos-570431	2355	0.2	2355	3	0	0.1	2355	16	0	0.3
neos-574665	506	0.1	506	3	0	0.1	506	16	0	0.3
neos-578379	14389	13.2	14389	3	0	13.4	14389	18	0	13.8

Appendix A. Experimental Data and Results

Table A.2 continued

Instance	SoPlex$_9$		SoPlex$_{50}$				SoPlex$_{250}$			
	iter	t	iter	R	R$_0$	t	iter	R	R$_0$	t
neos-582605	1067	0.1	1067	3	0	0.1	1067	15	0	0.1
neos-583731	602	0.0	602	0	0	0.0	602	0	0	0.0
neos-584146	603	0.1	603	3	0	0.1	603	15	0	0.1
neos-584851	612	0.0	612	3	0	0.1	612	16	0	0.1
neos-584866	11751	2.6	11751	3	0	2.8	11751	16	0	2.8
neos-585192	2614	0.5	2614	3	0	0.7	2614	17	0	0.7
neos-585467	1912	0.4	1912	3	0	0.2	1912	16	0	0.4
neos-593853	615	0.0	615	3	0	0.1	615	16	0	0.1
neos-595904	1276	0.2	1276	3	0	0.3	1276	16	0	0.3
neos-595905	390	0.0	390	3	0	0.1	390	16	0	0.1
neos-595925	521	0.0	521	3	0	0.1	521	16	0	0.1
neos-598183	1148	0.1	1148	3	0	0.1	1148	16	0	0.2
neos-603073	455	0.0	455	3	0	0.0	455	16	0	0.1
neos-611135	11914	4.7	11914	3	0	4.7	11914	16	0	5.1
neos-611838	2046	0.4	2046	3	0	0.2	2046	16	0	0.5
neos-612125	2002	0.4	2002	3	0	0.3	2002	16	0	0.7
neos-612143	2100	0.4	2100	3	0	0.3	2100	16	0	0.5
neos-612162	1828	0.3	1828	3	0	0.3	1828	16	0	0.7
neos-619167	6697	1.3	6697	3	0	1.1	6697	16	0	1.3
neos-631164	954	0.1	954	3	0	0.1	954	16	0	0.1
neos-631517	765	0.0	765	3	0	0.1	765	16	0	0.0
neos-631694	24940	5.2	24940	3	0	5.2	24940	16	0	5.2
neos-631709	70640	213.1	70640	3	0	213.0	70640	16	0	216.1
neos-631710	66455	923.9	66455	3	0	920.8	66455	16	0	922.3
neos-631784	26868	29.2	26868	3	0	29.5	26868	16	0	30.8
neos-632335	5592	1.0	5592	3	0	1.3	5592	16	0	1.7
neos-633273	5070	1.1	5070	3	0	1.1	5070	16	0	1.6
neos-641591	14066	6.7	14066	3	0	7.1	14066	16	0	7.6
neos-655508	119	0.1	119	0	0	0.1	119	0	0	0.0
neos-662469	14066	7.2	14066	3	0	7.0	14066	16	0	7.7
neos-686190	834	0.2	834	3	0	0.1	834	16	0	0.2
neos-691058	7206	1.3	7206	3	0	1.2	7206	17	0	1.4
neos-691073	6943	1.4	6943	3	0	1.3	6943	17	0	1.3
neos-693347	9600	2.1	9600	3	0	2.2	9600	17	0	2.5
neos-702280	12807	14.6	12807	3	0	15.3	12807	18	0	17.8
neos-709469	429	0.0	429	3	0	0.0	429	16	0	0.0
neos-717614	1049	0.0	1049	3	0	0.1	1049	16	0	0.2
neos-738098	25294	31.9	25294	3	0	31.8	25294	17	0	32.0
neos-775946	1720	0.6	1720	3	0	0.5	1720	16	0	0.6
neos-777800	1633	0.5	1633	3	0	0.3	1633	17	0	0.6
neos-780889	57556	184.5	57556	0	0	185.1	57556	0	0	185.1
neos-785899	411	0.1	411	3	0	0.1	411	16	0	0.1
neos-785912	1338	0.3	1338	3	0	0.3	1338	17	0	0.4
neos-785914	232	0.0	232	3	0	0.0	232	15	0	0.1
neos-787933	14008	31.7	14008	3	0	32.3	14008	15	0	47.2
neos-791021	10768	2.5	10768	3	0	2.4	10768	16	0	2.8
neos-796608	379	0.0	379	0	0	0.0	379	0	0	0.0
neos-799711	19400	3.0	19400	3	0	3.3	19400	16	0	4.6
neos-799838	11429	3.8	11429	3	0	3.8	11429	15	0	4.0
neos-801834	1782	0.3	1782	3	0	0.2	1782	16	0	0.6
neos-803219	1309	0.1	1309	3	0	0.1	1309	16	0	0.2
neos-803220	888	0.1	888	3	0	0.1	888	16	0	0.1
neos-806323	1246	0.1	1246	3	0	0.1	1246	16	0	0.2
neos-807454	12164	2.2	12164	3	0	2.3	12164	16	0	2.2
neos-807456	11272	1.6	11272	3	0	1.6	11272	17	0	1.3
neos-807639	1305	0.1	1305	3	0	0.1	1305	16	0	0.1
neos-807705	1259	0.1	1259	3	0	0.1	1259	17	0	0.3
neos-808072	8164	1.4	8164	3	0	1.4	8164	16	0	1.3
neos-808214	1236	0.2	1236	3	0	0.1	1236	17	0	0.2
neos-810286	13517	3.5	13517	3	0	3.5	13517	18	0	3.7
neos-810326	8031	1.2	8031	3	0	1.2	8031	17	0	1.3
neos-820146	249	0.0	249	3	0	0.0	249	15	0	0.0
neos-820157	569	0.0	569	3	0	0.0	569	16	0	0.1
neos-820879	1509	0.5	1509	3	0	0.7	1509	16	0	0.9
neos-824661	9912	5.2	9912	3	0	5.6	9912	16	0	5.6
neos-824695	4726	1.8	4726	3	0	1.8	4726	16	0	1.9
neos-825075	1062	0.1	1062	3	0	0.0	1062	17	0	0.1
neos-826224	5396	1.2	5396	3	0	1.4	5396	15	0	1.4
neos-826250	5747	1.3	5747	3	0	1.0	5747	16	0	1.1
neos-826650	8372	1.7	8372	3	0	1.9	8372	17	0	1.6
neos-826694	10754	2.6	10754	3	0	2.7	10754	16	0	2.8
neos-826812	10450	2.4	10450	3	0	2.4	10450	16	0	2.5

Table A.2 continued

Instance	SoPlex$_9$		SoPlex$_{50}$				SoPlex$_{250}$			
	iter	t	iter	R	R$_0$	t	iter	R	R$_0$	t
neos-826841	3738	0.8	3738	3	0	0.7	3738	15	0	0.8
neos-827015	18897	12.2	18897	3	0	12.8	18897	16	0	14.7
neos-827175	11439	3.2	11439	3	0	3.3	11439	16	0	3.8
neos-829552	13050	4.4	13050	3	0	4.8	13050	16	0	5.7
neos-830439	179	0.0	179	3	0	0.0	179	15	0	0.1
neos-831188	8374	1.1	8374	3	0	1.2	8374	16	0	1.2
neos-839838	6451	1.1	6451	3	0	1.4	6451	16	0	1.6
neos-839859	2312	0.4	2312	3	0	0.2	2312	16	0	0.6
neos-839894	21287	20.6	21287	3	0	20.8	21287	17	0	21.8
neos-841664	5975	0.7	5975	3	0	0.6	5975	16	0	0.9
neos-847051	2383	0.2	2383	3	0	0.3	2383	16	0	0.4
neos-847302	2146	0.3	2146	3	0	0.4	2146	17	0	0.2
neos-848150	1409	0.2	1409	3	0	0.2	1409	17	0	0.3
neos-848198	11363	2.1	11363	3	0	2.3	11363	15	0	2.6
neos-848589	1401	1.0	1401	3	0	1.9	1401	16	0	4.8
neos-848845	7295	1.4	7295	3	0	1.4	7295	17	0	1.5
neos-849702	5626	1.0	5626	3	0	1.0	5626	17	0	1.1
neos-850681	25783	5.9	25783	3	0	5.8	25783	17	0	6.2
neos-856059	580	0.1	580	0	0	0.1	580	0	0	0.1
neos-859770	294	0.3	294	3	0	0.3	294	16	0	0.6
neos-860244	90	0.1	90	3	0	0.1	90	15	0	0.3
neos-860300	349	0.2	349	3	0	0.3	349	16	0	0.8
neos-862348	1026	0.2	1026	3	0	0.2	1026	16	0	0.4
neos-863472	142	0.0	142	3	0	0.0	142	15	0	0.0
neos-872648	38712	176.3	38712	3	0	175.4	38712	15	0	180.9
neos-873061	30803	140.9	30803	3	0	142.5	30803	15	0	146.5
neos-876808	92322	98.6	92322	3	0	98.4	92322	16	0	100.0
neos-880324	232	0.0	232	3	0	0.0	232	16	0	0.0
neos-881765	424	0.0	424	3	0	0.1	424	16	0	0.0
neos-885086	3951	0.6	3951	3	0	0.9	3951	16	0	1.1
neos-885524	205	5.4	205	3	0	5.6	205	16	0	7.1
neos-886822	1834	0.3	1834	3	0	0.2	1834	18	0	0.4
neos-892255	1082	0.2	1082	3	0	0.2	1082	16	0	0.3
neos-905856	1228	0.1	1228	3	0	0.2	1228	16	0	0.2
neos-906865	792	0.1	792	3	0	0.1	792	16	0	0.1
neos-911880	319	0.0	319	3	0	0.0	319	16	0	0.1
neos-911970	258	0.0	258	3	0	0.0	258	16	0	0.1
neos-912015	1001	0.1	1001	3	0	0.2	1001	17	0	0.2
neos-912023	1095	0.1	1095	3	0	0.2	1095	16	0	0.3
neos-913984	4692	2.5	4692	3	0	2.8	4692	16	0	3.0
neos-914441	7787	1.9	7787	3	0	1.6	7787	16	0	2.0
neos-916173	862	0.2	862	3	0	0.3	862	17	0	1.2
neos-916792	1165	0.2	1165	3	0	0.6	1165	19	0	2.0
neos-930752	19648	6.0	19648	3	0	6.1	19648	17	0	6.1
neos-931517	10863	2.1	10863	3	0	2.2	10863	15	0	2.1
neos-931538	11171	2.4	11171	3	0	2.2	11171	16	0	2.6
neos-932721	30030	11.8	30030	3	0	11.7	30030	17	0	12.4
neos-932816	14154	4.1	14154	3	0	4.2	14154	17	0	5.0
neos-933364	1547	0.1	1547	3	0	0.1	1547	15	0	0.1
neos-933550	2301	0.5	2301	3	0	0.4	2301	16	0	0.6
neos-933562	4067	1.2	4067	3	0	1.1	4067	17	0	1.2
neos-933638	27131	18.3	27131	3	0	18.0	27131	17	0	18.8
neos-933815	1147	0.1	1147	3	0	0.0	1147	15	0	0.1
neos-933966	21146	11.1	21146	3	0	11.4	21146	17	0	12.2
neos-934184	1547	0.1	1547	3	0	0.1	1547	15	0	0.1
neos-934278	26638	16.5	26638	3	0	16.6	26638	17	0	17.2
neos-934441	29117	18.7	29117	3	0	18.6	29117	17	0	19.1
neos-934531	549	0.3	549	3	0	0.5	549	15	0	0.9
neos-935234	27636	16.9	27636	3	0	17.0	27636	17	0	17.4
neos-935348	28978	17.0	28978	3	0	16.8	28978	17	0	17.0
neos-935496	4000	1.1	4000	3	0	0.8	4000	17	0	1.3
neos-935627	30839	18.9	30839	3	0	18.6	30839	17	0	20.0
neos-935674	4015	1.1	4015	3	0	1.0	4015	17	0	1.1
neos-935769	24339	12.2	24339	3	0	12.5	24339	17	0	12.5
neos-936660	26498	14.6	26498	3	0	14.7	26498	17	0	15.1
neos-937446	26033	15.3	26033	3	0	15.2	26033	16	0	15.5
neos-937511	24828	13.7	24828	3	0	13.5	24828	16	0	14.0
neos-937815	33788	23.8	33788	3	0	23.8	33788	17	0	24.2
neos-941262	26859	15.1	26859	3	0	15.1	26859	18	0	15.6
neos-941313	46102	110.4	46102	3	0	112.8	46102	18	0	114.9
neos-941698	682	0.1	682	3	0	0.1	682	16	0	0.1
neos-941717	4805	0.7	4805	3	0	0.8	4805	17	0	0.7

Table A.2 continued

Instance	SoPlex$_9$		SoPlex$_{50}$				SoPlex$_{250}$			
	iter	t	iter	R	R$_0$	t	iter	R	R$_0$	t
neos-941782	1858	0.3	1858	3	0	0.2	1858	17	0	0.3
neos-942323	468	0.1	468	3	0	0.0	468	16	0	0.1
neos-942830	1444	0.2	1444	3	0	0.1	1444	16	0	0.2
neos-942886	336	0.0	336	3	0	0.1	336	16	0	0.1
neos-948126	30978	19.4	30978	3	0	19.5	30978	17	0	20.0
neos-948268	11614	3.8	11614	3	0	3.6	11614	17	0	3.9
neos-948346	4225	4.1	4225	3	0	4.4	4225	17	0	6.2
neos-950242	1897	0.4	1897	0	0	0.6	1897	0	0	0.5
neos-952987	531	0.3	531	3	0	0.6	531	16	0	1.3
neos-953928	1858	1.1	5020	4	1	2.8	5020	17	1	3.5
neos-954925	7761	14.9	*2098551*	*1*	*1*	*7200.0*	*2114187*	*1*	*1*	*7200.0*
neos-955215	850	0.0	850	3	0	0.0	850	15	0	0.1
neos-955800	882	0.1	882	0	0	0.0	882	0	0	0.0
neos-956971	6109	6.8	*4764547*	*1*	*1*	*7200.0*	*4744541*	*1*	*1*	*7200.0*
neos-957143	4374	4.2	*7409607*	*1*	*1*	*7200.0*	*7358729*	*1*	*1*	*7200.0*
neos-957270	515	0.2	515	3	0	0.3	515	15	0	0.4
neos-957323	3688	3.8	3688	3	0	4.3	3688	16	0	5.5
neos-957389	1299	0.3	1299	3	0	0.5	1299	16	0	0.6
neos-960392	12434	9.2	12434	3	0	9.8	12434	17	0	10.7
neos-983171	27536	16.1	27536	3	0	16.2	27536	19	0	16.7
neos-984165	30160	19.0	30160	3	0	19.1	30160	18	0	19.3
neos1	4509	12.5	4509	3	0	13.2	4509	16	0	15.1
neos13	2438	0.8	2438	3	0	0.9	2438	16	0	1.7
neos15	508	0.0	508	3	0	0.0	508	15	0	0.1
neos16	475	0.0	475	3	0	0.0	475	15	0	0.1
neos18	293	0.1	293	3	0	0.2	293	15	0	0.4
neos2	8434	34.4	8434	3	0	34.8	8434	16	0	37.6
neos6	6681	2.1	6681	3	0	2.1	6681	17	0	2.8
neos788725	603	0.1	603	3	0	0.0	603	17	0	0.1
neos808444	8106	4.1	8106	3	0	4.0	8106	16	0	4.3
neos858960	2	0.0	2	0	0	0.0	2	0	0	0.0
nesm	6164	0.6	6164	3	0	0.4	6164	18	0	0.8
net12	11945	2.9	11945	3	0	2.9	11945	16	0	3.4
netdiversion	26331	27.9	26331	3	0	27.8	26331	15	0	29.4
netlarge2	111407	1542.8	111407	0	0	1528.1	111407	0	0	1525.0
newdano	455	0.0	455	3	0	0.0	455	15	0	0.0
nl	14132	3.9	14132	3	0	3.9	14132	18	0	4.8
nobel-eu-DBE	2927	0.3	2927	3	0	0.1	2927	15	0	0.3
noswot	134	0.0	135	3	1	0.0	135	16	1	0.0
npmv07	168166	45.2	168166	4	0	47.6	168166	17	0	55.3
ns1111636	27016	78.9	27016	3	0	80.2	27016	16	0	99.7
ns1116954	15636	24.2	15636	3	0	24.4	15636	16	0	25.3
ns1208400	9117	1.7	9117	3	0	1.8	9117	17	0	1.9
ns1456591	2343	0.5	2343	3	0	0.6	2343	16	0	0.8
ns1606230	15157	6.2	15157	3	0	6.1	15157	17	0	6.1
ns1631475	101171	253.3	101171	3	0	256.6	101171	17	0	259.3
ns1644855	47488	109.7	47488	3	0	110.5	47488	18	0	112.8
ns1663818	1723	24.9	1723	3	0	25.8	1723	17	0	29.1
ns1685374	105867	779.2	105867	3	0	774.7	105867	18	0	774.3
ns1686196	157	0.1	157	3	0	0.1	157	16	0	0.2
ns1688347	178	0.1	178	3	0	0.1	178	16	0	0.2
ns1696083	498	0.3	498	3	0	0.3	498	16	0	0.6
ns1702808	58	0.0	58	3	0	0.0	58	15	0	0.0
ns1745726	182	0.1	182	3	0	0.1	182	15	0	0.3
ns1758913	27608	335.6	27608	3	0	335.8	27608	17	0	336.2
ns1766074	27	0.0	27	3	0	0.0	27	15	0	0.0
ns1769397	206	0.2	206	3	0	0.2	206	15	0	0.3
ns1778858	14685	5.4	14685	3	0	5.4	14685	17	0	5.9
ns1830653	2753	0.5	2753	3	0	0.7	2753	16	0	0.8
ns1856153	3231	0.7	3231	3	0	0.7	3231	15	0	1.0
ns1904248	42350	55.0	42365	4	1	55.9	42365	16	1	57.4
ns1905797	5721	2.5	5721	3	0	2.8	5721	16	0	3.8
ns1905800	2057	0.5	2057	3	0	0.4	2057	16	0	0.8
ns1952667	81	0.2	81	3	0	0.2	81	18	0	0.3
ns2017839	60283	118.1	60283	4	0	117.7	60283	17	0	120.4
ns2081729	115	0.0	115	3	0	0.0	115	15	0	0.0
ns2118727	23580	111.8	23580	3	0	112.7	23580	16	0	117.4
ns2122603	13837	3.5	13837	3	0	3.8	13837	16	0	4.0
ns2124243	94850	312.1	94850	3	0	313.0	94850	16	0	314.2
ns2137859	2864	5.2	2864	3	0	6.2	2864	16	0	10.6
ns4-pr3	10658	1.8	10658	3	0	1.8	10658	15	0	2.1
ns4-pr9	9965	1.5	9965	3	0	1.3	9965	15	0	1.6

Table A.2 continued

Instance	SoPlex$_9$ iter	t	SoPlex$_{50}$ iter	R	R_0	t	SoPlex$_{250}$ iter	R	R_0	t
ns894236	36452	16.2	36452	3	0	15.4	36452	17	0	15.7
ns894244	29128	21.5	29128	3	0	21.7	29128	17	0	22.1
ns894786	18747	14.5	18747	3	0	14.8	18747	16	0	15.6
ns894788	10998	1.5	10998	3	0	1.4	10998	16	0	1.7
ns903616	30158	22.4	30158	3	0	22.4	30158	16	0	23.1
ns930473	10736	8.2	10736	3	0	8.2	10736	17	0	8.6
nsa	1162	0.1	1162	3	0	0.1	1162	16	0	0.2
nsct1	4841	1.2	4841	3	0	1.4	4841	16	0	1.9
nsct2	6810	1.3	6810	3	0	1.5	6810	16	0	2.1
nsic1	254	0.0	254	3	0	0.0	254	16	0	0.0
nsic2	268	0.0	268	3	0	0.0	268	16	0	0.0
nsir1	3762	0.6	3762	3	0	1.0	3762	16	0	0.8
nsir2	3288	0.5	3288	3	0	0.7	3288	16	0	0.9
nsr8k	102631	402.4	102631	4	0	397.4	102631	22	0	398.9
nsrand-ipx	152	0.1	152	3	0	0.4	152	16	0	0.6
nu120-pr3	6187	1.1	6187	3	0	1.1	6187	16	0	1.1
nu60-pr9	5736	1.1	5736	3	0	1.2	5736	15	0	1.0
nug05	161	0.0	161	3	0	0.0	161	16	0	0.0
nug06	1117	0.1	1117	3	0	0.1	1117	17	0	0.1
nug07	7569	1.0	7569	3	0	0.8	7569	19	0	0.9
nug08	13541	2.0	13541	3	0	1.7	13541	19	0	2.0
nug08-3rd	43975	1840.1	43975	4	0	1869.3	43975	22	0	1861.6
nug12	99946	160.0	99946	4	0	159.6	99946	22	0	160.2
nw04	82	0.3	82	3	0	1.0	82	15	0	3.0
nw14	166	0.7	166	3	0	1.9	166	15	0	4.9
ofi	139069	407.6	139104	4	2	412.4	139104	17	2	424.2
opm2-z10-s2	21918	111.0	21918	3	0	110.7	21918	17	0	113.7
opm2-z11-s8	24349	162.2	24349	3	0	163.7	24349	16	0	167.4
opm2-z12-s14	28647	279.0	28647	3	0	283.9	28647	16	0	298.7
opm2-z12-s7	30797	313.1	30797	3	0	315.2	30797	16	0	310.4
opm2-z7-s2	10304	7.3	10304	3	0	7.4	10304	16	0	8.0
opt1217	143	0.0	143	3	0	0.0	143	15	0	0.0
orna1	1327	0.2	1327	4	0	0.2	1327	19	0	0.2
orna2	1539	0.2	1539	4	0	0.2	1539	18	0	0.5
orna3	1754	0.2	1754	4	0	0.3	1754	19	0	0.6
orna4	2708	0.2	2708	4	0	0.3	2708	18	0	0.6
orna7	2280	0.3	2280	4	0	0.3	2280	19	0	0.6
orswq2	148	0.0	148	3	0	0.0	148	16	0	0.0
osa-07	966	0.5	966	3	0	0.5	966	15	0	0.8
osa-14	2104	1.2	2104	3	0	1.4	2104	15	0	2.2
osa-30	5628	5.9	5628	3	0	6.3	5628	15	0	7.9
osa-60	13364	38.1	13364	3	0	39.1	13364	16	0	43.1
p0033	19	0.0	19	3	0	0.0	19	15	0	0.0
p0040	13	0.0	23	4	1	0.0	23	16	1	0.0
p010	13700	1.8	13700	3	0	2.2	13700	16	0	2.7
p0201	50	0.0	51	1	1	0.0	51	1	1	0.0
p0282	114	0.0	114	3	0	0.0	114	16	0	0.0
p0291	27	0.0	27	3	0	0.0	27	15	0	0.0
p05	7515	0.9	7515	3	0	0.8	7515	16	0	1.3
p0548	84	0.0	84	3	0	0.0	84	16	0	0.0
p100x588b	235	0.0	235	3	0	0.0	235	15	0	0.0
p19	315	0.0	315	3	0	0.0	315	17	0	0.1
p2756	73	0.0	73	3	0	0.0	73	15	0	0.1
p2m2p1m1p0n100	10	0.0	10	3	0	0.0	10	15	0	0.0
p6000	728	0.1	728	3	0	0.2	728	16	0	0.5
p6b	548	0.1	548	0	0	0.1	548	0	0	0.0
p80x400b	152	0.0	152	3	0	0.0	152	15	0	0.0
pcb1000	2684	0.3	2684	3	0	0.4	2684	16	0	0.6
pcb3000	7719	0.9	7719	3	0	1.0	7719	16	0	1.3
pds-02	2713	0.1	2713	0	0	0.1	2713	0	0	0.1
pds-06	10699	1.1	10699	2	0	1.1	10699	2	0	1.4
pds-10	15362	1.9	15362	0	0	1.9	15362	0	0	2.4
pds-100	661386	4676.7	661386	3	0	4662.1	661386	15	0	4667.9
pds-20	37294	13.1	37294	3	0	13.4	37294	15	0	14.2
pds-30	65338	58.4	65338	0	0	57.5	65338	0	0	58.3
pds-40	101787	159.1	101787	3	0	155.7	101787	15	0	157.2
pds-50	137692	289.9	137692	3	0	293.2	137692	16	0	302.2
pds-60	188842	563.5	188842	3	0	563.8	188842	15	0	567.6
pds-70	222389	759.3	222389	0	0	761.6	222389	0	0	760.2
pds-80	240224	823.9	240224	3	0	821.3	240224	16	0	824.8
pds-90	512992	3214.8	512992	3	0	3210.0	512992	16	0	3220.0
perold	5155	0.8	5155	4	0	0.6	5155	18	0	0.8

Table A.2 continued

Instance	SoPlex$_9$		SoPlex$_{50}$				SoPlex$_{250}$			
	iter	t	iter	R	R$_0$	t	iter	R	R$_0$	t
pf2177	9474	3.4	9474	3	0	3.3	9474	18	0	3.5
pg	2164	0.0	2164	3	0	0.1	2164	16	0	0.2
pg5_34	3400	0.1	3400	3	0	0.1	3400	16	0	0.3
pgp2	3713	0.1	3713	3	0	0.2	3713	16	0	0.5
pigeon-10	245	0.0	245	3	0	0.0	245	15	0	0.0
pigeon-11	237	0.0	237	3	0	0.0	237	15	0	0.0
pigeon-12	305	0.0	305	3	0	0.0	305	15	0	0.1
pigeon-13	288	0.0	288	3	0	0.0	288	15	0	0.1
pigeon-19	608	0.0	608	3	0	0.1	608	15	0	0.2
pilot	10227	3.6	10227	3	0	3.6	10227	19	0	4.4
pilot-ja	11126	1.7	11126	4	0	1.6	11126	18	0	1.6
pilot-we	5730	0.8	5730	4	0	0.8	5730	19	0	0.7
pilot4	1483	0.2	1483	4	0	0.2	1483	20	0	0.5
pilot87	15871	9.5	15871	4	0	9.5	15871	19	0	10.5
pilotnov	7134	0.9	7134	4	0	0.7	7134	21	0	0.8
pk1	101	0.0	101	3	0	0.0	101	16	0	0.0
pldd000b	1667	0.2	1667	3	0	0.1	1667	16	0	0.3
pldd001b	1662	0.2	1662	3	0	0.2	1662	16	0	0.6
pldd002b	1649	0.2	1649	3	0	0.3	1649	16	0	0.6
pldd003b	1657	0.2	1657	3	0	0.3	1657	16	0	0.3
pldd004b	1654	0.2	1654	3	0	0.2	1654	16	0	0.6
pldd005b	1655	0.2	1655	3	0	0.2	1655	16	0	0.6
pldd006b	1693	0.2	1693	3	0	0.3	1693	16	0	0.5
pldd007b	1714	0.2	1714	3	0	0.3	1714	16	0	0.6
pldd008b	1716	0.2	1716	3	0	0.2	1716	16	0	0.2
pldd009b	2258	0.3	2258	3	0	0.4	2258	16	0	0.7
pldd010b	2462	0.4	2462	3	0	0.5	2462	16	0	0.8
pldd011b	2413	0.4	2413	3	0	0.5	2413	16	0	0.4
pldd012b	2447	0.4	2447	3	0	0.5	2447	16	0	0.4
pltexpa2-16	1094	0.1	1094	3	0	0.0	1094	15	0	0.1
pltexpa2-6	411	0.0	411	3	0	0.0	411	15	0	0.1
pltexpa3_16	15975	4.1	15975	3	0	4.4	15975	15	0	5.9
pltexpa3_6	2741	0.3	2741	3	0	0.5	2741	15	0	0.5
pltexpa4_6	14762	3.5	14762	3	0	4.0	14762	15	0	5.2
pp08a	145	0.0	145	3	0	0.0	145	15	0	0.0
pp08aCUTS	224	0.0	224	3	0	0.0	224	16	0	0.0
primagaz	1802	0.4	1802	3	0	0.4	1802	16	0	0.7
problem	9	0.0	9	0	0	0.0	9	0	0	0.0
probportfolio	126	0.0	126	3	0	0.0	126	15	0	0.0
prod1	193	0.0	193	3	0	0.0	193	16	0	0.1
prod2	344	0.0	344	3	0	0.1	344	16	0	0.1
progas	3998	0.8	3998	3	0	0.8	3998	18	0	0.9
protfold	13331	2.8	13331	3	0	2.7	13331	17	0	2.6
pw-myciel4	2154	0.6	2154	3	0	0.7	2154	16	0	0.5
qap10	58606	30.2	58606	4	0	30.0	58606	22	0	30.2
qiu	1604	0.2	1604	3	0	0.2	1604	16	0	0.3
qiulp	1604	0.2	1604	3	0	0.1	1604	16	0	0.2
qnet1	789	0.1	789	3	0	0.1	789	16	0	0.2
qnet1_o	406	0.0	406	3	0	0.0	406	16	0	0.1
queens-30	13322	10.8	13322	3	0	13.5	13322	17	0	14.6
r05	7499	0.7	7499	3	0	0.9	7499	16	0	1.3
r80x800	188	0.0	188	3	0	0.0	188	15	0	0.1
rail01	349407	4435.1	349603	7	1	4461.4	349603	37	1	4453.6
rail2586	30254	675.5	30254	3	0	679.5	30254	18	0	717.2
rail4284	54982	1670.1	54982	3	0	1678.7	54982	18	0	1708.1
rail507	13033	11.2	13033	3	0	11.6	13033	18	0	13.5
rail516	6992	5.6	6992	3	0	5.9	6992	15	0	7.0
rail582	10608	9.2	10608	3	0	9.5	10608	16	0	10.8
ramos3	19446	58.7	19446	4	0	60.3	19446	21	0	60.5
ran10x10a	133	0.0	133	3	0	0.0	133	15	0	0.0
ran10x10b	149	0.0	149	3	0	0.0	149	15	0	0.0
ran10x10c	161	0.0	161	3	0	0.0	161	15	0	0.0
ran10x12	162	0.0	162	3	0	0.0	162	15	0	0.0
ran10x26	423	0.0	423	3	0	0.0	423	15	0	0.1
ran12x12	186	0.0	186	3	0	0.0	186	15	0	0.0
ran12x21	380	0.0	380	3	0	0.0	380	15	0	0.1
ran13x13	236	0.0	236	3	0	0.0	236	15	0	0.0
ran14x18	397	0.0	397	3	0	0.0	397	15	0	0.0
ran14x18-disj-8	1915	0.2	1916	4	1	0.3	1916	19	1	0.6
ran14x18.disj-8	1915	0.2	1916	4	1	0.3	1916	19	1	0.6
ran14x18.1	434	0.0	434	3	0	0.0	434	15	0	0.1
ran16x16	379	0.0	379	3	0	0.0	379	15	0	0.0

Table A.2 continued

Instance	SoPlex$_9$		SoPlex$_{50}$				SoPlex$_{250}$			
	iter	t	iter	R	R$_0$	t	iter	R	R$_0$	t
ran17x17	258	0.0	258	3	0	0.0	258	15	0	0.0
ran4x64	515	0.0	515	3	0	0.0	515	15	0	0.0
ran6x43	447	0.0	447	3	0	0.0	447	15	0	0.1
ran8x32	402	0.0	402	3	0	0.0	402	15	0	0.1
rat1	2870	0.8	2870	3	0	0.7	2870	16	0	1.4
rat5	3024	1.1	3024	3	0	1.4	3024	17	0	2.2
rat7a	11319	12.3	11319	3	0	13.7	11319	17	0	15.0
rd-rplusc-21	138	6.3	139	3	1	7.7	139	15	1	12.6
reblock166	3359	1.4	3359	3	0	1.3	3359	16	0	1.6
reblock354	12854	7.9	12854	3	0	8.2	12854	17	0	8.7
reblock420	9480	11.5	9480	3	0	11.6	9480	17	0	12.9
reblock67	972	0.1	972	3	0	0.1	972	16	0	0.4
recipe	40	0.0	40	3	0	0.0	40	15	0	0.0
refine	21	0.0	21	3	0	0.0	21	15	0	0.0
rentacar	6483	1.2	6483	3	0	1.1	6483	17	0	1.5
rgn	85	0.0	85	3	0	0.0	85	15	0	0.0
rlfddd	825	0.2	825	1	0	0.3	825	1	0	0.4
rlfdual	8652	1.7	8652	0	0	1.7	8652	0	0	1.8
rlfprim	5532	1.2	5532	0	0	1.2	5532	0	0	1.2
rlp1	248	0.0	248	3	0	0.0	248	16	0	0.0
rmatr100-p10	6260	1.6	6260	3	0	1.5	6260	16	0	1.7
rmatr100-p5	10885	3.1	10885	3	0	3.4	10885	16	0	3.1
rmatr200-p10	13646	11.7	13646	3	0	11.8	13646	17	0	12.6
rmatr200-p20	11166	7.7	11166	3	0	8.2	11166	17	0	8.1
rmatr200-p5	17650	18.8	17650	3	0	18.8	17650	17	0	19.4
rmine10	13871	20.9	13871	3	0	21.6	13871	17	0	22.7
rmine14	92846	1720.4	92846	3	0	1714.0	92846	17	0	1735.3
rmine6	2058	0.7	2058	3	0	0.6	2058	16	0	0.7
rocII-4-11	751	0.3	751	3	0	0.5	751	16	0	1.2
rocII-7-11	1075	0.5	1075	3	0	0.8	1075	16	0	2.1
rocII-9-11	1523	0.8	1523	3	0	1.1	1523	16	0	2.7
rococoB10-011000	4263	0.8	4263	3	0	0.7	4263	16	0	0.9
rococoC10-001000	2344	0.3	2344	3	0	0.2	2344	16	0	0.5
rococoC11-011100	10112	1.6	10112	3	0	1.8	10112	16	0	1.6
rococoC12-111000	11920	3.4	11920	3	0	3.2	11920	16	0	3.8
roll3000	2627	0.3	2627	3	0	0.2	2627	17	0	0.6
rosen1	898	0.2	898	3	0	0.1	898	17	0	0.3
rosen10	2401	1.0	2401	3	0	0.9	2401	17	0	1.2
rosen2	1627	0.6	1627	3	0	0.6	1627	17	0	1.0
rosen7	332	0.0	332	3	0	0.0	332	16	0	0.1
rosen8	662	0.1	662	3	0	0.2	662	16	0	0.2
rout	293	0.0	293	3	0	0.0	293	16	0	0.0
route	2239	1.1	2239	3	0	1.2	2239	16	0	2.0
roy	119	0.0	119	3	0	0.0	119	15	0	0.0
rvb-sub	512	0.8	512	3	0	1.4	512	16	0	4.1
satellites1-25	9023	4.7	9023	3	0	4.9	9023	19	0	5.2
satellites2-60	79705	373.8	79705	3	0	376.6	79705	19	0	381.0
satellites2-60-fs	65144	270.6	65144	4	0	271.8	65144	19	0	271.9
satellites3-40	190485	2657.9	190485	4	0	2654.6	190485	21	0	2666.7
satellites3-40-fs	199929	2345.5	199931	4	1	2342.0	199931	21	1	2355.8
sc105	99	0.0	99	3	0	0.0	99	16	0	0.0
sc205	217	0.0	217	3	0	0.0	217	16	0	0.0
sc205-2r-100	1109	0.1	1109	3	0	0.2	1109	15	0	0.2
sc205-2r-16	171	0.0	171	3	0	0.0	171	15	0	0.0
sc205-2r-1600	9189	4.7	9189	3	0	4.9	9189	15	0	5.2
sc205-2r-200	2195	0.4	2195	3	0	0.5	2195	15	0	0.5
sc205-2r-27	327	0.0	327	3	0	0.0	327	15	0	0.0
sc205-2r-32	331	0.0	331	3	0	0.0	331	15	0	0.1
sc205-2r-4	55	0.0	55	3	0	0.0	55	15	0	0.0
sc205-2r-400	4393	1.1	4393	3	0	1.3	4393	15	0	1.3
sc205-2r-50	621	0.1	621	3	0	0.0	621	16	0	0.1
sc205-2r-64	651	0.1	651	3	0	0.0	651	15	0	0.1
sc205-2r-8	100	0.0	100	3	0	0.0	100	15	0	0.0
sc205-2r-800	8729	3.4	8729	3	0	3.4	8729	15	0	4.0
sc50a	45	0.0	45	3	0	0.0	45	16	0	0.0
sc50b	49	0.0	49	3	0	0.0	49	16	0	0.0
scagr25	784	0.0	784	3	0	0.1	784	16	0	0.1
scagr7	178	0.0	178	3	0	0.0	178	16	0	0.0
scagr7-2b-16	717	0.0	717	3	0	0.1	717	16	0	0.1
scagr7-2b-4	189	0.0	189	3	0	0.0	189	16	0	0.0
scagr7-2b-64	11614	1.2	11614	3	0	1.6	11614	16	0	2.2
scagr7-2c-16	684	0.0	684	3	0	0.0	684	16	0	0.1

Table A.2 continued

	SoPlex$_9$		SoPlex$_{50}$				SoPlex$_{250}$			
Instance	iter	t	iter	R	R$_0$	t	iter	R	R$_0$	t
scagr7-2c-4	186	0.0	186	3	0	0.0	186	16	0	0.0
scagr7-2c-64	2727	0.2	2727	3	0	0.2	2727	16	0	0.4
scagr7-2r-108	4538	0.3	4538	3	0	0.4	4538	16	0	0.7
scagr7-2r-16	707	0.0	707	3	0	0.0	707	16	0	0.1
scagr7-2r-216	9082	0.8	9082	3	0	1.0	9082	16	0	1.7
scagr7-2r-27	1144	0.0	1144	3	0	0.0	1144	16	0	0.1
scagr7-2r-32	1335	0.1	1335	3	0	0.0	1335	16	0	0.1
scagr7-2r-4	185	0.0	185	3	0	0.0	185	16	0	0.0
scagr7-2r-432	16562	2.5	16562	3	0	2.8	16562	16	0	4.0
scagr7-2r-54	2299	0.1	2299	3	0	0.1	2299	16	0	0.5
scagr7-2r-64	2763	0.2	2763	3	0	0.2	2763	16	0	0.5
scagr7-2r-8	357	0.0	357	3	0	0.0	357	16	0	0.0
scagr7-2r-864	32972	12.0	32972	3	0	12.6	32972	16	0	15.2
scfxm1	447	0.0	447	3	0	0.0	447	17	0	0.1
scfxm1-2b-16	4522	0.4	4522	3	0	0.5	4522	18	0	0.9
scfxm1-2b-4	1135	0.1	1135	3	0	0.1	1135	17	0	0.1
scfxm1-2b-64	26349	8.8	26349	3	0	9.2	26349	19	0	11.4
scfxm1-2c-4	1071	0.1	1071	3	0	0.1	1071	16	0	0.1
scfxm1-2r-128	23788	7.7	23788	3	0	8.0	23788	19	0	10.1
scfxm1-2r-16	4672	0.5	4672	3	0	0.5	4672	18	0	0.6
scfxm1-2r-256	47280	30.8	47280	4	0	31.7	47280	20	0	36.2
scfxm1-2r-27	8158	0.8	8158	3	0	0.6	8158	17	0	1.2
scfxm1-2r-32	8924	1.0	8924	3	0	0.8	8924	19	0	1.5
scfxm1-2r-4	1203	0.1	1203	3	0	0.0	1203	17	0	0.1
scfxm1-2r-64	13133	1.8	13133	3	0	2.1	13133	18	0	3.1
scfxm1-2r-8	2252	0.2	2252	3	0	0.1	2252	18	0	0.2
scfxm1-2r-96	16931	3.7	16931	3	0	3.9	16931	19	0	5.5
scfxm2	1119	0.1	1119	3	0	0.1	1119	19	0	0.1
scfxm3	1992	0.1	1992	3	0	0.1	1992	18	0	0.4
scorpion	245	0.0	245	3	0	0.0	245	17	0	0.1
scrs8	608	0.0	608	3	0	0.0	608	16	0	0.1
scrs8-2b-16	88	0.0	88	3	0	0.0	88	16	0	0.1
scrs8-2b-4	22	0.0	22	3	0	0.0	22	15	0	0.0
scrs8-2b-64	296	0.0	296	3	0	0.0	296	16	0	0.1
scrs8-2c-16	91	0.0	91	3	0	0.0	91	16	0	0.0
scrs8-2c-32	179	0.0	179	3	0	0.0	179	16	0	0.1
scrs8-2c-4	22	0.0	22	3	0	0.0	22	15	0	0.0
scrs8-2c-64	358	0.0	358	3	0	0.1	358	16	0	0.2
scrs8-2c-8	45	0.0	45	3	0	0.0	45	15	0	0.0
scrs8-2r-128	543	0.1	543	3	0	0.1	543	16	0	0.3
scrs8-2r-16	96	0.0	96	3	0	0.0	96	16	0	0.1
scrs8-2r-256	1119	0.2	1119	3	0	0.3	1119	16	0	0.6
scrs8-2r-27	103	0.0	103	3	0	0.0	103	15	0	0.0
scrs8-2r-32	192	0.0	192	3	0	0.0	192	16	0	0.1
scrs8-2r-4	24	0.0	24	3	0	0.0	24	15	0	0.0
scrs8-2r-512	2532	0.4	2532	3	0	0.6	2532	16	0	1.1
scrs8-2r-64	384	0.0	384	3	0	0.0	384	16	0	0.1
scrs8-2r-64b	271	0.0	271	3	0	0.0	271	16	0	0.1
scrs8-2r-8	41	0.0	41	3	0	0.0	41	15	0	0.0
scsd1	97	0.0	97	3	0	0.0	97	15	0	0.1
scsd6	423	0.0	423	3	0	0.1	423	16	0	0.1
scsd8	1837	0.2	1837	3	0	0.3	1837	16	0	0.5
scsd8-2b-16	295	0.0	295	3	0	0.0	295	15	0	0.1
scsd8-2b-4	47	0.0	47	1	0	0.0	47	1	0	0.0
scsd8-2b-64	2056	0.1	2056	3	0	0.4	2056	16	0	1.0
scsd8-2c-16	198	0.0	198	3	0	0.0	198	15	0	0.0
scsd8-2c-4	47	0.0	47	1	0	0.0	47	1	0	0.0
scsd8-2c-64	2056	0.2	2056	3	0	0.4	2056	16	0	1.1
scsd8-2r-108	1114	0.1	1114	3	0	0.1	1114	16	0	0.7
scsd8-2r-16	231	0.0	231	3	0	0.0	231	15	0	0.0
scsd8-2r-216	2314	0.1	2314	3	0	0.3	2314	16	0	0.9
scsd8-2r-27	286	0.0	286	3	0	0.0	286	15	0	0.1
scsd8-2r-32	429	0.0	429	3	0	0.1	429	15	0	0.1
scsd8-2r-4	47	0.0	47	1	0	0.0	47	1	0	0.0
scsd8-2r-432	4530	0.2	4530	3	0	0.5	4530	16	0	1.8
scsd8-2r-54	600	0.1	600	3	0	0.1	600	15	0	0.2
scsd8-2r-64	1188	0.1	1188	3	0	0.1	1188	16	0	0.4
scsd8-2r-8	97	0.0	97	3	0	0.0	97	15	0	0.0
scsd8-2r-8b	97	0.0	97	3	0	0.0	97	15	0	0.0
sct1	15178	7.4	15178	4	0	7.6	15178	19	0	8.5
sct32	14051	3.8	14051	4	0	3.7	14051	21	0	4.7
sct5	6217	3.1	6217	3	0	3.5	6217	18	0	4.2

Table A.2 continued

Instance	SoPlex$_9$ iter	t	SoPlex$_{50}$ iter	R	R$_0$	t	SoPlex$_{250}$ iter	R	R$_0$	t
sctap1	262	0.0	262	3	0	0.0	262	16	0	0.0
sctap1-2b-16	323	0.0	323	3	0	0.0	323	15	0	0.1
sctap1-2b-4	83	0.0	83	3	0	0.0	83	15	0	0.0
sctap1-2b-64	4773	0.6	4773	3	0	0.7	4773	16	0	1.7
sctap1-2c-16	329	0.0	329	3	0	0.0	329	16	0	0.1
sctap1-2c-4	85	0.0	85	3	0	0.0	85	16	0	0.0
sctap1-2c-64	1146	0.1	1146	3	0	0.1	1146	15	0	0.3
sctap1-2r-108	2109	0.2	2109	3	0	0.4	2109	15	0	0.8
sctap1-2r-16	282	0.0	282	3	0	0.0	282	15	0	0.0
sctap1-2r-216	4248	0.5	4248	3	0	0.8	4248	16	0	1.4
sctap1-2r-27	536	0.0	536	3	0	0.0	536	15	0	0.1
sctap1-2r-32	559	0.0	559	3	0	0.0	559	15	0	0.1
sctap1-2r-4	84	0.0	84	3	0	0.0	84	15	0	0.0
sctap1-2r-480	9373	1.5	9373	3	0	1.9	9373	16	0	3.4
sctap1-2r-54	1069	0.1	1069	3	0	0.1	1069	15	0	0.2
sctap1-2r-64	1123	0.1	1123	3	0	0.2	1123	15	0	0.3
sctap1-2r-8	147	0.0	147	3	0	0.0	147	15	0	0.0
sctap1-2r-8b	161	0.0	161	3	0	0.0	161	15	0	0.0
sctap2	505	0.1	505	3	0	0.1	505	16	0	0.1
sctap3	604	0.1	604	3	0	0.1	604	15	0	0.1
seba	6	0.0	6	3	0	0.0	6	15	0	0.0
self	13495	48.6	13495	3	0	49.7	13495	18	0	51.8
set1ch	513	0.0	513	3	0	0.0	513	15	0	0.0
set3-10	2279	0.1	2279	3	0	0.2	2279	16	0	0.6
set3-15	2269	0.2	2269	3	0	0.2	2269	16	0	0.6
set3-20	2344	0.1	2344	3	0	0.2	2344	17	0	0.3
seymour	3858	1.0	3858	3	0	0.9	3858	16	0	0.9
seymour-disj-10	5700	1.8	5702	4	1	2.1	5702	18	1	2.6
seymour.disj-10	5700	1.7	5702	4	1	2.1	5702	18	1	2.7
seymourl	3858	0.9	3858	3	0	1.1	3858	16	0	0.9
sgpf5y6	153350	167.8	197667	3	1	221.6	197667	16	1	232.0
share1b	217	0.0	217	3	0	0.0	217	17	0	0.0
share2b	159	0.0	159	3	0	0.0	159	17	0	0.0
shell	595	0.0	595	1	0	0.0	595	1	0	0.0
ship04l	473	0.0	473	3	0	0.0	473	16	0	0.1
ship04s	383	0.0	383	3	0	0.0	383	16	0	0.1
ship08l	810	0.1	810	3	0	0.1	810	16	0	0.3
ship08s	513	0.0	513	3	0	0.1	513	16	0	0.2
ship12l	1085	0.1	1085	3	0	0.2	1085	16	0	0.5
ship12s	648	0.0	648	3	0	0.1	648	16	0	0.3
shipsched	2174	0.6	2174	3	0	0.9	2174	15	0	1.1
shs1023	174484	604.0	177407	7	5	618.0	177407	35	19	633.0
siena1	28441	19.0	28441	4	0	19.7	28441	24	0	20.7
sierra	640	0.0	640	3	0	0.0	640	16	0	0.2
sing2	37541	40.3	37541	3	0	40.9	37541	16	0	42.0
sing245	218115	1608.1	218115	3	0	1611.8	218115	17	0	1622.6
sing359	352087	5830.9	352087	3	0	5799.4	352087	18	0	5800.1
slptsk	4856	1.2	4856	3	0	1.8	4856	17	0	2.6
small000	557	0.0	557	3	0	0.0	557	16	0	0.1
small001	725	0.0	737	3	1	0.0	737	16	1	0.2
small002	834	0.0	850	3	1	0.1	850	17	1	0.2
small003	681	0.0	693	3	1	0.1	693	16	1	0.2
small004	465	0.0	476	3	1	0.1	476	16	1	0.2
small005	601	0.0	621	4	1	0.1	621	17	1	0.2
small006	518	0.0	540	3	1	0.1	540	16	1	0.2
small007	517	0.0	547	3	1	0.1	547	16	1	0.2
small008	489	0.0	511	3	1	0.1	511	16	1	0.2
small009	416	0.0	435	3	1	0.1	435	16	1	0.1
small010	328	0.0	342	3	1	0.1	342	16	1	0.1
small011	337	0.0	348	3	1	0.0	348	16	1	0.1
small012	287	0.0	287	3	0	0.1	287	16	0	0.1
small013	289	0.0	289	3	0	0.1	289	16	0	0.1
small014	341	0.0	341	3	0	0.0	341	16	0	0.1
small015	338	0.0	338	3	0	0.1	338	16	0	0.1
small016	338	0.0	338	3	0	0.0	338	16	0	0.1
south31	23654	14.9	23654	3	0	15.2	23654	16	0	16.6
sp97ar	6984	2.9	6986	4	1	3.3	6986	17	1	4.5
sp97ic	3102	0.8	3102	4	0	0.7	3102	18	0	0.8
sp98ar	10342	2.9	10349	4	1	3.1	10349	18	1	3.9
sp98ic	3276	0.7	3276	4	0	0.8	3276	17	0	1.5
sp98ir	2967	0.7	2970	4	1	0.5	2970	19	1	1.1
square15	208626	2424.8	208626	0	0	2424.6	208626	0	0	2438.8

Table A.2 continued

Instance	SoPlex$_9$		SoPlex$_{50}$				SoPlex$_{250}$			
	iter	t	iter	R	R$_0$	t	iter	R	R$_0$	t
stair	658	0.1	658	3	0	0.1	658	18	0	0.2
standata	50	0.0	50	3	0	0.0	50	15	0	0.0
standmps	189	0.0	189	3	0	0.0	189	16	0	0.1
stat96v1	275141	2283.3	*875464*	*4*	*4*	*6516.9*	*875464*	*16*	*16*	*6522.2*
stat96v4	144049	395.8	144049	4	0	402.2	144049	20	0	403.8
stat96v5	11838	18.4	13119	5	1	21.2	13119	20	1	25.3
stein27	32	0.0	32	3	0	0.0	32	16	0	0.0
stein45	59	0.0	59	3	0	0.0	59	16	0	0.0
stocfor1	108	0.0	108	3	0	0.0	108	16	0	0.0
stocfor2	1994	0.3	1994	3	0	0.2	1994	16	0	0.7
stocfor3	16167	6.2	16167	3	0	6.5	16167	16	0	7.6
stockholm	25891	20.3	26971	4	1	21.7	26971	17	1	23.4
stormG2_1000	732642	4648.3	732642	4	0	4679.4	732642	17	0	4728.4
stormg2-125	92103	45.6	92103	3	0	47.4	92103	16	0	50.6
stormg2-27	20754	2.4	20754	3	0	2.7	20754	16	0	3.1
stormg2-8	6863	0.6	6863	3	0	0.4	6863	16	0	0.5
stormg2_1000	732642	4660.6	732642	4	0	4685.5	732642	17	0	4750.8
stp3d	148139	1238.6	148139	3	0	1242.3	148139	18	0	1247.0
sts405	434	0.5	434	3	0	0.7	434	16	0	1.5
sts729	907	1.3	907	3	0	1.8	907	18	0	4.5
swath	127	0.1	127	3	0	0.1	127	16	0	0.2
sws	1150	0.3	1150	3	0	0.3	1150	16	0	0.9
t0331-4l	15487	14.7	15487	4	0	15.4	15487	19	0	18.6
t1717	12220	7.0	12220	4	0	7.6	12220	19	0	10.7
t1722	10612	3.3	10612	3	0	3.6	10612	17	0	4.7
tanglegram1	478	0.3	478	0	0	0.3	478	0	0	0.3
tanglegram2	236	0.1	236	0	0	0.0	236	0	0	0.1
testbig	8010	4.4	8010	3	0	4.5	8010	15	0	5.0
timtab1	20	0.0	20	3	0	0.0	20	15	0	0.0
timtab2	36	0.0	36	3	0	0.0	36	15	0	0.0
toll-like	717	0.1	717	0	0	0.1	717	0	0	0.0
tr12-30	696	0.0	696	3	0	0.0	696	15	0	0.1
transportmoment	9617	1.7	9617	4	0	1.9	9617	20	0	2.5
triptim1	68167	93.3	68167	3	0	92.8	68167	17	0	94.3
triptim2	204335	572.2	204335	3	0	567.7	204335	18	0	579.6
triptim3	89514	156.7	89514	3	0	161.5	89514	18	0	162.3
truss	21892	4.7	21892	3	0	4.6	21892	16	0	5.1
tuff	212	0.0	212	3	0	0.0	212	17	0	0.1
tw-myciel4	11588	2.6	11588	3	0	2.3	11588	15	0	2.4
uc-case11	39603	75.9	39603	3	0	75.3	39603	17	0	79.8
uc-case3	34156	65.4	34156	3	0	66.7	34156	17	0	66.7
uct-subprob	2988	0.5	2988	3	0	0.5	2988	16	0	0.7
ulevimin	125858	108.8	125858	3	0	108.6	125858	17	0	109.4
umts	5537	0.9	5549	5	2	0.7	5549	17	2	1.2
unitcal_7	21824	8.9	21824	3	0	9.1	21824	16	0	10.1
us04	338	0.2	338	3	0	0.5	338	16	0	1.2
usAbbrv-8-25_70	2434	0.1	2434	3	0	0.1	2434	16	0	0.1
van	11014	5.2	11014	3	0	6.0	11014	19	0	10.7
vpm1	130	0.0	130	3	0	0.0	130	15	0	0.0
vpm2	192	0.0	192	3	0	0.0	192	15	0	0.0
vpphard	11075	11.3	11075	3	0	11.5	11075	17	0	12.4
vpphard2	8024	32.4	8024	3	0	33.1	8024	16	0	39.6
vtp-base	75	0.0	75	3	0	0.0	75	16	0	0.0
wachplan	2033	0.5	2033	3	0	0.4	2033	16	0	0.4
watson_1	188103	366.3	188496	4	1	369.2	188496	17	1	386.2
watson_2	333044	1519.4	333083	4	1	1518.6	333083	17	1	1585.2
wide15	229488	2733.8	229488	0	0	2734.6	229488	0	0	2731.0
wnq-n100-mw99-14	764	5.0	764	3	0	5.8	764	15	0	8.7
wood1p	142	0.1	142	3	0	0.2	142	16	0	0.9
woodw	1832	0.6	1832	3	0	0.6	1832	16	0	1.0
world	70204	131.2	70320	15	12	132.1	70320	29	12	137.6
zed	31	0.0	31	3	0	0.0	31	16	0	0.0
zib54-UUE	1855	0.1	1855	3	0	0.1	1855	16	0	0.4

A.1.3 Exact Linear Programming

Table A.3: Detailed results comparing QSopt_ex's performance warm started from bases returned by SoPlex₉ and SoPlex₅₀. Entries corresponding to time or memory outs (all due to QSopt_ex) are printed in italics. See Table 3.1 for aggregated results.

iter — number of simplex iterations by SoPlex and QSopt_ex
prc — max. precision used by QSopt_ex
t_9/t_{50} — running time of SoPlex₉/SoPlex₉ (in seconds)
t_{ex} — running time of QSopt_ex (in seconds)
t — total running time (in seconds)
R_0 — number of refinements to final basis

Instance	SoPlex₉+QSopt_ex					R_0	SoPlex₅₀+QSopt_ex				
	iter	prc	t_9	t_{ex}	t		iter	prc	t_{50}	t_{ex}	t
10teams	1611	64	0.2	0.1	0.3	0	1611	64	0.3	0.1	0.4
16_n14	329933	64	376.0	1.6	377.6	0	329933	64	371.8	1.9	373.7
22433	1041	64	0.1	0.1	0.1	0	1041	64	0.1	0.0	0.1
23588	548	64	0.0	0.0	0.1	0	548	64	0.0	0.0	0.0
25fv47	4359	64	0.8	0.9	1.6	0	4359	64	0.7	0.7	1.3
30_70_45_095_100	16103	64	7.4	0.5	7.9	0	16103	64	7.6	0.3	7.9
30n20b8	269	64	0.3	0.2	0.4	0	269	64	0.3	0.1	0.4
50v-10	220	64	0.0	0.0	0.0	0	220	64	0.0	0.0	0.1
80bau3b	7797	64	1.0	0.2	1.2	0	7797	64	1.1	0.2	1.3
Test3	6948	64	3.3	1.9	5.2	0	6948	64	3.7	1.9	5.6
a1c1s1	1742	64	0.1	0.0	0.1	0	1742	64	0.1	0.0	0.1
aa01	11898	64	3.5	0.4	3.8	0	11898	64	3.6	0.4	4.0
aa03	7584	64	2.0	0.3	2.3	0	7584	64	1.9	0.3	2.1
aa3	7584	64	2.1	0.3	2.4	0	7584	64	2.0	0.1	2.1
aa4	4486	64	1.1	0.2	1.4	0	4486	64	1.2	0.2	1.4
aa5	9706	64	2.4	0.3	2.7	0	9706	64	2.6	0.2	2.9
aa6	4870	64	1.1	0.1	1.2	0	4870	64	1.2	0.1	1.4
acc-tight4	11142	64	2.6	0.5	3.1	0	11142	64	2.2	0.5	2.7
acc-tight5	10469	64	2.0	0.4	2.5	0	10469	64	2.2	0.2	2.4
acc-tight6	10672	64	2.2	0.4	2.6	0	10672	64	2.3	0.4	2.6
adlittle	88	64	0.0	0.0	0.0	0	88	64	0.0	0.0	0.0
afiro	16	64	0.0	0.0	0.0	0	16	64	0.0	0.0	0.0
aflow30a	396	64	0.0	0.0	0.1	0	396	64	0.0	0.0	0.0
aflow40b	1826	64	0.2	0.1	0.2	0	1826	64	0.1	0.1	0.2
agg	80	64	0.0	0.0	0.0	0	80	64	0.0	0.0	0.0
agg2	152	64	0.0	0.0	0.1	0	152	64	0.0	0.0	0.0
agg3	165	64	0.0	0.0	0.1	0	165	64	0.0	0.0	0.1
air02	95	64	0.1	0.1	0.2	0	95	64	0.1	0.1	0.2
air03	626	64	0.3	0.1	0.5	0	626	64	0.4	0.1	0.6
air04	11898	64	3.4	0.2	3.6	0	11898	64	3.4	0.3	3.8
air05	4486	64	1.2	0.2	1.4	0	4486	64	1.3	0.2	1.5
air06	7584	64	2.0	0.3	2.3	0	7584	64	2.0	0.3	2.2
aircraft	1912	64	0.4	0.2	0.6	0	1912	64	0.2	0.2	0.4
aligninq	1492	64	0.1	0.1	0.2	0	1492	64	0.2	0.1	0.3
app1-2	14617	64	7.2	2.1	9.4	1	14622	64	8.0	2.1	10.1
arki001	1570	128	0.2	2.1	2.3	1	1571	64	0.3	0.4	0.7
ash608gpia-3col	3123	64	0.2	0.3	0.6	0	3123	64	0.3	0.5	0.8
atlanta-ip	38285	128	34.8	3035.9	3070.7	1	38286	64	35.2	2.1	37.3
atm20-100	4586	128	0.4	20.2	20.6	1	4699	64	0.5	0.1	0.6
b2c1s1	2621	64	0.1	0.1	0.2	0	2621	64	0.1	0.1	0.2
bab1	5711	64	4.0	3.4	7.4	0	5711	64	4.7	3.3	8.0
bab3	697791	64	5263.4	14.6	5278.0	0	697791	64	5266.5	14.5	5281.0
bab5	29788	64	10.3	0.5	10.8	0	29788	64	10.5	0.4	11.0
bal8x12	102	64	0.0	0.0	0.0	0	102	64	0.0	0.0	0.0
bandm	474	64	0.0	0.1	0.1	0	474	64	0.0	0.1	0.1
bas1lp	2582	64	1.0	1.3	2.3	0	2582	64	1.1	1.3	2.4
baxter	12347	64	2.6	0.6	3.2	0	12347	64	2.8	0.3	3.2
bc	3759	64	1.3	1.6	2.8	0	3759	64	1.4	1.6	3.0
bc1	3759	64	1.3	1.6	2.9	0	3759	64	1.4	1.6	3.0
beaconfd	88	64	0.0	0.0	0.0	0	88	64	0.0	0.0	0.0
beasleyC3	1140	64	0.1	0.0	0.1	0	1140	64	0.1	0.0	0.1
bell3a	81	64	0.0	0.0	0.0	0	81	64	0.0	0.0	0.0
bell5	66	64	0.0	0.0	0.0	0	66	64	0.0	0.0	0.0
berlin_5_8_0	1017	64	0.0	0.1	0.1	0	1017	64	0.1	0.0	0.1
bg512142	2238	64	0.3	0.2	0.4	0	2238	64	0.1	0.2	0.3
biella1	16158	64	4.2	0.4	4.6	0	16158	64	4.2	0.3	4.5
bienst1	455	64	0.0	0.0	0.1	0	455	64	0.0	0.0	0.0

Table A.3 continued

Instance	SoPlex$_9$+QSopt_ex					SoPlex$_{50}$+QSopt_ex				
	iter	prc	t_9	t_{ex}	t	R_0	iter	prc	t_{50}	t_{ex}	t
bienst2	455	64	0.0	0.0	0.1	0	455	64	0.0	0.0	0.0
binkar10_1	1267	64	0.1	0.1	0.1	0	1267	64	0.1	0.0	0.1
bk4x3	16	64	0.0	0.0	0.0	0	16	64	0.0	0.0	0.0
blend	97	64	0.0	0.0	0.0	0	97	64	0.0	0.0	0.0
blend2	175	64	0.0	0.0	0.0	0	175	64	0.0	0.0	0.0
blp-ar98	368	64	0.2	0.3	0.4	0	368	64	0.5	0.3	0.8
blp-ic97	446	64	0.3	0.4	0.7	0	446	64	0.4	0.4	0.8
bnatt350	685	64	0.1	0.1	0.2	0	685	64	0.1	0.1	0.1
bnatt400	831	64	0.1	0.1	0.2	0	831	64	0.1	0.1	0.2
bnl1	1474	64	0.1	0.1	0.2	0	1474	64	0.2	0.1	0.2
bnl2	2718	64	0.4	0.1	0.6	0	2718	64	0.5	0.1	0.6
boeing1	458	64	0.0	0.0	0.1	0	458	64	0.0	0.0	0.1
boeing2	145	64	0.0	0.0	0.0	0	145	64	0.0	0.0	0.0
bore3d	100	64	0.0	0.0	0.0	0	100	64	0.0	0.0	0.1
brandy	482	64	0.0	0.1	0.1	0	482	64	0.0	0.1	0.1
buildingenergy	144595	64	827.1	18.9	846.0	0	144595	64	835.0	19.2	854.2
cap6000	815	64	0.2	0.2	0.4	0	815	64	0.3	0.2	0.5
capri	338	64	0.0	0.0	0.1	0	338	64	0.0	0.0	0.1
car4	10442	64	1.9	5.6	7.5	0	10442	64	2.0	5.5	7.5
cari	681	64	0.1	0.4	0.5	0	681	64	0.3	0.6	0.9
cep1	1399	64	0.1	0.1	0.1	0	1399	64	0.1	0.1	0.2
ch	10747	64	1.4	0.2	1.7	0	10747	64	1.5	0.2	1.7
circ10-3	10910	64	10.2	0.8	11.1	0	10910	64	10.4	0.8	11.2
co-100	783	64	1.2	3.7	4.9	0	783	64	2.3	3.7	5.9
co5	12067	64	3.2	1.2	4.4	0	12067	64	3.5	1.2	4.7
co9	19820	64	11.6	5.1	16.7	0	19820	64	11.7	5.3	17.0
complex	9592	64	2.4	2.0	4.3	1	9646	64	2.6	1.8	4.5
cont1	40707	64	*985.6*	*7200.0*	*8185.6*	0	40707	64	*1226.7*	*7200.0*	*8426.7*
cont4	40802	128	*2698.5*	*7200.0*	*9898.5*	0	40802	128	*2839.8*	*7200.0*	*10039.8*
core2536-691	41182	64	16.0	0.3	16.3	0	41182	64	16.8	0.3	17.1
core4872-1529	69516	64	69.1	3.7	72.8	0	69516	64	70.6	3.6	74.2
cov1075	3270	64	0.5	0.8	1.3	0	3270	64	0.4	0.6	1.0
cq5	12350	64	3.1	0.7	3.8	0	12350	64	3.1	0.8	4.0
cq9	17826	64	7.1	6.3	13.5	0	17826	64	7.2	6.3	13.5
cr42	999	64	0.1	0.1	0.1	0	999	64	0.1	0.0	0.1
cre-a	3555	64	0.3	0.1	0.4	0	3555	64	0.3	0.1	0.5
cre-b	12712	64	5.0	0.7	5.7	0	12712	64	5.4	0.7	6.1
cre-c	2842	64	0.3	0.1	0.4	0	2842	64	0.3	0.1	0.4
cre-d	9271	64	3.5	0.6	4.2	0	9271	64	3.9	0.6	4.5
crew1	1849	64	0.6	0.1	0.7	0	1849	64	0.4	0.0	0.4
csched007	5522	64	0.7	0.1	0.7	0	5522	64	0.4	0.0	0.4
csched008	3102	64	0.4	0.0	0.5	0	3102	64	0.4	0.0	0.4
csched010	6832	64	0.8	0.0	0.8	0	6832	64	0.8	0.0	0.8
cycle	920	64	0.1	0.1	0.2	0	920	64	0.1	0.1	0.1
czprob	1572	64	0.2	0.1	0.2	0	1572	64	0.1	0.1	0.2
d10200	2852	64	0.6	0.6	1.1	0	2852	64	0.5	0.6	1.1
d20200	9723	64	1.8	0.9	2.7	0	9723	64	1.9	1.0	2.8
d2q06c	13920	64	3.9	28.9	32.8	0	13920	64	3.7	28.8	32.5
d6cube	1179	64	0.4	0.2	0.6	0	1179	64	0.4	0.1	0.5
dano3_3	46251	64	28.8	7.8	36.6	0	46251	64	29.0	7.8	36.9
dano3_4	46251	64	28.9	7.9	36.9	0	46251	64	28.7	8.0	36.7
dano3_5	46251	64	30.5	8.1	38.6	0	46251	64	29.0	8.4	37.4
dano3mip	46251	64	28.8	8.2	36.9	0	46251	64	29.1	7.8	36.9
danoint	2763	64	0.3	0.1	0.4	0	2763	64	0.3	0.0	0.3
dbir1	14306	64	11.2	1.9	13.1	0	14306	64	11.4	1.9	13.3
dbir2	11641	64	4.9	2.2	7.1	0	11641	64	5.3	2.5	7.7
dc1c	19795	64	6.2	0.7	6.9	0	19795	64	6.2	0.5	6.7
dc1l	24825	64	21.1	1.1	22.2	0	24825	64	21.4	1.1	22.5
dcmulti	479	64	0.0	0.0	0.0	0	479	64	0.0	0.0	0.0
de063155	2546	64	0.3	0.6	0.9	3	2623	64	0.4	0.3	0.7
de080285	888	64	0.1	0.3	0.4	0	888	64	0.1	0.3	0.4
degen2	1325	64	0.1	0.1	0.2	0	1325	64	0.1	0.1	0.2
degen3	5832	64	0.9	0.2	1.1	0	5832	64	0.8	0.2	1.1
delf000	1675	64	0.4	0.6	1.0	0	1675	64	0.5	0.6	1.1
delf001	1703	64	0.4	0.6	1.0	0	1703	64	0.5	0.5	1.0
delf002	2009	64	0.4	0.7	1.1	0	2009	64	0.2	0.3	0.5
delf003	3118	128	0.6	7.9	8.6	1	3180	64	0.8	0.9	1.6
delf004	2624	128	0.6	6.7	7.3	1	2704	64	0.3	0.9	1.2
delf005	3226	128	0.7	7.5	8.3	1	3294	64	0.8	0.9	1.7
delf006	3066	128	0.6	8.8	9.3	1	3125	64	0.3	1.6	1.9
delf007	2757	128	0.5	9.5	10.0	1	2826	64	0.3	1.1	1.5
delf008	3393	128	0.7	9.8	10.5	2	3493	64	0.6	1.5	2.0

Table A.3 continued

Instance	iter	prc	t_9	t_{ex}	t	R_0	iter	prc	t_{50}	t_{ex}	t
		SoPlex$_9$+QSopt_ex						SoPlex$_{50}$+QSopt_ex			
delf009	3307	128	0.6	11.2	11.8	1	3383	64	0.8	1.7	2.6
delf010	3132	128	0.7	8.6	9.3	2	3209	64	0.4	1.5	1.9
delf011	3030	128	0.6	8.5	9.1	2	3104	64	0.4	0.8	1.2
delf012	2890	128	0.6	8.9	9.4	1	2963	64	0.5	1.1	1.6
delf013	3103	128	0.6	9.4	10.0	1	3184	64	0.5	1.4	1.9
delf014	4257	128	0.8	7.7	8.5	1	4322	64	0.9	0.9	1.8
delf015	3176	128	0.7	9.9	10.7	1	3242	64	0.4	0.9	1.3
delf017	3075	128	0.5	9.8	10.3	1	3131	64	0.4	1.0	1.4
delf018	3281	128	0.5	10.9	11.4	1	3335	64	0.4	0.6	1.0
delf019	3160	64	0.6	0.7	1.3	0	3160	64	0.6	0.5	1.1
delf020	3701	128	0.6	10.7	11.3	2	3784	64	0.6	0.5	1.1
delf021	3247	128	0.5	10.3	10.9	1	3331	64	0.3	0.7	1.1
delf022	3673	128	0.6	11.3	11.9	2	3757	64	0.6	0.5	1.1
delf023	4603	128	0.8	11.4	12.1	2	4747	64	0.6	0.8	1.4
delf024	3973	128	0.8	10.1	10.8	1	4150	64	0.7	1.5	2.1
delf025	4032	128	0.7	8.9	9.6	2	4152	64	0.8	1.0	1.8
delf026	3378	128	0.8	9.8	10.6	1	3495	64	0.5	0.8	1.3
delf027	3402	128	0.8	9.2	9.9	2	3519	64	0.5	0.6	1.2
delf028	3269	128	0.6	9.1	9.6	2	3425	64	0.4	0.7	1.1
delf029	2949	128	0.6	10.3	10.9	1	3085	64	0.5	0.9	1.3
delf030	3002	128	0.6	11.0	11.6	1	3135	64	0.6	0.9	1.6
delf031	2835	128	0.6	8.8	9.4	1	2969	64	0.5	0.8	1.3
delf032	2958	128	0.5	12.3	12.8	1	3091	64	0.6	0.8	1.4
delf033	2210	128	0.4	8.7	9.1	1	2344	64	0.3	1.0	1.3
delf034	2699	128	0.6	9.2	9.7	1	2832	64	0.4	0.8	1.2
delf035	2407	128	0.5	11.3	11.8	2	2545	64	0.4	0.7	1.1
delf036	2522	128	0.5	9.7	10.2	2	2655	64	0.4	0.7	1.1
deter0	3725	64	0.2	0.1	0.3	0	3725	64	0.2	0.1	0.3
deter1	9022	64	0.6	0.3	0.9	0	9022	64	0.8	0.1	0.9
deter2	11017	64	0.6	0.3	0.9	0	11017	64	0.8	0.3	1.2
deter3	12348	64	0.8	0.4	1.2	0	12348	64	1.0	0.2	1.3
deter4	5633	64	0.3	0.2	0.4	0	5633	64	0.2	0.1	0.3
deter5	9277	64	0.6	0.3	0.8	0	9277	64	0.8	0.3	1.1
deter6	7482	64	0.4	0.2	0.6	0	7482	64	0.3	0.1	0.5
deter7	11149	64	0.8	0.4	1.1	0	11149	64	0.8	0.3	1.1
deter8	6977	64	0.4	0.2	0.6	0	6977	64	0.3	0.2	0.5
df2177	1391	64	0.7	0.5	1.1	0	1391	64	0.7	0.5	1.1
dfl001	30478	64	18.5	1.4	19.9	0	30478	64	18.6	1.5	20.1
dfn-gwin-UUM	373	64	0.0	0.0	0.0	0	373	64	0.0	0.0	0.0
dg012142	11646	64	2.0	0.2	2.2	0	11646	64	1.9	0.1	2.0
disctom	13965	64	1.7	0.2	1.9	0	13965	64	1.7	0.2	1.9
disp3	490	64	0.0	0.1	0.1	0	490	64	0.0	0.1	0.1
dolom1	19382	64	7.3	0.5	7.9	0	19382	64	7.5	0.5	8.0
ds	13089	64	17.8	3.1	20.9	0	13089	64	19.0	3.1	22.1
ds-big	44732	64	604.9	25.3	630.2	0	44732	64	610.6	25.3	635.9
dsbmip	2179	64	0.1	0.1	0.2	0	2179	64	0.2	0.0	0.2
e18	12395	64	4.6	0.6	5.2	0	12395	64	4.8	0.8	5.6
e226	402	64	0.0	0.1	0.1	0	402	64	0.0	0.0	0.0
egout	96	64	0.0	0.0	0.0	0	96	64	0.0	0.0	0.0
eil33-2	308	64	0.1	0.2	0.3	0	308	64	0.2	0.1	0.4
eilA101-2	5338	64	20.6	1.8	22.4	0	5338	64	21.9	1.8	23.7
eilB101	1459	64	0.3	0.1	0.4	0	1459	64	0.4	0.1	0.4
enigma	44	64	0.0	0.0	0.0	0	44	64	0.0	0.0	0.0
enlight13	0	64	0.0	0.0	0.0	0	0	64	0.0	0.0	0.0
enlight14	0	64	0.0	0.0	0.0	0	0	64	0.0	0.0	0.0
enlight15	0	64	0.0	0.0	0.0	0	0	64	0.0	0.0	0.0
enlight16	0	64	0.0	0.0	0.0	0	0	64	0.0	0.0	0.0
enlight9	0	64	0.0	0.0	0.0	0	0	64	0.0	0.0	0.0
etamacro	766	128	0.0	0.5	0.6	1	771	64	0.1	0.0	0.1
ex10	115687	64	1791.6	41.6	1833.3	0	115687	64	1799.0	42.7	1841.8
ex1010-pi	19385	64	10.0	2.3	12.3	0	19385	64	10.1	2.3	12.4
ex3sta1	7713	64	4.8	83.6	88.4	0	7713	64	5.0	84.0	89.0
ex9	57559	64	349.2	12.4	361.6	0	57559	64	348.8	12.4	361.2
f2000	40611	64	60.6	46.9	107.5	0	40611	64	60.2	46.8	107.0
farm	0	64	0.0	0.0	0.0	0	0	64	0.0	0.0	0.0
fast0507	13213	64	11.4	0.7	12.1	0	13213	64	11.9	0.7	12.6
fffff800	855	64	0.1	0.1	0.1	0	855	64	0.0	0.0	0.1
fiball	2818	64	0.4	0.4	0.8	0	2818	64	0.7	0.6	1.2
fiber	278	64	0.0	0.0	0.1	0	278	64	0.0	0.0	0.0
finnis	524	64	0.0	0.0	0.1	0	524	64	0.0	0.0	0.0
fit1d	1006	64	0.1	0.1	0.1	0	1006	64	0.0	0.0	0.1
fit1p	3573	64	0.4	0.1	0.4	0	3573	64	0.4	0.1	0.5

Table A.3 continued

Instance	\multicolumn{5}{SoPlex$_9$+QSopt_ex}	\multicolumn{6}{SoPlex$_{50}$+QSopt_ex}
	iter	prc	t_9	t_{ex}	t	R_0	iter	prc	t_{50}	t_{ex}	t
fit2d	10781	64	1.9	0.5	2.4	0	10781	64	2.1	0.4	2.5
fit2p	16070	64	3.8	0.5	4.3	0	16070	64	4.0	0.2	4.3
fixnet6	184	64	0.0	0.0	0.0	0	184	64	0.0	0.0	0.0
flugpl	11	64	0.0	0.0	0.0	0	11	64	0.0	0.0	0.0
fome11	45759	128	38.5	2932.2	2970.7	1	46151	64	39.9	1.7	41.6
fome12	93828	128	108.4	5503.8	5612.2	1	95445	64	115.1	3.5	118.6
fome13	175861	128	309.8	7200.0	7509.8	1	177979	64	321.1	7.5	328.6
fome20	37294	64	13.1	1.0	14.1	0	37294	64	13.3	1.0	14.3
fome21	81051	64	77.4	2.3	79.7	0	81051	64	76.9	2.3	79.2
forplan	369	64	0.0	0.1	0.1	0	369	64	0.0	0.0	0.1
fxm2-16	6817	64	0.6	0.3	0.8	0	6817	64	0.8	0.1	1.0
fxm2-6	2277	64	0.1	0.1	0.3	0	2277	64	0.1	0.1	0.2
fxm3_16	48534	64	22.2	7200.0	7222.2	0	48534	64	22.7	7200.0	7222.7
fxm3_6	9986	64	0.8	7200.0	7200.8	0	9986	64	1.1	7200.0	7201.1
fxm4_6	25805	64	4.9	7200.0	7204.9	0	25805	64	5.7	7200.0	7205.7
g200x740i	721	64	0.0	0.0	0.0	0	721	64	0.0	0.0	0.0
gams10a	38	64	0.0	0.0	0.0	1	44	64	0.0	0.0	0.0
gams30a	146	128	0.0	0.1	0.1	1	184	64	0.0	0.0	0.0
ganges	1291	64	0.1	0.1	0.1	0	1291	64	0.1	0.1	0.2
ge	11171	128	2.4	94.9	97.3	1	11172	64	2.5	1.2	3.7
gen	384	64	0.0	0.0	0.1	0	384	64	0.0	0.0	0.0
gen1	11914	128	12.1	2143.3	2155.4	1	12282	64	12.4	297.3	309.7
gen2	12239	64	25.4	7200.0	7225.4	0	12239	64	26.4	7200.0	7226.4
gen4	14397	64	56.3	7200.0	7256.3	1	14823	64	57.2	7200.0	7257.2
ger50_17_trans	4819	64	1.3	0.3	1.6	0	4819	64	1.6	0.3	1.9
germanrr	8775	64	2.7	0.4	3.1	0	8775	64	2.8	0.4	3.2
germany50-DBM	8399	64	1.2	0.1	1.2	0	8399	64	0.9	0.1	1.0
gesa2	1118	64	0.0	0.1	0.1	0	1118	64	0.0	0.0	0.1
gesa2-o	653	64	0.0	0.0	0.1	0	653	64	0.0	0.0	0.1
gesa2_o	653	64	0.0	0.1	0.1	0	653	64	0.0	0.0	0.1
gesa3	974	64	0.1	0.1	0.1	0	974	64	0.1	0.1	0.1
gesa3_o	530	64	0.0	0.1	0.1	0	530	64	0.0	0.0	0.1
gfrd-pnc	664	64	0.0	0.0	0.1	0	664	64	0.1	0.0	0.1
glass4	73	64	0.0	0.0	0.0	0	73	64	0.0	0.0	0.0
gmu-35-40	316	64	0.0	0.0	0.1	0	316	64	0.0	0.0	0.1
gmu-35-50	359	64	0.0	0.0	0.1	0	359	64	0.1	0.0	0.1
gmut-75-50	6042	64	8.1	1.8	9.9	0	6042	64	8.5	1.8	10.3
gmut-77-40	4047	64	1.7	0.6	2.2	0	4047	64	2.0	0.4	2.4
go19	2606	64	0.2	0.6	0.8	0	2606	64	0.1	0.3	0.4
gr4x6	36	64	0.0	0.0	0.0	0	36	64	0.0	0.0	0.0
greenbea	18382	64	4.7	0.8	5.5	0	18382	64	4.8	0.6	5.5
greenbeb	11957	64	2.5	1.9	4.4	0	11957	64	2.5	1.8	4.3
grow15	2102	64	0.2	0.8	1.0	0	2102	64	0.1	0.7	0.8
grow22	3334	64	0.4	3.2	3.5	0	3334	64	0.2	3.2	3.4
grow7	1071	64	0.1	0.6	0.7	0	1071	64	0.0	0.3	0.3
gt2	19	64	0.0	0.0	0.0	0	19	64	0.0	0.0	0.0
hanoi5	7389	64	1.6	0.3	1.9	0	7389	64	1.7	0.3	2.0
haprp	1303	64	0.0	0.1	0.1	0	1303	64	0.0	0.1	0.1
harp2	323	64	0.0	0.1	0.1	0	323	64	0.0	0.1	0.1
i_n13	948541	64	2586.7	6.8	2593.5	0	948541	64	2571.1	6.8	2577.8
ic97_potential	339	64	0.0	0.0	0.0	0	339	64	0.0	0.0	0.0
iiasa	1562	64	0.1	0.1	0.1	0	1562	64	0.1	0.1	0.1
iis-100-0-cov	1279	64	0.5	0.2	0.7	0	1279	64	0.5	0.2	0.7
iis-bupa-cov	4196	64	1.1	0.3	1.5	0	4196	64	1.1	0.3	1.5
iis-pima-cov	3532	64	1.2	0.5	1.7	0	3532	64	1.1	0.5	1.6
israel	149	64	0.0	0.0	0.0	0	149	64	0.0	0.0	0.0
ivu06-big	*37753*	*64*	*3149.6*	*7200.0*	*10349.6*	*0*	*37753*	*64*	*3170.1*	*7200.0*	*10370.1*
ivu52	18983	64	107.6	8.4	116.0	0	18983	64	110.3	8.4	118.7
janos-us-DDM	1042	64	0.0	0.0	0.1	0	1042	64	0.0	0.0	0.1
jendrec1	11116	64	2.4	1.9	4.3	0	11116	64	2.8	1.9	4.7
k16x240	39	64	0.0	0.0	0.0	0	39	64	0.0	0.0	0.0
kb2	58	64	0.0	0.0	0.0	0	58	64	0.0	0.0	0.0
ken-07	2777	64	0.1	0.1	0.3	0	2777	64	0.2	0.1	0.2
ken-11	17739	64	2.7	0.5	3.3	0	17739	64	3.0	0.6	3.7
ken-13	41646	64	20.1	1.0	21.0	0	41646	64	20.5	1.4	21.9
ken-18	192333	64	500.8	5.7	506.5	0	192333	64	501.6	5.5	507.1
kent	1680	64	0.3	0.5	0.9	0	1680	64	0.5	0.5	1.0
khb05250	119	64	0.0	0.0	0.0	0	119	64	0.0	0.0	0.0
kl02	263	64	0.2	0.4	0.5	0	263	64	0.4	0.4	0.8
kleemin3	0	64	0.0	0.0	0.0	0	0	64	0.0	0.0	0.0
kleemin4	0	64	0.0	0.0	0.0	0	0	64	0.0	0.0	0.0
kleemin5	0	64	0.0	0.0	0.0	0	0	64	0.0	0.0	0.0

Table A.3 continued

Instance	iter	prc	t_9	t_{ex}	t	R_0	iter	prc	t_{50}	t_{ex}	t
		SoPlex$_9$+QSopt_ex						SoPlex$_{50}$+QSopt_ex			
kleemin6	0	64	0.0	0.0	0.0	0	0	64	0.0	0.0	0.0
kleemin7	0	64	0.0	0.0	0.0	0	0	64	0.0	0.0	0.0
kleemin8	0	64	0.0	0.0	0.0	0	0	64	0.0	0.0	0.0
l152lav	700	64	0.1	0.0	0.1	0	700	64	0.1	0.0	0.1
l9	746	64	0.1	0.3	0.4	0	746	64	0.1	0.3	0.5
large000	3748	64	0.6	0.8	1.5	0	3748	64	0.7	0.9	1.6
large001	7844	64	1.4	1.3	2.7	0	7844	64	1.4	1.1	2.6
large002	3866	128	0.8	17.6	18.4	2	4065	64	0.6	1.4	2.1
large003	4179	128	0.8	15.1	15.9	1	4254	64	0.7	1.3	2.0
large004	4647	128	1.0	13.4	14.4	2	4681	64	0.8	2.2	3.0
large005	4501	128	0.8	14.2	15.0	1	4567	64	0.8	1.2	1.9
large006	4864	128	0.8	15.2	16.0	2	4942	64	0.7	1.3	2.0
large007	5010	128	0.8	15.1	16.0	1	5094	64	0.9	1.7	2.6
large008	5203	128	0.8	15.0	15.9	1	5291	64	0.9	1.8	2.7
large009	4988	128	0.9	16.4	17.3	1	5074	64	1.1	2.3	3.4
large010	4639	128	0.7	14.7	15.4	1	4725	64	0.8	1.4	2.3
large011	5135	128	0.9	16.4	17.4	1	5221	64	0.9	1.5	2.4
large012	4924	128	0.8	17.5	18.3	1	5009	64	0.9	1.5	2.4
large013	4975	128	0.8	16.2	17.0	1	5062	64	0.7	0.9	1.6
large014	5082	128	0.9	14.8	15.6	1	5148	64	0.9	1.2	2.1
large015	4318	128	0.7	15.0	15.8	1	4380	64	0.6	1.1	1.7
large016	4571	128	0.6	13.4	14.0	1	4633	64	0.8	1.1	1.9
large017	3980	64	0.8	1.2	2.0	0	3980	64	0.8	1.1	1.9
large018	4459	64	0.8	1.0	1.8	0	4459	64	0.8	0.7	1.6
large019	4909	64	0.8	0.9	1.7	0	4909	64	0.8	0.6	1.4
large020	6984	128	1.1	16.1	17.2	2	7059	64	0.9	1.1	2.0
large021	6201	128	0.9	20.4	21.3	2	6288	64	0.9	1.0	2.0
large022	6907	128	0.9	20.0	20.9	2	6993	64	0.8	0.7	1.5
large023	4224	128	0.9	18.0	19.0	2	4398	64	0.7	1.3	2.0
large024	5788	128	1.0	17.3	18.3	2	5988	64	1.2	1.3	2.5
large025	4811	128	0.8	19.1	19.9	2	5000	64	1.0	2.2	3.2
large026	4196	128	0.8	18.4	19.3	2	4367	64	0.7	1.6	2.2
large027	4172	128	0.8	16.4	17.3	2	4353	64	0.8	0.9	1.7
large028	4691	128	1.1	19.1	20.1	1	4906	64	0.8	1.1	1.9
large029	4158	128	0.8	19.5	20.3	2	4372	64	0.6	1.7	2.4
large030	3732	128	0.6	18.7	19.3	2	3930	64	0.7	1.1	1.8
large031	3729	128	0.6	18.7	19.3	1	3931	64	0.6	2.0	2.6
large032	4851	128	0.9	21.6	22.4	1	5052	64	0.9	3.1	4.0
large033	3675	128	0.7	16.5	17.2	2	3877	64	0.5	1.2	1.7
large034	4009	128	0.7	21.1	21.8	2	4201	64	0.7	2.7	3.4
large035	3450	128	0.8	22.1	22.9	1	3655	64	0.7	3.0	3.6
large036	3111	128	0.6	21.5	22.1	2	3314	64	0.6	2.4	3.0
lectsched-1	7	64	1.1	0.7	1.8	0	7	64	1.3	0.7	2.0
lectsched-1-obj	963	64	1.2	0.8	2.0	0	963	64	1.5	0.8	2.2
lectsched-2	3	64	0.5	0.3	0.8	0	3	64	0.7	0.4	1.0
lectsched-3	7	64	0.9	0.9	1.8	0	7	64	1.1	0.6	1.7
lectsched-4-obj	174	64	0.3	0.3	0.6	0	174	64	0.2	0.1	0.4
leo1	862	64	0.3	0.2	0.5	0	862	64	0.5	0.2	0.7
leo2	1637	64	0.6	0.3	0.9	0	1637	64	0.9	0.5	1.4
liu	543	64	0.1	0.0	0.1	0	543	64	0.1	0.0	0.1
lo10	953870	64	5941.4	3.8	5945.2	0	953870	64	5931.8	3.8	5935.7
long15	229488	64	2753.1	8.6	2761.7	0	229488	64	2738.1	8.5	2746.6
lotfi	226	64	0.0	0.0	0.0	0	226	64	0.0	0.0	0.0
lotsize	1460	64	0.0	0.0	0.1	0	1460	64	0.1	0.0	0.1
lp22	38451	64	35.6	5.4	41.0	0	38451	64	35.8	5.3	41.1
lpl1	36759	64	59.4	1.7	61.1	0	36759	64	60.0	1.4	61.4
lpl2	1465	64	0.2	0.1	0.3	0	1465	64	0.1	0.1	0.2
lpl3	5040	64	1.1	0.2	1.3	0	5040	64	1.2	0.2	1.4
lrn	11450	128	2.6	167.8	170.4	1	11452	64	2.5	0.3	2.9
lrsa120	9787	64	2.5	1.6	4.1	1	9789	64	2.4	0.2	2.7
lseu	25	64	0.0	0.0	0.0	0	25	64	0.0	0.0	0.0
m100n500k4r1	174	64	0.0	0.0	0.1	0	174	64	0.0	0.0	0.0
macrophage	706	64	0.0	0.0	0.1	0	706	64	0.0	0.0	0.1
manna81	3018	64	0.1	0.1	0.2	0	3018	64	0.1	0.1	0.2
map06	23840	64	39.4	5.9	45.3	0	23840	64	40.6	5.7	46.3
map10	23747	64	40.6	5.5	46.1	0	23747	64	41.5	5.6	47.1
map14	23178	64	38.3	5.5	43.9	0	23178	64	39.4	5.6	44.9
map18	20964	64	33.6	5.5	39.1	0	20964	64	34.7	5.5	40.2
map20	19686	64	33.1	5.7	38.8	0	19686	64	30.9	5.9	36.8
markshare1	35	64	0.0	0.0	0.0	0	35	64	0.0	0.0	0.0
markshare2	43	64	0.0	0.0	0.0	0	43	64	0.0	0.0	0.0
markshare_5_0	24	64	0.0	0.0	0.0	0	24	64	0.0	0.0	0.0

Table A.3 continued

Instance	SoPlex$_9$+QSopt_ex					SoPlex$_{50}$+QSopt_ex				
	iter	prc	t_9	t_{ex}	t	R_0	iter	prc	t_{50}	t_{ex}	t
maros	1255	64	0.1	0.2	0.3	0	1255	64	0.2	0.1	0.3
maros-r7	7953	192	2.9	787.1	790.0	0	7953	192	3.5	785.2	788.7
mas74	224	64	0.0	0.0	0.0	0	224	64	0.0	0.0	0.0
mas76	132	64	0.0	0.0	0.0	0	132	64	0.0	0.0	0.0
maxgasflow	6737	64	0.6	0.2	0.8	0	6737	64	0.8	0.1	0.9
mc11	1239	64	0.1	0.0	0.1	0	1239	64	0.1	0.0	0.1
mcf2	2763	64	0.3	0.1	0.4	0	2763	64	0.1	0.0	0.2
mcsched	2546	64	0.3	0.1	0.4	0	2546	64	0.3	0.1	0.4
methanosarcina	655	64	0.1	0.2	0.3	0	655	64	0.2	0.2	0.3
mik-250-1-100-1	100	64	0.0	0.0	0.0	0	100	64	0.0	0.0	0.0
mine-166-5	1642	64	0.5	0.2	0.8	0	1642	64	0.5	0.2	0.8
mine-90-10	1948	64	0.5	0.2	0.7	0	1948	64	0.4	0.2	0.6
misc03	45	64	0.0	0.0	0.0	0	45	64	0.0	0.0	0.0
misc06	816	64	0.1	0.1	0.1	0	816	64	0.1	0.1	0.1
misc07	157	64	0.0	0.0	0.1	0	157	64	0.0	0.0	0.1
mitre	2451	64	0.3	0.2	0.5	0	2451	64	0.4	0.2	0.6
mkc	538	64	0.1	0.1	0.1	0	538	64	0.1	0.1	0.2
mkc1	538	64	0.1	0.1	0.2	0	538	64	0.0	0.1	0.1
mod008	27	64	0.0	0.0	0.0	0	27	64	0.0	0.0	0.0
mod010	1062	64	0.1	0.0	0.2	0	1062	64	0.2	0.0	0.2
mod011	6153	64	0.9	0.2	1.1	0	6153	64	0.8	0.2	1.0
mod2	58340	128	100.4	7200.0	7300.4	11	58534	64	103.8	4.3	108.1
model1	180	128	0.0	0.2	0.2	1	182	64	0.0	0.0	0.0
model10	44687	64	22.8	35.0	57.8	0	44687	64	23.2	34.9	58.1
model11	5273	64	1.0	0.5	1.5	0	5273	64	1.0	0.3	1.3
model2	3466	64	0.4	0.2	0.6	0	3466	64	0.2	0.2	0.4
model3	9208	64	1.3	1.5	2.8	0	9208	64	1.3	1.5	2.8
model4	15098	64	3.0	1.8	4.8	0	15098	64	2.8	1.8	4.6
model5	19877	64	3.8	1.5	5.3	0	19877	64	4.1	1.5	5.6
model6	14770	64	3.3	12.1	15.4	0	14770	64	3.2	12.1	15.3
model7	16070	64	4.0	8.1	12.1	0	16070	64	4.0	7.8	11.8
model8	2522	64	0.3	7200.0	7200.4	1	2538	64	0.4	7200.0	7200.4
model9	12397	64	2.3	1.8	4.0	0	12397	64	2.5	1.8	4.2
modglob	359	64	0.0	0.0	0.0	0	359	64	0.0	0.0	0.0
modszk1	653	64	0.0	0.1	0.1	0	653	64	0.0	0.0	0.1
momentum1	3305	64	0.8	4.1	4.9	1	3306	64	1.1	1.5	2.6
momentum2	45882	128	18.8	240.8	259.6	2	45950	64	19.7	2.6	22.2
momentum3	46184	128	189.5	7200.0	7389.5	1	46185	64	192.0	1833.5	2025.5
msc98-ip	9496	128	1.3	862.3	863.6	1	9549	64	1.7	0.6	2.3
mspp16	*52*	*64*	*19.3*	*7200.0*	*7219.3*	*0*	*52*	*64*	*20.0*	*7200.0*	*7220.0*
multi	59	64	0.0	0.0	0.0	0	59	64	0.0	0.0	0.0
mzzv11	37921	64	26.4	0.5	26.8	0	37921	64	26.2	0.3	26.6
mzzv42z	34373	64	17.6	0.4	18.1	0	34373	64	17.2	0.5	17.7
n15-3	43662	64	114.6	3.4	118.0	0	43662	64	115.3	3.4	118.7
n3-3	2965	64	0.6	0.2	0.8	0	2965	64	0.5	0.1	0.6
n3700	8698	64	1.6	0.2	1.9	0	8698	64	1.6	0.2	1.8
n3701	8106	64	1.3	0.2	1.5	0	8106	64	1.3	0.2	1.5
n3702	7987	64	1.2	0.2	1.5	0	7987	64	1.1	0.2	1.3
n3703	7397	64	1.1	0.2	1.3	0	7397	64	1.0	0.2	1.2
n3704	7325	64	1.1	0.2	1.2	0	7325	64	0.9	0.2	1.2
n3705	6974	64	1.3	0.2	1.5	0	6974	64	1.2	0.2	1.4
n3706	7305	64	1.1	0.2	1.3	0	7305	64	1.1	0.2	1.4
n3707	8681	64	1.2	0.2	1.4	0	8681	64	1.4	0.2	1.6
n3708	7133	64	1.6	0.2	1.8	0	7133	64	1.2	0.2	1.4
n3709	8030	64	1.5	0.2	1.7	0	8030	64	1.4	0.2	1.7
n370a	8608	64	1.2	0.2	1.5	0	8608	64	1.3	0.2	1.5
n370b	8553	64	1.4	0.2	1.6	0	8553	64	1.2	0.2	1.4
n370c	7273	64	1.1	0.2	1.3	0	7273	64	0.9	0.2	1.2
n370d	7273	64	1.1	0.2	1.3	0	7273	64	0.9	0.2	1.2
n370e	7852	64	1.2	0.2	1.4	0	7852	64	1.0	0.2	1.2
n3div36	306	64	0.4	0.4	0.8	0	306	64	0.7	0.4	1.1
n3seq24	4646	64	14.1	10.5	24.6	1	4720	64	15.9	5.7	21.7
n4-3	1341	64	0.2	0.0	0.2	0	1341	64	0.2	0.0	0.2
n9-3	3531	64	0.6	0.1	0.6	0	3531	64	0.5	0.1	0.6
nag	2476	64	0.2	0.1	0.2	0	2476	64	0.1	0.1	0.2
nemsafm	755	64	0.0	0.0	0.1	0	755	64	0.0	0.0	0.1
nemscem	932	64	0.1	0.1	0.1	0	932	64	0.1	0.1	0.1
nemsemm1	10608	64	3.6	2.6	6.2	0	10608	64	5.4	2.7	8.1
nemsemm2	10445	64	1.5	0.7	2.2	0	10445	64	1.9	0.6	2.6
nemspmm1	15049	64	4.3	1.7	6.0	0	15049	64	4.5	1.8	6.3
nemspmm2	15056	128	4.2	223.0	227.2	1	15058	64	4.4	8.6	13.0
nemswrld	35922	64	32.6	160.3	192.9	0	35922	64	33.7	160.3	194.0

Table A.3 continued

Instance	SoPlex$_9$+QSopt_ex						SoPlex$_{50}$+QSopt_ex				
	iter	prc	t_9	t_{ex}	t	R_0	iter	prc	t_{50}	t_{ex}	t
neos	89942	64	497.3	22.5	519.8	1	90253	64	497.8	13.1	511.0
neos-1053234	257	64	0.1	0.1	0.2	0	257	64	0.0	0.1	0.1
neos-1053591	821	64	0.0	0.0	0.1	0	821	64	0.0	0.0	0.0
neos-1056905	140	64	0.0	0.0	0.0	0	140	64	0.0	0.0	0.0
neos-1058477	48	64	0.0	0.1	0.1	0	48	64	0.0	0.0	0.1
neos-1061020	16447	64	7.3	0.4	7.7	0	16447	64	7.5	0.6	8.0
neos-1062641	873	128	0.0	0.4	0.5	1	902	64	0.0	0.0	0.0
neos-1067731	12132	64	2.4	0.2	2.7	0	12132	64	2.5	0.2	2.8
neos-1096528	115	64	47.5	20.5	68.0	0	115	64	54.0	20.3	74.4
neos-1109824	115	64	0.5	0.3	0.7	0	115	64	0.6	0.1	0.7
neos-1112782	622	64	0.1	0.1	0.1	0	622	64	0.0	0.0	0.1
neos-1112787	557	64	0.1	0.1	0.1	0	557	64	0.0	0.0	0.1
neos-1120495	96	64	0.2	0.2	0.4	0	96	64	0.3	0.2	0.5
neos-1121679	35	64	0.0	0.0	0.0	0	35	64	0.0	0.0	0.0
neos-1122047	5499	64	1.2	1.0	2.2	0	5499	64	1.5	0.8	2.3
neos-1126860	5216	64	0.9	0.5	1.4	0	5216	64	1.0	0.5	1.5
neos-1140050	13327	64	30.1	1115.7	1145.8	0	13327	64	32.6	1113.4	1146.0
neos-1151496	4930	64	0.8	0.1	0.9	0	4930	64	0.8	0.1	0.9
neos-1171448	4075	64	0.6	0.5	1.1	0	4075	64	0.5	0.3	0.8
neos-1171692	1290	64	0.1	0.1	0.2	0	1290	64	0.1	0.1	0.2
neos-1171737	2060	64	0.3	0.1	0.4	0	2060	64	0.2	0.1	0.3
neos-1173026	33	64	0.0	0.0	0.0	0	33	64	0.0	0.0	0.1
neos-1200887	294	64	0.0	0.1	0.1	0	294	64	0.0	0.0	0.1
neos-1208069	3008	64	0.6	0.1	0.7	0	3008	64	0.6	0.1	0.7
neos-1208135	2364	64	0.4	0.1	0.5	0	2364	64	0.4	0.1	0.6
neos-1211578	187	64	0.0	0.0	0.0	0	187	64	0.0	0.0	0.0
neos-1215259	6319	64	1.0	0.1	1.1	0	6319	64	1.0	0.1	1.1
neos-1215891	3541	64	0.9	0.2	1.1	0	3541	64	0.7	0.2	1.0
neos-1223462	13116	64	3.7	0.3	4.0	0	13116	64	3.6	0.2	3.9
neos-1224597	10673	64	1.8	0.1	1.9	0	10673	64	1.5	0.1	1.6
neos-1225589	438	64	0.0	0.0	0.1	0	438	64	0.0	0.0	0.1
neos-1228986	197	64	0.0	0.0	0.0	0	197	64	0.0	0.0	0.0
neos-1281048	2276	64	0.2	0.0	0.3	0	2276	64	0.3	0.1	0.3
neos-1311124	639	64	0.0	0.0	0.1	0	639	64	0.0	0.0	0.1
neos-1324574	6564	64	0.8	0.2	1.0	0	6564	64	0.9	0.2	1.1
neos-1330346	1983	64	0.3	0.1	0.4	0	1983	64	0.1	0.1	0.2
neos-1330635	141	64	0.0	0.1	0.1	0	141	64	0.0	0.1	0.1
neos-1337307	10964	64	2.1	0.3	2.3	0	10964	64	2.2	0.3	2.5
neos-1337489	187	64	0.0	0.0	0.0	0	187	64	0.0	0.0	0.0
neos-1346382	350	64	0.0	0.0	0.0	0	350	64	0.0	0.0	0.0
neos-1354092	12518	64	7.0	28.2	35.2	0	12518	64	7.1	28.2	35.4
neos-1367061	18400	64	23.5	1.4	24.9	0	18400	64	24.1	1.4	25.5
neos-1396125	4834	64	0.6	0.1	0.7	0	4834	64	0.7	0.0	0.8
neos-1407044	27160	64	31.7	71.6	103.3	0	27160	64	32.1	67.7	99.8
neos-1413153	1027	64	0.2	0.1	0.3	0	1027	64	0.2	0.0	0.2
neos-1415183	1763	64	0.3	0.1	0.4	0	1763	64	0.4	0.1	0.4
neos-1417043	13642	64	57.3	3.9	61.2	0	13642	64	57.2	3.9	61.1
neos-1420205	944	128	0.0	0.5	0.5	0	944	64	0.0	0.0	0.0
neos-1420546	102727	64	92.3	2.1	94.4	0	102727	64	92.2	2.1	94.3
neos-1420790	12930	64	2.5	0.4	2.9	0	12930	64	2.4	0.3	2.6
neos-1423785	19268	64	3.1	0.6	3.8	0	19268	64	3.6	0.5	4.0
neos-1425699	26	64	0.0	0.0	0.0	0	26	64	0.0	0.0	0.0
neos-1426635	350	64	0.0	0.0	0.0	0	350	64	0.0	0.0	0.0
neos-1426662	665	64	0.0	0.0	0.1	0	665	64	0.0	0.0	0.0
neos-1427181	591	64	0.0	0.0	0.0	0	591	64	0.0	0.0	0.0
neos-1427261	883	64	0.1	0.0	0.1	0	883	64	0.1	0.0	0.1
neos-1429185	466	64	0.0	0.0	0.0	0	466	64	0.0	0.0	0.0
neos-1429212	28835	64	220.5	10.5	231.0	0	28835	64	218.8	10.3	229.1
neos-1429461	399	64	0.0	0.0	0.0	0	399	64	0.0	0.0	0.0
neos-1430701	257	64	0.0	0.0	0.0	0	257	64	0.0	0.0	0.0
neos-1430811	32707	64	340.4	17.1	357.6	0	32707	64	338.3	17.3	355.6
neos-1436709	505	64	0.0	0.0	0.0	0	505	64	0.0	0.0	0.0
neos-1436713	835	64	0.0	0.0	0.1	0	835	64	0.1	0.0	0.1
neos-1437164	296	64	0.0	0.0	0.1	0	296	64	0.1	0.0	0.1
neos-1439395	312	64	0.0	0.0	0.0	0	312	64	0.0	0.0	0.1
neos-1440225	486	64	0.1	0.1	0.2	0	486	64	0.1	0.1	0.2
neos-1440447	239	64	0.0	0.0	0.0	0	239	64	0.0	0.0	0.0
neos-1440457	820	64	0.0	0.0	0.1	0	820	64	0.1	0.0	0.1
neos-1440460	359	64	0.0	0.0	0.0	0	359	64	0.0	0.0	0.0
neos-1441553	288	64	0.0	0.1	0.1	0	288	64	0.1	0.0	0.1
neos-1442119	607	64	0.0	0.0	0.1	0	607	64	0.1	0.0	0.1
neos-1442657	498	64	0.0	0.0	0.0	0	498	64	0.0	0.0	0.0

Table A.3 continued

Instance	SoPlex$_9$+QSopt_ex						SoPlex$_{50}$+QSopt_ex				
	iter	prc	t_9	t_{ex}	t	R_0	iter	prc	t_{50}	t_{ex}	t
neos-1445532	14103	64	1.7	0.2	1.9	0	14103	64	1.8	0.2	2.0
neos-1445738	15164	64	2.8	0.3	3.1	0	15164	64	2.9	0.2	3.1
neos-1445743	16103	64	3.2	0.2	3.4	0	16103	64	3.1	0.2	3.3
neos-1445755	15634	64	3.1	0.2	3.4	0	15634	64	3.2	0.2	3.4
neos-1445765	15960	64	3.3	0.2	3.6	0	15960	64	3.4	0.2	3.6
neos-1451294	10087	64	1.8	0.6	2.4	0	10087	64	1.6	0.6	2.2
neos-1456979	957	64	0.4	0.1	0.4	0	957	64	0.2	0.1	0.3
neos-1460246	276	64	0.0	0.0	0.0	0	276	64	0.0	0.0	0.1
neos-1460265	808	64	0.1	0.1	0.1	0	808	64	0.1	0.0	0.1
neos-1460543	9199	64	1.3	0.1	1.5	0	9199	64	1.2	0.1	1.4
neos-1460641	10168	64	1.1	0.1	1.2	0	10168	64	1.1	0.0	1.1
neos-1461051	418	64	0.0	0.0	0.1	0	418	64	0.1	0.0	0.1
neos-1464762	10563	64	1.2	0.1	1.3	0	10563	64	1.2	0.1	1.3
neos-1467067	673	64	0.0	0.0	0.0	0	673	64	0.0	0.0	0.0
neos-1467371	8604	64	1.1	0.1	1.2	0	8604	64	0.9	0.1	1.0
neos-1467467	8764	64	0.9	0.1	1.0	0	8764	64	0.9	0.1	1.0
neos-1480121	86	64	0.0	0.0	0.0	0	86	64	0.0	0.0	0.0
neos-1489999	835	64	0.1	0.0	0.1	0	835	64	0.1	0.0	0.1
neos-1516309	134	64	0.0	0.0	0.1	0	134	64	0.1	0.1	0.1
neos-1582420	2433	64	0.5	0.2	0.8	0	2433	64	0.7	0.2	0.9
neos-1593097	921	64	0.6	0.2	0.8	0	921	64	0.7	0.2	1.0
neos-1595230	518	64	0.0	0.0	0.1	0	518	64	0.0	0.0	0.0
neos-1597104	150	64	0.5	1.3	1.8	0	150	64	0.7	1.3	2.0
neos-1599274	357	64	0.1	0.1	0.2	0	357	64	0.1	0.1	0.1
neos-1601936	14758	64	5.3	0.5	5.9	0	14758	64	5.4	0.7	6.1
neos-1603512	814	64	0.1	0.0	0.1	0	814	64	0.1	0.0	0.1
neos-1603518	2362	64	0.4	0.1	0.5	0	2362	64	0.4	0.1	0.5
neos-1605061	23399	61	11.5	0.5	12.0	0	23399	64	11.5	0.8	12.2
neos-1605075	16980	64	8.8	0.8	9.6	0	16980	64	8.8	0.7	9.4
neos-1616732	314	64	0.0	0.0	0.0	0	314	64	0.0	0.0	0.0
neos-1620770	772	64	0.1	0.1	0.1	0	772	64	0.1	0.0	0.1
neos-1620807	222	64	0.0	0.0	0.0	0	222	64	0.0	0.0	0.0
neos-1622252	807	64	0.1	0.1	0.1	0	807	64	0.1	0.0	0.1
neos-430149	240	64	0.0	0.0	0.0	0	240	64	0.0	0.0	0.0
neos-476283	5536	64	3.8	15.9	19.7	0	5536	64	4.3	16.1	20.4
neos-480878	466	64	0.1	0.2	0.2	0	466	64	0.1	0.1	0.2
neos-494568	553	64	0.2	0.1	0.3	0	553	64	0.2	0.2	0.4
neos-495307	963	64	0.3	0.1	0.3	0	963	64	0.2	0.1	0.2
neos-498623	1034	64	0.3	0.3	0.7	1	1037	64	0.6	0.2	0.8
neos-501453	37	64	0.0	0.0	0.0	0	37	64	0.0	0.0	0.0
neos-501474	288	64	0.0	0.0	0.0	0	288	64	0.0	0.0	0.0
neos-503737	4883	64	0.8	0.1	0.9	0	4883	64	0.8	0.1	0.9
neos-504674	548	64	0.0	0.0	0.1	0	548	64	0.0	0.0	0.1
neos-504815	468	64	0.0	0.0	0.1	0	468	64	0.0	0.0	0.0
neos-506422	98	64	0.1	0.1	0.1	0	98	64	0.1	0.0	0.1
neos-506428	2092	64	1.0	1.7	2.7	0	2092	64	1.1	1.7	2.8
neos-512201	543	64	0.0	0.1	0.1	0	543	64	0.0	0.0	0.1
neos-520729	32688	64	19.9	0.5	20.4	0	32688	64	20.2	0.7	20.9
neos-522351	205	64	0.0	0.0	0.0	0	205	64	0.0	0.0	0.0
neos-525149	1124	64	0.7	5.7	6.5	0	1124	64	0.8	5.9	6.7
neos-530627	68	64	0.0	0.0	0.0	0	68	64	0.0	0.0	0.0
neos-538867	169	64	0.0	0.0	0.1	1	171	64	0.0	0.0	0.1
neos-538916	168	64	0.0	0.0	0.0	0	168	64	0.0	0.0	0.0
neos-544324	2629	64	1.2	2.9	4.0	0	2629	64	1.5	2.8	4.3
neos-547911	1895	64	0.3	1.0	1.3	0	1895	64	0.5	1.0	1.4
neos-548047	11782	64	1.6	0.2	1.7	0	11782	64	1.8	0.2	2.0
neos-548251	2037	64	0.1	0.0	0.1	0	2037	64	0.1	0.0	0.1
neos-551991	7378	64	1.1	0.2	1.4	0	7378	64	1.1	0.2	1.3
neos-555001	10198	64	1.0	0.1	1.1	0	10198	64	0.6	0.1	0.7
neos-555298	3995	64	0.2	0.1	0.4	0	3995	64	0.1	0.1	0.3
neos-555343	11043	64	1.1	0.1	1.2	0	11043	64	0.9	0.1	0.9
neos-555424	5895	64	0.4	0.1	0.5	0	5895	64	0.2	0.1	0.3
neos-555694	565	64	0.1	0.1	0.2	0	565	64	0.1	0.0	0.1
neos-555771	575	64	0.1	0.1	0.2	0	575	64	0.1	0.0	0.1
neos-555884	7183	64	0.6	0.1	0.8	0	7183	64	0.4	0.1	0.6
neos-555927	2202	64	0.1	0.1	0.2	0	2202	64	0.1	0.1	0.1
neos-565672	87785	64	259.8	7.5	267.2	0	87785	64	260.9	7.2	268.2
neos-565815	6971	64	2.3	0.4	2.8	0	6971	64	2.4	0.5	3.0
neos-570431	2355	64	0.2	0.1	0.3	0	2355	64	0.1	0.0	0.1
neos-574665	506	64	0.1	0.1	0.2	0	506	64	0.1	0.1	0.2
neos-578379	*14389*	64	*13.2*	*7200.0*	*7213.2*	0	*14389*	64	*13.4*	*7200.0*	*7213.4*
neos-582605	1067	64	0.1	0.0	0.1	0	1067	64	0.1	0.0	0.1

Table A.3 continued

Instance	iter	prc	t_9	t_{ex}	t	R_0	iter	prc	t_{50}	t_{ex}	t
			SoPlex$_9$+QSopt_ex						SoPlex$_{50}$+QSopt_ex		
neos-583731	602	64	0.0	0.0	0.0	0	602	64	0.0	0.0	0.0
neos-584146	603	64	0.1	0.0	0.1	0	603	64	0.1	0.0	0.1
neos-584851	612	64	0.0	0.0	0.1	0	612	64	0.1	0.0	0.1
neos-584866	11751	64	2.6	0.2	2.8	0	11751	64	2.8	0.1	2.9
neos-585192	2614	64	0.5	0.3	0.8	0	2614	64	0.7	0.3	1.0
neos-585467	1912	64	0.4	0.2	0.6	0	1912	64	0.2	0.2	0.4
neos-593853	615	64	0.0	0.1	0.1	0	615	64	0.1	0.0	0.1
neos-595904	1276	64	0.2	0.1	0.4	0	1276	64	0.3	0.1	0.4
neos-595905	390	64	0.0	0.0	0.1	0	390	64	0.1	0.0	0.1
neos-595925	521	64	0.0	0.1	0.1	0	521	64	0.1	0.0	0.1
neos-598183	1148	64	0.1	0.1	0.2	0	1148	64	0.1	0.1	0.2
neos-603073	455	64	0.0	0.0	0.0	0	455	64	0.0	0.0	0.1
neos-611135	11914	64	4.7	1.6	6.2	0	11914	64	4.7	1.6	6.2
neos-611838	2046	64	0.4	0.2	0.6	0	2046	64	0.2	0.2	0.4
neos-612125	2002	64	0.4	0.2	0.6	0	2002	64	0.3	0.2	0.5
neos-612143	2100	64	0.4	0.2	0.6	0	2100	64	0.3	0.2	0.5
neos-612162	1828	64	0.3	0.2	0.5	0	1828	64	0.3	0.2	0.5
neos-619167	6697	64	1.3	0.2	1.5	0	6697	64	1.1	0.2	1.3
neos-631164	954	64	0.1	0.0	0.1	0	954	64	0.1	0.0	0.1
neos-631517	765	64	0.0	0.0	0.1	0	765	64	0.1	0.0	0.1
neos-631694	24940	64	5.2	0.1	5.3	0	24940	64	5.2	0.1	5.3
neos-631709	70640	64	213.1	0.8	213.9	0	70640	64	213.0	0.8	213.9
neos-631710	66455	64	923.9	4.6	928.5	0	66455	64	920.8	4.6	925.4
neos-631784	26868	64	29.2	0.2	29.4	0	26868	64	29.5	0.3	29.8
neos-632335	5592	64	1.0	0.5	1.6	0	5592	64	1.3	0.3	1.6
neos-633273	5070	64	1.1	0.5	1.6	0	5070	64	1.1	0.5	1.6
neos-641591	14066	64	6.7	0.3	7.1	0	14066	64	7.1	0.3	7.4
neos-655508	119	64	0.1	0.3	0.4	0	119	64	0.1	0.2	0.3
neos-662469	14066	64	7.2	0.4	7.6	0	14066	64	7.0	0.3	7.3
neos-686190	834	64	0.2	0.1	0.3	0	834	64	0.1	0.1	0.2
neos-691058	7206	64	1.3	0.2	1.5	0	7206	64	1.2	0.1	1.3
neos-691073	6943	64	1.4	0.1	1.6	0	6943	64	1.3	0.1	1.3
neos-693347	9600	64	2.1	0.3	2.3	0	9600	64	2.2	0.3	2.5
neos-702280	12807	64	14.6	15.1	29.7	0	12807	64	15.3	15.0	30.3
neos-709469	429	64	0.0	0.0	0.1	0	429	64	0.0	0.0	0.0
neos-717614	1049	64	0.0	0.1	0.1	0	1049	64	0.1	0.0	0.1
neos-738098	25294	64	31.9	0.8	32.6	0	25294	64	31.8	0.7	32.6
neos-775946	1720	64	0.6	0.3	0.9	0	1720	64	0.5	0.3	0.9
neos-777800	1633	64	0.5	0.2	0.6	0	1633	64	0.3	0.1	0.4
neos-780889	57556	64	184.5	2.9	187.4	0	57556	64	185.1	3.0	188.1
neos-785899	411	64	0.1	0.1	0.1	0	411	64	0.1	0.1	0.2
neos-785912	1338	64	0.3	0.1	0.4	0	1338	64	0.3	0.1	0.4
neos-785914	232	64	0.0	0.0	0.1	0	232	64	0.0	0.0	0.1
neos-787933	14008	64	31.7	0.6	32.3	0	14008	64	32.3	0.6	32.9
neos-791021	10768	64	2.5	0.2	2.7	0	10768	64	2.4	0.2	2.7
neos-796608	379	64	0.0	0.0	0.0	0	379	64	0.0	0.0	0.0
neos-799711	19400	64	3.0	0.9	3.9	0	19400	64	3.3	0.9	4.2
neos-799838	11429	64	3.8	0.3	4.2	0	11429	64	3.8	0.2	4.0
neos-801834	1782	64	0.3	0.2	0.5	0	1782	64	0.2	0.1	0.3
neos-803219	1309	64	0.1	0.1	0.2	0	1309	64	0.1	0.0	0.1
neos-803220	888	64	0.1	0.0	0.1	0	888	64	0.1	0.0	0.1
neos-806323	1246	64	0.1	0.1	0.1	0	1246	64	0.1	0.1	0.2
neos-807454	12164	64	2.2	0.1	2.3	0	12164	64	2.3	0.1	2.4
neos-807456	11272	64	1.6	0.4	2.0	0	11272	64	1.6	0.4	2.1
neos-807639	1305	64	0.1	0.1	0.1	0	1305	64	0.1	0.0	0.1
neos-807705	1259	64	0.1	0.1	0.2	0	1259	64	0.1	0.1	0.2
neos-808072	8164	64	1.4	0.1	1.5	0	8164	64	1.4	0.1	1.5
neos-808214	1236	64	0.2	0.1	0.3	0	1236	64	0.1	0.1	0.2
neos-810286	13517	64	3.5	0.3	3.8	0	13517	64	3.5	0.3	3.8
neos-810326	8031	64	1.2	0.1	1.4	0	8031	64	1.2	0.1	1.3
neos-820146	249	64	0.0	0.0	0.0	0	249	64	0.0	0.0	0.0
neos-820157	569	64	0.0	0.0	0.1	0	569	64	0.0	0.0	0.0
neos-820879	1509	64	0.5	0.1	0.7	0	1509	64	0.7	0.1	0.8
neos-824661	9912	64	5.2	0.4	5.6	0	9912	64	5.6	0.4	6.0
neos-824695	4726	64	1.8	0.3	2.1	0	4726	64	1.8	0.1	1.9
neos-825075	1062	64	0.1	0.0	0.1	0	1062	64	0.0	0.0	0.1
neos-826224	5396	64	1.2	0.3	1.5	0	5396	64	1.4	0.4	1.8
neos-826250	5747	64	1.3	0.1	1.4	0	5747	64	1.0	0.1	1.1
neos-826650	8372	64	1.7	0.2	1.9	0	8372	64	1.9	0.1	1.9
neos-826694	10754	64	2.6	0.4	3.0	0	10754	64	2.7	0.3	3.1
neos-826812	10450	64	2.4	0.1	2.5	0	10450	64	2.4	0.1	2.6
neos-826841	3738	64	0.8	0.1	0.9	0	3738	64	0.7	0.1	0.7

Table A.3 continued

Instance	iter	prc	SoPlex$_9$+QSopt_ex t_9	t_{ex}	t	R_0	iter	prc	SoPlex$_{50}$+QSopt_ex t_{50}	t_{ex}	t
neos-827015	18897	64	12.2	0.6	12.8	0	18897	64	12.8	0.6	13.3
neos-827175	11439	64	3.2	0.5	3.7	0	11439	64	3.3	0.4	3.7
neos-829552	13050	64	4.4	0.5	4.8	0	13050	64	4.8	0.4	5.3
neos-830439	179	64	0.0	0.0	0.0	0	179	64	0.0	0.0	0.0
neos-831188	8374	64	1.1	0.1	1.2	0	8374	64	1.2	0.1	1.2
neos-839838	6451	64	1.1	0.5	1.6	0	6451	64	1.4	0.4	1.8
neos-839859	2312	64	0.4	0.1	0.5	0	2312	64	0.2	0.1	0.3
neos-839894	21287	64	20.6	1.2	21.8	0	21287	64	20.8	1.2	21.9
neos-841664	5975	64	0.7	0.1	0.8	0	5975	64	0.6	0.1	0.7
neos-847051	2383	64	0.2	0.2	0.4	0	2383	64	0.3	0.1	0.4
neos-847302	2146	64	0.3	0.1	0.5	0	2146	64	0.4	0.1	0.4
neos-848150	1409	64	0.2	0.1	0.3	0	1409	64	0.2	0.0	0.3
neos-848198	11363	64	2.1	0.1	2.2	0	11363	64	2.3	0.1	2.4
neos-848589	1401	64	1.0	3.8	4.8	0	1401	64	1.9	3.7	5.6
neos-848845	7295	64	1.4	0.2	1.7	0	7295	64	1.4	0.3	1.7
neos-849702	5626	64	1.0	0.3	1.2	0	5626	64	1.0	0.3	1.3
neos-850681	25783	64	5.9	0.2	6.0	0	25783	64	5.8	0.2	6.0
neos-856059	580	64	0.1	0.1	0.3	0	580	64	0.1	0.1	0.2
neos-859770	294	64	0.3	1.6	1.9	0	294	64	0.3	1.6	2.0
neos-860244	90	64	0.1	0.4	0.5	0	90	64	0.1	0.5	0.6
neos-860300	349	64	0.2	0.6	0.8	0	349	64	0.3	0.6	0.9
neos-862348	1026	64	0.2	0.3	0.5	0	1026	64	0.2	0.3	0.4
neos-863472	142	64	0.0	0.0	0.0	0	142	64	0.0	0.0	0.0
neos-872648	38712	64	176.3	1.8	178.1	0	38712	64	175.4	2.0	177.4
neos-873061	30803	64	140.9	1.7	142.6	0	30803	64	142.5	1.9	144.4
neos-876808	92322	64	98.6	3.1	101.7	0	92322	64	98.4	3.1	101.5
neos-880324	232	64	0.0	0.0	0.0	0	232	64	0.0	0.0	0.0
neos-881765	424	64	0.0	0.0	0.1	0	424	64	0.1	0.0	0.1
neos-885086	3951	64	0.6	0.5	1.2	0	3951	64	0.9	0.5	1.4
neos-885524	205	64	5.4	0.4	5.8	0	205	64	5.6	0.4	6.0
neos-886822	1834	64	0.3	0.1	0.4	0	1834	64	0.2	0.1	0.2
neos-892255	1082	64	0.2	0.1	0.3	0	1082	64	0.2	0.0	0.2
neos-905856	1228	64	0.1	0.1	0.2	0	1228	64	0.2	0.0	0.2
neos-906865	792	64	0.1	0.1	0.1	0	792	64	0.1	0.1	0.1
neos-911880	319	64	0.0	0.0	0.1	0	319	64	0.0	0.0	0.1
neos-911970	258	64	0.0	0.0	0.1	0	258	64	0.0	0.0	0.1
neos-912015	1001	64	0.1	0.1	0.2	0	1001	64	0.2	0.1	0.2
neos-912023	1095	64	0.1	0.1	0.2	0	1095	64	0.2	0.0	0.2
neos-913984	4692	64	2.5	0.2	2.7	0	4692	64	2.8	0.2	3.0
neos-914441	7787	64	1.9	0.4	2.2	0	7787	64	1.6	0.4	2.0
neos-916173	862	64	0.2	0.4	0.6	0	862	64	0.3	0.3	0.7
neos-916792	1165	64	0.2	0.4	0.6	0	1165	64	0.6	0.4	1.0
neos-930752	19648	64	6.0	0.3	6.3	0	19648	64	6.1	0.2	6.3
neos-931517	10863	64	2.1	0.3	2.4	0	10863	64	2.2	0.3	2.5
neos-931538	11171	64	2.4	0.3	2.7	0	11171	64	2.2	0.3	2.5
neos-932721	30030	64	11.8	0.5	12.3	0	30030	64	11.7	0.4	12.1
neos-932816	14154	64	4.1	1.1	5.2	0	14154	64	4.2	1.1	5.4
neos-933364	1547	64	0.1	0.0	0.1	0	1547	64	0.1	0.0	0.1
neos-933550	2301	64	0.5	0.2	0.7	0	2301	64	0.4	0.2	0.6
neos-933562	4067	64	1.2	1.6	2.8	0	4067	64	1.1	1.6	2.7
neos-933638	27131	64	18.3	0.5	18.8	0	27131	64	18.0	0.6	18.6
neos-933815	1147	64	0.1	0.0	0.1	0	1147	64	0.0	0.0	0.0
neos-933966	21146	64	11.1	0.6	11.7	0	21146	64	11.4	0.8	12.2
neos-934184	1547	64	0.1	0.0	0.1	0	1547	64	0.1	0.0	0.1
neos-934278	26638	64	16.5	0.9	17.4	0	26638	64	16.6	1.1	17.7
neos-934441	29117	64	18.7	0.8	19.6	0	29117	64	18.6	0.8	19.4
neos-934531	549	64	0.3	0.3	0.6	0	549	64	0.5	0.6	1.0
neos-935234	27636	64	16.9	0.9	17.9	0	27636	64	17.0	0.7	17.7
neos-935348	28978	64	17.0	0.6	17.6	0	28978	64	16.8	0.6	17.4
neos-935496	4000	64	1.1	0.6	1.7	0	4000	64	0.8	0.7	1.4
neos-935627	30839	64	18.9	0.5	19.4	0	30839	64	18.6	0.4	19.1
neos-935674	4015	64	1.1	0.6	1.7	0	4015	64	1.0	0.4	1.5
neos-935769	24339	64	12.2	0.4	12.6	0	24339	64	12.5	0.2	12.7
neos-936660	26498	64	14.6	0.8	15.4	0	26498	64	14.7	0.6	15.3
neos-937446	26033	64	15.3	0.3	15.6	0	26033	64	15.2	0.2	15.3
neos-937511	24828	64	13.7	0.5	14.2	0	24828	64	13.5	0.3	13.8
neos-937815	33788	64	23.8	1.2	24.9	0	33788	64	23.8	1.1	24.9
neos-941262	26859	64	15.1	0.9	16.0	0	26859	64	15.1	0.7	15.8
neos-941313	46102	64	110.4	2.2	112.6	0	46102	64	112.8	2.2	115.0
neos-941698	682	64	0.1	0.1	0.2	0	682	64	0.1	0.0	0.1
neos-941717	4805	64	0.7	0.2	0.9	0	4805	64	0.8	0.1	0.9
neos-941782	1858	64	0.3	0.1	0.5	0	1858	64	0.2	0.1	0.2

Table A.3 continued

Instance	SoPlex$_9$+QSopt_ex						SoPlex$_{50}$+QSopt_ex				
	iter	prc	t_9	t_{ex}	t	R_0	iter	prc	t_{50}	t_{ex}	t
neos-942323	468	64	0.1	0.1	0.1	0	468	64	0.0	0.0	0.1
neos-942830	1444	64	0.2	0.1	0.3	0	1444	64	0.1	0.0	0.2
neos-942886	336	64	0.0	0.1	0.1	0	336	64	0.1	0.0	0.1
neos-948126	30978	64	19.4	1.3	20.7	0	30978	64	19.5	1.4	21.0
neos-948268	11614	64	3.8	0.5	4.2	0	11614	64	3.6	0.5	4.1
neos-948346	4225	64	4.1	0.9	5.0	0	4225	64	4.4	0.9	5.4
neos-950242	1897	64	0.4	0.3	0.8	0	1897	64	0.6	0.2	0.8
neos-952987	531	64	0.3	0.2	0.5	0	531	64	0.6	0.2	0.8
neos-953928	1858	128	1.1	66.5	67.6	1	5020	64	2.8	0.4	3.2
neos-955215	850	64	0.0	0.0	0.1	0	850	64	0.0	0.0	0.0
neos-955800	882	64	0.1	0.1	0.2	0	882	64	0.0	0.0	0.1
neos-957270	515	64	0.2	0.5	0.7	0	515	64	0.3	0.5	0.9
neos-957323	3688	64	3.8	0.9	4.8	0	3688	64	4.3	1.0	5.3
neos-957389	1299	64	0.3	0.5	0.8	0	1299	64	0.5	0.5	1.0
neos-960392	12434	64	9.2	0.4	9.6	0	12434	64	9.8	0.4	10.2
neos-983171	27536	64	16.1	1.3	17.3	0	27536	64	16.2	1.2	17.4
neos-984165	30160	64	19.0	0.9	19.9	0	30160	64	19.1	1.0	20.1
neos1	4509	64	12.5	2.6	15.1	0	4509	64	13.2	2.7	15.9
neos13	2438	64	0.8	4.5	5.2	0	2438	64	0.9	4.7	5.6
neos15	508	64	0.0	0.0	0.0	0	508	64	0.0	0.0	0.1
neos16	475	64	0.0	0.0	0.0	0	475	64	0.0	0.0	0.0
neos18	293	64	0.1	0.1	0.2	0	293	64	0.2	0.1	0.3
neos2	8434	64	34.4	3.2	37.5	0	8434	64	34.8	3.2	38.0
neos6	6681	64	2.1	0.3	2.4	0	6681	64	2.1	0.3	2.4
neos788725	603	64	0.1	0.0	0.1	0	603	64	0.0	0.0	0.1
neos808444	8106	64	4.1	0.5	4.6	0	8106	64	4.0	0.6	4.6
neos858960	2	64	0.0	0.0	0.0	0	2	64	0.0	0.0	0.0
nesm	6164	64	0.6	0.2	0.7	0	6164	64	0.4	0.1	0.5
net12	11945	64	2.9	0.3	3.2	0	11945	64	2.9	0.4	3.3
netdiversion	26331	64	27.9	1.9	29.8	0	26331	64	27.8	1.9	29.7
netlarge2	111407	64	1542.8	17.2	1560.0	0	111407	64	1528.1	17.3	1545.4
newdano	455	64	0.0	0.0	0.1	0	455	64	0.0	0.0	0.0
nl	14132	64	3.9	0.8	4.7	0	14132	64	3.9	0.8	4.7
nobel-eu-DBE	2927	64	0.3	0.0	0.3	0	2927	64	0.1	0.0	0.2
noswot	134	128	0.0	0.1	0.1	1	135	64	0.0	0.0	0.0
npmv07	168166	64	45.2	4.9	50.1	0	168166	64	47.6	4.7	52.3
ns1111636	27016	64	78.9	2.2	81.1	0	27016	64	80.2	2.1	82.2
ns1116954	15636	64	24.2	1.6	25.9	0	15636	64	24.4	1.6	25.9
ns1208400	9117	64	1.7	0.3	2.0	0	9117	64	1.8	0.2	2.0
ns1456591	2343	64	0.5	0.3	0.8	0	2343	64	0.6	0.3	0.9
ns1606230	15157	64	6.2	0.6	6.8	0	15157	64	6.1	0.5	6.6
ns1631475	101171	64	253.3	0.8	254.1	0	101171	64	256.6	0.9	257.5
ns1644855	47488	64	109.7	47.1	156.8	0	47488	64	110.5	47.2	157.7
ns1663818	*1723*	64	24.9	7200.0	7224.9	*0*	*1723*	64	25.8	7200.0	7225.8
ns1685374	105867	64	779.2	426.5	1205.7	0	105867	64	774.7	426.3	1200.9
ns1686196	157	64	0.1	0.1	0.2	0	157	64	0.1	0.1	0.2
ns1688347	178	64	0.1	0.2	0.2	0	178	64	0.1	0.2	0.3
ns1696083	498	64	0.3	0.5	0.8	0	498	64	0.3	0.5	0.8
ns1702808	58	64	0.0	0.0	0.1	0	58	64	0.0	0.0	0.0
ns1745726	182	64	0.1	0.1	0.3	0	182	64	0.1	0.1	0.2
ns1758913	27608	64	335.6	13.4	349.0	0	27608	64	335.8	13.4	349.2
ns1766074	27	64	0.0	0.0	0.0	0	27	64	0.0	0.0	0.0
ns1769397	206	64	0.2	0.2	0.4	0	206	64	0.2	0.3	0.5
ns1778858	14685	64	5.4	0.5	5.8	0	14685	64	5.4	0.5	5.9
ns1830653	2753	64	0.5	0.2	0.8	0	2753	64	0.7	0.3	0.9
ns1856153	3231	64	0.7	0.5	1.2	0	3231	64	0.7	0.6	1.3
ns1904248	42350	128	55.0	2907.3	2962.3	1	42365	64	55.9	2.3	58.2
ns1905797	5721	64	2.5	0.8	3.3	0	5721	64	2.8	0.8	3.6
ns1905800	2057	64	0.5	0.2	0.7	0	2057	64	0.4	0.1	0.5
ns1952667	81	64	0.2	0.3	0.5	0	81	64	0.2	0.3	0.5
ns2017839	60283	64	118.1	2.7	120.8	0	60283	64	117.7	2.7	120.4
ns2081729	115	64	0.0	0.0	0.0	0	115	64	0.0	0.0	0.0
ns2118727	23580	64	111.8	13.3	125.1	0	23580	64	112.7	13.3	126.0
ns2122603	13837	64	3.5	0.6	4.1	0	13837	64	3.8	0.7	4.5
ns2124243	94850	64	312.1	2.7	314.8	0	94850	64	313.0	2.7	315.8
ns2137859	2864	64	5.2	5.2	10.3	0	2864	64	6.2	5.2	11.4
ns4-pr3	10658	64	1.8	0.1	1.8	0	10658	64	1.8	0.0	1.9
ns4-pr9	9965	64	1.5	0.1	1.5	0	9965	64	1.3	0.1	1.4
ns894236	36452	64	16.2	0.3	16.6	0	36452	64	15.4	0.4	15.8
ns894244	29128	64	21.5	0.9	22.4	0	29128	64	21.7	0.8	22.4
ns894786	18747	64	14.5	0.5	15.0	0	18747	64	14.8	0.6	15.4
ns894788	10998	64	1.5	0.1	1.6	0	10998	64	1.4	0.1	1.5

Table A.3 continued

Instance	iter	prc	t_9	t_{ex}	t	R_0	iter	prc	t_{50}	t_{ex}	t
			SoPlex$_9$+QSopt_ex						SoPlex$_{50}$+QSopt_ex		
ns903616	30158	64	22.4	0.6	22.9	0	30158	64	22.4	0.7	23.0
ns930473	10736	64	8.2	0.3	8.6	0	10736	64	8.2	0.3	8.6
nsa	1162	64	0.1	0.1	0.2	0	1162	64	0.1	0.0	0.1
nsct1	4841	64	1.2	1.4	2.6	0	4841	64	1.4	1.4	2.8
nsct2	6810	64	1.3	1.5	2.9	0	6810	64	1.5	1.5	3.0
nsic1	254	64	0.0	0.0	0.0	0	254	64	0.0	0.0	0.0
nsic2	268	64	0.0	0.0	0.0	0	268	64	0.0	0.0	0.0
nsir1	3762	64	0.6	0.4	1.0	0	3762	64	1.0	0.2	1.2
nsir2	3288	64	0.5	0.3	0.7	0	3288	64	0.7	0.4	1.0
nsr8k	102631	64	402.4	196.1	598.5	0	102631	64	397.4	196.8	594.2
nsrand-ipx	152	64	0.1	0.2	0.4	0	152	64	0.4	0.3	0.7
nu120-pr3	6187	64	1.1	0.2	1.2	0	6187	64	1.1	0.1	1.2
nu60-pr9	5736	64	1.1	0.1	1.3	0	5736	64	1.2	0.1	1.3
nug05	161	64	0.0	0.0	0.0	0	161	64	0.0	0.0	0.0
nug06	1117	64	0.1	0.1	0.1	0	1117	64	0.1	0.0	0.1
nug07	7569	64	1.0	0.1	1.1	0	7569	64	0.8	0.1	0.9
nug08	13541	64	2.0	0.2	2.2	0	13541	64	1.7	0.2	1.9
nug08-3rd	43975	64	1840.1	7200.0	9040.1	0	43975	64	1869.3	7200.0	9069.3
nug12	99946	64	160.0	13.7	173.7	0	99946	64	159.6	13.5	173.2
nw04	82	64	0.3	0.5	0.8	0	82	64	1.0	0.5	1.4
nw14	166	64	0.7	0.7	1.4	0	166	64	1.9	0.7	2.7
ofi	*139069*	*128*	*407.6*	*7200.0*	*7607.6*	*2*	139104	64	412.4	19.8	432.2
opm2-z10-s2	21918	64	111.0	11.3	122.3	0	21918	64	110.7	11.3	122.0
opm2-z11-s8	24349	64	162.2	15.7	177.9	0	24349	64	163.7	15.2	178.9
opm2-z12-s14	28647	64	279.0	32.9	312.0	0	28647	64	283.9	33.8	317.6
opm2-z12-s7	30797	64	313.1	37.7	350.8	0	30797	64	315.2	39.6	354.8
opm2-z7-s2	10304	64	7.3	1.7	9.0	0	10304	64	7.4	1.6	9.0
opt1217	143	64	0.0	0.0	0.0	0	143	64	0.0	0.0	0.0
orna1	1327	64	0.2	2.2	2.3	0	1327	64	0.2	2.3	2.5
orna2	1539	64	0.2	2.3	2.4	0	1539	64	0.2	2.1	2.4
orna3	1754	64	0.2	2.5	2.7	0	1754	64	0.3	2.0	2.3
orna4	2708	64	0.2	3.4	3.7	0	2708	64	0.3	3.6	3.9
orna7	2280	64	0.3	2.3	2.5	0	2280	64	0.3	1.9	2.2
orswq2	148	64	0.0	0.0	0.0	0	148	64	0.0	0.0	0.0
osa-07	966	64	0.5	0.2	0.7	0	966	64	0.5	0.2	0.7
osa-14	2104	64	1.2	0.5	1.7	0	2104	64	1.4	0.5	1.9
osa-30	5628	64	5.9	0.9	6.8	0	5628	64	6.3	0.9	7.2
osa-60	13364	64	38.1	2.1	40.1	0	13364	64	39.1	2.1	41.2
p0033	19	64	0.0	0.0	0.0	0	19	64	0.0	0.0	0.0
p0040	13	64	0.0	0.0	0.0	1	23	64	0.0	0.0	0.0
p010	13700	64	1.8	1.0	2.8	0	13700	64	2.2	1.0	3.2
p0201	50	64	0.0	0.0	0.0	1	51	64	0.0	0.0	0.0
p0282	114	64	0.0	0.0	0.0	0	114	64	0.0	0.0	0.0
p0291	27	64	0.0	0.0	0.0	0	27	64	0.0	0.0	0.0
p05	7515	64	0.9	0.6	1.5	0	7515	64	0.8	0.6	1.4
p0548	84	64	0.0	0.0	0.0	0	84	64	0.0	0.0	0.0
p100x588b	235	64	0.0	0.0	0.0	0	235	64	0.0	0.0	0.0
p19	315	64	0.0	0.0	0.1	0	315	64	0.0	0.0	0.0
p2756	73	64	0.0	0.0	0.1	0	73	64	0.0	0.0	0.0
p2m2p1m1p0n100	10	64	0.0	0.0	0.0	0	10	64	0.0	0.0	0.0
p6000	728	64	0.1	0.2	0.3	0	728	64	0.2	0.2	0.4
p6b	548	64	0.1	0.0	0.1	0	548	64	0.1	0.0	0.1
p80x400b	152	64	0.0	0.0	0.0	0	152	64	0.0	0.0	0.0
pcb1000	2684	64	0.3	0.4	0.7	0	2684	64	0.4	0.4	0.8
pcb3000	7719	64	0.9	1.2	2.0	0	7719	64	1.0	1.4	2.4
pds-02	2713	64	0.1	0.1	0.2	0	2713	64	0.1	0.1	0.1
pds-06	10699	64	1.1	0.2	1.3	0	10699	64	1.1	0.2	1.3
pds-10	15362	64	1.9	0.4	2.3	0	15362	64	1.9	0.4	2.3
pds-100	661386	64	4676.7	9.3	4685.9	0	661386	64	4662.1	9.1	4671.2
pds-20	37294	64	13.1	1.0	14.1	0	37294	64	13.4	1.0	14.4
pds-30	65338	64	58.4	0.9	59.3	0	65338	64	57.5	1.2	58.7
pds-40	101787	64	159.1	3.2	162.2	0	101787	64	155.7	3.0	158.7
pds-50	137692	64	289.9	3.6	293.5	0	137692	64	293.2	3.6	296.8
pds-60	188842	64	563.5	5.1	568.6	0	188842	64	563.8	5.3	569.1
pds-70	222389	64	759.3	6.5	765.8	0	222389	64	761.6	6.3	767.9
pds-80	240224	64	823.9	7.9	831.8	0	240224	64	821.3	8.0	829.3
pds-90	512992	64	3214.8	8.7	3223.5	0	512992	64	3210.0	8.8	3218.8
perold	5155	64	0.8	5.5	6.2	0	5155	64	0.6	4.9	5.6
pf2177	9474	64	3.4	0.6	4.0	0	9474	64	3.3	0.6	3.9
pg	2164	64	0.0	0.1	0.1	0	2164	64	0.1	0.1	0.1
pg5_34	3400	64	0.1	0.1	0.2	0	3400	64	0.1	0.1	0.2
pgp2	3713	64	0.1	0.2	0.3	0	3713	64	0.2	0.2	0.4

Table A.3 continued

Instance	iter	prc	t_9	t_{ex}	t	R_0	iter	prc	t_{50}	t_{ex}	t
		SoPlex$_9$+QSopt_ex						SoPlex$_{50}$+QSopt_ex			
pigeon-10	245	64	0.0	0.0	0.0	0	245	64	0.0	0.0	0.0
pigeon-11	237	64	0.0	0.0	0.0	0	237	64	0.0	0.0	0.0
pigeon-12	305	64	0.0	0.0	0.1	0	305	64	0.0	0.0	0.1
pigeon-13	288	64	0.0	0.0	0.1	0	288	64	0.0	0.0	0.1
pigeon-19	608	64	0.0	0.1	0.1	0	608	64	0.1	0.1	0.1
pilot	10227	64	3.6	137.1	140.7	0	10227	64	3.6	136.9	140.5
pilot-ja	11126	64	1.7	11.5	13.2	0	11126	64	1.6	11.4	12.9
pilot-we	5730	64	0.8	1.8	2.7	0	5730	64	0.8	1.6	2.4
pilot4	1483	64	0.2	1.2	1.3	0	1483	64	0.2	1.2	1.4
pilot87	15871	64	9.5	2274.2	2283.7	0	15871	64	9.5	2283.4	2292.9
pilotnov	7134	64	0.9	1.6	2.5	0	7134	64	0.7	1.4	2.1
pk1	101	64	0.0	0.0	0.0	0	101	64	0.0	0.0	0.0
pldd000b	1667	64	0.2	0.8	1.0	0	1667	64	0.1	0.9	1.1
pldd001b	1662	64	0.2	0.9	1.1	0	1662	64	0.2	0.5	0.7
pldd002b	1649	64	0.2	0.9	1.1	0	1649	64	0.3	0.6	0.9
pldd003b	1657	64	0.2	0.8	1.0	0	1657	64	0.3	0.6	0.9
pldd004b	1654	64	0.2	0.8	1.0	0	1654	64	0.2	0.6	0.8
pldd005b	1655	64	0.2	0.7	0.9	0	1655	64	0.2	0.6	0.9
pldd006b	1693	64	0.2	0.8	0.9	0	1693	64	0.3	0.5	0.7
pldd007b	1714	64	0.2	0.7	0.9	0	1714	64	0.3	0.5	0.8
pldd008b	1716	64	0.2	0.8	0.9	0	1716	64	0.2	0.5	0.8
pldd009b	2258	64	0.3	0.9	1.2	0	2258	64	0.4	0.8	1.2
pldd010b	2462	64	0.4	0.9	1.3	0	2462	64	0.5	0.9	1.4
pldd011b	2413	64	0.4	0.8	1.2	0	2413	64	0.5	0.5	0.9
pldd012b	2447	64	0.4	0.8	1.2	0	2447	64	0.5	0.8	1.2
pltexpa2-16	1094	64	0.1	0.1	0.1	0	1094	64	0.0	0.0	0.1
pltexpa2-6	411	64	0.0	0.0	0.1	0	411	64	0.0	0.0	0.0
pltexpa3_16	15975	64	4.1	0.8	4.9	0	15975	64	4.4	0.8	5.2
pltexpa3_6	2741	64	0.3	0.2	0.5	0	2741	64	0.5	0.1	0.6
pltexpa4_6	14762	64	3.5	0.5	4.1	0	14762	64	4.0	0.6	4.6
pp08a	145	64	0.0	0.0	0.0	0	145	64	0.0	0.0	0.0
pp08aCUTS	224	64	0.0	0.0	0.0	0	224	64	0.0	0.0	0.0
primagaz	1802	64	0.4	0.1	0.6	0	1802	64	0.4	0.1	0.5
problem	9	64	0.0	0.0	0.0	0	9	64	0.0	0.0	0.0
probportfolio	126	64	0.0	0.0	0.0	0	126	64	0.0	0.0	0.1
prod1	193	64	0.0	0.0	0.0	0	193	64	0.0	0.0	0.1
prod2	344	64	0.0	0.0	0.1	0	344	64	0.1	0.1	0.1
progas	3998	64	0.8	15.4	16.2	0	3998	64	0.8	15.5	16.3
protfold	13331	64	2.8	0.3	3.1	0	13331	64	2.7	0.3	3.0
pw-myciel4	2154	64	0.6	0.1	0.8	0	2154	64	0.7	0.1	0.8
qap10	58606	64	30.2	1.4	31.7	0	58606	64	30.0	1.2	31.3
qiu	1604	64	0.2	0.1	0.3	0	1604	64	0.2	0.0	0.3
qiulp	1604	64	0.2	0.1	0.2	0	1604	64	0.1	0.0	0.1
qnet1	789	64	0.1	0.0	0.1	0	789	64	0.1	0.0	0.1
qnet1_o	406	64	0.0	0.0	0.1	0	406	64	0.0	0.0	0.1
queens-30	13322	64	10.8	748.5	759.3	0	13322	64	13.5	753.5	767.0
r05	7499	64	0.7	0.5	1.1	0	7499	64	0.9	0.5	1.5
r80x800	188	64	0.0	0.0	0.0	0	188	64	0.0	0.0	0.0
rail01	349407	128	4435.1	6310.6	10745.7	1	349603	64	4461.4	1.9	4463.3
rail2586	*30254*	64	*675.5*	*7200.0*	*7875.5*	0	*30254*	64	*679.5*	*7200.0*	*7879.6*
rail4284	*54982*	64	*1670.1*	*7200.0*	*8870.1*	0	*54982*	64	*1678.7*	*7200.0*	*8878.7*
rail507	13033	64	11.2	0.7	12.0	0	13033	64	11.6	0.8	12.5
rail516	6992	64	5.6	0.5	6.1	0	6992	64	5.9	0.5	6.4
rail582	10608	64	9.2	0.6	9.8	0	10608	64	9.5	0.6	10.1
ramos3	19446	64	58.7	975.6	1034.3	0	19446	64	60.3	975.4	1035.7
ran10x10a	133	64	0.0	0.0	0.0	0	133	64	0.0	0.0	0.0
ran10x10b	149	64	0.0	0.0	0.0	0	149	64	0.0	0.0	0.0
ran10x10c	161	64	0.0	0.0	0.0	0	161	64	0.0	0.0	0.0
ran10x12	162	64	0.0	0.0	0.0	0	162	64	0.0	0.0	0.0
ran10x26	423	64	0.0	0.0	0.0	0	423	64	0.0	0.0	0.0
ran12x12	186	64	0.0	0.0	0.0	0	186	64	0.0	0.0	0.0
ran12x21	380	64	0.0	0.0	0.0	0	380	64	0.0	0.0	0.0
ran13x13	236	64	0.0	0.0	0.0	0	236	64	0.0	0.0	0.0
ran14x18	397	64	0.0	0.0	0.0	0	397	64	0.0	0.0	0.0
ran14x18-disj-8	1915	128	0.2	14.1	14.4	1	1916	64	0.3	4.8	5.2
ran14x18.disj-8	1915	128	0.2	14.3	14.6	1	1916	64	0.3	4.8	5.2
ran14x18_1	434	64	0.0	0.0	0.0	0	434	64	0.0	0.0	0.0
ran16x16	379	64	0.0	0.0	0.0	0	379	64	0.0	0.0	0.1
ran17x17	258	64	0.0	0.0	0.0	0	258	64	0.0	0.0	0.0
ran4x64	515	64	0.0	0.0	0.0	0	515	64	0.0	0.0	0.0
ran6x43	447	64	0.0	0.0	0.0	0	447	64	0.0	0.0	0.0
ran8x32	402	64	0.0	0.0	0.0	0	402	64	0.0	0.0	0.0

Table A.3 continued

Instance	iter	prc	t_9	t_{ex}	t	R_0	iter	prc	t_{50}	t_{ex}	t
			SoPlex$_9$+QSopt_ex						SoPlex$_{50}$+QSopt_ex		
rat1	2870	64	0.8	5.8	6.6	0	2870	64	0.7	5.8	6.5
rat5	3024	64	1.1	609.6	610.7	0	3024	64	1.4	611.8	613.1
rat7a	11319	64	12.3	7200.0	7212.3	0	11319	64	13.7	7200.0	7213.7
rd-rplusc-21	138	64	6.3	42.3	48.6	1	139	64	7.7	41.9	49.5
reblock166	3359	64	1.4	0.4	1.8	0	3359	64	1.3	0.5	1.7
reblock354	12854	64	7.9	0.7	8.6	0	12854	64	8.2	0.9	9.1
reblock420	9480	64	11.5	1.0	12.5	0	9480	64	11.6	1.1	12.7
reblock67	972	64	0.1	0.1	0.2	0	972	64	0.1	0.0	0.2
recipe	40	64	0.0	0.0	0.0	0	40	64	0.0	0.0	0.0
refine	21	64	0.0	0.0	0.0	0	21	64	0.0	0.0	0.0
rentacar	6483	64	1.2	0.3	1.5	0	6483	64	1.1	0.2	1.3
rgn	85	64	0.0	0.0	0.0	0	85	64	0.0	0.0	0.0
rlfddd	825	64	0.2	0.3	0.5	0	825	64	0.3	0.3	0.6
rlfdual	8652	64	1.7	0.6	2.3	0	8652	64	1.7	0.6	2.3
rlfprim	5532	64	1.2	0.4	1.6	0	5532	64	1.2	0.5	1.7
rlp1	248	64	0.0	0.0	0.0	0	248	64	0.0	0.0	0.0
rmatr100-p10	6260	64	1.6	0.2	1.8	0	6260	64	1.5	0.1	1.7
rmatr100-p5	10885	64	3.1	0.2	3.3	0	10885	64	3.4	0.2	3.6
rmatr200-p10	13646	64	11.7	0.5	12.2	0	13646	64	11.8	0.5	12.3
rmatr200-p20	11166	64	7.7	0.4	8.1	0	11166	64	8.2	0.5	8.7
rmatr200-p5	17650	64	18.8	0.6	19.4	0	17650	64	18.8	0.7	19.5
rmine10	13871	64	20.9	2.9	23.8	0	13871	64	21.6	3.0	24.7
rmine14	92846	64	1720.4	22.5	1742.9	0	92846	64	1714.0	22.6	1736.5
rmine6	2058	64	0.7	0.2	0.9	0	2058	64	0.6	0.2	0.8
rocII-4-11	751	64	0.3	0.9	1.1	0	751	64	0.5	0.9	1.3
rocII-7-11	1075	64	0.5	1.8	2.3	0	1075	64	0.8	1.8	2.5
rocII-9-11	1523	64	0.8	2.6	3.4	0	1523	64	1.1	2.5	3.6
rococoB10-011000	4263	64	0.8	0.1	0.9	0	4263	64	0.7	0.0	0.8
rococoC10-001000	2344	64	0.3	0.1	0.4	0	2344	64	0.2	0.0	0.2
rococoC11-011100	10112	64	1.6	0.1	1.8	0	10112	64	1.8	0.2	2.0
rococoC12-111000	11920	64	3.4	0.3	3.7	0	11920	64	3.2	0.3	3.5
roll3000	2627	64	0.3	0.1	0.5	0	2627	64	0.2	0.1	0.3
rosen1	898	64	0.2	0.2	0.3	0	898	64	0.1	0.1	0.2
rosen10	2401	64	1.0	0.6	1.6	0	2401	64	0.9	0.6	1.6
rosen2	1627	64	0.6	0.3	0.9	0	1627	64	0.6	0.2	0.8
rosen7	332	64	0.0	0.0	0.1	0	332	64	0.0	0.1	0.1
rosen8	662	64	0.1	0.1	0.2	0	662	64	0.2	0.1	0.3
rout	293	64	0.0	0.0	0.0	0	293	64	0.0	0.0	0.0
route	2239	64	1.1	0.4	1.5	0	2239	64	1.2	0.4	1.6
roy	119	64	0.0	0.0	0.0	0	119	64	0.0	0.0	0.0
rvb-sub	512	64	0.8	1.5	2.3	0	512	64	1.4	1.5	3.0
satellites1-25	9023	64	4.7	0.6	5.3	0	9023	64	4.9	0.5	5.4
satellites2-60	79705	64	373.8	2.6	376.4	0	79705	64	376.6	2.6	379.2
satellites2-60-fs	65144	64	270.6	7.0	277.6	0	65144	64	271.8	6.5	278.3
satellites3-40	190485	64	2657.9	26.6	2684.5	0	190485	64	2654.6	26.3	2680.9
satellites3-40-fs	199929	128	2345.5	4825.0	7170.5	1	199931	64	2342.0	34.6	2376.6
sc105	99	64	0.0	0.0	0.0	0	99	64	0.0	0.0	0.0
sc205	217	64	0.0	0.0	0.0	0	217	64	0.0	0.0	0.0
sc205-2r-100	1109	64	0.1	0.1	0.2	0	1109	64	0.2	0.0	0.2
sc205-2r-16	171	64	0.0	0.0	0.0	0	171	64	0.0	0.0	0.0
sc205-2r-1600	9189	64	4.7	0.3	5.0	0	9189	64	4.9	0.2	5.2
sc205-2r-200	2195	64	0.4	0.1	0.5	0	2195	64	0.5	0.0	0.5
sc205-2r-27	327	64	0.0	0.0	0.0	0	327	64	0.0	0.0	0.0
sc205-2r-32	331	64	0.0	0.0	0.0	0	331	64	0.0	0.0	0.0
sc205-2r-4	55	64	0.0	0.0	0.0	0	55	64	0.0	0.0	0.0
sc205-2r-400	4393	64	1.1	0.2	1.3	0	4393	64	1.3	0.2	1.5
sc205-2r-50	621	64	0.1	0.0	0.1	0	621	64	0.0	0.0	0.0
sc205-2r-8	651	64	0.1	0.0	0.1	0	651	64	0.0	0.0	0.1
sc205-2r-8	100	64	0.0	0.0	0.0	0	100	64	0.0	0.0	0.0
sc205-2r-800	8729	64	3.4	0.4	3.8	0	8729	64	3.4	0.5	3.9
sc50a	45	64	0.0	0.0	0.0	0	45	64	0.0	0.0	0.0
sc50b	49	64	0.0	0.0	0.0	0	49	64	0.0	0.0	0.0
scagr25	784	64	0.0	0.0	0.1	0	784	64	0.1	0.0	0.1
scagr7	178	64	0.0	0.0	0.0	0	178	64	0.0	0.0	0.0
scagr7-2b-16	717	64	0.0	0.0	0.1	0	717	64	0.1	0.0	0.1
scagr7-2b-4	189	64	0.0	0.0	0.0	0	189	64	0.0	0.0	0.0
scagr7-2b-64	11614	64	1.2	0.4	1.6	0	11614	64	1.6	0.4	2.0
scagr7-2c-16	684	64	0.0	0.0	0.1	0	684	64	0.0	0.0	0.1
scagr7-2c-4	186	64	0.0	0.0	0.0	0	186	64	0.0	0.0	0.0
scagr7-2c-64	2727	64	0.2	0.1	0.3	0	2727	64	0.2	0.1	0.3
scagr7-2r-108	4538	64	0.3	0.2	0.5	0	4538	64	0.4	0.2	0.6
scagr7-2r-16	707	64	0.0	0.0	0.1	0	707	64	0.0	0.0	0.1

Table A.3 continued

Instance	iter	prc	t_9	t_{ex}	t	R_0	iter	prc	t_{50}	t_{ex}	t
			SoPlex$_9$+QSopt_ex						SoPlex$_{50}$+QSopt_ex		
scagr7-2r-216	9082	64	0.8	0.3	1.1	0	9082	64	1.0	0.3	1.4
scagr7-2r-27	1144	64	0.0	0.1	0.1	0	1144	64	0.0	0.1	0.1
scagr7-2r-32	1335	64	0.1	0.1	0.1	0	1335	64	0.0	0.1	0.1
scagr7-2r-4	185	64	0.0	0.0	0.0	0	185	64	0.0	0.0	0.0
scagr7-2r-432	16562	64	2.5	0.5	3.0	0	16562	64	2.8	0.5	3.3
scagr7-2r-54	2299	64	0.1	0.1	0.2	0	2299	64	0.1	0.1	0.2
scagr7-2r-64	2763	64	0.2	0.1	0.2	0	2763	64	0.2	0.1	0.3
scagr7-2r-8	357	64	0.0	0.0	0.0	0	357	64	0.0	0.0	0.0
scagr7-2r-864	32972	64	12.0	1.0	13.0	0	32972	64	12.6	0.9	13.6
scfxm1	447	64	0.0	0.0	0.1	0	447	64	0.0	0.0	0.0
scfxm1-2b-16	4522	64	0.4	0.2	0.6	0	4522	64	0.5	0.2	0.7
scfxm1-2b-4	1135	64	0.1	0.1	0.1	0	1135	64	0.1	0.1	0.1
scfxm1-2b-64	26349	64	8.8	0.8	9.6	0	26349	64	9.2	0.8	10.0
scfxm1-2c-4	1071	64	0.1	0.1	0.1	0	1071	64	0.1	0.0	0.1
scfxm1-2r-128	23788	64	7.7	0.7	8.5	0	23788	64	8.0	0.7	8.7
scfxm1-2r-16	4672	64	0.5	0.2	0.7	0	4672	64	0.5	0.1	0.6
scfxm1-2r-256	47280	64	30.8	1.3	32.1	0	47280	64	31.7	1.3	33.0
scfxm1-2r-27	8158	64	0.8	0.3	1.1	0	8158	64	0.6	0.1	0.8
scfxm1-2r-32	8924	64	1.0	0.3	1.3	0	8924	64	0.8	0.3	1.1
scfxm1-2r-4	1203	64	0.1	0.1	0.1	0	1203	64	0.0	0.1	0.1
scfxm1-2r-64	13133	64	1.8	0.5	2.2	0	13133	64	2.1	0.5	2.6
scfxm1-2r-8	2252	64	0.2	0.1	0.2	0	2252	64	0.1	0.1	0.2
scfxm1-2r-96	16931	64	3.7	0.6	4.3	0	16931	64	3.9	0.6	4.5
scfxm2	1119	64	0.1	0.1	0.1	0	1119	64	0.1	0.1	0.1
scfxm3	1992	64	0.1	0.1	0.2	0	1992	64	0.1	0.1	0.2
scorpion	245	64	0.0	0.0	0.0	0	245	64	0.0	0.0	0.0
scrs8	608	64	0.0	0.0	0.1	0	608	64	0.0	0.0	0.1
scrs8-2b-16	88	64	0.0	0.0	0.0	0	88	64	0.0	0.0	0.0
scrs8-2b-4	22	64	0.0	0.0	0.0	0	22	64	0.0	0.0	0.0
scrs8-2b-64	296	64	0.0	0.1	0.1	0	296	64	0.0	0.0	0.1
scrs8-2c-16	91	64	0.0	0.0	0.0	0	91	64	0.0	0.0	0.0
scrs8-2c-32	179	64	0.0	0.0	0.1	0	179	64	0.0	0.0	0.0
scrs8-2c-4	22	64	0.0	0.0	0.0	0	22	64	0.0	0.0	0.0
scrs8-2c-64	358	64	0.0	0.1	0.1	0	358	64	0.1	0.0	0.1
scrs8-2c-8	45	64	0.0	0.0	0.0	0	45	64	0.0	0.0	0.0
scrs8-2r-128	543	64	0.1	0.1	0.2	0	543	64	0.1	0.1	0.2
scrs8-2r-16	96	64	0.0	0.0	0.0	0	96	64	0.0	0.0	0.0
scrs8-2r-256	1119	64	0.2	0.2	0.4	0	1119	64	0.3	0.2	0.5
scrs8-2r-27	103	64	0.0	0.0	0.0	0	103	64	0.0	0.0	0.0
scrs8-2r-32	192	64	0.0	0.0	0.0	0	192	64	0.0	0.0	0.1
scrs8-2r-4	24	64	0.0	0.0	0.0	0	24	64	0.0	0.0	0.0
scrs8-2r-512	2532	64	0.4	0.4	0.8	0	2532	64	0.6	0.5	1.0
scrs8-2r-64	384	64	0.0	0.1	0.1	0	384	64	0.0	0.1	0.1
scrs8-2r-64b	271	64	0.0	0.1	0.1	0	271	64	0.0	0.1	0.1
scrs8-2r-8	41	64	0.0	0.0	0.0	0	41	64	0.0	0.0	0.0
scsd1	97	64	0.0	0.0	0.0	0	97	64	0.0	0.0	0.0
scsd6	423	64	0.0	0.0	0.1	0	423	64	0.1	0.1	0.1
scsd8	1837	64	0.2	0.1	0.3	0	1837	64	0.3	0.0	0.3
scsd8-2b-16	295	64	0.0	0.1	0.1	0	295	64	0.0	0.0	0.1
scsd8-2b-4	47	64	0.0	0.0	0.0	0	47	64	0.0	0.0	0.0
scsd8-2b-64	2056	64	0.1	0.3	0.4	0	2056	64	0.4	0.4	0.8
scsd8-2c-16	198	64	0.0	0.0	0.1	0	198	64	0.0	0.0	0.1
scsd8-2c-4	47	64	0.0	0.0	0.0	0	47	64	0.0	0.0	0.0
scsd8-2c-64	2056	64	0.2	0.4	0.6	0	2056	64	0.4	0.3	0.8
scsd8-2r-108	1114	64	0.1	0.2	0.3	0	1114	64	0.1	0.2	0.4
scsd8-2r-16	231	64	0.0	0.0	0.1	0	231	64	0.0	0.0	0.0
scsd8-2r-216	2314	64	0.1	0.4	0.5	0	2314	64	0.3	0.3	0.6
scsd8-2r-27	286	64	0.0	0.1	0.1	0	286	64	0.0	0.0	0.1
scsd8-2r-32	429	64	0.0	0.1	0.1	0	429	64	0.1	0.0	0.1
scsd8-2r-4	47	64	0.0	0.0	0.0	0	47	64	0.0	0.0	0.0
scsd8-2r-432	4530	64	0.2	0.5	0.8	0	4530	64	0.5	0.5	1.1
scsd8-2r-54	600	64	0.1	0.1	0.2	0	600	64	0.1	0.1	0.2
scsd8-2r-64	1188	64	0.1	0.1	0.2	0	1188	64	0.1	0.1	0.2
scsd8-2r-8	97	64	0.0	0.0	0.0	0	97	64	0.0	0.0	0.0
scsd8-2r-8b	97	64	0.0	0.0	0.0	0	97	64	0.0	0.0	0.0
sct1	15178	64	7.4	5.3	12.7	0	15178	64	7.6	5.3	12.9
sct32	14051	64	3.8	1.0	4.8	0	14051	64	3.7	0.8	4.6
sct5	6217	64	3.1	0.5	3.6	0	6217	64	3.5	0.6	4.1
sctap1	262	64	0.0	0.0	0.0	0	262	64	0.0	0.0	0.1
sctap1-2b-16	323	64	0.0	0.0	0.0	0	323	64	0.0	0.0	0.0
sctap1-2b-4	83	64	0.0	0.0	0.0	0	83	64	0.0	0.0	0.0
sctap1-2b-64	4773	64	0.6	0.3	0.9	0	4773	64	0.7	0.3	1.0

Table A.3 continued

Instance	SoPlex$_9$ +QSopt_ex					SoPlex$_{50}$ +QSopt_ex					
	iter	prc	t_9	t_{ex}	t	R_0	iter	prc	t_{50}	t_{ex}	t
sctap1-2c-16	329	64	0.0	0.0	0.0	0	329	64	0.0	0.0	0.0
sctap1-2c-4	85	64	0.0	0.0	0.0	0	85	64	0.0	0.0	0.0
sctap1-2c-64	1146	64	0.1	0.1	0.1	0	1146	64	0.1	0.1	0.2
sctap1-2r-108	2109	64	0.2	0.1	0.4	0	2109	64	0.4	0.1	0.5
sctap1-2r-16	282	64	0.0	0.0	0.0	0	282	64	0.0	0.0	0.0
sctap1-2r-216	4248	64	0.5	0.3	0.8	0	4248	64	0.8	0.3	1.1
sctap1-2r-27	536	64	0.0	0.0	0.1	0	536	64	0.0	0.0	0.1
sctap1-2r-32	559	64	0.0	0.0	0.1	0	559	64	0.0	0.0	0.1
sctap1-2r-4	84	64	0.0	0.0	0.0	0	84	64	0.0	0.0	0.0
sctap1-2r-480	9373	64	1.5	0.6	2.2	0	9373	64	1.9	0.6	2.5
sctap1-2r-54	1069	64	0.1	0.1	0.2	0	1069	64	0.1	0.1	0.2
sctap1-2r-64	1123	64	0.1	0.1	0.2	0	1123	64	0.2	0.1	0.2
sctap1-2r-8	147	64	0.0	0.0	0.0	0	147	64	0.0	0.0	0.0
sctap1-2r-8b	161	64	0.0	0.0	0.0	0	161	64	0.0	0.0	0.0
sctap2	505	64	0.1	0.0	0.1	0	505	64	0.1	0.0	0.1
sctap3	604	64	0.1	0.0	0.1	0	604	64	0.1	0.0	0.1
seba	6	64	0.0	0.0	0.0	0	6	64	0.0	0.0	0.0
self	13495	64	48.6	6101.6	6150.2	0	13495	64	49.7	6095.1	6144.8
set1ch	513	64	0.0	0.0	0.0	0	513	64	0.0	0.0	0.0
set3-10	2279	64	0.1	0.1	0.3	0	2279	64	0.2	0.1	0.4
set3-15	2269	64	0.2	0.1	0.3	0	2269	64	0.2	0.1	0.4
set3-20	2344	64	0.1	0.1	0.3	0	2344	64	0.2	0.1	0.4
seymour	3858	64	1.0	0.2	1.2	0	3858	64	0.9	0.2	1.2
seymour-disj-10	5700	128	1.8	63.8	65.6	1	5702	64	2.1	8.6	10.7
seymour.disj-10	5700	128	1.7	63.8	65.5	1	5702	64	2.1	8.4	10.5
seymourl	3858	64	0.9	0.2	1.1	0	3858	64	1.1	0.1	1.2
sgpf5y6	*153350*	*128*	*167.8*	*7200.0*	*7367.8*	1	197667	64	221.6	7.1	228.7
share1b	217	64	0.0	0.0	0.0	0	217	64	0.0	0.0	0.0
share2b	159	64	0.0	0.0	0.0	0	159	64	0.0	0.0	0.0
shell	595	64	0.0	0.0	0.1	0	595	64	0.0	0.0	0.0
ship04l	473	64	0.0	0.0	0.1	0	473	64	0.0	0.0	0.1
ship04s	383	64	0.0	0.0	0.1	0	383	64	0.0	0.1	0.1
ship08l	810	64	0.1	0.1	0.1	0	810	64	0.1	0.1	0.2
ship08s	513	64	0.0	0.1	0.1	0	513	64	0.1	0.0	0.1
ship12l	1085	64	0.1	0.1	0.2	0	1085	64	0.2	0.1	0.2
ship12s	648	64	0.0	0.1	0.1	0	648	64	0.1	0.1	0.2
shipsched	2174	64	0.6	0.2	0.8	0	2174	64	0.9	0.3	1.2
shs1023	*174484*	*128*	*604.0*	*7200.0*	*7804.0*	5	177407	64	618.0	7.5	625.4
siena1	28441	64	19.0	3.2	22.2	0	28441	64	19.7	3.2	22.9
sierra	640	64	0.0	0.1	0.1	0	640	64	0.0	0.0	0.1
sing2	37541	64	40.3	0.7	41.0	0	37541	64	40.9	0.7	41.6
sing245	218115	64	1608.1	7.0	1615.1	0	218115	64	1611.8	6.9	1618.7
sing359	*352087*	*64*	*5830.9*	*7200.0*	*13030.9*	*0*	*352087*	*64*	*5799.4*	*7200.0*	*12999.4*
slptsk	4856	64	1.2	12.8	14.0	0	4856	64	1.8	13.0	14.8
small000	557	64	0.0	0.1	0.1	0	557	64	0.0	0.0	0.1
small001	725	128	0.0	0.8	0.9	1	737	64	0.0	0.0	0.1
small002	834	128	0.0	1.0	1.1	1	850	64	0.1	0.1	0.2
small003	681	128	0.0	0.8	0.8	1	693	64	0.1	0.1	0.1
small004	465	128	0.0	0.8	0.8	1	476	64	0.1	0.1	0.1
small005	601	128	0.0	0.9	1.0	1	621	64	0.1	0.1	0.2
small006	518	128	0.0	0.8	0.8	1	540	64	0.1	0.1	0.1
small007	517	128	0.0	1.0	1.1	1	547	64	0.1	0.1	0.1
small008	489	128	0.0	0.9	0.9	1	511	64	0.1	0.1	0.1
small009	416	128	0.0	0.9	0.9	1	435	64	0.1	0.1	0.1
small010	328	128	0.0	0.8	0.8	1	342	64	0.1	0.1	0.1
small011	337	128	0.0	0.8	0.9	1	348	64	0.0	0.0	0.1
small012	287	64	0.0	0.1	0.1	0	287	64	0.1	0.1	0.1
small013	289	64	0.0	0.1	0.1	0	289	64	0.1	0.1	0.1
small014	341	64	0.0	0.1	0.1	0	341	64	0.0	0.1	0.1
small015	338	64	0.0	0.1	0.1	0	338	64	0.1	0.1	0.1
small016	338	64	0.0	0.1	0.1	0	338	64	0.1	0.1	0.1
south31	23654	64	14.9	2.4	17.3	0	23654	64	15.2	2.5	17.8
sp97ar	6984	64	2.9	1.3	4.2	1	6986	64	3.3	0.5	3.8
sp97ic	3102	64	0.8	0.5	1.3	0	3102	64	0.7	0.6	1.3
sp98ar	10342	64	2.9	1.4	4.2	1	10349	64	3.1	0.5	3.6
sp98ic	3276	64	0.7	0.6	1.3	0	3276	64	0.8	0.6	1.4
sp98ir	2967	64	0.7	0.5	1.2	1	2970	64	0.5	0.2	0.7
square15	208626	64	2424.8	8.4	2433.2	0	208626	64	2424.6	8.5	2433.0
stair	658	64	0.1	4.0	4.1	0	658	64	0.1	4.0	4.1
standata	50	64	0.0	0.0	0.0	0	50	64	0.0	0.0	0.0
standmps	189	64	0.0	0.0	0.0	0	189	64	0.0	0.0	0.0
stat96v4	144049	64	395.8	3588.8	3984.6	0	144049	64	402.2	3590.3	3992.5

Table A.3 continued

Instance	SoPlex$_9$ + QSopt_ex					R_0	SoPlex$_{50}$ + QSopt_ex				
	iter	prc	t_9	t_{ex}	t		iter	prc	t_{50}	t_{ex}	t
stat96v5	*11838*	*128*	*18.4*	*7200.0*	*7218.4*	*1*	*13119*	*64*	*21.2*	*7200.0*	*7221.2*
stein27	32	64	0.0	0.0	0.0	0	32	64	0.0	0.0	0.0
stein45	59	64	0.0	0.0	0.0	0	59	64	0.0	0.0	0.0
stocfor1	108	64	0.0	0.0	0.0	0	108	64	0.0	0.0	0.0
stocfor2	1994	64	0.3	0.1	0.4	0	1994	64	0.2	0.1	0.4
stocfor3	16167	64	6.2	0.5	6.7	0	16167	64	6.5	0.6	7.1
stockholm	25891	64	20.3	11.6	32.0	1	26971	64	21.7	0.8	22.5
stormG2_1000	*732642*	*64*	*4648.3*	*7200.0*	*11848.3*	*0*	*732642*	*64*	*4679.4*	*7200.0*	*11879.5*
stormg2-125	92103	64	45.6	1.9	47.5	0	92103	64	47.4	1.9	49.3
stormg2-27	20754	64	2.4	0.5	2.9	0	20754	64	2.7	0.4	3.1
stormg2-8	6863	64	0.6	0.2	0.8	0	6863	64	0.4	0.2	0.6
stormg2_1000	*732642*	*64*	*4660.6*	*7200.0*	*11860.6*	*0*	*732642*	*64*	*4685.5*	*7200.0*	*11885.5*
stp3d	148139	64	1238.6	7.2	1245.8	0	148139	64	1242.3	7.1	1249.5
sts405	434	64	0.5	0.5	1.1	0	434	64	0.7	0.4	1.1
sts729	907	64	1.3	1.5	2.7	0	907	64	1.8	1.5	3.2
swath	127	64	0.1	0.1	0.2	0	127	64	0.1	0.1	0.2
sws	1150	64	0.3	0.4	0.7	0	1150	64	0.3	0.3	0.7
t0331-4l	15487	64	14.7	8.2	22.9	0	15487	64	15.4	8.2	23.7
t1717	12220	64	7.0	2.0	9.0	0	12220	64	7.6	2.0	9.6
t1722	10612	64	3.3	0.6	3.9	0	10612	64	3.6	0.6	4.2
tanglegram1	478	64	0.3	1.4	1.6	0	478	64	0.3	1.4	1.8
tanglegram2	236	64	0.1	0.1	0.2	0	236	64	0.0	0.1	0.2
testbig	8010	64	4.4	0.2	4.6	0	8010	64	4.5	0.3	4.8
timtab1	20	64	0.0	0.0	0.0	0	20	64	0.0	0.0	0.0
timtab2	36	64	0.0	0.0	0.0	0	36	64	0.0	0.0	0.0
toll-like	717	64	0.1	0.0	0.1	0	717	64	0.1	0.1	0.1
tr12-30	696	64	0.0	0.0	0.0	0	696	64	0.0	0.0	0.0
transportmoment	9617	64	1.7	0.7	2.4	0	9617	64	1.9	0.6	2.4
triptim1	68167	64	93.3	1.2	94.5	0	68167	64	92.8	1.2	94.1
triptim2	204335	64	572.2	4.3	576.4	0	204335	64	567.7	4.3	572.0
triptim3	89514	64	156.7	2.7	159.4	0	89514	64	161.5	2.6	164.1
truss	21892	64	4.7	0.2	4.9	0	21892	64	4.6	0.1	4.7
tuff	212	64	0.0	0.0	0.1	0	212	64	0.0	0.0	0.1
tw-myciel4	11588	64	2.6	0.1	2.7	0	11588	64	2.3	0.1	2.3
uc-case11	39603	64	75.9	1.1	77.0	0	39603	64	75.3	1.1	76.4
uc-case3	34156	64	65.4	1.2	66.6	0	34156	64	66.7	1.2	67.9
uct-subprob	2988	64	0.5	0.1	0.6	0	2988	64	0.5	0.0	0.6
ulevimin	125858	64	108.8	1.0	109.8	0	125858	64	108.6	1.1	109.6
umts	5537	64	0.9	0.5	1.5	2	5549	64	0.7	0.2	0.8
unitcal_7	21824	64	8.9	0.7	9.5	0	21824	64	9.1	0.8	9.9
us04	338	64	0.2	0.2	0.4	0	338	64	0.5	0.2	0.7
usAbbrv-8-25_70	2434	64	0.1	0.1	0.2	0	2434	64	0.1	0.1	0.1
van	11014	64	5.2	1.8	6.9	0	11014	64	6.0	1.7	7.7
vpm1	130	64	0.0	0.0	0.0	0	130	64	0.0	0.0	0.0
vpm2	192	64	0.0	0.0	0.0	0	192	64	0.0	0.0	0.0
vpphard	11075	64	11.3	1.0	12.3	0	11075	64	11.5	1.0	12.5
vpphard2	8024	64	32.4	4.1	36.5	0	8024	64	33.1	4.2	37.2
vtp-base	75	64	0.0	0.0	0.0	0	75	64	0.0	0.0	0.0
wachplan	2033	64	0.5	0.2	0.7	0	2033	64	0.4	0.1	0.5
watson_1	*188103*	*128*	*366.3*	*7200.0*	*7566.3*	*1*	*188496*	*64*	*369.2*	*10.6*	*379.8*
watson_2	*333044*	*128*	*1519.4*	*7200.0*	*8719.4*	*1*	*333083*	*64*	*1518.6*	*19.5*	*1538.1*
wide15	229488	64	2733.8	8.5	2742.3	0	229488	64	2734.6	8.7	2743.3
wnq-n100-mw99-14	764	64	5.0	13.3	18.2	0	764	64	5.8	13.2	19.0
wood1p	142	64	0.1	0.8	1.0	0	142	64	0.2	0.8	1.1
woodw	1832	64	0.6	0.2	0.8	0	1832	64	0.6	0.1	0.7
world	70204	128	131.2	7200.0	7331.2	12	70320	64	132.1	4.5	136.6
zed	31	64	0.0	0.0	0.0	0	31	64	0.0	0.0	0.0
zib54-UUE	1855	64	0.1	0.1	0.2	0	1855	64	0.1	0.1	0.2

Table A.4: Detailed results comparing exact SoPlex with rational factorization, exact SoPlex with rational reconstruction, and QSopt.ex. Entries corresponding to unsolved instances are printed in italics. See Tables 3.2 and 3.3 for aggregated results.

iter — number of simplex iterations
prc — max. precision used by QSopt.ex
R — number of refinements
#fac — number of rational factorizations
t_{fac} — time for rational factorization and LU solves (in seconds)
#rec — number of rational reconstructions
t_{rec} — time for rational reconstructions (in seconds)
t — total running time (in seconds)
dlcm — size ($\log_{10}$) of least common multiple of denominators in solution (two values if runs differ)

Instance	QSopt.ex			SoPlex$_{fac}$						SoPlex$_{rec}$					
	iter	prc	t	iter	R	#fac	t_{fac}	t	dlcm	iter	R	#rec	t_{rec}	t	dlcm
10teams	547	64	0.28	1611	2	1	0.01	0.28	14	1611	1	2	0.01	0.26	14
16_n14	280582	64	1405.51	329933	0	0	0.00	366.25	1	329933	0	0	0.00	368.41	1
22433	425	64	0.09	1041	2	1	0.00	0.07	6	1041	2	2	0.00	0.07	6
23588	256	64	0.06	548	2	1	0.01	0.04	7	548	1	2	0.01	0.04	7
25fv47	1942	64	1.12	4359	2	1	0.07	0.78	541	4359	83	19	2.00	3.81	541
30_70_45_095_100	15504	64	11.90	16103	2	1	0.05	7.59	27	16103	3	4	0.12	7.62	27
30n20b8	302	64	0.20	269	2	1	0.01	0.58	10	269	0	1	0.01	0.48	10
50v-10	318	128	0.10	220	2	1	0.00	0.03	16	220	2	3	0.00	0.04	16
80bau3b	5449	64	0.87	7797	2	1	0.01	0.89	199	7797	21	12	0.25	1.64	199
Test3	35222	64	32.23	6948	2	1	0.08	3.92	863	6948	69	18	1.19	15.12	863
a1c1s1	3304	64	0.55	1742	2	1	0.02	0.10	5	1742	0	1	0.01	0.07	5
aa01	8452	64	6.20	11898	2	1	0.01	3.57	6	11898	1	2	0.04	3.54	6
aa03	8843	64	6.39	7584	2	1	0.00	2.14	2	7584	0	1	0.01	1.96	2
aa3	5793	64	4.33	7584	2	1	0.00	1.98	2	7584	0	1	0.01	2.04	2
aa4	2173	64	1.55	4486	2	1	0.00	1.24	6	4486	1	2	0.02	1.22	6
aa5	4739	64	3.04	9706	2	1	0.00	2.39	2	9706	0	1	0.01	2.56	2
aa6	4164	64	2.27	4870	2	1	0.00	1.32	3	4870	1	1	0.01	1.21	3
acc-tight4	3265	64	2.84	11142	2	1	0.02	2.63	21	11142	2	3	0.01	2.40	21
acc-tight5	3440	64	2.80	10469	2	1	0.02	2.28	17	10469	2	3	0.03	2.10	17
acc-tight6	3387	64	2.86	10672	2	1	0.01	2.22	13	10672	1	2	0.01	2.22	13
adlittle	84	64	0.02	88	2	1	0.00	0.00	30	88	3	4	0.00	0.01	30
afiro	18	64	0.00	16	2	1	0.00	0.00	9	16	1	2	0.00	0.00	9
aflow30a	434	64	0.04	396	2	1	0.00	0.03	17	396	1	2	0.00	0.01	17
aflow40b	1297	64	0.21	1826	2	1	0.01	0.22	15	1826	1	2	0.02	0.22	15
agg	108	64	0.04	80	2	1	0.00	0.03	82	80	7	7	0.02	0.05	82
agg2	157	64	0.04	152	2	1	0.00	0.03	72	152	3	4	0.01	0.04	72
agg3	157	64	0.04	165	2	1	0.01	0.04	74	165	4	5	0.01	0.04	74
air02	134	64	0.19	95	2	1	0.00	0.14	1	95	0	1	0.01	0.10	1
air03	470	64	0.61	626	2	0	0.00	0.42	2	626	1	1	0.02	0.37	2
air04	8452	64	5.99	11898	2	1	0.01	3.85	6	11898	1	2	0.04	3.32	6
air05	2391	64	1.54	4486	2	1	0.00	1.07	6	4486	1	2	0.02	1.14	6
air06	8843	64	6.25	7584	2	1	0.01	2.11	2	7584	0	1	0.01	1.92	2
aircraft	4570	64	0.91	1912	2	1	0.00	0.42	10	1912	1	1	0.02	0.41	10

Table A.4 continued

Instance	QSopt_ex			SoPlex_fac						SoPlex_rec					
	iter	prc	t	iter	R	#fac	t_{fac}	t	dlcm	iter	R	#rec	t_{rec}	t	dlcm
aligninq	600	64	0.25	1492	2	1	0.01	0.20	12	1492	1	2	0.01	0.20	12
appl1-2	14696	64	28.21	14622	3	1	0.28	8.31	31	14622	4	5	0.73	8.83	31
arki001	1697	192	1.85	1571	3	1	0.08	0.43	961	1571	100	20	0.81	2.57	961
ash608gpia-3col	5535	64	13.52	3123	2	1	0.09	0.73	1	3123	0	1	0.02	0.26	1
atlanta-ip	83029	128	3066.61	38286	3	1	0.46	35.86	316	38286	39	15	5.86	46.59	316
atm20-100	4970	128	1.88	4699	3	1	0.03	0.59	1176	4699	7	7	0.23	0.78	1176
b2c1s1	4089	64	0.89	2621	2	1	0.02	0.16	6	2621	0	2	0.01	0.12	6
bab1	11781	64	60.15	5711	2	1	0.10	4.89	7	5711	1	2	0.25	4.62	7
bab3	61415	64	270.39	697791	2	1	0.14	5248.65	10	697791	1	2	0.66	5264.91	10
bab5	8313	64	2.72	29788	2	1	0.03	10.58	7	29788	1	2	0.07	10.32	7
bal8x12	111	64	0.00	102	2	1	0.00	0.00	4	102	0	1	0.00	0.00	4
bandm	295	64	0.14	474	2	1	0.02	0.08	183	474	26	13	0.31	0.58	183
bas1lp	789	64	1.52	2582	2	1	0.08	1.22	61	2582	9	8	0.18	1.40	61
baxter	8303	64	3.76	12347	2	1	0.09	3.11	781	12347	14	10	0.53	3.88	781
bc	564	64	2.12	3759	2	1	0.43	1.89	274	3759	39	15	1.84	3.92	274
bc1	564	64	2.12	3759	2	1	0.43	1.87	274	3759	39	15	1.88	3.94	274
beaconfd	109	64	0.02	88	2	1	0.00	0.02	7	88	0	1	0.01	0.02	7
beasleyC3	1151	64	0.11	1140	2	1	0.01	0.13	3	1140	0	1	0.01	0.11	3
bell3a	111	64	0.02	81	2	1	0.00	0.01	10	81	1	2	0.00	0.00	10
bell5	83	64	0.02	66	2	1	0.00	0.00	11	66	1	2	0.00	0.00	11
berlin_5_8_0	1504	64	0.28	1017	2	1	0.01	0.06	31	1017	1	2	0.02	0.06	31
bg512142	1455	64	0.52	2238	2	1	0.06	0.37	267	2238	32	14	0.76	1.22	267
biella1	5426	64	3.15	16158	2	1	0.02	4.42	21	16158	3	4	0.06	4.47	21
bienst1	613	64	0.11	455	2	1	0.00	0.03	11	455	1	2	0.00	0.03	11
bienst2	613	64	0.12	455	2	1	0.01	0.03	11	455	1	2	0.00	0.03	11
binkar10_1	1581	64	0.13	1267	2	1	0.00	0.08	40	1267	4	5	0.06	0.16	40
bk4x3	24	64	0.00	16	2	1	0.00	0.00	2	16	0	1	0.00	0.00	2
blend	97	64	0.02	97	2	1	0.00	0.00	46	97	5	6	0.00	0.02	46
blend2	184	64	0.03	175	2	1	0.01	0.02	23	175	3	4	0.00	0.01	23
blp-ar98	469	64	0.47	368	2	1	0.01	0.59	12	368	1	2	0.04	0.31	12
blp-ic97	290	64	0.39	446	2	1	0.00	0.59	9	446	1	2	0.04	0.31	9
bnatt350	750	64	0.16	685	2	1	0.02	0.16	24	685	2	3	0.03	0.13	24
bnatt400	1073	64	0.25	831	2	1	0.03	0.20	32	831	3	4	0.06	0.22	32
bnl1	1149	64	0.31	1474	2	1	0.01	0.20	109	1474	11	9	0.13	0.40	109
bnl2	2816	64	0.94	2718	2	1	0.02	0.52	118	2718	14	10	0.19	0.78	118
boeing1	397	64	0.07	458	2	1	0.00	0.05	67	458	7	7	0.03	0.09	67
boeing2	133	64	0.01	145	2	1	0.00	0.01	23	145	2	3	0.00	0.01	23
bore3d	174	64	0.04	100	2	1	0.00	0.03	98	100	11	9	0.08	0.11	98
brandy	303	64	0.09	482	2	1	0.00	0.04	86	482	11	9	0.08	0.17	86
buildingenergy	121087	432	*memout*	144595	2	1	0.83	830.37	22	144595	2	3	0.95	833.58	22
cap6000	859	64	0.25	815	2	1	0.01	0.27	8	815	1	2	0.02	0.26	8
capri	340	64	0.06	338	2	1	0.00	0.03	152	338	21	12	0.16	0.29	152
car4	1982	64	6.50	10442	2	1	2.52	4.77	258	10442	47	16	2.31	6.03	258
cari	649	64	0.49	681	2	1	0.01	0.43	21	681	3	4	0.05	0.23	21
cep1	2105	64	0.23	1399	2	1	0.00	0.08	9	1399	1	2	0.03	0.10	9
ch	6316	128	1.83	10747	2	1	0.02	1.47	280	10747	17	11	0.22	2.00	280

Table A.4 continued

instance	QSopt_ex			SoPlex_fac						SoPlex_rec					
	iter	prc	t	iter	R	#fac	t_{fac}	t	dlcm	iter	R	#rec	t_{rec}	t	dlcm
circ10-3	2864	64	9.43	10910	2	1	0.14	10.59	22	10910	3	4	0.13	10.58	22
co-100	1278	64	5.79	783	2	1	0.00	2.47	291	783	7	7	1.76	5.26	291
co5	7974	128	7.17	12067	2	1	0.19	3.89	972	12067	121	21	5.89	21.57	972
co9	19376	128	47.42	19820	2	1	0.76	13.51	1745	19820	212	24	32.28	107.35	1745
complex	5074	64	5.90	9646	3	1	0.28	2.89	10	9646	7	7	0.17	2.68	39
cont1	0	64	7200.00	40707	3	1	5991.02	7216.51	—	40707	606	29	3971.55	7201.85	—
cont4	35671	288	7200.00	40802	3	1	4361.53	7203.80	—	40802	369	27	1775.20	5508.69	2306
core2536-691	7407	64	7.05	41182	2	1	0.01	16.13	8	41182	1	2	0.03	15.91	8
core4872-1529	29433	64	84.78	69516	2	1	0.29	70.12	72	69516	11	9	1.78	72.39	72
cov1075	297	64	0.13	3270	2	1	0.06	0.63	3	3270	0	1	0.01	0.48	3
cq5	7401	128	4.78	12350	2	1	0.11	3.39	755	12350	83	19	3.16	11.80	755
cq9	15864	128	31.56	17826	2	1	0.64	8.82	1368	17826	146	22	21.04	54.60	1368
cr42	1436	64	0.24	999	2	1	0.02	0.13	50	999	5	6	0.19	0.34	50
cre-a	2990	64	0.70	3555	2	1	0.02	0.35	11	3555	1	2	0.04	0.34	11
cre-b	17561	64	15.01	12712	2	1	0.02	5.27	16	12712	1	2	0.07	5.19	16
cre-c	2856	64	0.76	2842	2	1	0.02	0.32	18	2842	2	3	0.04	0.34	18
cre-d	16562	64	14.42	9271	2	1	0.02	3.90	11	9271	1	2	0.11	3.82	11
crew1	360	64	0.35	1849	2	1	0.00	0.55	5	1849	0	1	0.01	0.59	5
csched007	548	64	0.12	5522	2	1	0.01	0.70	9	5522	1	2	0.01	0.62	9
csched008	317	64	0.08	3102	2	1	0.00	0.40	7	3102	1	2	0.02	0.39	7
csched010	586	64	0.13	6832	2	1	0.00	0.69	8	6832	0	1	0.01	0.71	8
cycle	646	128	0.61	920	2	1	0.01	0.13	71	920	7	7	0.01	0.18	71
czprob	1014	64	0.28	1572	2	1	0.01	0.20	37	1572	4	5	0.03	0.25	37
d10200	1085	64	0.81	2852	2	1	0.02	0.57	14	2852	2	3	0.03	0.64	14
d20200	2399	64	1.92	9723	2	1	0.06	2.00	27	9723	4	5	0.10	2.01	27
d2q06c	6004	128	63.56	13920	2	1	1.25	6.48	1998	13920	307	26	36.00	75.83	1998
d6cube	12373	64	4.72	1179	2	1	0.00	0.42	17	1179	2	3	0.07	0.53	17
dano3_3	22239	128	51.03	46251	2	1	0.19	29.55	218	46251	32	14	1.63	31.84	218
dano3_4	22239	128	51.07	46251	2	1	0.18	29.35	218	46251	32	14	1.64	31.86	218
dano3_5	22239	128	51.01	46251	2	1	0.19	29.22	218	46251	32	14	1.64	31.86	218
dano3mip	22239	128	51.15	46251	2	1	0.18	29.39	218	46251	32	14	1.67	31.90	218
danoint	1041	64	0.35	2763	2	1	0.01	0.34	77	2763	11	9	0.12	0.51	77
dbir1	14173	64	3.36	14306	2	1	0.08	11.59	6	14306	0	1	0.04	11.63	6
dbir2	9899	64	3.17	11641	2	1	0.04	5.45	306	11641	0	1	0.08	5.29	306
dc1c	7230	64	6.71	19795	2	1	0.03	6.37	29	19795	4	5	0.11	6.43	29
dc11	12240	64	32.21	24825	2	1	0.03	21.46	36	24825	5	6	0.43	21.87	36
dcmulti	418	64	0.03	479	5	1	0.00	0.02	9	479	0	1	0.01	0.02	9
de063155	2140	192	2.41	2623	2	1	0.14	0.67	1802	2623	255	25	4.79	11.33	1802
de063157	2277	128	1.66	144457346	2	1	0.03	7200.00	—	145289366	255	25	3.27	7200.00	—
de080285	1892	128	1.66	888	2	1	0.10	0.25	1525	888	212	24	3.71	7.48	1525
degen2	599	64	0.13	1325	2	1	0.01	0.13	4	1325	0	1	0.01	0.12	4
degen3	2106	64	1.22	5832	2	1	0.02	0.98	3	5832	0	1	0.00	0.79	3
delf000	3522	128	2.72	1675	2	1	0.05	0.60	4284	1675	532	29	24.35	77.46	4284
delf001	4029	128	1.97	1703	2	1	0.11	0.69	4269	1703	532	29	26.70	82.64	4269
delf002	3210	128	2.34	2009	2	1	0.14	0.81	4401	2009	532	29	29.07	83.89	4401
delf003	3512	128	2.40	3180	3	1	0.17	1.09	4728	3180	639	30	63.43	142.94	4728

Table A.4 continued

instance	QSoptEx			SoPlex_fac						SoPlex_rec					
	iter	prc	t	iter	R	#fac	t_{fac}	t	dlcm	iter	R	#rec	t_{rec}	t	dlcm
delf004	2503	128	2.66	2704	3	1	0.21	1.11	4844	2704	639	30	64.03	138.77	4844
delf005	2975	128	2.52	3294	3	1	0.17	1.04	4660	3294	639	30	61.48	136.10	4660
delf006	2971	128	4.59	3125	3	1	0.24	1.16	4641	3125	639	30	75.55	153.76	4641
delf007	3549	128	3.21	2826	3	1	0.23	1.13	4484	2826	639	30	80.32	159.71	4484
delf008	2961	128	3.36	3493	4	1	0.24	1.20	4631	3493	639	30	69.09	151.61	4631
delf009	3126	128	4.24	3383	3	1	0.22	1.25	4582	3383	639	30	73.28	153.23	4582
delf010	3392	128	3.09	3209	4	1	0.19	1.16	4588	3209	639	30	66.47	147.02	4588
delf011	3295	128	2.60	3104	4	1	0.15	1.00	4372	3104	639	30	56.06	137.61	4372
delf012	3348	128	2.86	2963	3	1	0.17	1.07	4397	2963	639	30	62.12	143.78	4397
delf013	3301	128	3.38	3184	3	1	0.23	1.17	4549	3184	639	30	71.55	154.71	4549
delf014	2934	128	2.56	4322	3	1	0.15	1.11	4489	4322	639	30	59.36	139.58	4489
delf015	3154	128	2.80	3242	3	1	0.15	0.98	4359	3242	639	30	57.48	139.66	4359
delf017	3108	128	2.54	3131	3	1	0.18	0.98	4435	3131	639	30	58.93	137.37	4435
delf018	3340	128	2.13	3335	3	1	0.09	0.85	4249	3335	639	30	41.78	121.52	4249
delf019	3138	128	2.01	3160	2	1	0.16	0.96	4296	3160	639	30	38.31	116.93	4296
delf020	4877	128	2.55	3784	4	1	0.11	0.89	4365	3784	639	30	50.36	135.79	4365
delf021	5483	128	2.66	3331	3	1	0.13	0.90	4316	3331	639	30	50.07	136.35	4316
delf022	5452	128	2.39	3757	4	1	0.12	0.93	4345	3757	639	30	49.13	132.96	4345
delf023	4713	128	2.72	4747	4	1	0.21	1.34	4571	4747	639	30	56.30	142.75	4571
delf024	4743	128	3.42	4150	3	1	0.37	1.33	4907	4150	639	30	90.80	178.87	4907
delf025	3128	128	2.37	4152	4	1	0.23	1.22	4783	4152	639	30	65.10	151.80	4783
delf026	3377	128	2.44	3495	3	1	0.25	1.16	4849	3495	639	30	67.13	146.79	4849
delf027	3422	128	2.04	3519	4	1	0.20	1.09	4805	3519	639	30	55.60	134.54	4805
delf028	3313	128	2.14	3425	4	1	0.24	1.23	4823	3425	639	30	60.22	143.47	4823
delf029	4459	128	2.20	3085	3	1	0.24	1.17	4849	3085	639	30	58.70	140.05	4849
delf030	4311	128	2.72	3135	3	1	0.23	1.15	4825	3135	639	30	60.84	141.38	4825
delf031	3267	128	2.10	2969	3	1	0.19	0.89	4757	2969	639	30	55.06	137.31	4757
delf032	4722	128	2.52	3091	3	1	0.22	1.04	4756	3091	639	30	61.48	143.16	4756
delf033	3451	128	2.10	2344	3	1	0.33	1.06	4770	2344	639	30	56.98	137.65	4770
delf034	3212	128	2.46	2832	3	1	0.28	1.26	4826	2832	639	30	60.54	142.12	4826
delf035	3153	128	2.45	2545	4	1	0.33	1.14	4767	2545	639	30	60.32	145.23	4767
delf036	3183	128	2.22	2655	4	1	0.20	1.01	4721	2655	639	30	62.74	143.67	4721
deter0	3598	128	0.64	3725	2	1	0.01	0.25	235	3725	9	8	0.13	0.58	235
deter1	10172	64	2.03	9022	2	1	0.02	0.73	314	9022	9	8	0.25	1.10	314
deter2	11937	128	2.99	11017	2	1	0.03	0.91	455	11017	9	8	0.33	1.54	455
deter3	14182	64	3.19	12348	2	1	0.03	0.91	375	12348	9	8	0.35	1.73	451
deter4	6547	64	1.39	5633	2	1	0.02	0.43	214	5633	5	6	0.23	0.73	214
deter5	9956	64	2.01	9277	2	1	0.02	0.77	309	9277	9	8	0.22	1.07	309
deter6	7953	64	1.60	7482	2	1	0.01	0.51	278	7482	9	8	0.21	1.07	278
deter7	12074	128	3.25	11149	2	1	0.02	0.89	425	11149	9	8	0.29	1.38	425
deter8	7008	64	1.11	6977	2	1	0.01	0.61	287	6977	11	9	0.23	1.13	287
df2177	1020	64	0.96	1391	2	1	0.03	0.77	37	1391	5	6	0.06	0.79	37
dfl001	28510	128	88.41	30478	2	1	0.04	18.69	22	30478	3	4	0.13	18.76	22
dfn-gwin-UUM	157	64	0.01	373	0	1	0.00	0.03	4	373	0	1	0.00	0.02	4
dg012142	3212	64	1.38	11646	2	1	0.03	2.07	173	11646	17	11	0.41	2.55	173
disctom	9281	64	4.35	13965	2	1	0.00	1.82	13	13965	1	2	0.02	1.69	13

Table A.4 continued

Instance	QSoptEx iter	prc	t	SoPlex_fac iter	R	#fac	t_fac	t	dlcm	SoPlex_rec iter	R	#rec	t_rec	t	dlcm
disp3	1275	64	0.10	490	2	1	0.00	0.06	67	490	1	2	0.00	0.05	67
dolom1	10481	64	13.82	19382	2	1	0.03	7.52	25	19382	4	5	0.14	7.80	25
ds	14982	64	58.48	13089	2	1	0.09	19.35	56	13089	4	5	0.55	19.74	56
ds-big	42933	128	*memout*	44732	2	1	0.33	611.55	38	44732	7	7	7.23	623.53	38
dsbmip	914	64	0.11	2179	2	1	0.00	0.18	54	2179	7	7	0.04	0.24	54
e18	10492	64	11.81	12395	2	1	0.06	4.74	7	12395	0	1	0.03	4.79	7
e226	302	64	0.09	402	2	1	0.01	0.05	117	402	14	10	0.08	0.18	117
egout	84	64	0.02	96	2	1	0.00	0.00	7	96	0	1	0.01	0.01	7
eil33-2	137	64	0.27	308	2	1	0.00	0.26	8	308	1	2	0.08	0.31	8
eilA101-2	752	64	5.92	5338	2	1	0.01	21.86	16	5338	2	3	0.81	22.60	16
eilB101	460	64	0.32	1459	2	1	0.01	0.37	17	1459	2	3	0.04	0.40	17
enigma	80	64	0.00	44	2	1	0.00	0.00	6	44	0	1	0.00	0.00	6
enlight13	0	64	0.00	0	0	0	0.00	0.00	1	0	0	0	0.00	0.00	1
enlight14	0	64	0.00	0	0	0	0.00	0.01	1	0	0	0	0.00	0.00	1
enlight15	0	64	0.00	0	0	0	0.00	0.01	1	0	0	0	0.00	0.00	1
enlight16	0	64	0.01	0	0	0	0.00	0.00	1	0	0	0	0.00	0.00	1
enlight9	3	64	0.00	0	0	0	0.00	0.00	1	0	0	0	0.00	0.00	1
etamacro	1214	128	0.59	771	3	1	0.00	0.07	87	771	9	8	0.07	0.16	87
ex10	0	64	7200.00	115687	2	1	3.02	1806.24	145	115687	21	12	3.27	1813.26	145
ex1010-pi	11898	64	20.93	19385	2	1	0.53	10.80	78	19385	11	9	0.44	11.04	78
ex3sta1	7712	64	133.34	7713	2	1	94.63	101.24	1311	7713	212	24	113.88	164.74	1311
ex9	156024	64	4617.26	57559	2	1	1.39	350.35	120	57559	17	11	1.76	353.90	120
f2000	34138	64	142.68	40611	2	1	21.44	81.98	187	40611	32	14	2.89	64.53	187
farm	3	64	0.00	0	0	0	0.00	0.00	1	0	0	0	0.00	0.00	1
fast0507	3299	64	7.45	13213	2	1	0.01	11.65	10	13213	0	2	0.06	11.58	10
ffff800	205	64	0.06	855	2	1	0.01	0.08	156	855	11	9	0.09	0.23	156
fiball	7698	64	1.69	2818	2	1	0.02	0.69	18	2818	1	2	0.04	0.72	18
fiber	228	64	0.04	278	2	1	0.00	0.04	5	278	1	2	0.00	0.03	5
finnis	481	64	0.06	524	2	1	0.00	0.04	65	524	4	5	0.03	0.07	65
fit1d	535	64	0.15	1006	2	1	0.00	0.10	10	1006	1	2	0.01	0.09	10
fit1p	742	64	0.19	3573	2	1	0.01	0.40	10	3573	1	2	0.01	0.39	10
fit2d	6064	64	5.22	10781	2	1	0.00	2.04	14	10781	2	3	0.07	1.95	14
fit2p	15468	64	5.91	16070	2	1	0.04	4.12	14	16070	1	2	0.06	4.07	14
fixnet6	296	64	0.02	184	2	1	0.01	0.02	3	184	0	1	0.00	0.01	3
flugpl	22	64	0.02	11	2	1	0.00	0.00	6	11	0	1	0.00	0.00	6
fome11	51915	128	161.13	46151	5	2	0.13	40.15	22	46151	7	8	0.41	41.53	22
fome12	108076	128	389.61	95445	5	2	0.31	115.90	22	95445	7	8	0.85	118.21	22
fome13	706407	128	3036.68	179979	5	2	0.65	322.41	22	179979	7	8	1.73	323.49	22
fome20	39628	64	20.83	37294	2	1	0.11	13.37	1	37294	0	1	0.07	13.10	1
fome21	81439	64	57.26	81051	2	1	0.22	78.67	3	81051	1	1	0.15	77.24	3
forplan	133	64	0.05	369	2	1	0.00	0.05	69	369	9	8	0.04	0.05	69
fxm2-16	5159	64	0.89	6817	2	1	0.01	0.73	154	6817	9	8	0.21	1.12	154
fxm2-6	2034	64	0.35	2277	2	1	0.02	0.23	112	2277	9	9	0.26	0.69	112
fxm3-16	0	64	7200.00	48534	2	1	0.26	22.82	158	48534	11	9	2.69	27.70	158
fxm3.6	0	64	7200.00	9986	2	1	0.04	1.08	183	9986	11	9	0.35	1.79	183
fxm4.6	0	64	7200.00	25805	2	1	0.16	5.68	136	25805	11	9	1.80	8.96	136

Table A.4 continued

	QSopt_ex			SoPlex_fac						SoPlex_rec					
Instance	iter	prc	t	iter	R	#fac	t_{fac}	t	dlcm	iter	R	#rec	t_{rec}	t	dlcm
g200x740i	716	64	0.05	721	2	1	0.00	0.05	3	721	0	1	0.00	0.04	3
gams10a	21	128	0.05	44	3	1	0.00	0.00	28	44	3	4	0.00	0.00	28
gams30a	69	128	0.07	184	3	1	0.00	0.02	28	184	3	4	0.01	0.02	28
ganges	1548	64	0.15	1291	2	1	0.01	0.10	31	1291	2	3	0.04	0.13	31
ge	7786	128	8.42	11172	3	1	0.11	2.70	1324	11172	176	23	11.87	35.88	1324
gen	921	64	0.06	384	3	1	0.00	0.03	16	384	1	2	0.00	0.02	16
gen1	10125	64	7200.00	12282	2	1	166.71	183.99	2918	12282	532	29	117.81	227.71	2918
gen2	5027	64	7200.00	12239	2	1	5934.53	5978.35	5383	12239	921	32	997.85	1571.22	5383
gen4	41388	64	7200.00	14823	4	1	7151.58	7210.24	—	14823	921	32	1374.38	2016.87	5625
ger50_17_trans	479	128	0.96	4819	2	1	0.02	1.57	10	4819	0	2	0.10	1.54	10
germanrr	3489	64	2.36	8775	2	1	0.02	2.96	7	8775	1	2	0.05	2.76	7
germany50-DBM	3289	64	0.69	8399	2	1	0.01	1.10	2	8399	0	1	0.00	1.17	2
gesa2	1411	64	0.11	1118	2	1	0.01	0.08	51	1118	2	3	0.01	0.07	71
gesa2-o	1254	64	0.09	653	2	1	0.00	0.06	46	653	2	3	0.02	0.06	46
gesa2_o	1254	64	0.09	653	2	1	0.00	0.05	46	653	9	3	0.01	0.06	46
gesa3	1327	64	0.11	974	2	1	0.01	0.08	90	974	9	8	0.07	0.20	90
gesa3_o	1268	64	0.11	530	2	1	0.01	0.05	88	530	1	8	0.04	0.14	88
gfrd-pnc	852	64	0.08	664	2	1	0.00	0.04	17	664	0	2	0.02	0.05	17
glass4	45	64	0.01	73	2	1	0.00	0.01	4	73	5	1	0.00	0.00	4
gmu-35-40	1123	64	0.09	316	2	1	0.00	0.04	40	316	5	6	0.03	0.06	40
gmu-35-50	1462	64	0.12	359	2	1	0.00	0.06	39	359	7	6	0.02	0.10	39
gmut-75-50	38867	64	24.46	6042	2	1	0.01	8.81	55	6042	7	7	0.43	9.55	55
gmut-77-40	14740	64	3.38	4047	2	1	0.00	1.85	55	4047	1	7	0.16	2.40	55
go19	779	64	0.67	2606	2	1	0.09	0.40	7	2606	0	2	0.02	0.27	7
gr4x6	45	64	0.00	36	2	1	0.00	0.00	4	36	0	1	0.00	0.00	4
greenbea	8493	64	3.71	18382	2	1	0.07	4.98	732	18382	100	20	2.47	11.18	732
greenbeb	4505	64	3.72	11957	2	1	0.14	2.94	917	11957	121	21	4.36	12.01	917
grow15	456	64	2.73	2102	2	1	0.44	0.81	1357	2102	146	22	3.27	5.42	1357
grow22	725	64	1.40	3334	2	1	0.71	1.35	1743	3334	255	25	10.83	17.36	1743
grow7	185	64	0.40	1071	2	1	0.14	0.27	602	1071	83	19	0.98	1.59	602
gt2	75	64	0.01	19	2	1	0.00	0.01	7	19	0	1	0.00	0.00	7
hanoi5	5929	64	1.85	7389	2	1	0.05	1.73	4	7389	1	1	0.01	1.53	4
haprp	1758	64	0.20	1303	2	1	0.01	0.08	274	1303	39	15	0.53	0.89	299
harp2	539	64	0.09	323	2	1	0.00	0.07	167	323	9	8	0.01	0.16	167
i_n13	62497	64	56.53	948541	1	0	0.00	2546.11	1	948541	8	1	0.61	2547.13	1
ic97-potential	448	64	0.02	339	2	1	0.00	0.03	2	339	1	1	0.00	0.01	2
iiasa	1612	64	0.12	1562	2	1	0.07	0.09	25	1562	0	2	0.03	0.06	25
iis-100-0-cov	480	64	0.70	1279	2	1	0.06	0.59	16	1279	1	3	0.02	0.52	16
iis-bupa-cov	850	64	0.94	4196	2	1	0.05	1.08	21	4196	2	3	0.04	1.23	21
iis-pima-cov	1268	64	1.80	3532	2	1	0.01	1.35	15	3532	2	3	0.02	1.29	15
israel	144	64	0.03	149	2	1	0.01	0.03	35	149	4	5	0.02	0.05	35
ivu06-big	_0_	_64_	_memout_	37753	2	1	0.62	3224.36	63	37753	9	8	39.21	3267.91	63
ivu52	28882	128	219.04	18983	2	1	0.07	111.42	45	18983	7	7	5.92	119.42	45
janos-us-DDM	888	64	0.04	1042	0	0	0.00	0.03	3	1042	0	0	0.00	0.05	3
jendrec1	10855	64	6.85	11116	2	1	0.71	3.99	447	11116	69	18	11.52	25.67	447
k16x240	62	64	0.01	39	2	1	0.00	0.01	4	39	0	1	0.00	0.00	4

Table A.4 continued

instance	QSopt_ex			SoPlex_fac						SoPlex_rec					
	iter	prc	t	iter	R	#fac	t_{fac}	t	dlcm	iter	R	#rec	t_{rec}	t	dlcm
kb2	56	64	0.01	58	2	1	0.00	0.00	42	58	5	6	0.01	0.01	42
ken-07	1756	64	0.23	2777	2	1	0.00	0.20	7	2777	1	2	0.05	0.21	7
ken-11	12075	64	3.09	17739	2	1	0.05	2.82	7	17739	1	2	0.12	2.99	7
ken-13	28652	64	23.34	41646	2	1	0.13	20.50	7	41646	1	2	0.27	20.48	7
ken-18	115170	64	351.80	192333	2	1	0.48	505.33	7	192333	1	2	1.04	502.76	7
kent	8342	64	1.81	1680	2	1	0.07	0.64	7	1680	1	2	0.09	0.50	7
khb05250	210	64	0.02	119	2	1	0.00	0.01	4	119	0	1	0.00	0.01	4
kl02	191	64	0.42	263	2	1	0.00	0.32	2	263	0	1	0.03	0.21	2
kleemin3	2	64	0.00	0	0	0	0.00	0.00	1	0	0	0	0.00	0.00	1
kleemin4	2	64	0.00	0	0	0	0.00	0.00	1	0	0	0	0.00	0.00	1
kleemin5	2	64	0.00	0	0	0	0.00	0.00	1	0	0	0	0.00	0.00	1
kleemin6	2	64	0.01	0	0	0	0.00	0.00	1	0	0	0	0.00	0.00	1
kleemin7	2	64	0.01	0	0	0	0.00	0.00	1	0	0	0	0.00	0.00	1
kleemin8	2	64	0.02	0	1	1	0.00	0.00	3	0	0	1	0.00	0.09	3
l152lav	373	64	0.10	700	2	1	0.00	0.05	3	700	0	1	0.00	0.09	3
l30	67237	128	7200.00	3293091	4	2	2.96	7200.02	—	3273677	255	26	37.87	7200.02	—
l9	689	64	0.36	746	2	1	0.04	0.22	243	746	39	15	0.62	1.09	243
large000	5623	128	3.78	3748	2	1	0.10	1.10	5415	3748	767	31	57.97	215.38	5415
large001	3857	128	2.60	7844	2	1	0.24	1.90	5913	7844	767	31	84.07	237.59	5913
large002	4059	128	5.71	4065	4	1	0.38	1.67	5794	4065	767	31	127.89	307.08	5794
large003	4627	128	5.10	4254	3	1	0.31	1.47	5393	4254	767	31	119.85	290.91	5393
large004	4073	128	4.83	4681	4	1	0.35	1.70	5890	4681	921	32	144.56	382.26	5890
large005	4302	128	3.58	4567	3	1	0.20	1.36	5318	4567	767	31	100.30	257.66	5318
large006	4484	128	4.22	4942	4	1	0.24	1.36	5361	4942	767	31	106.22	271.86	5361
large007	4516	128	4.32	5094	3	1	0.24	1.54	5401	5094	767	31	103.79	270.23	5401
large008	4263	128	5.15	5291	3	1	0.24	1.49	5467	5291	767	31	112.04	282.13	5467
large009	4255	128	6.13	5074	3	1	0.25	1.43	5396	5074	767	31	112.39	279.33	5396
large010	4907	128	4.32	4725	3	1	0.24	1.53	5464	4725	767	31	111.74	278.80	5464
large011	4360	128	4.26	5221	3	1	0.23	1.31	5356	5221	767	31	102.33	272.17	5356
large012	4600	128	4.50	5009	3	1	0.26	1.45	5429	5009	767	31	106.47	274.26	5429
large013	4552	128	3.88	5062	3	1	0.22	1.32	5257	5062	767	31	95.08	263.42	5257
large014	4600	128	3.82	5148	3	1	0.21	1.34	5366	5148	767	31	97.92	263.13	5366
large015	4524	128	4.40	4380	3	1	0.19	1.05	5325	4380	767	31	95.48	264.99	5325
large016	4371	128	4.75	4633	3	1	0.21	1.18	5365	4633	767	31	103.67	269.78	5365
large017	4493	128	3.75	3980	2	1	0.18	1.20	5308	3980	767	31	85.79	252.85	5308
large018	4834	128	3.13	4459	2	1	0.12	1.14	5130	4459	767	31	67.20	232.57	5130
large019	4825	128	2.79	4909	4	1	0.14	1.22	5117	4909	639	30	57.76	178.01	5117
large020	6784	128	3.66	7059	4	1	0.17	1.51	5212	7059	767	31	83.08	254.98	5212
large021	6964	128	3.69	6288	4	1	0.16	1.33	5131	6288	639	30	67.76	191.73	5131
large022	6976	128	4.03	6993	4	1	0.16	1.55	5131	6993	639	30	67.56	191.63	5131
large023	6792	128	3.65	4398	4	1	0.26	1.31	5584	4398	767	31	88.12	260.96	5584
large024	6514	128	4.69	5988	4	1	0.38	1.90	5825	5988	767	31	116.35	286.58	5825
large025	6761	128	6.01	5000	4	1	0.41	1.76	6062	5000	921	32	159.86	396.72	6062
large026	6470	128	4.52	4367	4	1	0.37	1.69	6047	4367	767	31	123.99	290.86	6047
large027	4815	128	3.45	4353	4	1	0.29	1.53	5717	4353	639	30	83.51	196.02	5717
large028	6677	128	4.33	4906	3	1	0.38	1.64	5901	4906	639	30	99.68	218.75	5901

Table A.4 continued

instance	QSoptEx			SoPlex_fac						SoPlex_rec					
	iter	prc	t	iter	R	#fac	t_{fac}	t	dlcm	iter	R	#rec	t_{rec}	t	dlcm
large029	6475	128	4.29	4372	4	1	0.41	1.57	6120	4372	767	31	125.18	283.39	6120
large030	6695	128	3.90	3930	4	1	0.36	1.56	6022	3930	639	30	95.71	207.59	6022
large031	6605	128	4.81	3931	3	1	0.36	1.38	6115	3931	639	30	95.94	207.22	6115
large032	6397	128	7.31	5052	3	1	0.43	1.78	6209	5052	639	30	100.56	210.57	6209
large033	6290	128	3.56	3877	4	1	0.32	1.38	6070	3877	639	30	91.75	205.72	6070
large034	6544	128	6.85	4201	4	1	0.41	1.70	6210	4201	639	30	102.84	214.38	6210
large035	6746	128	7.68	3655	3	1	0.81	2.08	6573	3655	767	31	157.41	292.99	6573
large036	6419	128	6.40	3314	4	1	0.36	1.43	6067	3314	639	30	98.50	209.12	6067
lectsched-1	17318	64	20.59	7	2	1	0.24	1.59	4	7	0	1	0.05	1.18	4
lectsched-1-obj	17831	64	21.23	963	2	1	0.25	1.78	4	963	0	1	0.06	1.49	4
lectsched-2	10228	64	4.87	3	2	1	0.16	0.79	4	3	0	1	0.03	0.52	4
lectsched-3	16294	64	12.43	7	2	1	0.21	1.36	4	7	0	1	0.05	0.98	4
lectsched-4-obj	3287	64	0.53	174	2	1	0.05	0.45	4	174	0	1	0.01	0.31	4
leo1	313	64	0.37	862	2	1	0.01	0.55	18	862	2	3	0.02	0.31	18
leo2	563	64	0.89	1637	2	1	0.01	0.84	18	1637	2	3	0.05	0.84	18
liu	395	64	0.09	543	2	1	0.01	0.11	11	543	0	1	0.01	0.08	11
lo10	88280	64	27.06	953870	0	0	0.00	5902.34	1	953870	0	0	0.00	5918.75	1
long15	0	64	7200.00	229488	0	0	0.00	2728.13	1	229488	0	0	0.00	2736.62	1
lotfi	141	64	0.03	226	2	1	0.00	0.02	16	226	2	3	0.01	0.02	16
lotsize	1785	64	0.06	1460	2	1	0.01	0.08	20	1460	0	1	0.01	0.06	20
lp22	28599	64	48.96	38451	2	1	0.49	36.22	131	38451	21	12	1.43	37.97	131
lpl1	601521	64	6097.85	36759	2	1	0.10	61.17	39	36759	4	5	0.59	61.51	39
lpl2	1365	64	0.22	1465	2	1	0.01	0.29	3	1465	1	1	0.01	0.26	3
lpl3	5543	64	1.47	5040	2	1	0.03	1.55	3	5040	0	1	0.02	1.28	3
lrn	10217	64	5.20	11452	3	1	0.09	2.85	668	11452	69	18	2.93	9.19	668
lrsa120	7483	64	2.67	9789	3	1	0.05	2.67	12	9789	1	2	0.04	2.56	12
lseu	54	64	0.00	25	2	1	0.00	0.00	9	25	1	1	0.00	0.00	9
m100n500k4r1	164	64	0.05	174	2	1	0.00	0.02	11	174	1	2	0.00	0.02	11
macrophage	979	64	0.09	706	0	0	0.00	0.04	1	706	0	0	0.00	0.03	1
manna81	4282	64	0.18	3018	0	0	0.00	0.15	1	3018	0	0	0.00	0.16	1
map06	28459	64	601.02	23840	2	1	0.36	40.96	15	23840	2	3	0.52	40.81	15
map10	30505	64	643.73	23747	2	1	0.39	42.05	13	23747	2	3	0.40	41.71	13
map14	31300	64	672.97	23178	2	1	0.39	39.83	13	23178	2	3	0.46	39.44	13
map18	33110	64	713.15	20964	2	1	0.39	34.84	11	20964	1	2	0.28	34.25	11
map20	33358	64	724.49	19686	2	1	0.38	31.43	12	19686	1	2	0.31	30.60	12
markshare1	33	64	0.01	35	2	1	0.00	0.00	11	35	1	2	0.00	0.00	11
markshare2	42	64	0.02	43	2	1	0.00	0.00	14	43	1	2	0.00	0.00	14
markshare_5_0	27	64	0.01	24	2	1	0.00	0.00	9	24	1	2	0.00	0.00	9
maros	1544	64	0.50	1255	2	1	0.02	0.21	247	1255	32	14	0.35	0.90	247
maros-r7	35693	192	794.02	7953	2	1	33.66	58.45	7076	7953	1106	33	1004.62	2166.63	7076
mas74	107	64	0.04	224	2	1	0.01	0.03	99	224	14	10	0.02	0.07	99
mas76	82	64	0.03	132	2	1	0.00	0.02	69	132	9	8	0.02	0.04	69
maxgasflow	6526	64	0.67	6737	2	1	0.03	0.67	12	6737	1	2	0.08	0.73	12
mc11	1710	64	0.12	1239	2	1	0.01	0.14	3	1239	0	1	0.01	0.12	3
mcf2	1041	64	0.36	2763	2	1	0.02	0.34	77	2763	11	9	0.14	0.51	77
mcsched	2232	64	0.75	2546	2	1	0.02	0.29	7	2546	1	2	0.03	0.30	7

Table A.4 continued

Instance	QsoptEx			SoPlex_fac						SoPlex_rec					
	iter	prc	t	iter	R	#fac	t_fac	t	dlcm	iter	R	#rec	t_rec	t	dlcm
methanosarcina	3554	64	3.15	655	0	0	0.00	0.19	1	655	0	0	0.00	0.18	1
mik-250-1-100-1	101	64	0.02	100	2	1	0.00	0.03	10	100	0	1	0.01	0.02	10
mine-166-5	1598	64	1.03	1642	2	1	0.02	0.51	18	1642	2	3	0.05	0.62	18
mine-90-10	1589	64	1.00	1948	1	1	0.03	0.61	19	1948	2	3	0.08	0.66	19
misc03	73	64	0.00	45	0	0	0.00	0.00	1	45	0	0	0.00	0.00	1
misc06	1152	128	0.32	816	2	1	0.00	0.10	37	816	4	5	0.03	0.14	37
misc07	153	64	0.03	157	2	1	0.00	0.03	2	157	0	1	0.00	0.02	2
mitre	7135	64	0.37	2451	2	1	0.01	0.39	31	2451	1	2	0.03	0.35	31
mkc	451	64	0.15	538	2	1	0.00	0.10	20	538	0	1	0.01	0.07	20
mkc1	451	64	0.14	538	2	1	0.00	0.10	20	538	0	1	0.00	0.07	20
mod008	38	64	0.01	27	2	1	0.00	0.01	15	27	1	2	0.01	0.01	15
mod010	502	64	0.14	1062	2	1	0.00	0.17	2	1062	0	1	0.01	0.16	2
mod011	2318	64	0.52	6153	1	1	0.01	0.88	202	6153	4	5	0.13	1.12	202
mod2	77757	192	306.08	58534	13	2	1.73	104.55	3021	58534	369	27	106.74	523.17	3021
model1	423	128	0.16	182	3	1	0.00	0.02	69	182	9	8	0.05	0.08	69
model10	56934	128	1446.45	44687	2	1	1.20	27.13	2677	44687	369	27	56.43	201.27	2677
model11	6537	64	3.29	5273	2	1	0.05	0.96	645	5273	83	19	4.28	9.07	645
model2	908	64	0.34	3466	2	1	0.02	0.45	268	3466	39	15	0.29	0.95	268
model3	3440	64	2.40	9208	2	1	0.08	1.51	707	9208	100	20	2.39	6.77	707
model4	3794	128	5.12	15098	2	1	0.15	3.46	1062	15098	146	22	4.82	15.65	1062
model5	5264	128	5.02	19877	2	1	0.10	4.41	1197	19877	100	20	2.98	14.85	1197
model6	3503	128	23.52	14770	2	1	0.47	4.46	1715	14770	255	25	23.53	45.14	1715
model7	17401	128	178.41	16070	2	1	0.33	5.21	1313	16070	176	23	12.45	33.74	1313
model8	0	64	7200.00	2538	3	1	0.02	0.49	155	2538	21	12	0.55	1.34	155
model9	5601	128	6.44	12397	2	1	0.24	3.00	1928	12397	255	25	12.06	44.33	1944
modglob	232	64	0.02	359	2	1	0.00	0.02	35	359	3	4	0.01	0.03	35
modszk1	887	64	0.30	653	2	1	0.01	0.09	351	653	39	15	0.80	1.24	351
momentum1	7178	64	15.77	3306	3	1	0.32	1.67	1197	3306	121	21	11.38	26.40	1197
momentum2	11540	128	61.38	45950	4	1	0.78	20.78	783	45950	32	14	5.24	30.05	783
momentum3	21824	128	5676.87	46185	3	1	96.85	377.77	5496	46185	639	30	2892.66	4810.12	5496
msc98-ip	103661	128	697.14	9549	3	1	0.07	1.64	23	9549	3	4	0.10	1.62	23
mspp16	0	64	memout	52	2	1	0.27	25.82	3	52	0	1	0.12	19.45	3
multi	103	64	0.02	59	2	1	0.00	0.01	27	59	2	3	0.00	0.01	27
mzzv11	14798	64	15.17	37921	2	1	0.04	26.31	19	37921	1	2	0.02	26.23	19
mzzv42z	4449	64	1.96	34373	2	1	0.05	17.27	26	34373	2	3	0.04	17.53	26
n15-3	43797	64	153.04	43662	2	1	0.12	115.06	3	43662	0	1	0.14	114.43	3
n3-3	2803	64	1.01	2965	2	1	0.01	0.61	2	2965	1	1	0.01	0.60	2
n3700	5641	64	1.23	8698	2	1	0.02	1.35	120	8698	5	6	0.30	1.85	120
n3701	5644	64	1.26	8106	2	1	0.01	1.36	112	8106	4	5	0.21	1.63	112
n3702	5660	64	1.33	7987	2	1	0.01	1.41	129	7987	5	6	0.29	1.70	129
n3703	5695	64	1.49	7397	2	1	0.02	1.25	109	7397	4	5	0.22	1.63	146
n3704	5693	64	1.47	7325	2	1	0.02	1.11	89	7325	4	5	0.21	1.47	89
n3705	5701	64	1.19	6974	2	1	0.02	1.22	141	6974	7	7	0.45	1.78	141
n3706	5727	64	1.30	7305	2	1	0.02	1.31	109	7305	7	7	0.44	1.75	109
n3707	5671	64	1.40	8681	2	1	0.02	1.31	137	8681	7	7	0.40	1.92	137
n3708	5674	64	1.22	7133	2	1	0.02	1.51	118	7133	5	6	0.32	1.83	118

Table A.4 continued

Instance	QSopt_ex			SoPlex_fac						SoPlex_rec					
	iter	prc	t	iter	R	#fac	t_{fac}	t	dlcm	iter	R	#rec	t_{rec}	t	dlcm
n3709	5709	64	1.45	8030	2	1	0.02	1.48	110	8030	5	6	0.28	1.93	110
n370a	5685	64	1.26	8608	2	1	0.02	1.35	119	8608	5	6	0.32	1.79	119
n370b	5677	64	1.29	8553	2	1	0.02	1.52	103	8553	5	6	0.29	1.67	103
n370c	5716	64	1.38	7273	2	1	0.02	1.31	129	7273	5	6	0.27	1.60	129
n370d	5716	64	1.41	7273	2	1	0.02	1.11	129	7273	5	6	0.38	1.80	129
n370e	5708	64	1.29	7852	2	1	0.02	1.39	114	7852	4	5	0.23	1.51	114
n3div36	89	64	0.54	306	2	1	0.00	0.68	7	306	0	1	0.04	0.47	7
n3seq24	2532	64	26.13	4720	3	1	0.01	16.21	16	4720	3	4	0.19	16.19	16
n4-3	1286	64	0.31	1341	2	1	0.00	0.19	2	1341	0	1	0.01	0.17	2
n9-3	2896	64	0.78	3531	2	1	0.00	0.65	2	3531	0	1	0.01	0.61	2
nag	2304	64	1.51	2476	2	1	0.05	0.28	8	2476	2	1	0.02	0.19	8
nemsafm	1053	64	0.05	755	2	1	0.00	0.04	19	755	2	3	0.01	0.06	19
nemscem	1295	64	0.11	932	2	1	0.01	0.08	53	932	5	6	0.04	0.16	53
nemsemm1	8857	64	3.85	10608	2	1	0.04	5.46	200	10608	11	9	0.68	10.39	200
nemsemm2	13367	64	1.99	10445	2	1	0.04	1.95	731	10445	39	15	0.78	8.65	731
nemspmm1	10729	64	7.42	15049	2	1	0.16	5.03	753	15049	121	21	4.80	21.96	753
nemspmm2	10852	128	27.69	15058	3	1	0.44	5.85	1012	15058	146	22	13.61	39.94	1012
nemswrld	53676	128	588.06	35922	3	1	2.70	41.01	1586	35922	255	25	72.04	232.45	1586
neos	159091	64	1351.76	90253	3	1	1.57	497.27	4	90253	1	2	0.71	498.33	4
neos-1053234	432	64	0.17	257	2	1	0.01	0.11	69	257	7	7	0.02	0.15	69
neos-1053591	455	64	0.06	821	2	1	0.00	0.06	11	821	1	2	0.01	0.04	11
neos-1056905	83	64	0.02	140	2	1	0.00	0.01	3	140	0	1	0.00	0.01	3
neos-1058477	86	64	0.06	48	2	1	0.00	0.05	10	48	1	2	0.00	0.04	10
neos-1061020	21111	64	32.00	16447	2	1	0.04	7.29	4	16447	1	2	0.06	7.36	4
neos-1062641	866	128	0.62	902	3	1	0.01	0.07	19	902	2	3	0.02	0.07	19
neos-1067731	25031	64	22.35	12132	2	1	0.02	2.61	2	12132	0	1	0.02	2.63	2
neos-1096528	265	64	25.87	115	2	1	0.35	49.52	1	115	0	1	0.14	51.96	1
neos-1109824	239	64	0.42	115	2	1	0.02	0.08	1	115	0	1	0.01	0.47	1
neos-1112782	92	64	0.07	622	2	1	0.00	0.08	13	622	3	3	0.01	0.07	13
neos-1112787	82	64	0.06	557	2	1	0.00	0.07	13	557	2	3	0.02	0.09	13
neos-1120495	224	64	0.28	96	0	0	0.00	0.29	1	96	0	0	0.00	0.28	1
neos-1121679	33	64	0.01	35	2	1	0.00	0.00	11	35	1	2	0.00	0.00	11
neos-1122047	9983	64	7.89	5499	2	1	0.19	1.69	10	5499	1	2	0.09	1.45	10
neos-1126860	5954	64	5.89	5216	2	1	0.11	1.07	10	5216	1	2	0.06	1.07	10
neos-1140050	3738	128	2337.04	13327	2	1	69.60	106.22	1399	13327	369	27	507.78	779.98	1399
neos-1151496	847	64	0.41	4930	2	1	0.00	0.87	11	4930	1	2	0.00	0.81	11
neos-1171448	3360	64	0.83	4075	1	0	0.00	0.69	3	4075	0	1	0.01	0.73	3
neos-1171692	1123	64	0.35	1290	2	1	0.04	0.22	4	1290	0	1	0.01	0.19	4
neos-1171737	1550	64	0.34	2060	2	1	0.04	0.35	1	2060	0	1	0.00	0.28	1
neos-1173026	187	64	0.07	33	2	1	0.00	0.03	10	33	1	2	0.00	0.02	10
neos-1200887	334	64	0.09	294	2	1	0.02	0.05	5	294	1	1	0.01	0.03	5
neos-1208069	853	64	0.46	3008	2	1	0.02	0.61	29	3008	3	4	0.02	0.61	29
neos-1208135	796	64	0.45	2364	2	1	0.01	0.44	28	2364	3	4	0.02	0.45	28
neos-1211578	211	64	0.02	187	2	1	0.01	0.02	5	187	1	1	0.00	0.00	5
neos-1215259	2185	64	1.12	6319	2	1	0.00	0.93	23	6319	3	4	0.01	1.05	23
neos-1215891	2649	64	0.76	3541	2	1	0.01	1.01	25	3541	3	4	0.04	1.03	25

Table A.4 continued

Instance	QSopt_ex iter	prc	t	SoPlex_fac R	#fac	iter	t_{fac}	t	dlcm	SoPlex_rec R	#rec	iter	t_{rec}	t	dlcm
neos-1223462	1566	64	0.48	2	1	13116	0.01	3.89	50	7	7	13116	0.08	4.02	50
neos-1224597	979	64	0.22	2	1	10673	0.00	1.87	12	1	2	10673	0.00	1.89	12
neos-1225589	52	64	0.03	2	1	438	0.00	0.01	7	0	1	438	0.00	0.02	7
neos-1228986	236	64	0.02	2	1	197	0.00	0.01	6	0	1	197	0.00	0.00	6
neos-1281048	759	64	0.22	2	1	2276	0.01	0.28	3	0	1	2276	0.00	0.30	3
neos-1311124	641	64	0.03	2	1	639	0.02	0.05	1	0	1	639	0.00	0.03	1
neos-1324574	2433	64	1.51	2	1	6564	0.01	0.95	5	0	1	6564	0.00	0.93	5
neos-1330346	1197	64	0.78	2	1	1983	0.02	0.33	5	1	2	1983	0.01	0.28	5
neos-1330635	475	64	0.09	2	1	141	0.01	0.04	11	0	1	141	0.01	0.03	11
neos-1337307	5960	64	2.16	2	1	10964	0.03	2.24	7	0	1	10964	0.01	2.29	7
neos-1337489	211	64	0.03	2	1	187	0.00	0.01	5	0	1	187	0.00	0.01	5
neos-1346382	347	64	0.01	2	1	350	0.00	0.03	5	0	1	350	0.01	0.02	5
neos-1354092	9807	64	62.25	2	1	12518	0.33	7.36	49	7	7	12518	0.18	7.87	49
neos-1367061	8419	64	62.75	2	1	18400	0.20	24.04	34	5	6	18400	0.32	24.68	34
neos-1396125	3057	64	0.95	2	1	4834	0.01	0.73	18	1	2	4834	0.00	0.69	18
neos-1407044	36222	64	245.36	2	1	27160	1.14	33.07	95	14	10	27160	0.98	33.85	95
neos-1413153	966	64	0.29	2	1	1027	0.00	0.24	82	7	7	1027	0.07	0.34	82
neos-1415183	1137	64	0.35	2	1	1763	0.01	0.39	85	7	7	1763	0.07	0.49	85
neos-1417043	3104	64	4.62	0	0	13642	0.00	57.71	1	0	0	13642	0.00	57.17	1
neos-1420205	246	128	0.13	2	1	944	0.00	0.01	10	1	2	944	0.00	0.01	10
neos-1420546	53753	128	103.82	2	1	102727	0.33	93.99	514	83	19	102727	38.89	137.57	514
neos-1420790	4330	64	1.28	2	1	12930	0.03	2.70	298	47	16	12930	3.19	6.18	298
neos-1423785	16005	64	10.53	2	1	19268	0.09	3.43	14	1	2	19268	0.10	3.46	14
neos-1425699	78	64	0.01	2	1	26	0.00	0.00	7	1	2	26	0.00	0.00	7
neos-1426635	347	64	0.01	2	1	350	0.00	0.01	5	0	1	350	0.01	0.02	5
neos-1426662	563	64	0.05	2	1	665	0.01	0.06	1	0	1	665	0.00	0.04	1
neos-1427181	559	64	0.04	0	0	591	0.00	0.03	1	0	0	591	0.00	0.03	1
neos-1427261	738	64	0.06	2	1	883	0.02	0.08	1	0	1	883	0.00	0.06	1
neos-1429185	462	64	0.04	2	1	466	0.00	0.03	1	0	1	466	0.01	0.03	1
neos-1429212	30099	64	184.72	2	1	28835	0.60	220.36	221	26	13	28835	2.21	227.11	221
neos-1429461	417	64	0.03	2	1	399	0.00	0.03	1	0	1	399	0.00	0.01	1
neos-1430701	257	64	0.02	2	1	257	0.01	0.02	4	0	1	257	0.00	0.01	4
neos-1430811	32043	64	192.03	2	1	32707	0.76	340.95	307	39	15	32707	4.63	355.88	307
neos-1436709	533	64	0.04	2	1	505	0.00	0.04	2	0	1	505	0.00	0.03	2
neos-1436713	867	64	0.08	2	1	835	0.01	0.08	1	0	1	835	0.00	0.04	1
neos-1437164	314	64	0.08	2	1	296	0.00	0.06	38	2	3	296	0.02	0.06	38
neos-1439395	303	64	0.02	2	1	312	0.00	0.03	5	1	1	312	0.00	0.02	5
neos-1440225	392	64	0.31	2	1	486	0.02	0.12	15	2	3	486	0.02	0.11	15
neos-1440447	248	64	0.04	2	1	239	0.01	0.02	4	0	1	239	0.01	0.02	4
neos-1440457	695	64	0.05	2	1	820	0.01	0.07	2	0	1	820	0.01	0.04	2
neos-1440460	387	64	0.03	2	1	359	0.01	0.03	2	0	1	359	0.00	0.01	2
neos-1441553	243	64	0.09	2	1	288	0.00	0.06	40	2	3	288	0.01	0.06	40
neos-1442119	531	64	0.04	2	1	607	0.01	0.05	2	0	1	607	0.01	0.03	2
neos-1442657	417	64	0.03	2	1	498	0.01	0.04	1	0	1	498	0.00	0.02	1
neos-1445532	4306	64	0.54	2	1	14103	0.00	1.79	28	3	4	14103	0.02	1.79	28
neos-1445738	4914	64	0.74	2	1	15164	0.01	2.92	16	2	3	15164	0.02	3.08	16

Table A.4 continued

Instance	QSopt_ex			SoPlex$_{fac}$						SoPlex$_{rec}$					
	iter	prc	t	iter	R	#fac	t_{fac}	t	dlcm	iter	R	#rec	t_{rec}	t	dlcm
neos-1445743	5089	64	0.69	16103	2	1	0.00	3.38	20	16103	2	3	0.02	3.29	20
neos-1445755	5023	64	0.58	15634	2	1	0.00	3.11	21	15634	2	3	0.04	3.31	21
neos-1445765	5044	64	0.71	15960	2	1	0.01	3.29	20	15960	2	3	0.03	3.44	20
neos-1451294	2005	64	1.50	10087	2	1	0.03	1.80	16	10087	2	3	0.02	1.67	16
neos-1456979	1140	64	0.46	957	2	1	0.00	0.45	4	957	0	1	0.01	0.40	4
neos-1460246	227	64	0.05	276	2	1	0.00	0.03	16	276	2	3	0.00	0.03	16
neos-1460265	676	64	0.11	808	2	1	0.01	0.09	7	808	0	1	0.00	0.05	7
neos-1460543	2065	64	1.01	9199	2	1	0.02	1.39	41	9199	7	7	0.05	1.50	41
neos-1460641	993	64	0.29	10168	2	1	0.01	1.29	8	10168	1	2	0.00	1.28	8
neos-1461051	433	64	0.16	418	2	1	0.03	0.11	3	418	0	1	0.01	0.05	3
neos-1464762	1211	64	0.38	10563	2	1	0.00	1.27	18	10563	2	3	0.01	1.34	18
neos-1467067	639	64	0.02	673	2	1	0.01	0.03	1	673	1	1	0.00	0.03	1
neos-1467371	1099	64	0.35	8604	2	1	0.01	0.90	37	8604	4	5	0.00	1.13	37
neos-1467467	1207	64	0.43	8764	2	1	0.00	1.00	25	8764	3	4	0.00	0.93	25
neos-1480121	111	64	0.00	86	2	1	0.00	0.01	2	86	0	1	0.00	0.01	2
neos-1489999	942	64	0.12	835	2	1	0.00	0.08	8	835	0	1	0.00	0.06	8
neos-1516309	135	64	0.06	134	2	1	0.00	0.08	1	134	1	1	0.01	0.06	1
neos-1582420	3951	64	1.30	2433	2	1	0.00	0.50	16	2433	1	2	0.00	0.56	16
neos-1593097	478	64	0.65	921	2	1	0.01	0.58	145	921	21	12	0.10	1.60	145
neos-1595230	589	64	0.16	518	2	1	0.00	0.04	2	518	0	1	0.01	0.03	2
neos-1597104	140	64	2.20	150	2	1	0.12	0.89	1	150	0	1	0.03	0.55	1
neos-1599274	300	64	0.25	357	2	1	0.00	0.15	2	357	0	1	0.00	0.13	2
neos-1601936	5844	64	4.73	14758	2	1	0.03	5.55	17	14758	2	3	0.02	5.43	17
neos-1603512	419	64	0.13	814	2	1	0.00	0.12	3	814	0	1	0.01	0.11	3
neos-1603518	1388	64	0.77	2362	2	1	0.01	0.46	8	2362	1	2	0.02	0.44	8
neos-1603965	16126	128	33.92	1109741	4	7	0.06	1034.90	—	1109741	4	5	0.11	*1029.31*	—
neos-1605061	7474	64	9.28	23399	2	1	0.04	11.56	28	23399	4	5	0.06	11.69	28
neos-1605075	6575	64	6.81	16980	2	1	0.09	8.92	39	16980	5	6	0.15	9.03	39
neos-1616732	492	64	0.07	314	0	0	0.00	0.03	1	314	0	0	0.00	0.02	1
neos-1620770	812	64	0.73	772	0	0	0.00	0.06	1	772	0	0	0.00	0.07	1
neos-1620807	229	64	0.04	222	0	0	0.00	0.01	1	222	0	0	0.00	0.01	1
neos-1622252	898	64	0.41	807	0	0	0.00	0.06	1	807	0	0	0.00	0.07	1
neos-430149	273	64	0.05	240	2	1	0.00	0.02	9	240	1	1	0.00	0.01	9
neos-476283	2758	64	25.17	5536	2	1	0.05	4.92	19	5536	2	3	0.25	4.51	19
neos-480878	445	64	0.25	466	2	1	0.02	0.17	15	466	1	2	0.02	0.13	15
neos-494568	867	64	0.48	553	2	1	0.00	0.30	5	553	0	1	0.00	0.20	5
neos-495307	75	64	0.10	963	2	1	0.00	0.42	5	963	1	1	0.02	0.34	5
neos-498623	1416	64	0.58	1037	3	1	0.00	0.66	8	1034	1	1	0.00	0.49	8
neos-501453	77	64	0.02	37	2	1	0.00	0.00	10	37	1	2	0.00	0.01	10
neos-501474	224	64	0.04	288	2	1	0.00	0.01	22	288	2	3	0.00	0.02	22
neos-503737	983	64	0.60	4883	2	1	0.01	0.84	11	4883	1	2	0.00	0.98	11
neos-504674	723	64	0.07	548	2	1	0.01	0.05	11	548	1	2	0.00	0.04	11
neos-504815	593	64	0.06	468	2	1	0.00	0.03	11	468	1	2	0.00	0.03	11
neos-506422	763	64	0.47	98	2	1	0.02	0.12	3	98	0	1	0.00	0.06	3
neos-506428	1503	64	5.38	2092	0	0	0.00	1.34	1	2092	0	0	0.00	1.04	1
neos-512201	731	64	0.07	543	2	1	0.01	0.04	11	543	1	2	0.00	0.03	11

Table A.4 continued

Instance	QSoptEx			SoPlex$_{fac}$						SoPlex$_{rec}$					
	iter	prc	t	R	iter	#fac	t$_{fac}$	t	dlcm	R	iter	#rec	t$_{rec}$	t	dlcm
neos-520729	5943	64	1.25	2	32688	1	0.10	20.19	4	0	32688	1	0.08	20.04	4
neos-522351	213	64	0.02	2	205	1	0.00	0.02	4	0	205	1	0.00	0.02	4
neos-525149	400	64	7.32	0	1124	0	0.00	0.82	1	0	1124	0	0.00	0.78	1
neos-530627	95	64	0.02	3	68	1	0.00	0.00	25	2	68	3	0.00	0.00	25
neos-538867	78	64	0.02	2	171	1	0.00	0.04	3	1	171	2	0.00	0.03	3
neos-538916	140	64	0.03	2	168	1	0.01	0.04	3	0	168	1	0.00	0.03	3
neos-544324	534	64	4.71	2	2629	1	0.02	1.61	10	1	2629	2	0.05	1.38	10
neos-547911	424	64	1.41	2	1895	1	0.02	0.52	9	1	1895	2	0.02	0.42	9
neos-548047	3191	64	2.19	2	11782	1	0.02	1.67	12	1	11782	2	0.01	1.57	12
neos-548251	1357	64	0.43	2	2037	1	0.01	0.13	2	0	2037	1	0.01	0.09	2
neos-551991	2011	64	1.34	2	7378	1	0.01	0.87	8	1	7378	1	0.01	1.23	8
neos-555001	3598	64	0.79	2	10198	1	0.00	0.92	11	1	10198	2	0.01	0.78	11
neos-555298	2073	64	0.17	2	3995	1	0.01	0.29	10	1	3995	2	0.02	0.26	10
neos-555343	3229	64	0.59	2	11043	1	0.01	1.22	10	1	11043	2	0.01	1.13	10
neos-555424	3253	64	0.33	2	5895	1	0.02	0.47	14	1	5895	2	0.02	0.46	14
neos-555694	560	64	0.24	2	565	1	0.01	0.19	4	0	565	1	0.01	0.15	4
neos-555771	435	64	0.11	2	575	1	0.01	0.19	6	0	575	1	0.01	0.15	6
neos-555884	2838	64	0.38	2	7183	1	0.01	0.66	16	1	7183	2	0.03	0.64	16
neos-555927	1203	64	0.09	2	2202	1	0.00	0.11	11	1	2202	2	0.00	0.11	11
neos-565672	80796	64	136.17	2	87785	1	0.32	261.62	1119	69	87785	18	7.56	289.53	1119
neos-565815	2462	64	3.13	2	6971	1	0.04	2.47	5	0	6971	1	0.01	2.27	5
neos-570431	837	64	0.36	2	2355	1	0.02	0.27	6	1	2355	1	0.01	0.24	6
neos-574665	690	64	0.18	2	506	1	0.02	0.15	44	4	506	5	0.03	0.18	44
neos-578379	0	64	7200.00	2	14389	1	0.07	13.61	19	2	14389	3	0.05	13.53	19
neos-582605	692	64	0.10	2	1067	1	0.01	0.11	4	0	1067	1	0.00	0.09	4
neos-583731	522	64	0.04	0	602	0	0.00	0.01	1	0	602	0	0.00	0.02	1
neos-584146	526	64	0.09	2	603	1	0.00	0.06	4	0	603	1	0.01	0.05	4
neos-584851	429	64	0.04	2	612	1	0.00	0.05	6	0	612	1	0.00	0.04	6
neos-584866	4819	64	2.07	2	11751	1	0.02	2.56	9	1	11751	2	0.02	2.77	9
neos-585192	1699	128	2.07	2	2614	1	0.02	0.60	31	4	2614	5	0.11	0.80	31
neos-585467	1387	64	0.44	2	1912	1	0.03	0.49	31	4	1912	5	0.13	0.60	31
neos-593853	795	64	0.08	2	615	1	0.00	0.03	51	2	615	3	0.01	0.06	51
neos-595904	1677	64	0.45	2	1276	1	0.01	0.30	147	11	1276	9	0.09	0.53	147
neos-595905	554	64	0.09	2	390	1	0.01	0.05	78	5	390	6	0.02	0.06	78
neos-595925	551	64	0.10	2	521	1	0.00	0.06	86	5	521	6	0.02	0.10	86
neos-598183	891	64	0.15	2	1148	1	0.00	0.16	100	5	1148	6	0.01	0.19	100
neos-603073	903	64	0.09	2	455	1	0.00	0.05	5	0	455	1	0.01	0.04	5
neos-611135	984	64	2.11	2	11914	1	0.01	4.81	11	1	11914	2	0.01	4.69	11
neos-611838	2611	64	0.78	2	2046	1	0.01	0.42	10	1	2046	2	0.04	0.50	10
neos-612125	2762	64	0.93	2	2002	1	0.01	0.48	10	1	2002	2	0.05	0.47	10
neos-612143	2824	64	1.00	2	2100	1	0.00	0.45	10	1	2100	2	0.04	0.49	10
neos-612162	2491	64	0.65	2	1828	1	0.01	0.39	12	1	1828	2	0.04	0.37	12
neos-619167	13105	128	28.96	2	6697	1	0.02	1.16	65	9	6697	8	0.22	1.61	65
neos-631164	360	128	0.15	2	954	1	0.00	0.08	81	9	954	8	0.00	0.15	81
neos-631517	323	64	0.05	2	765	1	0.00	0.05	69	7	765	7	0.03	0.09	69
neos-631694	568	64	0.29	2	24940	1	0.00	5.14	4	0	24940	1	0.01	5.37	4

Table A.4 continued

Instance	QSopt_ex			SoPlex_fac						SoPlex_rec					
	iter	prc	t	iter	R	#fac	t_fac	t	dlcm	iter	R	#rec	t_rec	t	dlcm
neos-631709	5661	64	15.55	70640	2	1	0.06	212.38	6	70640	0	1	0.05	212.42	6
neos-631710	96015	64	2133.75	66455	2	1	0.24	920.08	7	66455	0	1	0.18	914.82	7
neos-631784	3243	64	4.82	26868	2	1	0.03	29.69	4	26868	0	1	0.02	29.36	4
neos-632335	3594	64	1.31	5592	2	1	0.05	1.36	10	5592	1	2	0.05	1.26	10
neos-633273	2548	64	0.88	5070	2	1	0.06	1.32	8	5070	1	2	0.04	1.27	8
neos-641591	5742	64	4.12	14066	2	1	0.01	6.92	9	14066	1	2	0.06	7.02	9
neos-655508	1445	64	0.35	119	0	0	0.00	0.09	1	119	0	0	0.00	0.09	1
neos-662469	5742	64	4.10	14066	2	1	0.00	6.95	9	14066	1	2	0.06	6.84	9
neos-686190	497	64	0.31	834	2	1	0.01	0.28	3	834	0	1	0.01	0.26	3
neos-691058	2033	64	1.24	7206	2	1	0.02	1.43	15	7206	2	3	0.02	1.47	15
neos-691073	2363	64	1.35	6943	2	1	0.02	1.35	13	6943	1	2	0.02	1.49	13
neos-693347	1530	64	1.35	9600	2	1	0.03	2.34	16	9600	2	3	0.02	2.23	16
neos-702280	1591	64	25.08	12807	2	1	3.68	20.34	96	12807	14	10	8.26	25.44	96
neos-709469	175	64	0.04	429	2	1	0.00	0.04	11	429	1	2	0.00	0.04	11
neos-717614	2133	64	0.35	1049	2	1	0.00	0.09	44	1049	2	3	0.02	0.08	44
neos-738098	21092	64	50.03	25294	2	1	0.06	31.43	27	25294	4	5	0.09	31.70	27
neos-775946	872	64	0.45	1720	2	1	0.01	0.61	7	1720	0	1	0.00	0.57	7
neos-777800	2067	64	3.98	1633	2	1	0.01	0.53	21	1633	2	3	0.01	0.56	21
neos-780889	29192	64	94.03	57556	0	0	0.00	184.15	1	57556	0	0	0.00	184.76	1
neos-785899	401	64	0.16	411	2	1	0.00	0.08	7	411	0	1	0.00	0.07	7
neos-785912	291	64	0.11	1338	2	1	0.01	0.31	14	1338	1	2	0.01	0.30	14
neos-785914	157	64	0.04	232	2	1	0.00	0.04	4	232	0	1	0.00	0.03	4
neos-787933	2165	64	0.75	14008	2	1	0.01	31.79	3	14008	0	1	0.11	31.35	3
neos-791021	2241	64	0.89	10768	2	1	0.01	2.38	6	10768	0	1	0.00	2.38	6
neos-796608	233	64	0.00	379	0	0	0.00	0.01	1	379	0	0	0.00	0.01	1
neos-799711	16520	128	267.39	19400	2	1	0.13	3.38	29	19400	2	3	0.17	3.32	29
neos-799838	4562	64	2.54	11429	2	1	0.01	3.69	2	11429	0	1	0.01	3.87	2
neos-801834	2246	64	0.78	1782	2	1	0.01	0.42	4	1782	0	1	0.01	0.36	4
neos-803219	386	64	0.14	1309	2	1	0.01	0.14	15	1309	2	3	0.03	0.16	15
neos-803220	291	64	0.11	888	2	1	0.01	0.07	13	888	1	2	0.02	0.07	13
neos-806323	416	64	0.14	1246	2	1	0.01	0.11	15	1246	2	3	0.02	0.10	15
neos-807454	1592	64	0.97	12164	2	1	0.00	2.19	4	12164	1	1	0.01	2.33	4
neos-807456	1913	64	1.24	11272	2	1	0.04	1.70	38	11272	5	6	0.07	1.65	38
neos-807639	1115	64	0.22	1305	2	1	0.02	0.13	16	1305	2	3	0.03	0.13	16
neos-807705	427	64	0.17	1259	2	1	0.02	0.13	49	1259	7	7	0.15	0.28	49
neos-808072	2914	64	1.39	8164	2	1	0.01	1.31	2	8164	0	1	0.01	1.46	2
neos-808214	800	64	0.40	1236	2	1	0.01	0.23	7	1236	1	1	0.00	0.21	7
neos-810286	3446	64	2.87	13517	2	1	0.01	3.66	14	13517	2	3	0.02	3.57	14
neos-810326	2686	64	1.26	8031	2	1	0.01	1.41	5	8031	1	2	0.02	1.40	5
neos-820146	155	64	0.01	249	2	1	0.00	0.03	2	249	0	1	0.00	0.01	2
neos-820157	345	64	0.04	569	2	1	0.01	0.06	4	569	0	1	0.00	0.05	4
neos-820879	997	64	0.65	1509	2	1	0.00	0.55	4	1509	0	1	0.02	0.60	4
neos-824661	129814	64	193.83	9912	2	1	0.01	5.55	5	9912	0	1	0.02	5.44	5
neos-824695	57655	64	45.26	4726	2	1	0.01	1.94	4	4726	1	1	0.01	1.68	4
neos-825075	738	64	0.11	1062	2	1	0.01	0.08	3	1062	0	1	0.00	0.07	3
neos-826224	24146	64	17.89	5396	2	1	0.03	1.41	4	5396	0	1	0.01	1.38	4

Table A.4 continued

Instance	QSoptex			SoPlex_fac						SoPlex_rec					
	iter	prc	t	iter	R	#fac	t_{fac}	t	dlcm	iter	R	#rec	t_{rec}	t	dlcm
neos-826250	34785	64	15.42	5747	2	1	0.01	1.14	4	5747	0	1	0.01	1.08	4
neos-826650	2080	64	0.88	8372	2	1	0.01	1.66	10	8372	1	2	0.01	1.79	10
neos-826694	36110	64	12.34	10754	2	1	0.02	2.77	4	10754	0	1	0.00	2.76	4
neos-826812	20724	64	7.34	10450	2	1	0.03	2.43	4	10450	0	1	0.00	2.51	4
neos-826841	3710	64	1.32	3738	2	1	0.01	0.83	5	3738	0	1	0.01	0.84	5
neos-827015	21599	128	28.75	18897	2	1	0.04	12.62	14	18897	1	2	0.16	12.60	14
neos-827175	21692	64	15.62	11439	2	1	0.03	3.38	6	11439	0	1	0.01	3.05	6
neos-829552	12356	64	8.84	13050	2	1	0.02	4.79	14	13050	1	2	0.09	4.76	14
neos-830439	279	64	0.03	179	2	1	0.00	0.03	4	179	0	1	0.00	0.02	4
neos-831188	4594	64	1.82	8374	2	1	0.00	1.20	11	8374	1	2	0.02	1.01	11
neos-839838	13573	64	13.82	6451	2	1	0.03	1.35	159	6451	11	9	0.45	1.91	159
neos-839859	2975	64	1.05	2312	2	1	0.02	0.46	61	2312	5	6	0.10	0.57	61
neos-839894	25749	128	172.00	21287	2	1	0.14	21.01	273	21287	32	14	4.09	27.35	273
neos-841664	6269	64	3.14	5975	2	1	0.02	0.75	8	5975	0	1	0.03	0.86	8
neos-847051	2635	64	0.61	2383	2	1	0.02	0.31	377	2383	47	16	0.24	1.37	377
neos-847302	1256	64	0.75	2146	2	1	0.01	0.38	8	2146	1	2	0.01	0.37	8
neos-848150	581	64	0.27	1409	2	1	0.01	0.24	13	1409	1	2	0.00	0.22	13
neos-848198	1078	64	0.11	11363	2	1	0.00	2.32	4	11363	0	1	0.01	2.25	4
neos-848589	1830	64	4.70	1401	2	1	0.02	1.78	9	1401	1	2	0.17	1.54	9
neos-848845	894	64	0.66	7295	2	1	0.02	1.44	22	7295	3	4	0.03	1.27	22
neos-849702	756	64	0.60	5626	2	1	0.02	1.24	26	5626	3	4	0.04	1.09	26
neos-850681	1649	64	0.95	25783	2	1	0.01	5.89	57	25783	7	7	0.05	6.16	57
neos-856059	827	64	0.48	580	2	1	0.00	0.19	1	580	0	0	0.00	0.19	1
neos-859770	332	64	1.92	294	0	0	0.01	0.40	2	294	0	1	0.00	0.27	2
neos-860244	101	64	0.48	90	2	1	0.00	0.16	2	90	0	1	0.00	0.11	2
neos-860300	817	64	1.02	349	2	1	0.03	0.34	13	349	1	2	0.03	0.27	13
neos-862348	773	64	0.39	1026	2	1	0.01	0.32	9	1026	1	2	0.01	0.31	9
neos-863472	142	64	0.04	142	2	1	0.00	0.03	9	142	1	2	0.00	0.02	9
neos-872648	36405	64	228.81	38712	2	1	0.17	176.12	11	38712	1	2	0.26	175.98	11
neos-873061	38182	64	214.00	30803	2	1	0.15	142.58	11	30803	1	2	0.24	142.51	11
neos-876808	67949	64	337.69	92322	2	1	0.16	99.42	476	92322	14	10	0.36	100.29	476
neos-880324	84	64	0.02	232	2	1	0.00	0.01	13	232	1	2	0.00	0.01	13
neos-881765	244	64	0.05	424	2	1	0.00	0.05	3	424	0	1	0.01	0.05	3
neos-885086	2743	64	0.71	3951	2	1	0.14	0.86	12	3951	1	2	0.02	0.69	12
neos-885524	109	64	0.51	205	2	1	0.01	5.57	87	205	4	5	0.06	6.37	87
neos-886822	869	64	0.38	1834	2	1	0.02	0.36	140	1834	21	12	0.45	0.90	140
neos-892255	1780	64	0.79	1082	2	1	0.00	0.22	13	1082	1	2	0.01	0.22	13
neos-905856	727	64	0.25	1228	2	1	0.01	0.17	7	1228	0	1	0.01	0.16	7
neos-906865	911	64	0.26	792	2	1	0.01	0.07	7	792	1	2	0.01	0.07	7
neos-911880	111	64	0.03	319	2	1	0.00	0.02	55	319	7	7	0.01	0.05	55
neos-911970	87	64	0.03	258	2	1	0.01	0.02	63	258	5	6	0.01	0.04	63
neos-912015	546	64	0.26	1001	2	1	0.01	0.17	10	1001	1	2	0.02	0.16	10
neos-912023	713	64	0.22	1095	2	1	0.00	0.18	3	1095	0	1	0.00	0.15	3
neos-913984	1080	64	0.25	4692	2	1	0.01	2.70	6	4692	0	1	0.02	2.60	6
neos-914441	19336	128	66.35	7787	2	1	0.03	1.93	65	7787	7	7	0.08	1.95	65
neos-916173	1134	64	0.58	862	2	1	0.01	0.60	53	862	7	7	0.11	0.88	53

Table A.4 continued

Instance	QSopt_ex			$SoPlex_{fac}$						$SoPlex_{rec}$					
	iter	prc	t	iter	R	#fac	t_{fac}	t	dlcm	iter	R	#rec	t_{rec}	t	dlcm
neos-916792	1007	64	0.80	1165	2	1	0.01	0.69	58	1165	9	8	0.34	1.46	58
neos-930752	10686	64	4.14	19648	2	1	0.02	5.94	4	19648	0	1	0.01	5.94	4
neos-931517	5943	64	2.11	10863	2	1	0.01	2.09	3	10863	0	1	0.01	2.12	3
neos-931538	8640	64	3.87	11171	2	1	0.01	2.33	4	11171	1	1	0.01	2.32	4
neos-932721	5644	64	3.96	30030	2	1	0.03	11.85	12	30030	1	2	0.03	11.86	12
neos-932816	8082	64	12.54	14154	2	1	0.06	4.38	11	14154	1	2	0.05	4.16	11
neos-933364	1040	64	0.13	1547	2	1	0.00	0.13	5	1547	0	1	0.01	0.11	5
neos-933550	1676	64	0.94	2301	2	1	0.02	0.48	7	2301	0	1	0.01	0.44	7
neos-933562	3774	64	7.12	4067	2	1	0.03	1.11	13	4067	1	2	0.02	1.26	13
neos-933638	18071	64	26.87	27131	2	1	0.03	18.10	9	27131	1	2	0.04	18.10	9
neos-933815	971	64	0.09	1147	2	1	0.00	0.07	4	1147	0	1	0.01	0.06	4
neos-933966	14934	64	16.47	21146	2	1	0.03	11.40	8	21146	1	2	0.03	11.29	8
neos-934184	1040	64	0.11	1547	2	1	0.00	0.11	5	1547	0	1	0.00	0.11	5
neos-934278	22924	64	36.12	26638	2	1	0.05	16.51	9	26638	1	2	0.03	16.46	9
neos-934441	23396	64	39.46	29117	2	1	0.05	18.70	8	29117	1	2	0.02	19.12	8
neos-934531	732	64	0.94	549	2	1	0.05	0.57	7	549	0	1	0.02	0.34	7
neos-935234	17470	64	22.80	27636	2	1	0.05	17.66	9	27636	1	2	0.03	16.99	9
neos-935348	17440	64	20.04	28978	2	1	0.04	16.94	5	28978	0	1	0.01	16.63	5
neos-935496	3267	64	3.42	4000	2	1	0.02	1.21	12	4000	1	2	0.01	1.19	12
neos-935627	19394	64	25.89	30839	2	1	0.04	18.97	5	30839	0	1	0.01	18.95	5
neos-935674	3872	64	4.35	4015	2	1	0.03	1.00	14	4015	1	2	0.02	1.19	14
neos-935769	11869	64	8.77	24339	2	1	0.01	12.30	9	24339	1	2	0.02	12.39	9
neos-936660	13779	64	11.96	26498	2	1	0.02	14.82	9	26498	1	1	0.01	14.58	9
neos-937446	13859	64	14.80	26033	2	1	0.03	15.17	5	26033	0	1	0.01	15.11	5
neos-937511	13425	64	11.28	24828	2	1	0.02	13.76	7	24828	0	1	0.02	14.01	7
neos-937815	20525	64	36.04	33788	2	1	0.04	23.66	8	33788	1	1	0.01	23.57	8
neos-941262	13781	64	13.28	26859	2	1	0.03	15.17	8	26859	2	2	0.01	15.21	8
neos-941313	60767	64	106.09	46102	2	1	0.05	110.64	12	46102	0	1	0.11	109.75	12
neos-941698	754	64	0.36	682	2	1	0.00	0.14	12	682	1	2	0.01	0.12	12
neos-941717	1048	64	0.59	4805	2	1	0.00	0.85	13	4805	1	2	0.00	0.80	13
neos-941782	981	64	0.62	1858	2	1	0.02	0.41	13	1858	1	2	0.01	0.40	13
neos-942323	537	64	0.17	468	2	1	0.01	0.09	10	468	1	1	0.01	0.05	10
neos-942830	664	64	0.32	1444	2	1	0.02	0.28	7	1444	0	1	0.00	0.25	7
neos-942886	293	64	0.10	336	3	1	0.00	0.06	5	336	0	1	0.00	0.03	5
neos-948126	16328	64	21.10	30978	2	1	0.04	19.52	13	30978	1	2	0.03	19.66	13
neos-948268	1911	64	1.11	11614	1	1	0.03	3.78	18	11614	2	3	0.01	3.81	18
neos-948346	4723	64	10.77	4225	2	1	0.02	4.43	72	4225	9	8	0.19	5.35	72
neos-950242	1427	64	2.99	1897	0	0	0.01	0.47	1	1897	0	0	0.00	0.56	1
neos-952987	1980	64	0.84	531	2	1	0.00	0.55	34	531	2	3	0.06	0.60	34
neos-953928	10399	128	119.35	5020	3	1	0.01	2.81	15	5020	2	3	0.04	2.74	15
neos-954925	15578	128	1072.75	*2104607*	*1*	*0*	*0.00*	*7200.00*	—	2088602	1	1	*0.01*	7200.00	—
neos-955215	736	64	0.05	850	2	1	0.00	0.05	3	850	0	1	0.01	0.04	3
neos-955800	512	64	0.63	882	0	0	0.00	0.10	1	882	1	0	0.00	0.11	1
neos-956971	11492	128	103.65	*4738324*	*1*	*0*	*0.00*	*7200.00*	—	*4738601*	1	1	*0.01*	7200.00	—
neos-957143	10400	128	120.98	*7403652*	*1*	*0*	*0.00*	*7200.00*	—	*7387864*	1	1	*0.02*	7200.00	—
neos-957270	367	64	0.50	515	2	1	0.02	0.28	9	515	1	2	0.00	0.21	9

Table A.4 continued

Instance	QSopt.ex			SoPlex_fac						SoPlex_rec					
	iter	prc	t	iter	R	#fac	t_{fac}	t	dlcm	iter	R	#rec	t_{rec}	t	dlcm
neos-957323	7738	128	56.97	3688	2	1	0.01	4.18	25	3688	2	3	0.14	4.22	25
neos-957389	1847	64	1.79	1299	2	1	0.03	0.43	14	1299	1	2	0.01	0.36	14
neos-960392	2253	64	1.19	12434	2	1	0.02	9.47	22	12434	2	3	0.05	9.29	22
neos-983171	13259	64	14.53	27536	2	1	0.04	16.29	14	27536	2	3	0.05	16.28	14
neos-984165	15405	64	20.60	30160	2	1	0.04	18.91	8	30160	1	2	0.02	18.87	8
neos1	7934	64	163.28	4509	2	1	0.39	13.58	5	4509	0	1	0.06	12.65	5
neos13	2050	64	6.90	2438	2	1	0.07	1.03	21	2438	3	4	0.11	1.08	21
neos15	499	64	0.03	508	2	1	0.00	0.02	12	508	1	2	0.00	0.02	12
neos16	497	64	0.08	475	2	1	0.01	0.04	2	475	0	1	0.01	0.03	2
neos18	1719	64	0.79	293	2	1	0.04	0.20	2	293	0	1	0.01	0.11	2
neos2	11960	64	271.12	8434	2	1	0.53	35.83	5	8434	1	1	0.06	34.94	5
neos6	1009	64	0.86	6681	2	1	0.01	2.24	10	6681	1	2	0.01	2.10	10
neos788725	611	64	0.13	603	2	1	0.00	0.07	5	603	1	1	0.00	0.05	5
neos808444	2317	64	0.40	8106	2	1	0.03	3.93	56	8106	7	7	0.15	4.31	56
neos858960	5	64	0.00	2	0	0	0.00	0.01	1	2	0	0	0.00	0.00	1
nesm	1925	128	0.60	6164	2	1	0.00	0.64	249	6164	32	14	0.34	1.48	249
net12	1872	64	1.42	11945	2	1	0.06	3.06	5	11945	0	1	0.01	2.95	5
netdiversion	26291	64	43.09	26331	2	1	0.23	28.32	1	26331	0	1	0.08	27.80	1
netlarge2	207282	64	4955.57	111407	0	0	0.00	1523.88	1	111407	0	0	0.00	1532.00	1
newdano	613	64	0.12	455	2	1	0.00	0.04	11	455	1	2	0.01	0.04	11
nl	13736	64	10.09	14132	2	1	0.10	4.31	792	14132	100	20	4.91	18.63	792
nobel-eu-DBE	1057	64	0.17	2927	2	1	0.00	0.32	7	2927	0	1	0.00	0.30	7
noswot	398	128	0.52	135	3	1	0.00	0.01	119	135	9	8	0.01	0.03	263
npmv07	129008	128	118.18	168166	2	1	0.26	47.36	1844	168166	14	10	4.71	58.08	1844
ns1111636	18278	64	30.01	27016	2	1	0.06	80.52	3	27016	2	1	0.10	78.27	3
ns1116954	43175	64	452.46	15636	2	1	0.12	25.13	22	15636	0	3	0.08	24.77	22
ns1208400	1742	64	1.51	9117	2	1	0.03	1.73	40	9117	5	6	0.06	1.90	40
ns1456591	1972	64	1.02	2343	2	1	0.01	0.59	11	2343	1	2	0.00	0.55	11
ns1606230	8356	64	7.51	15157	2	1	0.04	6.11	23	15157	3	4	0.05	6.10	23
ns1631475	38116	64	89.37	101171	2	1	0.09	256.97	81	101171	11	9	0.90	258.30	81
ns1644855	41374	64	1204.90	47488	2	1	1.45	112.24	11	47488	1	2	0.09	110.28	11
ns1663818	0	64	memout	1723	2	1	0.17	27.67	15	1723	1	2	0.15	25.27	15
ns1685374	67409	128	2531.77	105867	2	1	43.64	819.49	906	105867	146	22	109.59	942.72	906
ns1686196	301	64	0.27	157	2	1	0.00	0.17	4	157	0	1	0.00	0.13	4
ns1688347	800	128	2.30	178	2	1	0.01	0.11	4	178	0	1	0.00	0.08	4
ns1696083	233	64	0.54	498	2	1	0.02	0.38	8	498	0	1	0.01	0.29	8
ns1702808	341	64	0.07	58	2	1	0.00	0.02	22	58	1	2	0.01	0.55	22
ns1745726	325	64	0.35	182	2	1	0.01	0.20	5	182	0	1	0.00	0.14	5
ns1758913	8964	64	119.31	27608	2	1	0.51	332.24	958	27608	32	14	1.28	340.06	958
ns1766074	52	64	0.00	27	2	1	0.00	0.01	2	27	0	1	0.00	0.00	2
ns1769397	151	64	0.34	206	2	1	0.01	0.25	5	206	0	1	0.00	0.20	5
ns1778858	9270	64	5.77	14685	2	1	0.06	5.70	21	14685	3	4	0.14	5.77	21
ns1830653	3477	64	2.63	2753	2	1	0.02	0.69	6	2753	0	1	0.00	0.71	6
ns1856153	3638	64	0.63	3231	2	1	0.05	0.89	5	3231	0	1	0.01	0.82	5
ns1904248	24597	128	388.94	42365	3	1	0.45	55.68	304	42365	7	7	1.23	57.65	304
ns1905797	1176	64	1.52	5721	2	1	0.02	2.83	11	5721	1	2	0.06	2.68	11

Table A.4 continued

Instance	QSoptLex			SoPlex$_{fac}$						SoPlex$_{rec}$					
	iter	prc	t	iter	R	#fac	t$_{fac}$	t	dlcm	iter	R	#rec	t$_{rec}$	t	dlcm
ns1905800	637	64	0.38	2057	2	1	0.01	0.68	9	2057	1	2	0.00	0.64	9
ns1952667	49	64	0.39	81	2	1	0.00	0.30	18	81	2	3	0.01	0.26	18
ns2017839	55935	128	297.59	60283	2	1	0.16	117.71	152	60283	14	10	0.96	124.19	152
ns2081729	186	64	0.04	115	2	1	0.01	0.03	4	115	0	1	0.00	0.02	4
ns2118727	42709	64	498.46	23580	2	1	0.39	113.13	361	23580	47	16	11.24	146.90	361
ns2122603	82054	128	222.62	13837	2	1	0.06	3.74	818	13837	57	17	2.76	9.14	818
ns2124243	52571	64	134.48	94850	2	1	0.25	312.87	3	94850	0	1	0.11	309.60	3
ns2137859	1330	64	7.43	2864	2	1	0.15	6.50	58	2864	3	4	0.21	6.43	58
ns4-pr3	2537	64	0.78	10658	2	1	0.01	1.97	4	10658	0	1	0.01	1.78	4
ns4-pr9	2222	64	0.61	9965	2	1	0.00	1.56	3	9965	0	1	0.01	1.46	3
ns894236	18867	64	21.20	36452	2	1	0.04	15.65	36	36452	4	5	0.16	15.80	36
ns894244	68088	128	1153.08	29128	2	1	0.05	21.57	37	29128	5	6	0.18	21.85	37
ns894786	71777	64	158.92	18747	2	1	0.06	14.69	35	18747	4	5	0.25	14.99	35
ns894788	3324	64	1.30	10998	2	1	0.01	1.54	9	10998	1	2	0.02	1.44	9
ns903616	87763	128	229.76	30158	2	1	0.05	22.48	35	30158	4	5	0.19	22.77	35
ns930473	2467	64	2.28	10736	2	1	0.03	8.22	89	10736	14	10	0.18	8.85	89
nsa	775	64	0.17	1162	2	1	0.01	0.12	10	1162	1	2	0.01	0.13	10
nsct1	14168	64	2.42	4841	2	1	0.06	1.49	6	4841	0	1	0.03	1.27	6
nsct2	11722	64	1.73	6810	2	1	0.06	1.67	175	6810	0	1	0.04	1.38	175
nsic1	377	64	0.04	254	2	1	0.00	0.02	7	254	0	1	0.00	0.00	7
nsic2	262	64	0.03	268	2	1	0.00	0.02	87	268	0	1	0.00	0.01	87
nsir1	3925	64	0.53	3762	2	1	0.02	0.66	6	3762	0	1	0.00	0.74	6
nsir2	2608	64	0.53	3288	2	1	0.01	0.63	272	3288	0	1	0.01	0.52	272
nsr8k	94271	64	840.57	102631	2	1	6.87	407.12	210	102631	39	15	23.58	425.61	210
nsrand-ipx	164	64	0.27	152	2	1	0.00	0.33	3	152	0	1	0.01	0.16	3
nu120-pr3	2948	64	0.77	6187	2	1	0.00	1.21	6	6187	0	1	0.01	1.28	6
nu60-pr9	2061	64	0.53	5736	2	1	0.00	1.05	3	5736	0	1	0.01	1.17	3
nug05	174	64	0.02	161	2	1	0.01	0.02	1	161	0	1	0.00	0.01	1
nug06	603	64	0.20	1117	2	1	0.01	0.11	1	1117	0	1	0.00	0.10	1
nug07	1787	64	0.72	7569	2	1	0.01	0.98	5	7569	0	1	0.00	0.85	5
nug08	3224	64	1.74	13541	2	1	0.01	2.06	3	13541	0	1	0.00	1.94	3
nug08-3rd	45325	64	7200.00	43975	2	1	4923.93	6787.18	520	43975	100	20	41.86	1926.23	520
nug12	424355	64	1031.11	99946	2	1	0.94	160.68	28	99946	5	6	0.20	160.17	28
nw04	172	64	1.46	82	2	1	0.00	0.92	1	82	0	1	0.09	0.51	1
nw14	311	64	2.30	166	2	1	0.01	1.56	1	166	0	1	0.14	1.01	1
ofi	82959	128	memout	139104	4	1	1.18	414.87	13601	139104	867	31	621.49	7203.60	—
opm2-z10-s2	36173	64	552.91	21918	2	1	0.64	112.12	21	21918	2	3	0.56	111.30	21
opm2-z11-s8	65225	64	1899.38	24349	2	1	0.93	163.92	18	24349	2	3	0.69	163.19	18
opm2-z12-s14	26629	64	756.19	28647	2	1	1.54	294.42	21	28647	3	4	1.40	290.82	21
opm2-z12-s7	49729	64	2235.30	30797	2	1	1.71	316.75	23	30797	3	4	1.41	309.29	23
opm2-z7-s2	9582	64	23.15	10304	2	1	0.13	7.67	23	10304	3	4	0.18	7.56	23
opt1217	63	64	0.01	143	2	1	0.00	0.01	4	143	0	1	0.00	0.02	4
orna1	1275	64	2.47	1327	2	1	0.75	1.40	3744	1327	639	30	77.57	111.46	3744
orna2	1360	64	2.49	1539	2	1	0.91	1.57	3744	1539	639	30	82.24	116.39	3744
orna3	1524	64	2.73	1754	2	1	0.75	1.47	3785	1754	639	30	80.85	114.84	3785
orna4	1468	64	3.77	2708	2	1	0.85	1.58	3709	2708	532	29	63.00	88.13	3709

Table A.4 continued

Instance	QSoptEx			SoPlex_fac						SoPlex_rec					
	iter	prc	t	iter	R	#fac	t_{fac}	t	dlcm	iter	R	#rec	t_{rec}	t	dlcm
orna7	2022	64	2.52	2280	2	1	0.95	1.69	3752	2280	639	30	78.68	112.24	3752
orswq2	124	64	0.02	148	2	1	0.01	0.01	319	148	39	15	0.20	0.26	319
osa-07	622	64	0.55	966	2	1	0.01	0.58	9	966	1	2	0.03	0.39	9
osa-14	1371	64	1.47	2104	2	1	0.01	1.41	9	2104	1	2	0.07	1.34	9
osa-30	2753	64	4.29	5628	2	1	0.01	6.34	9	5628	1	2	0.14	6.14	9
osa-60	5550	64	16.00	13364	2	1	0.03	39.06	9	13364	1	2	0.31	38.71	9
p0033	31	64	0.00	19	3	1	0.00	0.00	5	19	0	1	0.00	0.00	5
p0040	37	64	0.00	23	2	1	0.00	0.00	7	23	1	2	0.00	0.00	7
p010	12417	64	3.26	13700	1	0	0.12	2.11	16	13700	2	3	0.17	2.22	16
p0201	169	64	0.02	51	2	1	0.00	0.01	1	51	1	1	0.00	0.01	1
p0282	87	64	0.02	114	2	1	0.00	0.01	10	114	2	2	0.00	0.01	10
p0291	39	64	0.02	27	2	1	0.00	0.01	7	27	1	1	0.00	0.00	7
p05	6224	64	1.13	7515	2	1	0.05	0.86	16	7515	2	3	0.10	1.00	16
p0548	391	64	0.03	84	2	1	0.00	0.02	9	84	0	1	0.00	0.01	9
p100x588b	283	64	0.02	235	2	1	0.00	0.03	4	235	0	1	0.01	0.02	4
p19	419	64	0.06	315	2	1	0.00	0.05	180	315	21	12	0.06	0.26	180
p2756	192	64	0.05	73	2	1	0.00	0.04	7	73	0	1	0.00	0.03	7
p2m2p1m1p0n100	10	64	0.00	10	2	1	0.00	0.00	4	10	0	1	0.00	0.00	4
p6000	711	64	0.20	728	2	1	0.00	0.24	8	728	1	2	0.03	0.22	8
p6b	625	64	0.09	548	0	0	0.00	0.06	1	548	0	0	0.00	0.07	1
p80x400b	210	64	0.01	152	2	1	0.01	0.02	4	152	0	0	0.00	0.01	4
pcb1000	3729	64	0.82	2684	2	1	0.04	0.41	14	2684	1	2	0.04	0.37	14
pcb3000	9920	64	2.53	7719	2	1	0.05	1.16	23	7719	3	4	0.21	1.26	23
pds-02	2817	64	0.20	2713	0	0	0.00	0.15	1	2713	0	0	0.00	0.16	1
pds-06	9644	64	1.19	10699	0	0	0.00	1.30	1	10699	0	1	0.02	1.27	1
pds-10	16627	64	2.97	15362	0	0	0.00	2.26	1	15362	0	0	0.00	1.89	1
pds-100	276385	64	1910.01	661386	2	1	0.52	4660.75	3	661386	0	1	0.45	4654.37	3
pds-20	39628	64	20.75	37294	1	1	0.11	13.38	1	37294	0	1	0.07	13.29	1
pds-30	77469	64	113.80	65338	0	0	0.00	57.05	1	65338	0	0	0.00	56.67	1
pds-40	111375	64	248.29	101787	2	1	0.23	156.35	2	101787	0	1	0.17	154.93	2
pds-50	165089	64	581.03	137692	2	1	0.29	290.09	2	137692	0	1	0.25	289.63	2
pds-60	221330	64	1179.96	188842	1	1	0.33	566.15	2	188842	0	1	0.29	564.43	2
pds-70	241925	64	1347.88	222389	0	0	0.00	754.76	1	222389	0	0	0.00	757.40	1
pds-80	274550	64	1894.39	240224	2	1	0.46	815.33	2	240224	0	1	0.40	816.31	2
pds-90	286273	64	2100.68	512992	2	1	0.49	3220.25	1	512992	0	1	0.46	3219.84	1
perold	2070	64	7.35	5155	2	1	0.83	1.83	1336	5155	176	23	6.95	11.00	1336
pf2177	856	64	1.51	9474	2	1	0.05	3.50	37	9474	5	6	0.08	3.45	37
p8	1576	64	0.05	2164	2	1	0.00	0.07	150	2164	1	1	0.01	0.06	150
pg5_34	1525	64	0.06	3400	2	1	0.00	0.15	126	3400	1	2	0.02	0.14	126
pgp2	12072	128	5.64	3713	1	1	0.03	0.27	7	3713	0	1	0.03	0.18	7
pigeon-10	208	64	0.03	245	2	1	0.02	0.04	4	245	0	1	0.00	0.01	4
pigeon-11	225	64	0.04	237	2	1	0.01	0.04	4	237	0	1	0.00	0.02	4
pigeon-12	214	64	0.04	305	2	1	0.01	0.06	3	305	0	1	0.00	0.03	3
pigeon-13	238	64	0.05	288	2	1	0.01	0.05	5	288	0	1	0.01	0.04	5
pigeon-19	484	64	0.14	608	2	1	0.03	0.13	4	608	0	1	0.00	0.07	4
pilot	4911	128	291.66	10227	2	1	23.30	31.19	2291	10227	369	27	101.87	166.02	2291

Table A.4 continued

instance	QSopt.ex			SoPlex_fac						SoPlex_rec					
	iter	prc	t	iter	R	#fac	t_{fac}	t	dlcm	iter	R	#rec	t_{rec}	t	dlcm
pilot-ja	7118	128	21.90	11126	2	1	0.74	3.05	1562	11126	212	24	12.66	23.00	1562
pilot-we	1764	128	2.92	5730	2	1	0.46	1.66	1902	5730	307	26	16.84	29.16	1902
pilot4	993	64	1.43	1483	2	1	0.81	1.16	1068	1483	176	23	5.28	8.29	1068
pilot87	14093	128	4655.05	15871	2	1	411.38	444.09	5187	15871	921	32	877.09	1436.26	5187
pilotnov	1745	64	4.94	7134	2	1	0.14	1.34	1181	7134	212	24	5.85	10.01	1181
pk1	64	64	0.01	101	2	1	0.00	0.01	26	101	3	4	0.00	0.04	26
pldd000b	2373	64	1.63	1667	2	1	0.12	0.52	2895	1667	443	28	28.60	56.38	2895
pldd001b	2417	64	1.51	1662	2	1	0.12	0.49	2860	1662	443	28	27.57	54.46	2860
pldd002b	2428	64	1.57	1649	2	1	0.14	0.52	2873	1649	443	28	27.42	54.28	2873
pldd003b	2139	64	1.27	1657	2	1	0.12	0.50	2878	1657	443	28	27.60	54.33	2878
pldd004b	2489	64	1.39	1654	2	1	0.14	0.49	2889	1654	443	28	28.31	55.52	2889
pldd005b	1994	64	1.36	1655	2	1	0.14	0.51	2889	1655	443	28	27.68	54.78	2889
pldd006b	2238	128	1.88	1693	2	1	0.15	0.52	2914	1693	443	28	27.85	54.81	2914
pldd007b	2317	64	1.47	1714	2	1	0.15	0.54	2911	1714	443	28	28.13	55.07	2911
pldd008b	2430	128	2.08	1716	2	1	0.15	0.50	2923	1716	443	28	28.14	55.27	2923
pldd009b	2353	128	1.86	2258	2	1	0.16	0.70	2945	2258	443	28	28.37	55.04	2945
pldd010b	2502	128	2.08	2462	2	1	0.16	0.79	2939	2462	443	28	29.01	56.43	2939
pldd011b	2412	128	1.88	2413	2	1	0.08	0.61	2933	2413	443	28	27.81	54.78	2933
pldd012b	2049	128	1.81	2447	2	1	0.14	0.66	2921	2447	443	28	27.86	54.62	2921
pltexpa2-16	785	64	0.09	1094	2	1	0.01	0.10	14	1094	1	2	0.02	0.10	14
pltexpa2-6	308	64	0.04	411	2	1	0.00	0.03	14	411	1	2	0.02	0.04	14
pltexpa3.16	12495	128	4.09	15975	2	1	0.07	4.40	14	15975	1	2	0.18	4.36	14
pltexpa3.6	1948	128	0.55	2741	2	1	0.02	0.38	14	2741	1	2	0.06	0.37	14
pltexpa4.6	45726	128	139.86	14762	2	1	0.08	3.93	14	14762	1	2	0.17	3.88	14
pp08a	132	64	0.00	145	2	1	0.00	0.01	5	145	0	1	0.00	0.01	5
pp08aCUTS	275	64	0.03	224	2	1	0.01	0.01	23	224	2	3	0.00	0.01	23
primagaz	3095	64	0.64	1802	2	1	0.01	0.47	2	1802	0	1	0.02	0.41	2
problem	21	64	0.00	9	0	0	0.00	0.00	1	9	0	0	0.00	0.00	1
probportfolio	291	64	0.07	126	2	1	0.00	0.01	7	126	1	1	0.00	0.00	7
prod1	66	64	0.02	193	2	1	0.00	0.03	37	193	4	5	0.01	0.02	37
prod2	110	64	0.04	344	2	1	0.01	0.05	46	344	5	6	0.02	0.07	46
progas	1401	64	17.32	3998	2	1	3.15	6.32	7896	3998	1328	34	366.04	560.92	7896
profold	1912	64	1.20	13331	2	1	0.01	2.94	28	13331	0	4	0.02	2.95	28
pw-myciel4	936	64	0.57	2154	2	1	0.04	0.71	7	2154	0	1	0.00	0.61	7
qap10	13196	64	15.36	58606	2	1	0.07	30.25	11	58606	2	3	0.04	30.20	11
qiu	1563	128	0.99	1604	2	1	0.01	0.24	14	1604	1	2	0.02	0.22	14
qiulp	1563	128	0.99	1604	2	1	0.00	0.23	14	1604	1	2	0.03	0.25	14
qnet1	473	64	0.07	789	2	1	0.00	0.08	101	789	1	2	0.00	0.08	101
qnet1_o	468	64	0.03	406	2	1	0.01	0.04	85	406	0	1	0.00	0.02	85
queens-30	950	64	753.61	13322	2	1	35.84	49.41	34	13322	4	5	0.57	14.29	34
r05	6434	64	1.45	7499	2	1	0.12	0.87	16	7499	2	3	0.13	0.94	16
r80x800	334	64	0.02	188	2	1	0.00	0.03	4	188	0	1	0.00	0.02	4
rail01	152557	128	936.08	349603	5	2	0.35	4440.46	23	349603	4	6	1.43	4429.14	23
rail2586	0	64	memout	30254	2	1	0.16	683.26	51	30254	7	7	4.29	692.71	51
rail4284	0	64	memout	54982	2	1	0.19	1667.00	51	54982	7	7	4.40	1691.87	51
rail507	3350	64	8.38	13033	2	1	0.00	11.60	11	13033	1	2	0.06	11.42	11

Table A.4 continued

Instance	QSopt_ex			SoPlex_fac						SoPlex_rec					
	iter	prc	t	R	iter	#fac	t_fac	t	dlcm	iter	R	#rec	t_rec	t	dlcm
rail516	1472	64	2.25	2	6992	1	0.00	5.81	3	6992	0	1	0.03	5.66	3
rail582	3832	64	6.58	2	10608	1	0.01	9.48	7	10608	0	1	0.05	9.31	7
ramos3	10154	64	1111.63	2	19446	1	387.97	448.59	50	19446	9	8	0.38	60.55	50
ran10x10a	159	64	0.02	2	133	1	0.00	0.00	6	133	0	1	0.00	0.01	6
ran10x10b	157	64	0.01	2	149	1	0.00	0.00	6	149	0	1	0.00	0.00	6
ran10x10c	155	64	0.01	2	161	1	0.00	0.01	7	161	0	1	0.00	0.00	7
ran10x12	183	64	0.01	2	162	1	0.00	0.01	5	162	0	1	0.00	0.01	5
ran10x26	386	64	0.04	2	423	1	0.01	0.02	10	423	1	2	0.00	0.02	10
ran12x12	209	64	0.03	2	186	1	0.00	0.01	9	186	0	1	0.01	0.01	9
ran12x21	357	64	0.03	2	380	1	0.00	0.02	10	380	0	1	0.00	0.01	10
ran13x13	245	64	0.01	2	236	1	0.00	0.01	5	236	0	1	0.00	0.01	5
ran14x18	335	64	0.03	2	397	1	0.00	0.02	10	397	0	1	0.00	0.02	10
ran14x18-disj-8	669	128	9.75	3	1916	1	0.56	1.01	603	1916	100	20	3.90	5.74	603
ran14x18.disj-8	669	128	10.45	3	1916	1	0.57	0.92	603	1916	100	20	4.03	6.06	603
ran14x18.1	334	64	0.03	2	434	1	0.01	0.03	10	434	0	1	0.00	0.02	10
ran16x16	382	64	0.04	2	379	1	0.00	0.02	10	379	1	2	0.00	0.02	10
ran17x17	361	64	0.04	2	258	1	0.00	0.01	12	258	0	1	0.00	0.01	12
ran4x64	464	64	0.02	2	515	1	0.01	0.03	11	515	0	1	0.00	0.02	11
ran6x43	396	64	0.03	2	447	1	0.00	0.02	11	447	0	1	0.00	0.02	11
ran8x32	380	64	0.03	2	402	1	0.00	0.02	10	402	0	1	0.00	0.02	10
rat1	4943	64	10.16	2	2870	1	1.17	3.71	2068	2870	307	26	83.93	126.41	2068
rat5	3929	64	615.03	2	3024	1	228.13	245.29	4344	3024	639	30	494.35	751.16	4344
rat7a	5340	64	7200.00	3	11319	7	7326.09	7339.40	—	11319	767	31	1544.37	2339.70	5025
rd-plusc-21	374	64	55.89	3	139	1	0.70	8.73	28	139	2	3	0.55	7.85	28
reblock166	3512	64	4.49	2	3359	1	0.06	1.39	19	3359	7	3	0.07	1.42	19
reblock354	10689	64	13.68	2	12854	1	0.08	8.26	51	12854	7	7	0.43	8.78	51
reblock420	10155	64	46.38	2	9480	1	0.17	11.74	19	9480	2	3	0.23	12.23	19
reblock67	945	64	0.43	2	972	1	0.02	0.22	19	972	2	3	0.06	0.24	19
recipe	51	64	0.00	2	40	1	0.00	0.01	3	40	2	1	0.00	0.00	3
refine	30	64	0.00	2	21	1	0.00	0.00	3	21	0	1	0.00	0.00	3
rentacar	1440	64	0.82	2	6483	1	0.02	1.13	29	6483	3	4	0.05	1.36	29
rgn	57	64	0.01	2	85	1	0.00	0.01	17	85	0	2	0.00	0.00	17
rlfddd	482	64	0.39	1	825	1	0.00	0.24	1	825	0	1	0.01	0.25	1
rlfdual	4677	64	2.08	0	8652	0	0.00	1.72	1	8652	0	0	0.00	1.78	1
rlfprim	6908	64	18.66	0	5532	0	0.00	1.19	1	5532	0	0	0.00	1.19	1
rlp1	288	64	0.01	2	248	1	0.00	0.01	4	248	0	1	0.00	0.01	4
rmatr100-p10	2326	64	1.74	2	6260	1	0.01	1.62	9	6260	1	2	0.01	1.62	9
rmatr100-p5	3030	64	2.31	2	10885	1	0.02	3.32	8	10885	1	2	0.01	3.27	8
rmatr200-p10	5399	64	24.53	2	13646	1	0.05	11.86	19	13646	2	3	0.11	11.99	19
rmatr200-p20	3281	64	15.98	2	11166	1	0.04	7.78	14	11166	1	2	0.05	7.64	14
rmatr200-p5	9855	64	33.92	2	17650	1	0.06	18.80	22	17650	3	4	0.30	19.06	22
rmine10	31242	64	134.18	2	13871	1	0.23	21.42	173	13871	26	13	3.18	27.03	173
rmine14	167422	128	5172.52	2	92846	1	1.18	1720.88	202	92846	32	14	15.52	1750.58	202
rmine6	2283	64	1.34	2	2058	1	0.01	0.66	70	2058	9	8	0.12	0.83	70
rocII-4-11	547	64	1.23	2	751	1	0.18	0.78	44	751	4	5	0.16	0.67	44
rocII-7-11	922	64	2.77	2	1075	1	0.19	1.20	23	1075	2	3	0.12	0.82	23

Table A.4 continued

	QSopt_ex			SoPlex_fac						SoPlex_rec					
Instance	iter	prc	t	R	iter	#fac	t_{fac}	t	dlcm	iter	R	#rec	t_{rec}	t	dlcm
rocI-9-11	1102	64	3.96	2	1523	1	0.27	1.38	49	1523	4	5	0.43	1.75	49
rococoB10-011000	630	64	0.13	2	4263	1	0.01	0.78	19	4263	2	3	0.01	0.90	19
rococoC10-001000	608	64	0.11	2	2344	1	0.00	0.36	17	2344	1	2	0.02	0.33	17
rococoC11-011100	922	64	0.23	2	10112	1	0.00	1.71	25	10112	2	3	0.01	1.71	25
rococoC12-111000	716	64	0.35	2	11920	1	0.01	3.44	28	11920	3	4	0.03	3.52	28
roll3000	1024	64	0.50	2	2627	1	0.03	0.44	14	2627	1	2	0.02	0.46	14
rosen1	520	64	0.35	2	898	1	0.03	0.33	159	898	21	12	0.13	0.74	159
rosen10	1694	64	1.37	2	2401	1	0.04	1.01	283	2401	39	15	0.22	2.46	283
rosen2	993	64	0.84	2	1627	1	0.01	0.75	309	1627	47	16	0.26	2.09	309
rosen7	227	64	0.07	2	332	1	0.02	0.07	62	332	9	8	0.03	0.13	62
rosen8	434	64	0.24	2	662	1	0.00	0.21	114	662	17	11	0.07	0.52	114
rout	250	64	0.05	2	293	1	0.02	0.02	14	293	1	2	0.00	0.02	14
route	6696	64	4.17	2	2239	1	0.00	1.29	17	2239	1	2	0.02	1.18	17
roy	158	64	0.01	2	119	1	0.00	0.00	6	119	0	1	0.00	0.01	6
rvb-sub	769	128	10.08	2	512	1	0.00	1.59	19	512	2	3	0.35	1.66	19
satellites1-25	7082	64	3.32	2	9023	1	0.05	4.87	16	9023	1	2	0.02	4.82	16
satellites2-60	37088	64	46.01	2	79705	1	0.12	376.43	37	79705	4	5	0.31	380.78	37
satellites2-60-fs	28885	64	30.83	2	65144	1	0.10	272.46	20	65144	2	2	0.09	271.40	20
satellites3-40	76068	64	261.33	3	190485	1	0.48	2659.25	52	190485	7	7	1.12	2661.34	52
satellites3-40-fs	68491	64	176.10	2	199931	1	0.41	2344.48	55	199931	7	7	1.14	2351.08	55
sc105	92	64	0.01	2	99	1	0.00	0.01	12	99	1	2	0.00	0.00	12
sc205	191	64	0.03	2	217	1	0.01	0.01	21	217	2	3	0.01	0.02	21
sc205-2r-100	1149	64	0.25	2	1109	1	0.00	0.17	4	1109	0	1	0.01	0.15	4
sc205-2r-16	157	64	0.01	2	171	1	0.05	0.01	4	171	1	1	0.00	0.01	4
sc205-2r-1600	2046	64	0.95	2	9189	1	0.02	4.88	2	9189	0	1	0.03	4.91	2
sc205-2r-200	2220	64	0.81	2	2195	1	0.01	0.56	4	2195	1	1	0.01	0.47	4
sc205-2r-27	336	64	0.02	2	327	1	0.00	0.02	4	327	0	1	0.00	0.02	4
sc205-2r-32	301	64	0.03	2	331	1	0.00	0.03	4	331	0	1	0.00	0.02	4
sc205-2r-4	54	64	0.00	2	55	1	0.00	0.00	4	55	0	1	0.00	0.00	4
sc205-2r-400	4334	64	1.86	2	4393	1	0.02	1.16	4	4393	0	1	0.01	1.02	4
sc205-2r-50	666	64	0.08	2	621	1	0.00	0.07	5	621	0	1	0.00	0.05	5
sc205-2r-64	589	64	0.08	2	651	1	0.00	0.08	4	651	0	1	0.00	0.07	4
sc205-2r-8	92	64	0.01	2	100	1	0.01	0.01	4	100	0	1	0.00	0.00	4
sc205-2r-800	6110	64	4.76	2	8729	1	0.04	3.51	7	8729	0	1	0.02	3.43	7
sc50a	45	64	0.02	2	45	1	0.00	0.00	4	45	0	1	0.00	0.00	4
sc50b	50	64	0.00	2	49	1	0.00	0.00	4	49	0	1	0.00	0.00	4
scagr25	581	64	0.07	2	784	1	0.01	0.06	24	784	3	4	0.03	0.08	24
scagr7	148	64	0.02	2	178	1	0.00	0.01	10	178	1	2	0.00	0.01	10
scagr7-2b-16	660	64	0.07	2	717	1	0.01	0.04	7	717	1	2	0.00	0.04	7
scagr7-2b-4	187	64	0.02	2	189	1	0.00	0.01	7	189	0	1	0.00	0.00	7
scagr7-2b-64	10601	64	2.38	2	11614	1	0.04	1.48	10	11614	1	2	0.09	1.57	10
scagr7-2c-16	668	64	0.07	2	684	1	0.00	0.04	7	684	1	2	0.02	0.04	7
scagr7-2c-4	189	64	0.02	2	186	1	0.00	0.01	7	186	1	2	0.01	0.01	7
scagr7-2c-64	2632	64	0.38	2	2727	1	0.01	0.25	10	2727	1	2	0.05	0.25	10
scagr7-2r-108	4200	64	0.54	2	4538	1	0.03	0.43	8	4538	1	2	0.08	0.43	8
scagr7-2r-16	666	64	0.07	2	707	1	0.00	0.04	7	707	1	2	0.00	0.05	7

Table A.4 continued

Instance	QSopt.ex			SoPlex_fac						SoPlex_rec					
	iter	prc	t	iter	R	#fac	t_{fac}	t	dlcm	iter	R	#rec	t_{rec}	t	dlcm
scag7-2r-216	8933	64	1.84	9082	2	1	0.03	1.01	8	9082	1	2	0.07	0.81	8
scag7-2r-27	1090	64	0.12	1144	2	1	0.01	0.08	10	1144	1	2	0.02	0.09	10
scag7-2r-32	1316	64	0.13	1335	2	1	0.01	0.08	10	1335	1	2	0.03	0.08	10
scag7-2r-4	191	64	0.02	185	2	1	0.00	0.01	7	185	0	1	0.00	0.01	7
scag7-2r-432	17565	128	6.92	16562	2	1	0.07	2.75	8	16562	1	2	0.15	2.78	8
scag7-2r-54	2194	64	0.30	2299	2	1	0.01	0.19	10	2299	1	2	0.04	0.19	10
scag7-2r-64	2646	64	0.36	2763	2	1	0.01	0.25	10	2763	1	2	0.05	0.25	10
scag7-2r-8	355	64	0.03	357	2	1	0.01	0.03	7	357	1	2	0.01	0.02	7
scag7-2r-864	35438	128	28.98	32972	2	1	0.16	12.80	9	32972	1	2	0.30	12.96	9
scfxm1	403	64	0.06	447	2	1	0.00	0.05	92	447	9	8	0.07	0.12	92
scfxm1-2b-16	1954	64	0.67	4522	2	1	0.03	0.53	185	4522	26	13	0.55	1.50	185
scfxm1-2b-4	517	64	0.14	1135	2	1	0.01	0.10	130	1135	21	12	0.32	0.55	130
scfxm1-2b-64	15274	128	13.93	26349	2	1	0.17	9.50	212	26349	32	14	3.97	17.47	212
scfxm1-2c-4	514	64	0.12	1071	2	1	0.01	0.09	128	1071	17	11	0.22	0.45	128
scfxm1-2r-128	15242	128	13.76	23788	2	1	0.19	8.18	235	23788	32	14	3.85	15.88	235
scfxm1-2r-16	1903	64	0.77	4672	2	1	0.05	0.62	185	4672	26	13	0.54	1.43	185
scfxm1-2r-256	29104	128	42.25	47280	2	1	0.34	31.79	235	47280	39	15	8.39	50.69	235
scfxm1-2r-27	2999	64	0.96	8158	2	1	0.04	0.98	160	8158	21	12	0.61	2.09	160
scfxm1-2r-32	3780	64	1.31	8924	2	1	0.04	1.04	210	8924	32	14	1.06	3.21	210
scfxm1-2r-4	517	64	0.13	1203	2	1	0.01	0.11	127	1203	17	11	0.23	0.45	127
scfxm1-2r-64	7491	128	4.27	13133	2	1	0.08	2.20	214	13133	26	13	1.67	5.27	214
scfxm1-2r-8	994	64	0.30	2252	2	1	0.02	0.24	159	2252	26	13	0.40	0.87	159
scfxm1-2r-96	11316	128	8.33	16931	2	1	0.13	4.13	212	16931	32	14	2.89	10.12	212
scfxm2	802	64	0.11	1119	2	1	0.01	0.11	143	1119	17	11	0.16	0.41	143
scfxm3	1245	64	0.20	1992	2	1	0.01	0.20	138	1992	14	10	0.21	0.51	138
scorpion	391	64	0.04	245	2	1	0.01	0.03	33	245	2	3	0.02	0.03	33
scrs8	560	64	0.10	608	2	1	0.00	0.06	93	608	9	8	0.09	0.16	93
scrs8-2b-16	152	64	0.03	88	2	1	0.00	0.01	10	88	1	2	0.01	0.02	10
scrs8-2b-4	61	64	0.02	22	2	1	0.00	0.00	9	22	1	2	0.00	0.01	9
scrs8-2b-64	431	64	0.08	296	2	1	0.01	0.05	12	296	1	2	0.01	0.04	12
scrs8-2c-16	158	64	0.03	91	2	1	0.01	0.02	10	91	1	2	0.01	0.01	10
scrs8-2c-32	296	64	0.05	179	2	1	0.00	0.02	10	179	1	2	0.01	0.03	10
scrs8-2c-4	61	64	0.02	22	2	1	0.00	0.01	9	22	1	2	0.00	0.00	9
scrs8-2c-64	587	64	0.09	358	2	1	0.00	0.06	10	358	1	2	0.01	0.06	10
scrs8-2c-8	81	64	0.02	45	2	1	0.01	0.01	8	45	1	2	0.01	0.01	8
scrs8-2r-128	901	64	0.18	543	2	1	0.01	0.12	13	543	1	2	0.02	0.12	13
scrs8-2r-16	169	64	0.03	96	2	1	0.00	0.02	10	96	1	2	0.00	0.00	10
scrs8-2r-256	1777	64	0.49	1119	2	1	0.03	0.33	14	1119	1	2	0.06	0.32	14
scrs8-2r-27	154	64	0.04	103	2	1	0.01	0.02	15	103	1	2	0.00	0.01	15
scrs8-2r-32	328	64	0.04	192	2	1	0.01	0.03	10	192	1	2	0.02	0.03	10
scrs8-2r-4	68	64	0.01	24	2	1	0.00	0.00	9	24	1	2	0.00	0.01	9
scrs8-2r-512	3995	64	0.94	2532	2	1	0.03	0.60	14	2532	1	2	0.06	0.50	14
scrs8-2r-64	649	64	0.10	384	2	1	0.01	0.07	11	384	1	2	0.01	0.05	11
scrs8-2r-64b	455	64	0.08	271	2	1	0.01	0.06	13	271	1	2	0.01	0.06	13
scrs8-2r-8	60	64	0.02	41	2	1	0.00	0.00	13	41	1	2	0.01	0.01	13
scsd1	120	128	0.08	97	2	1	0.00	0.01	40	97	5	6	0.02	0.04	40

Table A.4 continued

Instance	QSopt.ex			SoPlex_fac						SoPlex_rec					
	iter	prc	t	iter	R	#fac	t_{fac}	t	dlcm	iter	R	#rec	t_{rec}	t	dlcm
scsd6	792	128	1.02	423	2	1	0.01	0.05	73	423	9	8	0.08	0.17	73
scsd8	1135	64	0.25	1837	2	1	0.00	0.28	36	1837	3	4	0.03	0.30	36
scsd8-2b-16	242	64	0.06	295	2	1	0.00	0.04	10	295	0	1	0.00	0.03	10
scsd8-2b-4	72	64	0.01	47	1	0	0.00	0.01	2	47	0	1	0.00	0.01	2
scsd8-2b-64	16622	128	9.65	2056	2	1	0.01	0.31	6	2056	1	1	0.02	0.20	6
scsd8-2c-16	229	128	0.09	198	2	1	0.00	0.04	14	198	0	2	0.01	0.03	14
scsd8-2c-4	72	64	0.01	47	1	0	0.00	0.01	2	47	0	1	0.01	0.01	2
scsd8-2c-64	6001	128	3.42	2056	2	1	0.00	0.27	6	2056	1	1	0.02	0.25	6
scsd8-2r-108	5725	128	2.11	1114	2	1	0.00	0.25	8	1114	1	2	0.05	0.19	8
scsd8-2r-16	655	128	0.19	231	2	1	0.00	0.03	18	231	1	2	0.01	0.03	18
scsd8-2r-216	11034	128	5.33	2314	2	1	0.00	0.32	12	2314	1	2	0.04	0.32	12
scsd8-2r-27	1041	128	0.32	286	2	1	0.00	0.05	15	286	1	2	0.00	0.05	15
scsd8-2r-32	1522	128	0.45	429	2	1	0.00	0.07	14	429	1	2	0.01	0.07	14
scsd8-2r-4	72	64	0.01	47	1	0	0.00	0.00	2	47	0	1	0.00	0.01	2
scsd8-2r-432	14089	128	7.59	4530	2	1	0.01	0.55	8	4530	1	2	0.10	0.48	8
scsd8-2r-54	2618	128	0.93	600	2	1	0.00	0.12	14	600	1	2	0.03	0.11	14
scsd8-2r-64	3451	128	1.04	1188	2	1	0.00	0.19	26	1188	2	3	0.06	0.21	26
scsd8-2r-8	271	64	0.03	97	2	1	0.00	0.01	10	97	0	1	0.00	0.01	10
scsd8-2r-8b	271	64	0.02	97	2	1	0.00	0.01	10	97	0	1	0.00	0.01	10
sct1	6843	64	17.52	15178	2	1	0.36	8.28	305	15178	47	16	2.00	13.24	305
sct32	6741	128	11.65	14051	2	1	0.06	3.94	165	14051	26	13	0.59	5.77	165
sct5	7450	128	24.59	6217	2	1	0.02	3.35	114	6217	17	11	0.15	4.18	114
sctap1	149	64	0.02	262	2	1	0.00	0.03	8	262	1	2	0.00	0.02	8
sctap1-2b-16	249	64	0.02	323	2	1	0.00	0.04	5	323	0	1	0.01	0.03	5
sctap1-2b-4	66	64	0.00	83	2	1	0.00	0.01	5	83	0	1	0.00	0.01	5
sctap1-2b-64	9626	64	0.74	4773	2	1	0.03	0.80	5	4773	0	1	0.04	0.59	5
sctap1-2c-16	265	64	0.03	329	2	1	0.01	0.04	5	329	0	1	0.01	0.03	5
sctap1-2c-4	70	64	0.00	85	2	1	0.00	0.01	5	85	0	1	0.00	0.01	5
sctap1-2c-64	906	64	0.11	1146	2	1	0.02	0.17	5	1146	0	1	0.02	0.13	5
sctap1-2r-108	1483	64	0.25	2109	2	1	0.02	0.42	5	2109	0	1	0.03	0.29	5
sctap1-2r-16	216	64	0.02	282	2	1	0.00	0.03	7	282	0	1	0.00	0.03	7
sctap1-2r-216	7072	64	0.66	4248	2	1	0.03	0.65	5	4248	0	1	0.03	0.66	5
sctap1-2r-27	376	64	0.04	536	2	1	0.01	0.07	5	536	0	1	0.01	0.05	5
sctap1-2r-32	428	64	0.05	559	2	1	0.01	0.09	7	559	0	1	0.00	0.05	7
sctap1-2r-4	57	64	0.00	84	2	1	0.01	0.01	7	84	0	1	0.00	0.01	7
sctap1-2r-480	14802	64	1.33	9373	2	1	0.06	1.95	7	9373	0	1	0.07	1.64	7
sctap1-2r-54	711	64	0.10	1069	2	1	0.01	0.16	5	1069	0	1	0.02	0.11	5
sctap1-2r-64	854	64	0.13	1123	2	1	0.02	0.17	5	1123	0	1	0.02	0.15	5
sctap1-2r-8	111	64	0.01	147	2	1	0.00	0.02	7	147	0	1	0.00	0.01	7
sctap1-2r-8b	128	64	0.01	161	2	1	0.00	0.01	5	161	0	1	0.01	0.01	5
sctap2	311	64	0.04	505	2	1	0.01	0.06	7	505	0	1	0.00	0.05	7
sctap3	506	64	0.06	604	2	1	0.00	0.08	5	604	0	1	0.01	0.06	5
seba	160	64	0.02	6	2	1	0.00	0.03	2	6	0	1	0.00	0.02	2
self	24544	192	7200.00	13495	2	1	3548.11	3700.19	6992	13495	533	29	8208.91	8683.34	—
set1ch	507	64	0.03	513	2	1	0.02	0.02	98	513	2	3	0.02	0.03	98
set3-10	5531	64	1.86	2279	2	1	0.02	0.25	49	2279	1	2	0.03	0.23	49

Table A.4 continued

Instance	QsoptEx			SoPlex$_{fac}$						SoPlex$_{rec}$					
	iter	prc	t	iter	R	#fac	t$_{fac}$	t	dlcm	iter	R	#rec	t$_{rec}$	t	dlcm
set3-15	5512	64	1.79	2269	2	1	0.01	0.26	66	2269	4	5	0.10	0.38	66
set3-20	5631	64	2.01	2344	2	1	0.03	0.24	66	2344	3	4	0.11	0.33	66
seymour	2800	64	1.86	3858	2	1	0.02	1.07	15	3858	1	2	0.02	0.92	15
seymour-disj-10	3146	128	30.86	5702	3	1	2.24	5.48	675	5702	100	20	55.55	65.57	675
seymour.disj-10	3146	128	30.56	5702	3	1	2.24	5.49	675	5702	100	20	54.43	64.33	675
seymourl	2800	64	1.77	3858	2	1	0.03	0.99	15	3858	1	2	0.03	0.86	15
sgpf5y6	53669	128	memout	197667	3	1	0.81	223.21	15	197667	2	3	2.63	226.25	15
share1b	166	64	0.03	217	2	1	0.01	0.02	81	217	9	8	0.01	0.05	81
share2b	82	64	0.01	159	2	1	0.00	0.01	35	159	4	5	0.01	0.02	35
shell	620	64	0.06	595	1	0	0.00	0.03	1	595	0	1	0.01	0.04	1
ship04l	375	64	0.06	473	2	1	0.00	0.05	11	473	1	2	0.00	0.04	11
ship04s	364	64	0.05	383	2	1	0.00	0.04	16	383	1	2	0.01	0.04	16
ship08l	634	64	0.13	810	2	1	0.00	0.10	17	810	2	3	0.03	0.14	17
ship08s	568	64	0.08	513	2	1	0.01	0.06	17	513	2	3	0.03	0.07	17
ship12l	1086	64	0.22	1085	2	1	0.01	0.15	16	1085	2	3	0.04	0.16	16
ship12s	980	64	0.06	648	2	1	0.00	0.09	16	648	2	3	0.03	0.10	16
shipsched	9188	64	5.19	2174	2	2	0.12	0.92	6	2174	0	1	0.03	0.62	6
shs1023	192325	128	memout	177407	7	2	0.68	624.28	5476	177407	921	33	106.67	3202.52	5476
siena1	16144	64	31.16	28441	2	1	0.18	19.54	76	28441	14	10	1.16	21.20	76
sierra	694	64	0.08	640	1	1	0.00	0.06	8	640	1	2	0.01	0.04	8
sing2	61677	64	37.90	37541	2	1	0.09	40.60	47	37541	2	3	0.18	42.00	47
sing245	0	64	7200.00	218115	2	1	0.59	1614.49	238	218115	26	13	11.28	1653.59	238
sing359	0	64	7200.00	352087	2	1	1.32	5842.03	503	352087	57	17	56.30	5958.76	503
slptsk	3209	64	14.77	4856	2	1	6.63	9.92	1336	4856	176	23	50.71	87.90	1336
small000	678	128	0.24	557	3	1	0.01	0.07	920	557	121	21	0.85	2.06	920
small001	715	128	0.23	737	3	1	0.01	0.09	958	737	121	21	0.85	1.96	958
small002	718	128	0.39	850	3	1	0.03	0.15	1033	850	146	22	1.55	3.16	1033
small003	760	128	0.26	693	3	1	0.01	0.09	906	693	121	21	1.00	2.23	906
small004	866	128	0.24	476	3	1	0.02	0.08	872	476	121	21	0.87	2.10	872
small005	1075	128	0.25	621	3	1	0.02	0.10	1004	621	121	21	0.86	1.98	1004
small006	699	128	0.30	540	3	1	0.01	0.08	1021	540	100	20	0.99	1.93	1021
small007	776	128	0.27	547	3	1	0.02	0.09	1021	547	100	20	1.08	1.92	1021
small008	804	128	0.34	511	3	1	0.02	0.09	989	511	121	21	1.15	2.09	989
small009	739	128	0.24	435	3	1	0.01	0.08	977	435	121	21	1.09	2.05	977
small010	732	128	0.22	342	3	1	0.01	0.07	942	342	100	20	0.82	1.56	942
small011	693	128	0.24	348	3	1	0.01	0.06	899	348	100	20	0.75	1.49	899
small012	704	128	0.25	287	2	1	0.01	0.06	928	287	100	20	0.80	1.59	928
small013	691	128	0.24	289	2	1	0.01	0.04	945	289	100	20	0.68	1.36	945
small014	679	128	0.23	341	2	1	0.01	0.06	939	341	100	20	0.75	1.49	939
small015	662	128	0.26	338	2	1	0.01	0.06	951	338	100	20	0.67	1.47	951
small016	677	128	0.28	338	2	1	0.01	0.06	931	338	100	20	0.66	1.44	931
south31	48550	64	40.17	23654	2	1	0.47	16.27	375	23654	47	16	29.65	52.07	375
sp97ar	1268	64	1.52	6986	2	1	0.00	3.26	16	6986	3	4	0.13	3.41	16
sp97ic	965	64	0.97	3102	2	1	0.02	0.79	43	3102	7	7	0.12	0.86	43
sp98ar	3058	64	2.37	10349	3	1	0.03	3.33	46	10349	7	7	0.13	3.31	46
sp98ic	1333	64	1.07	3276	2	1	0.03	0.96	34	3276	4	5	0.07	1.03	34

Table A.4 continued

Instance	QSopt_ex			SoPlex_fac						SoPlex_rec					
	iter	prc	t	iter	R	#fac	t_{fac}	t	dlcm	iter	R	#rec	t_{rec}	t	dlcm
sp98ir	933	64	0.65	2970	3	1	0.01	0.75	21	2970	3	4	0.03	0.73	21
square15	0	64	7200.00	208626	0	0	0.00	2421.78	1	208626	0	0	0.00	2420.72	1
stair	519	64	3.92	658	2	1	0.77	0.97	698	658	100	20	2.99	3.94	698
standata	70	64	0.01	50	2	1	0.00	0.02	4	50	0	1	0.01	0.01	4
standmps	367	64	0.02	189	2	1	0.00	0.03	4	189	0	1	0.00	0.01	4
stat96v1	137191	288	7200.00	875464	3	1	0.01	6509.88	—	875464	2	3	0.09	6538.90	—
stat96v4	131330	128	7200.00	144049	2	1	177.58	1811.18	22797	144049	922	32	5618.49	8046.75	—
stat96v5	24501	128	memout	13119	3	1	120.78	681.55	21766	13119	1823	35	1078.48	7200.95	—
stein27	40	64	0.00	32	2	1	0.00	0.00	1	32	0	1	0.00	0.00	1
stein45	67	64	0.01	59	2	1	0.00	0.01	1	59	0	1	0.00	0.00	1
stocfor1	79	64	0.01	108	2	1	0.00	0.00	58	108	7	7	0.02	0.03	58
stocfor2	1688	64	0.49	1994	2	1	0.02	0.35	175	1994	26	13	0.68	1.30	175
stocfor3	13766	64	10.37	16167	2	1	0.11	6.51	256	16167	32	14	5.19	14.83	256
stockholm	59789	64	106.73	26971	3	1	0.21	22.01	9	26971	2	3	0.27	21.67	9
stormG2_1000	500001	64	7200.00	732642	2	1	2.00	4699.38	7	732642	1	2	3.39	4672.98	7
stormg2-125	82031	64	54.71	92103	2	1	0.25	46.72	6	92103	1	2	0.44	46.58	6
stormg2-27	16590	64	2.25	20754	2	1	0.04	2.67	6	20754	1	1	0.04	2.42	6
stormg2-8	4849	64	0.62	6863	2	1	0.03	0.82	6	6863	0	1	0.01	0.53	6
stormg2_1000	500001	64	7200.00	732642	2	1	1.99	4678.21	7	732642	1	2	3.42	4666.89	7
stp3d	135704	64	1877.91	148139	2	1	0.32	1242.21	20	148139	3	4	1.29	1243.94	20
sts405	1481	64	3.81	434	2	1	0.10	0.60	7	434	0	1	0.05	0.56	7
sts729	2817	64	25.35	907	2	1	0.39	2.19	15	907	2	3	0.21	1.87	15
swath	224	64	0.20	127	2	1	0.00	0.14	9	127	1	2	0.01	0.11	9
sws	6149	64	0.51	1150	2	1	0.05	0.56	7	1150	1	2	0.06	0.32	7
t0331-4l	6953	64	33.50	15487	2	1	1.42	17.96	93	15487	14	10	1.18	18.56	93
t1717	5499	64	23.24	12220	2	1	0.16	7.77	47	12220	7	7	0.36	8.45	47
t1722	2391	64	4.57	10612	2	1	0.03	3.59	29	10612	4	5	0.11	3.82	29
tanglegram1	446	64	2.20	478	0	0	0.00	0.32	1	478	0	0	0.00	0.31	1
tanglegram2	168	64	0.17	236	0	0	0.00	0.08	1	236	0	0	0.00	0.10	1
testbig	5679	64	4.69	8010	2	1	0.04	4.74	4	8010	1	1	0.03	4.43	4
timtab1	44	64	0.01	20	2	1	0.00	0.01	22	20	3	3	0.00	0.00	22
timtab2	66	64	0.00	36	2	1	0.00	0.01	2	36	1	1	0.00	0.01	2
toll-like	798	64	0.09	717	0	0	0.00	0.04	1	717	0	0	0.00	0.05	1
tr12-30	663	64	0.02	696	2	1	0.00	0.04	12	696	1	1	0.00	0.02	12
transportmoment	11768	64	5.74	9617	2	1	0.05	1.76	139	9617	21	12	2.29	4.81	139
triptim1	63662	64	236.43	68167	2	1	0.13	93.23	9	68167	0	1	0.07	92.83	9
triptim2	394399	64	7200.00	204335	2	1	0.29	570.10	10	204335	1	2	0.13	566.72	10
triptim3	732299	128	7200.00	89514	2	1	0.24	157.44	12	89514	1	2	0.11	162.06	12
truss	5384	64	2.20	21892	2	1	0.01	4.48	21	21892	2	3	0.05	4.76	21
tuff	156	64	0.05	212	2	1	0.01	0.03	80	212	7	7	0.04	0.06	80
tw-myciel4	2446	64	2.13	11588	2	1	0.02	2.32	4	11588	0	1	0.01	2.75	4
uc-case11	78755	128	223.53	39603	2	1	0.28	76.60	592	39603	39	15	3.77	85.09	592
uc-case3	49132	64	55.24	34156	2	1	0.27	65.05	332	34156	21	12	2.04	70.02	332
uct-subprob	874	64	0.53	2988	2	1	0.01	0.51	6	2988	0	1	0.00	0.51	6
ulevimin	21430	64	18.02	125858	2	1	0.05	108.77	356	125858	39	15	2.05	117.87	356
umts	4144	64	3.85	5549	4	1	0.02	1.00	24	5549	5	6	0.03	1.06	24

Table A.4 continued

Instance	QSopt.ex			SoPlex_fac						SoPlex_rec					
	iter	prc	t	iter	R	#fac	t_{fac}	t	dlcm	iter	R	#rec	t_{rec}	t	dlcm
unitcal.7	10802	64	2.71	21824	2	1	0.12	9.21	174	21824	17	11	0.85	10.87	174
us04	465	64	0.97	338	2	1	0.00	0.50	1	338	0	1	0.03	0.34	1
usAbbrv-8-25_70	3274	64	0.76	2434	2	1	0.02	0.16	27	2434	1	2	0.02	0.13	27
van	8953	64	11.74	11014	2	1	0.56	6.65	19	11014	2	3	0.51	6.40	19
vpm1	142	64	0.00	130	2	1	0.00	0.01	4	130	0	1	0.00	0.01	4
vpm2	156	64	0.00	192	2	1	0.00	0.01	9	192	0	1	0.00	0.01	9
vpphard	7479	64	15.13	11075	2	1	0.06	11.62	11	11075	1	2	0.05	11.47	11
vpphard2	4737	64	11.37	8024	2	1	0.22	33.39	5	8024	0	1	0.11	32.67	5
vtp-base	167	64	0.02	75	2	1	0.00	0.01	27	75	3	4	0.01	0.02	27
wachplan	3917	64	2.15	2033	2	1	0.01	0.45	13	2033	1	2	0.00	0.43	13
watson.1	98859	128	memout	188496	3	1	0.86	370.16	7045	188496	829	31	506.53	7213.99	—
watson.2	434602	128	7200.00	333083	3	1	1.38	1522.05	1439	333083	212	24	153.67	2582.82	1439
wide15	0	64	7200.00	229488	0	0	0.00	2736.49	1	229488	0	0	0.00	2785.90	1
wnq-n100-mw99-14	3021	64	71.31	764	2	1	0.35	6.76	3	764	1	1	0.18	5.13	3
wood1p	356	64	0.99	142	2	1	0.06	0.67	212	142	21	12	0.67	1.71	212
woodw	1570	64	0.57	1832	2	2	0.00	0.60	23	1832	2	3	0.01	0.61	23
world	128384	192	548.95	70320	14	1	1.96	136.07	2790	70320	443	28	138.48	683.09	2790
zed	60	64	0.02	31	2	1	0.01	0.01	14	31	1	2	0.00	0.00	14
zib54-UUE	1807	64	0.26	1855	2	1	0.02	0.18	5	1855	0	1	0.01	0.11	5

A.1.4 Accurate Flux Balance Analysis

Table A.5: Detailed results comparing floating-point LP solvers and iterative refinement on ME models parametrized by growth rate of Eschericchia coli under four different nutritional environments. See Section 4.2 for a discussion of the results.

t — total running time (in seconds)
t_{fac} — time for rational factorization and LU solves (in seconds)
t_{inf} — time for infeasibility box computation (in seconds)
κ — condition number of basis returned by $CPLEX_0$
obj. val. — exact objective value rounded to seven decimal digits
obj. dev. — relative deviation from exact objective value
opt. dist. — relative distance to optimal hyperplane in $\|\cdot\|_\infty$

Instance	SoPlex_fac (exact)			CPLEX_9 (double)				SoPlex_12 (80-bit)			SoPlex_25 (iter. refinement)			
	t	t_{fac}	obj. val.	t	κ	obj. dev.	opt. dist.	t	obj. dev.	opt. dist.	t	t_{inf}	obj. dev.	opt. dist.
Acetate														
1.0050000	314.3	190.8	4.231843e-04	20.1	2.5e+13	+2.1e-04	4.7e-08	31.1	+3.6e-08	6.0e-10	47.3	—	-1.2e-16	5.2e-17
1.0205469	322.8	196.7	1.885586e-04	20.8	2.5e+13	+4.8e-04	4.8e-08	39.9	+7.4e-08	5.1e-10	48.5	—	-9.2e-17	timeout
1.0283203	311.6	186.7	5.509611e-05	19.8	2.5e+13	+1.7e-03	4.9e-08	37.0	cycling	—	46.8	—	-5.7e-17	1.6e-17
1.0302637	305.0	183.5	2.467286e-05	22.2	2.2e+13	+3.6e-03	4.7e-08	33.9	+6.0e-07	8.8e-12	43.4	—	-1.3e-16	2.6e-17
1.0312353	297.8	173.7	8.806807e-06	20.7	2.5e+13	+1.0e-02	4.7e-08	43.3	+1.5e-06	4.5e-10	48.8	—	-1.5e-16	timeout
1.0317212	308.6	186.1	5.676483e-06	19.3	2.4e+13	+1.5e-02	4.7e-08	37.7	+2.9e-06	8.9e-12	43.0	—	-9.6e-17	4.3e-17
1.0318426	316.2	193.0	3.953857e-06	21.7	2.5e+13	+2.2e-02	4.4e-08	37.9	+3.8e-06	8.9e-12	45.7	—	-1.4e-16	5.6e-17
1.0319034	305.6	185.6	2.884513e-06	20.0	2.5e+13	+2.9e-02	4.8e-08	36.5	+4.8e-06	9.0e-12	44.5	—	-1.1e-18	3.3e-17
1.0319186	302.2	184.1	9.379177e-08	21.1	2.5e+13	+9.4e-01	4.9e-08	35.0	+1.8e-04	8.9e-12	42.6	—	-4.2e-17	4.5e-17
1.0319224	310.2	192.0	5.273213e-08	20.1	2.5e+13	+1.7e+00	4.9e-08	36.4	+3.2e-04	8.9e-12	43.1	—	-5.2e-17	5.7e-17
1.0319233	314.2	183.6	infeasible	21.5	2.7e+16	∞	—	24.4	infeasible	—	81.3	30.8	infeasible	—
1.0319243	313.9	187.5	infeasible	20.4	2.5e+13	∞	—	23.2	infeasible	—	81.8	33.9	infeasible	—
1.0319262	309.3	182.0	infeasible	20.4	2.5e+13	∞	—	22.9	infeasible	—	76.6	27.7	infeasible	—
1.0319338	320.7	188.7	infeasible	20.1	—	infeasible	—	21.7	infeasible	—	90.0	40.7	infeasible	—
1.0319641	311.5	184.9	infeasible	22.9	—	infeasible	—	24.6	infeasible	—	89.6	37.7	infeasible	—
1.0322070	326.2	199.4	infeasible	19.5	—	infeasible	—	20.6	infeasible	—	46.3	26.4	infeasible	—
1.0360938	335.2	202.7	infeasible	20.1	—	infeasible	—	21.6	infeasible	—	50.1	29.4	infeasible	—
1.0671875	331.5	197.8	infeasible	23.0	—	infeasible	—	24.3	infeasible	—	54.5	31.4	infeasible	—
1.1293750	473.4	203.5	infeasible	26.0	—	infeasible	—	27.8	infeasible	—	52.0	25.8	infeasible	—
1.2537500	331.4	189.7	infeasible	26.3	—	infeasible	—	27.6	infeasible	—	58.8	31.8	infeasible	—
1.5025000	622.1	191.6	infeasible	25.5	—	infeasible	—	316.8	singular	—	326.2	18.8	infeasible	—
D-Glucose														
1.0050000	313.8	184.2	3.685742e-03	22.8	2.4e+13	+2.3e-05	4.8e-08	36.9	+4.7e-09	1.9e-09	51.3	—	-1.1e-16	5.0e-17
1.2537500	321.6	198.0	7.574653e-04	21.8	2.3e+13	+1.3e-04	5.1e-08	39.1	cycling	—	47.2	—	-9.6e-17	3.7e-17
1.2848438	320.8	191.4	2.872281e-04	21.7	2.2e+13	+3.4e-04	5.5e-08	43.9	+3.6e-08	2.8e-11	47.1	—	-6.9e-17	4.6e-17
1.3003906	300.8	178.9	3.225480e-05	19.0	2.3e+13	+3.1e-03	5.0e-08	39.1	cycling	—	45.6	—	-1.5e-16	3.8e-17
1.3023340	326.0	198.9	5.367941e-06	20.1	2.3e+13	+1.7e-02	4.0e-08	38.1	+2.2e-06	4.9e-10	46.0	—	-1.3e-17	4.5e-17
1.3028198	320.8	195.9	4.837611e-06	20.9	2.3e+13	+1.9e-02	4.2e-08	42.8	cycling	—	47.3	—	-1.5e-16	4.5e-17
1.3028806	312.1	190.2	8.511367e-07	19.1	2.2e+13	+1.2e-01	5.0e-08	38.1	+1.4e-05	4.9e-10	44.2	—	-1.6e-17	5.0e-17
1.3028844	309.4	184.2	8.499246e-07	21.6	2.2e+13	+1.1e-01	4.2e-08	46.8	cycling	—	45.5	—	-8.1e-17	4.1e-17
1.3028853	302.4	179.0	8.293278e-07	22.0	2.3e+13	+1.1e-01	4.3e-08	41.3	cycling	—	47.4	—	-3.9e-17	5.6e-17
1.3028863	306.0	186.6	infeasible	18.1	—	infeasible	—	18.9	infeasible	—	47.0	28.6	infeasible	—
1.3028882	321.8	184.7	infeasible	20.5	—	infeasible	—	22.1	infeasible	—	49.4	28.8	infeasible	—
1.3028957	321.9	192.4	infeasible	20.7	—	infeasible	—	21.2	infeasible	—	49.3	28.7	infeasible	—

Table A.5 continued

Instance	SoPlex$_\text{fac}$ (exact)			CPLEX$_9$ (double)				SoPlex$_{12}$ (80-bit)			SoPlex$_{25}$ (iter. refinement)			
	t	t_fac	obj. val.	t	κ	obj. dev.	opt. dist.	t	obj. dev.	opt. dist.	t	t_inf	obj. dev.	opt. dist.
1.3029109	326.3	188.5	infeasible	19.8	—	infeasible	—	21.2	infeasible	—	56.4	35.8	infeasible	—
1.3029413	320.0	186.3	infeasible	21.1	—	infeasible	—	22.1	infeasible	—	52.6	31.4	infeasible	—
1.3030627	313.1	180.7	infeasible	21.5	—	infeasible	—	23.2	infeasible	—	44.0	21.8	infeasible	—
1.3033057	331.0	197.1	infeasible	20.8	—	infeasible	—	22.4	infeasible	—	42.6	21.1	infeasible	—
1.3042773	330.5	197.8	infeasible	21.8	—	infeasible	—	23.0	infeasible	—	59.7	38.0	infeasible	—
1.3081641	326.6	192.3	infeasible	21.0	—	infeasible	—	22.7	infeasible	—	42.6	21.2	infeasible	—
1.3159375	316.6	184.4	infeasible	21.0	—	infeasible	—	23.3	infeasible	—	55.7	33.4	infeasible	—
1.3781250	316.7	185.6	infeasible	21.7	—	infeasible	—	22.6	infeasible	—	42.1	19.8	infeasible	—
1.5025000	335.3	179.6	infeasible	21.3	—	infeasible	—	58.9	infeasible	—	55.6	33.5	infeasible	—
Fumarate														
1.0050000	319.0	186.4	5.451522e-03	22.4	2.0e+16	+1.6e-05	3.7e-07	41.8	cycling	—	53.1	—	-1.5e-16	4.0e-17
1.5025000	318.6	189.8	1.025725e-04	21.4	2.1e+13	+5.8e-04	4.9e-08	41.2	+6.7e-08	4.8e-10	47.5	—	-9.3e-17	timeout
1.5063867	298.7	174.7	4.324266e-05	21.4	1.9e+13	+1.4e-03	4.2e-08	41.6	+2.8e-07	7.8e-12	47.1	—	-3.9e-17	5.4e-17
1.5083301	328.1	200.1	1.062037e-05	25.9	2.1e+13	+4.7e-03	4.9e-08	45.9	+7.5e-07	4.2e-10	50.1	—	-6.3e-17	5.5e-17
1.5088159	300.6	176.1	9.646003e-06	20.3	1.9e+13	+6.5e-03	5.0e-08	41.6	+1.1e-06	6.1e-12	46.9	—	-8.9e-17	3.9e-17
1.5090588	312.8	190.6	2.228012e-06	21.1	1.9e+16	+2.5e-02	4.0e-08	34.7	+6.3e-06	9.6e-12	44.7	—	-1.3e-16	5.3e-17
1.5091803	307.3	185.0	8.372383e-07	20.4	1.9e+13	+5.9e-02	3.3e-08	35.2	+1.7e-05	9.6e-12	43.3	—	-6.4e-17	2.0e-17
1.5091955	314.1	188.9	3.602409e-07	21.6	1.9e+13	+1.6e-01	3.8e-08	37.9	+3.4e-05	7.9e-12	47.0	—	-1.2e-16	4.7e-17
1.5092031	317.6	189.1	2.956074e-07	19.9	1.9e+13	+2.2e-01	1.0e-07	39.5	cycling	—	50.0	—	-1.3e-16	4.0e-17
1.5092040	316.8	187.7	2.950558e-07	23.3	1.9e+13	+2.1e-01	4.4e-08	40.8	+3.5e-05	4.2e-10	50.8	—	-1.3e-16	4.7e-17
1.5092050	553.4	184.8	infeasible	18.8	—	infeasible	—	20.8	infeasible	—	331.3	42.8	infeasible	—
1.5092069	314.3	184.5	infeasible	20.1	—	infeasible	—	21.2	infeasible	—	93.2	44.6	infeasible	—
1.5092107	322.8	191.9	infeasible	19.7	—	infeasible	—	20.7	infeasible	—	93.8	44.4	infeasible	—
1.5092410	329.1	193.7	infeasible	18.3	—	infeasible	—	19.4	infeasible	—	100.1	46.0	infeasible	—
1.5093018	330.8	194.5	infeasible	21.1	—	infeasible	—	22.2	infeasible	—	39.0	17.6	infeasible	—
1.5102734	323.8	187.0	infeasible	20.6	—	infeasible	—	21.9	infeasible	—	49.5	28.5	infeasible	—
1.5180469	327.3	192.2	infeasible	21.8	—	infeasible	—	22.9	infeasible	—	45.2	23.2	infeasible	—
1.5335938	320.4	185.7	infeasible	23.7	—	infeasible	—	24.2	infeasible	—	59.0	35.3	infeasible	—
1.5646875	874.3	197.7	infeasible	340.6	—	infeasible	—	579.0	singular	—	581.9	13.4	infeasible	—
1.6268750	321.9	182.3	infeasible	25.1	—	infeasible	—	27.3	infeasible	—	55.6	30.6	infeasible	—
1.7512500	336.7	181.1	infeasible	31.0	—	infeasible	—	32.1	infeasible	—	52.9	21.8	infeasible	—
Pyruvate														
1.0050000	314.8	188.3	5.012927e-03	18.6	1.8e+13	+1.7e-05	4.0e-08	35.1	+3.3e-09	1.9e-11	48.1	—	-1.8e-17	4.8e-17
1.2537500	304.2	180.2	2.777866e-03	17.7	1.8e+13	+3.2e-05	2.4e-08	39.4	+2.8e-09	3.8e-12	45.8	—	-4.6e-17	4.9e-17
1.3781250	337.1	199.8	1.141272e-03	19.6	1.8e+13	+6.8e-05	7.6e-08	39.3	+7.3e-09	1.5e-09	57.9	—	-1.3e-17	5.6e-17
1.4403125	310.2	184.7	1.965934e-04	20.2	1.8e+13	+2.5e-04	8.3e-08	33.2	+5.6e-08	1.1e-09	45.8	—	-6.9e-17	3.7e-17
1.4480859	322.5	198.5	7.363938e-05	20.0	1.9e+13	+7.1e-04	2.2e-08	45.1	+1.2e-07	3.7e-12	45.4	—	-4.6e-17	3.4e-17
1.4519727	321.6	196.2	1.704581e-05	19.5	2.2e+13	+3.1e-03	2.8e-08	43.4	+5.2e-07	9.7e-10	45.3	—	-7.4e-17	5.6e-17
1.4529443	309.3	184.1	7.805980e-07	18.7	1.8e+13	+7.1e-02	8.4e-08	36.5	+1.1e-05	9.6e-10	47.6	—	-2.3e-17	5.7e-17
1.4529595	317.3	193.4	7.073102e-07	18.5	1.8e+13	+7.8e-02	2.8e-08	37.1	+1.3e-05	3.7e-12	46.7	—	-3.3e-17	4.3e-17
1.4529633	296.9	173.4	7.046134e-07	18.2	1.8e+13	+7.6e-02	8.4e-08	34.7	+1.6e-05	4.8e-12	44.6	—	-1.0e-17	3.2e-17
1.4529652	323.7	200.0	6.804141e-07	18.6	1.9e+13	+8.3e-02	2.8e-08	33.0	+1.5e-05	9.6e-10	47.0	—	-1.5e-16	3.2e-17
1.4529662	318.1	195.1	6.804141e-07	17.6	1.8e+13	+7.8e-02	8.4e-08	39.8	+1.3e-05	9.6e-10	44.6	—	-1.4e-16	3.2e-17
1.4529671	320.3	188.7	infeasible	16.5	—	infeasible	—	17.2	infeasible	—	98.5	49.1	infeasible	—
1.4529747	328.8	194.5	infeasible	17.8	—	infeasible	—	19.1	infeasible	—	91.6	40.8	infeasible	—
1.4530051	331.5	202.5	infeasible	19.4	—	infeasible	—	21.6	infeasible	—	42.5	22.8	infeasible	—

Table A.5 continued

Instance	SoPlex$_{fac}$ (exact)			CPLEX$_9$ (double)				SoPlex$_{12}$ (80-bit)			SoPlex$_{25}$ (iter. refinement)			
	t	t_{fac}	obj. val.	t	κ	obj. dev.	opt. dist.	t	obj. dev.	opt. dist.	t	t_{inf}	obj. dev.	opt. dist.
1.4530658	314.0	182.3	infeasible	20.4	—	infeasible	—	21.3	infeasible	—	52.9	32.5	infeasible	—
1.4531873	311.8	179.1	infeasible	18.7	—	infeasible	—	30.7	infeasible	—	37.8	18.6	infeasible	—
1.4534302	328.9	192.2	infeasible	19.4	—	infeasible	—	20.8	infeasible	—	50.3	30.5	infeasible	—
1.4539160	318.2	184.0	infeasible	19.9	—	infeasible	—	34.5	infeasible	—	53.1	32.9	infeasible	—
1.4558594	328.5	198.3	infeasible	18.9	—	infeasible	—	20.0	infeasible	—	38.6	19.5	infeasible	—
1.4714063	327.0	197.6	infeasible	19.4	—	infeasible	—	21.1	infeasible	—	45.9	25.9	infeasible	—
1.5025000	323.8	191.9	infeasible	21.7	—	infeasible	—	23.3	infeasible	—	49.5	27.3	infeasible	—

A.2 Mixed-Integer Nonlinear Programming

A.2.1 Optimization-Based Bound Tightening

Table A.6: Detailed results comparing filtering strategies for OBBT at the root node on test set INT. See Table 6.1 for aggregated results.

lp — number of LPs solved during OBBT
lp_{filt} — number of LPs solved during OBBT's filtering
iter — number of LP iterations during OBBT
$iter_{filt}$ — number of LP iterations of OBBT's filtering LPs
b_{obbt} — number of bounds tightened by OBBT
lvb — number of LVBs found by OBBT

	greedy (base)						greedy+filt. off						greedy+filt. aggr.					
Instance	lp	lp_{filt}	iter	$iter_{filt}$	b_{obbt}	lvb	lp	lp_{filt}	iter	$iter_{filt}$	b_{obbt}	lvb	lp	lp_{filt}	iter	$iter_{filt}$	b_{obbt}	lvb
4stufen	54	9	649	3	8	31	84	0	771	0	8	31	43	21	1087	338	8	30
alan	6	0	33	0	5	5	6	0	33	0	5	5	5	2	43	14	5	5
batch0812_nc	76	9	1723	1	40	54	104	0	1716	0	40	60	65	25	2045	352	40	55
batch0812	51	2	672	0	34	44	68	0	679	0	34	52	46	17	720	177	34	45
batchdes	15	0	104	0	9	11	18	0	104	0	9	12	16	8	150	46	9	14
batch_nc	56	10	1947	3	20	36	66	0	1395	0	20	40	43	18	2051	528	20	40
batch	36	0	638	0	16	27	44	0	638	0	16	30	26	10	680	236	16	24
batchs101006m	86	1	1710	1	19	32	98	0	1662	0	19	33	45	28	1816	582	19	25
batchs121208m	105	0	7776	0	23	39	118	0	8006	0	23	40	88	31	10115	2558	23	48
batchs151208m	108	0	3504	0	23	48	124	0	3335	0	23	49	54	43	3514	1795	23	45
batchs201210m	119	2	4635	0	23	47	134	0	4774	0	23	53	64	44	6004	3075	23	52
bchoco05	84	6	1848	12	37	80	118	0	1884	0	37	82	85	27	1972	332	37	78
bchoco06	142	42	6622	90	6	115	192	0	6374	0	6	121	137	50	6798	569	6	112
bchoco08	747	172	117650	2584	60	590	1228	0	72343	0	12	341	755	161	63019	4448	13	254
blend029	36	5	469	4	2	22	52	0	437	0	2	22	24	13	436	289	2	18
blend146	108	13	3729	53	0	54	152	0	4199	0	0	48	69	38	2512	1427	0	43
blend480	142	26	5335	78	0	88	216	0	4954	0	0	88	111	53	4025	1578	0	86
blend531	118	15	2311	48	0	57	186	0	2848	0	0	65	80	42	2270	1115	0	65
blend718	90	6	2087	20	1	43	148	0	2091	0	1	48	66	39	2309	974	1	55
blend721	104	17	2903	95	0	58	152	0	2734	0	0	60	75	47	3178	1698	0	61
blend852	145	14	3981	55	0	72	216	0	4259	0	0	76	94	42	3042	1432	0	70
carton7	39	1	1644	2	4	16	46	0	1881	0	4	16	31	31	2616	2015	4	11
carton9	60	7	1945	19	11	21	62	0	2343	0	11	19	55	28	3739	2005	11	21
casctanks	452	53	22032	22	185	352	692	0	23294	0	185	360	413	118	26492	4349	185	359
cecil.13	108	2	3864	2	43	79	112	0	3864	0	43	79	78	29	6698	3381	43	78
chp_partload	1561	297	146966	137	399	995	1926	0	148278	0	399	1001	1475	378	137553	12581	399	1003
clay0203h	264	140	3895	100	0	235	288	0	4530	0	0	227	247	137	6162	3528	0	233
clay0203m	12	0	135	0	0	0	12	0	135	0	0	0	3	6	160	125	0	0
clay0204h	357	176	5427	121	0	321	384	0	5303	0	0	317	345	172	7806	3355	0	328
clay0204m	15	0	158	0	0	1	16	0	158	0	0	1	6	6	196	175	0	1
clay0205h	489	267	16457	162	0	430	624	0	7621	0	0	441	456	384	12136	7768	0	381
clay0205m	19	0	394	0	0	3	20	0	394	0	0	3	7	7	391	249	0	1
clay0303h	454	249	4629	16	0	332	656	0	4498	0	0	326	322	331	4191	4161	0	315
clay0303m	12	0	174	0	0	1	12	0	174	0	0	1	1	8	133	133	0	1
clay0304h	589	323	9242	82	0	425	880	0	6392	0	0	433	444	448	16901	16052	0	414
clay0304m	16	0	236	0	0	1	16	0	236	0	0	1	5	11	195	195	0	4

Table A.6 continued

Instance	greedy (base)						greedy+filt. off						greedy+filt. aggr.					
	lp	lp$_{filt}$	iter	iter$_{filt}$	b$_{obbt}$	lvb	lp	lp$_{filt}$	iter	iter$_{filt}$	b$_{obbt}$	lvb	lp	lp$_{filt}$	iter	iter$_{filt}$	b$_{obbt}$	lvb
clay0305h	757	413	15207	269	0	558	1096	0	8800	0	0	562	550	529	26717	22846	0	532
clay0305m	20	1	288	1	0	0	20	0	288	0	0	0	2	9	280	280	0	1
crudeoil_lee1_05	91	34	3787	70	32	50	128	0	3830	0	33	52	86	49	4794	1581	34	52
crudeoil_lee1_06	117	48	8147	150	21	59	160	0	7376	0	22	60	104	61	8782	2291	21	64
crudeoil_lee1_07	135	50	10008	116	25	72	192	0	9692	0	25	74	116	66	11580	3242	25	78
crudeoil_lee1_08	176	69	12957	140	28	95	224	0	12791	0	30	94	150	83	15563	3383	32	96
crudeoil_lee1_09	192	83	18964	201	36	104	256	0	16841	0	36	107	173	97	18111	4484	35	100
crudeoil_lee1_10	195	64	18502	161	42	115	280	0	20539	0	44	119	187	105	26933	7243	43	117
crudeoil_lee2_05	158	48	20051	86	39	86	222	0	20192	0	39	92	139	74	26227	5963	39	87
crudeoil_lee2_06	212	64	30629	130	55	120	294	0	33766	0	55	115	189	89	44592	10164	55	119
crudeoil_lee2_07	246	81	46873	261	64	139	356	0	53951	0	64	150	208	85	59249	10089	64	142
crudeoil_lee2_08	304	122	62279	373	74	172	418	0	63449	0	74	178	256	113	74801	17040	74	174
crudeoil_lee2_09	342	117	99020	303	88	210	480	0	94021	0	88	202	298	132	115932	19841	88	202
crudeoil_lee2_10	378	123	110685	434	97	209	542	0	102334	0	97	236	341	151	108023	16769	97	229
crudeoil_lee3_05	289	91	18775	172	120	181	436	0	20743	0	120	200	268	121	23691	5248	120	192
crudeoil_lee3_06	385	125	61367	385	173	256	582	0	59409	0	173	278	386	172	67223	11979	173	257
crudeoil_lee3_07	473	146	65455	388	226	340	728	0	65310	0	226	327	464	180	67697	9114	226	335
crudeoil_lee3_08	556	157	154527	420	287	400	874	0	147336	0	287	403	548	192	155060	15434	287	406
crudeoil_lee3_09	653	177	199420	465	341	472	1020	0	192312	0	341	497	645	215	198085	12122	341	465
crudeoil_lee3_10	743	212	264413	897	397	550	1166	0	271953	0	397	577	721	233	258742	15557	397	537
crudeoil_lee4_05	194	49	22919	137	75	126	294	0	21797	0	75	138	176	73	32441	8285	75	134
crudeoil_lee4_06	247	55	54350	102	100	170	370	0	55073	0	100	176	233	83	57108	7687	100	165
crudeoil_lee4_07	311	85	74047	180	117	204	446	0	70032	0	117	199	275	102	92844	15771	117	201
crudeoil_lee4_08	351	82	95270	146	126	228	522	0	95480	0	126	239	299	92	100573	8046	126	231
crudeoil_lee4_09	417	102	132726	268	150	264	598	0	122505	0	150	267	360	121	142073	13099	150	272
crudeoil_lee4_10	430	107	157458	330	157	295	674	0	146931	0	157	298	382	122	146034	10315	157	289
crudeoil_li01	95	6	2553	8	36	32	120	0	2454	0	36	35	91	22	3669	1056	36	41
crudeoil_li02	32	6	1550	2	13	28	34	0	1550	0	13	29	30	19	2036	730	13	25
crudeoil_li03	254	31	12160	13	21	33	288	0	13074	0	21	27	175	81	19071	9094	21	27
crudeoil_li05	236	22	7467	229	13	22	288	0	7215	0	13	22	107	61	11176	6545	13	21
crudeoil_li06	255	25	11183	16	14	25	288	0	11466	0	14	25	176	67	14803	5176	14	24
crudeoil_li11	257	26	14927	16	19	21	288	0	14820	0	19	21	174	69	20897	7808	19	20
crudeoil_li21	261	31	31300	362	17	11	288	0	32063	0	17	14	188	63	29243	10207	17	13
csched1a	16	1	104	0	3	7	22	0	104	0	3	7	8	9	162	90	3	8
csched1	15	1	130	1	1	4	22	0	130	0	1	4	7	11	163	95	1	5
csched2a	125	28	6817	12	10	70	172	0	6128	0	10	71	109	101	7454	5651	10	66
csched2	152	64	1306	55	1	63	172	0	1481	0	1	66	105	109	1622	1390	1	64
densitymod	316	0	184553	0	315	315	420	0	190555	0	315	315	315	7	188504	3427	315	315
du-opt5	17	0	240	0	4	5	36	0	248	0	4	6	5	9	311	157	4	5
du-opt	17	0	184	0	4	5	40	0	203	0	4	0	5	9	282	176	4	5
edgecross10-060	52	0	2082	10	0	0	56	0	1996	0	0	0	37	43	1368	1368	0	0
edgecross14-039	161	4	7492	0	0	0	168	0	7492	0	0	0	95	100	5040	5040	0	0
elf	11	2	286	5	0	5	12	0	286	0	0	5	7	10	314	281	0	6
eniplac	65	2	369	5	0	23	94	0	383	0	0	23	27	30	418	394	0	21
enpro48pb	42	0	1204	0	10	14	58	0	1341	0	10	14	26	16	1860	773	10	14
enpro56pb	39	0	1667	0	8	17	48	0	1667	0	8	17	32	5	1823	350	8	14

Table A.6 continued

Instance	greedy (base)						greedy+ filt. off						greedy+filt. aggr.					
	lp	lp_{filt}	iter	$iter_{filt}$	b_{obbt}	lvb	lp	lp_{filt}	iter	$iter_{filt}$	b_{obbt}	lvb	lp	lp_{filt}	iter	$iter_{filt}$	b_{obbt}	lvb
ethanolh	159	30	2729	28	5	73	228	0	2248	0	5	85	143	63	2184	828	5	74
ethanolm	81	10	700	9	7	37	124	0	711	0	7	43	58	34	805	443	7	40
ex1224	8	0	67	0	4	6	8	0	67	0	4	6	8	6	130	70	4	6
ex1233	75	1	191	0	16	26	136	0	289	0	16	33	64	33	624	520	16	36
ex1243	42	4	85	4	0	5	56	0	86	0	0	4	33	23	41	41	0	15
ex1244	59	3	200	0	2	12	80	0	256	0	2	10	48	25	286	200	2	10
ex1252a	32	14	209	7	1	17	58	0	217	0	1	20	24	26	271	262	1	15
ex1252	32	17	500	20	3	16	58	0	287	0	3	25	28	29	413	361	3	21
ex1263a	22	1	112	1	12	16	40	0	108	0	12	15	20	5	102	34	12	15
ex1263	26	2	362	1	11	15	40	0	405	0	11	17	23	7	345	145	11	14
ex1264a	23	0	90	0	9	9	40	0	108	0	9	12	17	5	106	44	9	12
ex1264	24	1	198	0	8	13	40	0	199	0	8	14	17	9	250	140	8	14
ex1265a	40	0	231	0	12	16	60	0	292	0	12	20	27	5	303	83	12	14
ex1265	36	3	251	2	6	15	60	0	410	0	6	16	23	22	216	182	6	17
ex1266a	52	0	223	0	21	25	84	0	234	0	21	23	41	9	213	78	21	23
ex1266	55	1	770	1	19	26	84	0	913	0	19	26	43	25	1382	1016	19	34
ex4	8	0	238	0	4	5	10	0	238	0	4	9	5	6	273	119	4	5
fac1	23	11	70	14	0	8	32	0	93	0	0	9	11	15	115	90	0	9
fac2	48	21	156	25	0	37	108	0	191	0	0	33	46	23	164	58	0	35
fac3	69	17	338	16	0	36	108	0	473	0	0	29	55	14	360	35	0	36
feedtray	372	57	8258	71	241	115	412	0	7902	0	241	104	396	136	10509	3443	241	174
fin2bb	84	0	2975	0	61	61	84	0	2975	0	61	61	84	4	3943	1131	61	61
flay02h	3	0	74	0	2	2	4	0	61	0	2	2	2	11	195	181	2	2
flay02m	4	0	61	0	2	2	4	0	61	0	2	2	2	3	67	60	2	2
flay03h	6	0	213	0	3	4	6	0	213	0	3	4	4	14	718	509	3	4
flay03m	6	0	72	0	3	3	6	0	72	0	3	3	3	3	87	68	3	3
flay04h	8	0	234	0	4	1	8	0	234	0	4	1	9	24	314	252	4	0
flay04m	8	0	80	0	0	4	8	0	80	0	0	4	4	8	106	106	0	6
flay05h	10	0	1071	0	0	6	10	0	1071	0	0	6	6	4	1051	626	0	6
flay05m	9	0	103	0	0	5	10	0	103	0	0	5	5	3	151	113	0	5
flay06h	12	0	1776	0	0	5	12	0	1776	0	0	5	6	4	1206	1095	0	4
flay06m	12	0	253	0	0	2	12	0	253	0	0	2	6	10	207	207	0	6
fo7_2	27	4	101	4	0	1	28	0	103	0	0	1	17	7	183	80	0	1
fo7.ar2_1	20	0	98	0	0	14	28	0	118	0	0	14	14	17	159	86	0	14
fo7.ar25_1	19	0	149	0	0	4	28	0	129	0	0	4	9	17	203	108	0	4
fo7.ar3_1	19	0	103	0	0	1	28	0	89	0	0	1	7	15	156	86	0	1
fo7.ar4_1	17	0	94	0	0	5	28	0	96	0	0	7	10	21	128	83	0	5
fo7.ar5_1	23	1	107	1	0	6	28	0	138	0	0	6	17	19	185	107	0	6
fo7	20	1	183	2	0	1	28	0	224	0	0	1	12	3	336	182	0	1
fo8.ar2_1	22	0	99	0	0	16	32	0	91	0	0	16	16	19	146	78	0	16
fo8.ar25_1	25	0	123	0	0	7	32	0	138	0	0	7	12	17	223	109	0	7
fo8.ar3_1	18	0	106	0	0	4	32	0	129	0	0	4	10	18	159	88	0	4
fo8.ar4_1	18	2	114	0	0	7	32	0	120	0	0	8	15	23	221	130	0	11
fo8.ar5_1	25	2	122	0	0	6	32	0	130	0	0	6	16	24	172	86	0	6
fo8	26	0	240	0	0	1	32	0	297	0	0	1	15	6	338	150	0	1
fo9.ar2_1	25	0	58	0	0	18	36	0	63	0	0	18	18	19	178	75	0	18

Table A.6 continued

instance	greedy (base)						greedy+filt. off						greedy+filt. aggr.					
	lp	lp$_{filt}$	iter	iter$_{filt}$	b$_{obbr}$	lvb	lp	lp$_{filt}$	iter	iter$_{filt}$	b$_{obbr}$	lvb	lp	lp$_{filt}$	iter	iter$_{filt}$	b$_{obbr}$	lvb
fo9.ar25_1	24	0	104	0	0	7	36	0	116	0	0	7	11	18	223	115	0	7
fo9.ar3_1	25	0	160	0	0	4	36	0	134	0	0	4	10	27	306	159	0	4
fo9.ar4_1	24	1	178	0	0	7	36	0	178	0	0	8	17	21	311	147	0	7
fo9.ar5_1	30	2	153	0	0	8	36	0	146	0	0	8	20	27	214	97	0	8
fo9	31	0	266	0	0	0	36	0	318	0	0	0	12	6	368	232	0	0
fuel	11	1	192	1	4	8	12	0	227	0	4	8	9	4	225	76	4	9
fuzzy	123	19	5363	5	80	83	158	0	6291	0	80	83	118	31	10540	5868	80	84
gams01	236	8	120856	13	118	141	282	0	121454	0	118	144	240	16	143900	13548	118	139
gasprod_sarawak01	37	11	511	4	6	32	42	0	577	0	6	33	36	21	729	228	6	32
gasprod_sarawak16	572	149	30238	70	113	494	672	0	30937	0	113	543	528	271	29203	3985	113	483
gasprod_sarawak81	2899	813	265274	608	337	2443	3402	0	253840	0	337	2712	2715	1425	258950	24616	337	2456
gastrans	59	2	589	0	35	32	64	0	589	0	35	34	47	12	753	359	35	34
gear2	7	4	44	3	0	4	10	0	65	0	0	5	7	9	71	32	0	4
gear3	6	2	17	0	1	5	10	0	34	0	1	6	8	11	45	26	1	5
gear4	6	2	30	1	2	2	8	0	40	0	2	2	8	10	51	31	2	4
gear	6	2	17	0	1	5	10	0	34	0	1	6	8	11	45	26	1	5
genpooling_lee1	26	6	130	0	0	3	40	0	87	0	0	3	18	15	161	119	0	5
genpooling_lee2	29	3	88	4	0	9	48	0	130	0	0	11	14	12	140	114	0	9
genpooling_meyer04	34	1	1231	11	0	7	56	0	1481	0	0	4	16	7	921	559	0	10
genpooling_meyer10	135	5	6396	19	0	49	260	0	3783	0	0	52	94	14	2498	927	0	55
genpooling_meyer15	270	26	4818	92	0	106	540	0	5344	0	0	107	134	141	4078	3765	0	98
ghg_1veh	52	14	457	7	3	19	78	0	455	0	3	19	33	23	706	387	3	20
ghg_2veh	113	44	1918	21	27	59	150	0	2036	0	27	63	107	60	2741	1184	27	64
ghg_3veh	191	65	4903	45	46	107	256	0	4921	0	46	112	179	80	5614	1556	46	115
gkocis	4	0	19	0	2	3	4	0	19	0	2	3	3	2	26	15	2	3
heatexch_gen1	115	32	610	42	0	58	168	0	731	0	0	67	100	88	826	619	0	64
heatexch_gen3	747	112	1868	114	0	170	1050	0	2025	0	0	221	340	229	1984	1806	0	202
heatexch_spec1	94	31	373	40	1	21	120	0	438	0	1	31	46	51	506	506	1	33
heatexch_spec2	84	16	458	26	10	39	180	0	480	0	10	66	62	49	904	666	10	43
heatexch_spec3	442	86	1643	210	0	89	692	0	2844	0	0	116	201	162	2532	2189	0	90
heatexch_trigen	112	31	944	24	16	59	224	0	1001	0	16	92	70	49	805	363	16	55
hybriddynamic_var	52	1	381	1	22	25	58	0	379	0	22	25	33	10	440	151	22	25
hydroenergy1	154	13	1271	14	13	78	184	0	1231	0	13	93	118	59	1845	1109	13	76
hydroenergy2	284	33	2135	23	13	108	368	0	2217	0	13	164	169	93	3541	2402	13	98
hydroenergy3	480	89	5007	57	13	274	644	0	5005	0	13	365	360	197	9246	5012	13	259
jit1	8	0	67	0	6	6	8	0	8	0	6	6	8	2	133	74	6	6
johnall	7378	0	258635	0	7297	7318	7606	0	255913	0	7297	7319	7378	2	275737	4479	7297	7318
kport20	172	27	718	42	8	32	222	0	690	0	8	34	52	50	945	769	8	35
kport40	310	46	3692	34	0	94	470	0	3603	0	0	103	239	175	3959	1988	0	103
lip	8	0	103	0	8	7	8	0	103	0	8	7	8	2	103	0	8	7
m3	9	0	27	0	0	2	12	0	27	0	0	2	6	5	54	31	0	3
m6	18	1	103	1	0	3	24	0	103	0	0	3	15	8	238	110	0	3
m7_ar2_1	14	1	84	0	0	4	28	0	102	0	0	4	11	14	134	82	0	4
m7_ar25_1	18	1	99	0	0	3	28	0	105	0	0	3	11	14	141	81	0	3
m7_ar3_1	22	2	124	0	0	6	28	0	124	0	0	6	18	24	209	100	0	6
m7_ar4_1	19	1	153	1	0	7	28	0	153	0	0	7	17	26	255	156	0	7

Table A.6 continued

Instance	greedy (base)						greedy+ filt. off						greedy+filt. aggr.					
	lp	lp_{filt}	iter	$iter_{filt}$	b_{obtr}	lvb	lp	lp_{filt}	iter	$iter_{filt}$	b_{obtr}	lvb	lp	lp_{filt}	iter	$iter_{filt}$	b_{obbt}	lvb
m7.ar5.1	23	3	124	0	0	6	28	0	124	0	0	6	21	26	282	123	0	5
m7	16	0	139	0	0	1	28	0	132	0	0	1	12	3	176	91	0	1
meanvarxsc	10	0	46	0	0	8	14	0	46	0	0	10	7	12	42	38	0	7
milinfract	494	0	3759	0	0	1	1002	0	3800	0	0	1	2	61	7543	7543	0	2
minlphix	23	4	36	5	6	8	28	0	44	0	6	8	22	10	96	72	6	8
multiplants.mtg1a	56	6	8610	2	25	41	64	0	8610	0	25	41	55	24	7940	2311	25	47
multiplants.mtg1c	48	8	13566	3	19	27	64	0	10497	0	19	27	42	22	20697	10534	19	23
multiplants.mtg5	70	5	1654	6	33	44	80	0	1936	0	33	50	64	30	2504	1050	33	52
multiplants.stg1a	145	23	8303	34	8	48	234	0	9286	0	8	47	118	78	7042	5157	8	52
multiplants.stg1b	215	70	32291	109	8	37	266	0	38048	0	8	28	150	61	45667	14351	8	89
multiplants.stg1c	126	19	10159	29	33	41	202	0	15472	0	8	56	93	57	15392	10551	0	54
multiplants.stg1	212	75	31854	9483	8	91	330	0	23178	0	8	90	190	79	36124	19710	8	106
multiplants.stg6	217	42	48181	113	9	87	306	0	49075	0	9	82	172	74	101441	34886	9	88
ndcc12	836	5	19748	4	97	319	1172	0	19314	0	97	327	621	21	20442	1720	97	326
ndcc12persp	134	0	2002	0	4	76	184	0	1801	0	4	68	131	6	1807	929	4	80
ndcc13	837	1	20898	0	210	405	1176	0	20675	0	210	441	634	22	22759	3354	210	409
ndcc13persp	129	0	2067	0	72	86	168	0	1854	0	72	88	123	14	2200	1407	72	93
ndcc14	1138	4	37164	4	82	462	1620	0	39805	0	82	433	856	25	40173	6632	82	478
ndcc14persp	160	1	3123	1	2	88	216	0	2970	0	2	87	160	8	3728	1623	2	97
ndcc15	880	1	46475	0	87	361	1220	0	45696	0	87	363	664	31	50879	3580	87	382
ndcc16	1430	1	51437	1	60	535	2040	0	56198	0	60	512	1058	16	59065	6539	60	518
ndcc16persp	177	0	4552	0	0	58	240	0	3954	0	0	65	176	4	4424	1236	0	66
netmod_dol1	10	2	18	0	0	4	12	0	17	0	0	3	6	5	20	7	0	4
netmod_dol2	10	4	3547	5	0	2	12	0	3754	0	0	3	6	8	7705	6098	0	4
netmod_kar1	8	2	14	2	0	1	8	0	14	0	0	1	8	4	14	2	0	1
netmod_kar2	8	2	14	2	0	1	8	0	14	0	0	1	8	4	14	2	0	1
no7_ar2.1	20	0	113	0	0	14	28	0	117	0	0	14	14	17	167	99	0	14
no7_ar25.1	19	0	114	0	0	4	28	0	93	0	0	4	8	17	222	118	0	4
no7_ar3.1	27	0	137	0	0	1	28	0	137	0	0	1	15	18	250	121	0	1
no7_ar4.1	23	0	144	0	0	6	28	0	142	0	0	7	15	21	241	129	0	7
no7_ar5.1	22	2	188	0	0	7	28	0	188	0	0	7	19	20	285	142	0	6
nous1	54	10	285	13	6	34	76	0	300	0	6	39	45	37	351	189	6	34
nous2	61	19	381	20	7	37	76	0	368	0	7	41	55	43	344	167	7	44
nuclear104	2234	39	1764	16	29	29	4362	0	2005	0	10	29	2307	130	1683	461	0	29
nuclear14a	618	162	202427	89805	10	373	784	0	198348	0	379	379	591	209	227582	190963	10	376
nuclear14b	613	95	66415	4155	203	368	784	0	57354	0	203	383	605	122	83825	28942	203	380
nuclear14	417	0	325	0	0	0	820	0	654	0	0	0	410	6	958	61	0	0
nuclear25a	597	148	237521	137860	7	372	816	0	243772	0	7	376	594	216	243450	232252	7	372
nuclear25b	763	155	104529	2182	215	409	816	0	88376	0	215	428	767	179	123659	27889	215	411
nuclear25	435	1	386	1	0	0	856	0	947	0	0	0	429	5	810	49	0	0
nuclear49a	1390	361	94723	36372	8	809	1782	0	56127	0	8	775	1306	396	227466	173248	8	798
nuclear49b	1665	311	295948	9365	458	916	1782	0	326493	0	458	940	1865	581	530195	255470	458	1076
nuclear49	951	10	673	5	0	7	1866	0	2600	0	0	0	942	13	2327	98	0	7
nuclearva	189	2	132	2	0	6	366	0	201	0	0	0	185	6	290	20	0	0
nuclearvb	188	1	121	1	0	0	366	0	130	0	0	0	184	5	284	21	0	0
nuclearvc	188	1	121	1	0	0	366	0	150	0	0	0	184	5	274	21	0	0

Table A.6 continued

Instance	greedy (base)						greedy+filt. off						greedy+filt. aggr.					
	lp	lp_{filt}	iter	$iter_{filt}$	b_{obbt}	lvb	lp	lp_{filt}	iter	$iter_{filt}$	b_{obbt}	lvb	lp	lp_{filt}	iter	$iter_{filt}$	b_{obbt}	lvb
nuclearvd	188	1	135	1	0	0	366	0	255	0	0	0	184	5	203	21	0	0
nuclearve	188	1	135	1	0	0	366	0	255	0	0	0	184	5	203	21	0	0
nuclearvf	188	1	135	1	0	0	366	0	255	0	0	0	184	5	203	21	0	0
nvs01	11	1	110	1	3	5	14	0	110	0	3	5	11	10	213	107	3	5
nvs02	5	0	21	0	0	0	10	0	20	0	0	0	1	4	39	25	0	0
nvs05	25	1	111	1	2	7	42	0	120	0	2	7	20	6	205	74	2	6
nvs06	9	0	27	0	5	7	18	0	42	0	5	9	8	10	41	30	5	8
nvs08	5	0	36	0	1	1	6	0	36	0	1	1	2	7	48	33	1	1
nvs10	4	0	5	0	4	4	4	0	5	0	4	4	4	2	11	17	4	4
nvs11	6	0	36	0	6	6	6	0	36	0	6	6	6	4	41	25	6	6
nvs12	8	0	51	0	8	8	8	0	51	0	8	8	8	2	63	20	8	8
nvs13	10	0	53	0	8	8	10	0	53	0	8	8	10	2	85	22	8	8
nvs15	4	0	20	0	2	2	4	0	20	0	2	2	3	2	31	20	2	2
nvs16	13	4	42	4	0	0	14	0	51	0	0	0	11	16	65	59	0	0
nvs17	14	0	112	0	11	11	12	0	112	0	11	11	13	2	134	29	11	11
nvs18	12	0	90	0	10	10	16	0	90	0	10	10	11	2	108	23	10	10
nvs19	16	0	176	0	12	12	16	0	176	0	12	12	13	5	207	63	12	12
nvs20	43	0	1660	0	33	33	64	0	1751	0	33	35	34	11	1967	532	33	33
nvs21	5	0	5	0	0	2	5	0	11	0	0	5	3	7	5	5	0	2
nvs23	17	0	249	0	13	13	18	0	249	0	13	13	14	3	272	36	13	13
nvs24	20	0	349	0	12	12	20	0	349	0	12	12	16	3	358	59	12	12
o7_2	23	0	118	0	0	0	28	0	123	0	0	0	13	6	200	118	0	0
o7_ar2_1	20	0	173	0	0	14	28	0	172	0	0	14	14	15	290	147	0	14
o7_ar25_1	26	0	143	0	0	4	28	0	143	0	0	4	15	14	254	110	0	4
o7_ar3_1	27	0	145	0	0	1	28	0	151	0	0	1	14	14	286	135	0	1
o7_ar4_1	23	0	182	0	0	6	28	0	162	0	0	7	14	20	271	120	0	6
o7_ar5_1	25	2	117	0	0	6	28	0	117	0	0	6	19	20	239	117	0	6
o7	21	0	83	0	0	0	28	0	102	0	0	0	8	4	137	87	0	0
o8_ar4_1	22	1	232	1	0	8	32	0	216	0	0	8	14	24	285	162	0	8
o9_ar4_1	27	0	211	0	0	6	36	0	194	0	0	8	17	29	318	143	0	7
oil2	1081	0	10787	0	839	1073	1098	0	10799	0	839	1078	1169	100	12697	2017	839	1163
oil	1370	104	11555	52	863	616	1430	0	12384	0	863	618	1335	190	12863	3843	863	627
ortez	51	12	922	21	28	67	66	0	1022	0	28	39	37	17	1245	386	28	36
pooling_epa1	128	57	5547	34	32	95	170	0	6009	0	32	78	112	73	6177	1475	32	67
pooling_epa2	201	74	8024	86	96	283	296	0	10156	0	96	109	168	102	7774	1926	96	92
pooling_epa3	620	194	91900	235	10	283	984	0	101028	0	10	299	430	252	98390	14640	10	290
portfol_card	17	0	25	0	10	17	18	0	24	0	10	18	19	22	120	74	10	19
portfol_classical050_1	77	0	536	0	59	61	100	0	536	0	59	61	77	2	600	72	59	61
portfol_classical200_2	355	0	3041	0	263	271	400	0	3041	0	263	271	356	2	3248	120	263	271
portfol_robust050_34	178	0	1453	0	168	172	200	0	1453	0	168	172	185	16	1767	297	168	180
portfol_robust100_09	358	0	3493	0	334	337	402	0	3493	0	334	337	357	3	3484	127	334	337
portfol_robust200_03	752	0	6259	0	703	708	802	0	6259	0	703	708	751	2	6544	166	703	708
portfol_shortfall050_68	200	11	605	0	188	200	200	0	605	0	188	200	200	13	641	50	188	200
portfol_shortfall100_04	404	0	2237	0	404	404	404	0	2237	0	404	404	404	6	2255	199	404	404
portfol_shortfall200_05	804	0	5178	0	804	804	804	0	5178	0	804	804	804	2	5182	89	804	804
primary	117	9	1664	2	47	68	192	0	1664	0	47	73	117	26	2137	660	47	66

Table A.6 continued

Instance	greedy (base)						greedy+filt. off						greedy+filt. aggr.					
	lp	lp_filt	iter	iter_filt	b_obbt	lvb	lp	lp_filt	iter	iter_filt	b_obbt	lvb	lp	lp_filt	iter	iter_filt	b_obbt	lvb
procsel	2	0	6	0	2	2	4	0	15	0	2	3	2	2	6	6	2	2
product2	255	2	2123	1	157	211	318	0	2065	0	157	213	231	23	3318	1671	157	213
product	215	31	2819	20	0	32	326	0	2877	0	0	47	37	47	5047	4997	0	27
qapw	450	0	22954	0	450	450	450	0	22954	0	450	450	450	2	23189	212	450	450
ravempb	41	0	976	0	3	12	56	0	976	0	3	12	22	12	1958	1121	3	11
risk2bpb	6	0	285	0	3	3	6	0	285	0	3	3	3	3	241	223	3	3
routingdelay_bigm	3941	37	426	8	46	174	4110	0	427	0	46	267	3606	142	450	441	46	176
routingdelay_proj	4080	20	351	1	0	135	4246	0	355	0	0	239	3743	142	338	338	0	133
rsyn0805h	18	0	31	0	3	7	24	0	31	0	3	5	8	7	50	50	3	5
rsyn0805m02h	27	4	351	2	4	17	36	0	352	0	4	23	23	13	626	389	4	20
rsyn0805m02m	11	0	116	0	0	4	12	0	116	0	0	4	7	9	99	91	0	5
rsyn0805m03m	18	1	269	1	0	11	18	0	269	0	0	11	14	14	446	312	0	10
rsyn0805m04m	23	1	499	1	0	13	24	0	499	0	0	13	16	12	792	394	0	15
rsyn0805m	5	0	43	0	0	3	6	0	35	0	0	1	2	6	40	40	0	2
rsyn0810m02m	22	0	159	0	0	9	24	0	207	0	0	8	11	9	268	219	0	10
rsyn0810m03m	33	2	470	3	0	16	36	0	437	0	0	13	16	21	433	396	0	10
rsyn0810m04m	45	4	872	5	0	19	48	0	869	0	0	17	26	15	1100	601	0	17
rsyn0810m	9	0	62	0	0	3	12	0	63	0	0	2	6	10	100	94	0	5
rsyn0815m02m	39	3	589	3	0	15	48	0	592	0	0	16	22	19	1045	783	0	13
rsyn0815m03m	61	1	2026	0	2	21	72	0	1917	0	2	20	39	16	2737	1359	2	20
rsyn0815m04m	83	0	2744	0	3	27	96	0	2322	0	3	22	45	21	4672	3035	3	31
rsyn0815m	19	2	139	3	0	4	24	0	157	0	0	2	11	9	340	162	0	8
rsyn0820m02m	50	2	862	1	0	21	60	0	821	0	0	21	31	15	1152	612	0	18
rsyn0820m03m	68	4	1476	10	2	24	90	0	1467	0	2	21	46	15	2368	1175	2	28
rsyn0820m04m	102	1	1917	1	3	29	120	0	1846	0	3	33	58	34	3956	2616	3	33
rsyn0820m	26	0	190	1	0	7	30	0	193	0	0	7	13	5	288	110	3	7
rsyn0830m02m	62	4	830	3	3	25	84	0	794	0	3	23	42	15	1367	774	3	29
rsyn0830m03m	98	8	1428	9	0	34	126	0	1398	0	0	37	59	27	2662	1476	2	41
rsyn0830m04m	136	7	3120	7	4	53	168	0	2719	0	4	48	84	19	3204	1303	4	50
rsyn0830m	29	0	288	0	0	13	42	0	303	0	0	14	21	10	411	237	0	11
rsyn0840m02m	91	3	1115	8	0	36	116	0	1093	0	0	38	61	22	1697	1074	0	34
rsyn0840m03m	145	7	2006	15	1	51	174	0	2281	0	1	53	95	33	3386	1823	1	56
rsyn0840m04m	190	6	3313	6	3	53	232	0	3596	0	3	57	112	22	3766	1479	3	68
rsyn0840m	42	0	366	0	1	15	58	0	369	0	1	16	29	14	607	369	1	17
sep1	10	0	61	0	6	8	10	0	61	0	6	8	10	3	63	24	6	8
sepasequ_complex	653	101	12396	102	53	210	984	0	13587	0	53	267	331	145	18849	4202	53	223
sfacloc1_2_80	162	15	1415	13	8	42	198	0	1385	0	8	39	109	60	1893	1031	8	40
sfacloc1_2_90	152	24	1441	28	8	23	184	0	1563	0	8	28	102	64	1428	819	8	29
sfacloc1_2_95	153	31	1828	61	2	16	178	0	1471	0	2	16	86	94	2080	1560	2	18
sfacloc1_3_80	232	34	2585	38	0	53	306	0	2833	0	0	55	158	90	3226	1755	0	57
sfacloc1_3_90	229	46	3472	87	0	35	300	0	4144	0	0	44	145	96	4583	2089	0	43
sfacloc1_3_95	225	56	2068	90	1	40	312	0	2430	0	1	44	164	174	3136	2398	1	50
sfacloc1_4_80	313	36	4018	33	0	102	426	0	3493	0	0	104	195	157	4233	2966	0	75
sfacloc1_4_90	293	33	4758	37	0	58	436	0	4832	0	0	60	173	125	5487	2377	0	47
sfacloc1_4_95	313	73	3261	114	0	49	426	0	3193	0	0	42	239	209	3452	2292	0	57
sfacloc2_2_80	125	11	1903	10	6	40	150	0	1941	0	0	36	69	74	1506	1090	0	40

Table A.6 continued

Instance	greedy (base)						greedy+filt. off						greedy+filt. aggr.					
	lp	lp$_{filt}$	iter	iter$_{filt}$	b$_{obbt}$	lvb	lp	lp$_{filt}$	iter	iter$_{filt}$	b$_{obbt}$	lvb	lp	lp$_{filt}$	iter	iter$_{filt}$	b$_{obbt}$	lvb
sfacloc2.2.90	121	6	1613	8	0	35	150	0	1211	0	0	33	70	75	1332	1118	0	41
sfacloc2.2.95	63	6	1059	4	0	18	80	0	1001	0	0	14	35	40	696	630	0	19
sfacloc2.3.80	60	0	1463	0	0	51	120	0	1610	0	0	50	54	3	1666	395	0	54
sfacloc2.3.90	60	0	1087	0	0	48	120	0	1312	0	0	49	54	6	1324	684	0	54
sfacloc2.3.95	32	0	996	0	0	23	64	0	777	0	0	25	29	3	898	188	0	24
sfacloc2.4.80	75	0	2130	0	0	64	150	0	2448	0	0	67	69	6	3315	480	0	67
sfacloc2.4.90	75	0	1531	0	0	69	150	0	1815	0	0	67	69	3	1522	452	0	69
sfacloc2.4.95	40	0	1107	0	0	33	80	0	1255	0	0	33	37	3	1415	317	0	33
slay04h	14	1	636	5	4	9	16	0	636	0	4	9	11	12	814	751	4	9
slay04m	16	1	231	1	3	4	16	0	231	0	3	4	7	10	299	225	3	4
slay05h	18	0	973	0	5	11	20	0	973	0	5	11	19	19	1594	1486	5	19
slay05m	19	0	390	0	0	0	20	0	390	0	0	0	5	6	322	299	0	0
slay06h	24	0	1912	1	6	18	24	0	1912	0	6	18	27	30	4077	3857	6	24
slay06m	22	1	508	0	0	0	24	0	508	0	0	0	0	6	345	345	0	0
slay07h	28	0	3397	0	3	20	28	0	3397	0	3	20	25	23	7459	6995	3	22
slay07m	27	0	653	0	0	0	28	0	653	0	0	0	1	7	560	560	3	1
slay08h	32	1	2707	1	4	27	32	0	2707	0	4	27	34	33	5506	5323	4	34
slay08m	32	1	1191	1	0	1	32	0	1191	0	0	1	0	6	620	620	0	0
slay09h	36	1	5282	0	5	28	36	0	5282	0	5	28	36	38	12896	12731	5	33
slay09m	30	0	1419	0	0	0	36	0	1419	0	0	0	1	7	879	879	0	1
slay10h	40	2	5700	0	5	35	40	0	5700	0	5	35	42	45	30797	30142	5	39
slay10m	38	1	1797	1	5	0	40	0	1797	0	5	0	0	6	1412	1412	5	0
smallinvDAXr1b010-011	13	0	87	0	8	8	16	0	87	0	8	8	12	2	83	19	8	8
smallinvDAXr1b020-022	15	0	76	0	8	8	16	0	76	0	8	8	12	3	85	23	8	8
smallinvDAXr1b050-055	14	0	88	0	12	12	16	0	88	0	12	12	13	2	90	11	12	12
smallinvDAXr1b100-110	15	0	106	0	13	13	16	0	106	0	13	13	14	2	116	22	13	13
smallinvDAXr1b150-165	34	0	131	0	26	28	60	0	131	0	26	28	30	7	138	20	26	28
smallinvDAXr1b200-220	34	0	131	0	26	28	60	0	131	0	26	28	30	7	140	21	26	28
smallinvDAXr2b010-011	14	0	88	0	7	7	16	0	90	0	7	7	10	4	89	28	7	7
smallinvDAXr2b020-022	15	0	92	0	9	9	16	0	92	0	9	9	12	3	97	20	9	9
smallinvDAXr2b050-055	34	0	160	0	27	27	60	0	158	0	27	27	30	5	178	24	27	27
smallinvDAXr2b100-110	14	0	93	0	12	12	16	0	93	0	12	12	13	2	107	14	12	12
smallinvDAXr2b150-165	15	0	85	0	13	13	18	0	85	0	13	13	14	2	103	21	13	13
smallinvDAXr2b200-220	14	0	92	0	13	13	16	0	92	0	13	13	13	2	100	13	13	13
smallinvDAXr3b010-011	15	0	81	0	7	7	16	0	81	0	7	7	12	3	96	25	7	7
smallinvDAXr3b020-022	14	0	100	0	10	10	16	0	100	0	10	10	13	2	128	26	10	10
smallinvDAXr3b050-055	15	0	95	0	13	13	16	0	95	0	13	13	13	3	107	18	13	13
smallinvDAXr3b100-110	14	0	110	0	13	13	16	0	110	0	13	13	13	2	131	21	13	13
smallinvDAXr3b150-165	14	0	101	0	13	13	16	0	101	0	13	13	13	2	111	17	13	13
smallinvDAXr3b200-220	16	0	99	0	13	13	18	0	99	0	13	13	14	2	112	19	13	13
smallinvDAXr4b010-011	14	0	105	0	7	7	16	0	105	0	7	7	12	2	99	26	7	7
smallinvDAXr4b020-022	15	0	113	0	11	11	16	0	113	0	11	11	13	2	134	22	11	11
smallinvDAXr4b050-055	15	0	136	0	13	13	16	0	136	0	11	11	13	3	154	30	11	11
smallinvDAXr4b100-110	14	0	109	0	13	13	16	0	109	0	13	13	13	2	122	15	13	13
smallinvDAXr4b150-165	13	0	97	0	13	13	14	0	97	0	13	13	13	2	100	11	13	13
smallinvDAXr4b200-220	15	0	102	0	13	13	18	0	102	0	13	13	14	2	115	16	13	13

Table A.6 continued

Instance	greedy (base)						greedy+ filt. off						greedy+filt. aggr.					
	lp	lp_{filt}	iter	$iter_{\mathrm{filt}}$	b_{obbt}	lvb	lp	lp_{filt}	iter	$iter_{\mathrm{filt}}$	b_{obbt}	lvb	lp	lp_{filt}	iter	$iter_{\mathrm{filt}}$	b_{obbt}	lvb
smallinvDAXr5b010-011	13	0	101	0	8	8	16	0	101	0	8	8	11	3	148	48	8	8
smallinvDAXr5b020-022	15	0	128	0	11	11	16	0	126	0	11	11	13	3	131	23	11	11
smallinvDAXr5b050-055	14	0	99	0	13	13	16	0	99	0	13	13	13	2	116	16	13	13
smallinvDAXr5b100-110	16	0	99	0	9	9	18	0	99	0	9	9	14	2	111	14	9	9
smallinvDAXr5b150-165	15	0	98	0	13	13	16	0	98	0	13	13	13	3	105	16	13	13
smallinvDAXr5b200-220	15	0	98	0	13	13	18	0	98	0	13	13	14	2	117	20	13	13
smallinvSNPr1b010-011	114	0	963	0	68	72	198	0	925	0	68	72	98	19	1242	350	68	72
smallinvSNPr1b020-022	112	0	1045	0	67	79	200	0	1025	0	67	79	99	14	1231	210	67	79
smallinvSNPr1b050-055	107	0	1101	0	39	81	200	0	1106	0	39	81	99	8	1180	130	39	81
smallinvSNPr1b100-110	111	0	1232	0	26	79	200	0	1239	0	26	79	99	12	1386	215	26	79
smallinvSNPr1b150-165	112	0	1176	0	31	84	200	0	1173	0	31	84	99	9	1212	118	31	84
smallinvSNPr1b200-220	111	0	1175	0	33	84	200	0	1150	0	33	84	99	15	1370	239	33	84
smallinvSNPr2b010-011	113	0	781	0	75	76	198	0	754	0	75	76	98	16	975	254	75	76
smallinvSNPr2b020-022	111	0	1000	0	47	63	200	0	1074	0	47	63	100	11	1147	166	47	63
smallinvSNPr2b050-055	110	0	1237	0	36	82	200	0	1240	0	36	82	99	11	1390	201	36	82
smallinvSNPr2b100-110	110	0	1106	0	30	81	200	0	1137	0	30	81	99	15	1375	273	30	81
smallinvSNPr2b150-165	109	0	1371	0	22	85	200	0	1350	0	22	85	99	10	1535	188	22	85
smallinvSNPr2b200-220	112	0	1338	0	28	84	200	0	1343	0	28	84	99	17	1570	295	28	84
smallinvSNPr3b010-011	112	0	755	0	68	69	198	0	760	0	68	69	98	14	917	204	68	69
smallinvSNPr3b020-022	109	0	839	0	61	72	200	0	871	0	61	72	99	12	959	178	61	72
smallinvSNPr3b050-055	109	0	997	0	26	73	200	0	979	0	26	73	99	13	1141	205	26	73
smallinvSNPr3b100-110	110	0	1268	0	32	87	200	0	1239	0	32	87	98	11	1378	172	32	87
smallinvSNPr3b150-165	110	0	1151	0	21	85	200	0	1160	0	21	85	99	14	1347	205	21	85
smallinvSNPr3b200-220	114	0	1143	0	13	78	200	0	1126	0	13	78	99	14	1332	233	13	78
smallinvSNPr4b010-011	110	0	719	0	56	57	198	0	732	0	56	57	98	15	941	225	56	57
smallinvSNPr4b020-022	106	0	876	0	43	61	200	0	890	0	43	61	99	15	1063	227	43	61
smallinvSNPr4b050-055	110	0	1062	0	35	80	200	0	1070	0	35	80	99	13	1148	182	35	80
smallinvSNPr4b100-110	110	0	1197	0	18	82	200	0	1219	0	18	82	99	11	1345	179	18	82
smallinvSNPr4b150-165	109	0	1024	0	18	88	200	0	1062	0	18	88	99	15	1148	192	18	88
smallinvSNPr4b200-220	110	0	1308	0	20	89	200	0	1313	0	20	89	99	11	1478	204	20	89
smallinvSNPr5b010-011	110	0	620	0	66	67	196	0	618	0	66	67	97	14	769	169	66	67
smallinvSNPr5b020-022	110	0	798	0	44	60	198	0	847	0	44	60	98	14	1005	193	44	60
smallinvSNPr5b050-055	108	0	747	0	12	65	200	0	712	0	12	65	99	12	838	174	12	65
smallinvSNPr5b100-110	109	0	709	0	35	89	200	0	710	0	35	89	99	13	877	204	35	89
smallinvSNPr5b150-165	110	0	883	0	15	90	200	0	881	0	15	90	99	11	1038	206	15	90
smallinvSNPr5b200-220	111	0	1044	0	34	92	200	0	1055	0	34	92	98	14	1157	185	34	92
space25a	64	19	81	0	0	28	86	0	106	0	0	35	57	34	64	64	0	29
space25	44	112	80	128	0	2	52	0	80	0	0	2	30	4	87	62	0	0
space960	2494	0	804391	0	12	0	3394	0	769976	0	12	0	417	389	120398	101315	12	0
spectra2	36	8	817	14	26	26	60	0	819	0	26	27	26	5	940	201	26	26
sporttournament14	118	2	496	0	0	2	140	0	425	0	0	2	6	14	390	390	0	1
spring	23	3	117	3	7	12	24	0	117	0	7	12	24	7	178	61	7	13
squfl010-025	449	57	14415	65	0	160	500	0	13166	0	0	157	250	12	13753	841	0	163
squfl010-025persp	705	5	89169	2	267	424	1000	0	89133	0	267	422	580	91	89984	4551	267	421
squfl010-040	726	0	38553	0	0	273	800	0	36779	0	0	257	408	65	31586	2590	0	260
squfl010-040persp	1214	78	325905	105	508	712	1600	0	326615	0	508	706	1130	88	332883	6274	508	710

Table A.6 continued

Instance	greedy (base)						greedy+filt. off						greedy+filt. aggr.					
	lp	lp$_{filt}$	iter	iter$_{filt}$	b$_{obbt}$	lvb	lp	lp$_{filt}$	iter	iter$_{filt}$	b$_{obbt}$	lvb	lp	lp$_{filt}$	iter	iter$_{filt}$	b$_{obbt}$	lvb
squfl010-080	1391	32	36181	15	0	493	1600	0	37146	0	0	489	505	525	9573	9573	0	451
squfl010-080persp	2809	136	610351	144	800	1327	3200	0	621749	0	800	1310	1646	230	627460	27848	800	1336
squfl015-060	1718	26	86294	5	0	577	1800	0	96280	0	0	566	930	96	85533	6147	0	556
squfl015-060persp	2909	713	1081371	1657	900	1391	3600	0	1080039	0	900	1423	1878	212	1049697	18371	900	1419
squfl015-080	2236	53	93262	32	0	752	2400	0	92729	0	0	751	1185	133	80626	6777	0	724
squfl015-080persp	3860	322	1377896	380	1200	2008	4800	0	1473240	0	1200	1982	2479	222	1451764	33174	1200	1989
squfl020-040	1551	18	74615	6	0	492	1600	0	68652	0	0	502	824	77	66903	2902	0	510
squfl020-040persp	2265	252	881474	438	1109	1459	3200	0	895950	0	1109	1460	2017	267	921810	20486	1109	1465
squfl020-050	1906	54	138898	8	0	639	2000	0	156945	0	0	627	1012	93	143945	4975	0	632
squfl020-050persp	2877	192	1529480	250	1277	1896	4000	0	1501841	0	1277	1894	2578	263	1526331	52421	1277	1898
squfl020-150persp	9480	1279	4623861	2121	3000	5102	12000	0	4832813	0	3000	5121	6206	402	4896913	109949	3000	5285
squfl025-025	1188	2	59358	3	0	361	1250	0	61586	0	0	374	619	20	53480	1219	0	394
squfl025-025persp	1628	132	616521	276	919	1165	2500	0	603827	0	919	1142	1522	209	608157	8951	919	1162
squfl025-030	1482	13	76202	11	0	441	1500	0	78134	0	0	448	778	40	76950	1847	0	438
squfl025-030persp	1953	240	734069	418	788	1278	3000	0	747281	0	788	1298	1691	219	741719	13266	788	1282
squfl025-040	1939	24	199743	9	0	644	2000	0	184886	0	0	631	1009	85	168250	3927	0	605
squfl025-040persp	2791	302	1024115	419	1353	1845	4000	0	1028077	0	1353	1860	2201	239	1013467	20961	1353	1840
squfl030-100persp	8672	2314	3607947	4100	3000	5138	12000	0	3709486	0	3000	4975	6059	892	3773481	129782	3000	5066
squfl030-150persp	14300	3119	6118433	4888	4500	7839	18000	0	6219033	0	4500	7674	9827	1690	6200935	66619	4500	7780
squfl040-080	5885	159	568778	94	0	1989	6400	0	545366	0	0	1955	3164	152	552823	9776	0	1992
squfl040-080persp	9821	3614	2989762	6934	3200	5337	12800	0	2999354	0	3200	5349	6385	1217	3009751	71812	3200	5490
sssd08-04	14	3	147	0	4	10	16	0	161	0	4	12	10	15	231	129	4	10
sssd08-04persp	27	9	341	19	4	15	32	0	279	0	4	12	24	27	408	260	4	16
sssd12-05	18	4	294	0	0	13	20	0	360	0	0	14	14	16	331	142	0	13
sssd12-05persp	36	11	512	23	5	20	40	0	482	0	5	12	27	28	629	364	5	16
sssd15-04	16	5	293	1	4	12	16	0	293	0	4	12	13	17	375	171	4	12
sssd15-04persp	31	10	424	23	0	15	32	0	468	0	0	10	25	27	625	374	0	17
sssd15-06	20	4	503	0	6	14	24	0	526	0	6	16	17	18	611	223	6	14
sssd15-06persp	39	9	705	19	8	22	48	0	747	0	8	20	32	34	874	498	8	22
sssd15-08	26	6	652	1	8	18	32	0	746	0	8	21	21	20	773	247	8	18
sssd15-08persp	49	12	925	17	8	24	64	0	987	0	8	23	43	45	1370	823	8	30
sssd16-07	24	6	484	1	7	17	28	0	499	0	7	19	19	19	574	208	7	17
sssd16-07persp	48	14	774	22	7	25	56	0	739	0	7	22	35	37	995	560	7	24
sssd18-06	20	6	561	2	6	14	24	0	595	0	6	17	16	17	668	233	6	14
sssd18-06persp	40	13	737	19	6	19	48	0	843	0	6	22	34	39	1160	709	6	24
sssd18-08	28	6	840	0	8	20	32	0	783	0	8	22	24	24	991	360	8	20
sssd18-08persp	54	17	1136	31	8	33	64	0	1134	0	8	30	48	48	1376	801	8	34
sssd20-04	14	5	283	4	0	10	16	0	311	0	0	11	11	14	394	184	0	10
sssd20-04persp	27	7	449	17	4	12	32	0	454	0	4	16	22	25	628	428	4	14
sssd20-08	26	7	675	2	8	18	32	0	634	0	8	21	21	20	774	242	8	18
sssd20-08persp	52	12	915	28	8	29	64	0	923	0	8	23	35	36	1259	733	8	22
sssd22-08	28	7	789	1	8	20	32	0	822	0	8	22	22	19	920	207	8	20
sssd22-08persp	55	17	1029	35	8	26	64	0	949	0	8	30	42	44	1460	843	8	27
sssd25-04	14	4	399	1	4	10	16	0	424	0	4	11	11	14	418	172	4	10
sssd25-04persp	28	8	501	17	4	17	32	0	518	0	4	14	22	23	669	371	4	14
sssd25-08	26	6	814	2	8	18	32	0	831	0	8	22	21	20	897	291	8	18

Table A.6 continued

Instance	greedy (base)						greedy+filt. off						greedy+filt. aggr.					
	lp	lp$_{filt}$	iter	iter$_{filt}$	b$_{obbt}$	lvb	lp	lp$_{filt}$	iter	iter$_{filt}$	b$_{obbt}$	lvb	lp	lp$_{filt}$	iter	iter$_{filt}$	b$_{obbt}$	lvb
sssd25-08persp	51	13	1097	25	8	28	64	0	1023	0	8	26	42	43	1388	774	8	29
st_e29	8	0	67	0	4	6	8	0	67	0	4	6	8	6	130	70	4	6
st_e31	7	0	28	0	2	2	10	0	36	0	2	2	5	4	32	14	2	2
st_e32	63	2	373	2	17	23	94	0	381	0	17	23	39	10	573	244	17	23
st_e36	13	3	111	1	4	3	16	0	110	0	4	5	12	7	213	94	4	3
st_e38	10	1	47	0	8	7	14	0	47	0	8	7	11	6	67	19	8	7
st_e40	23	4	166	2	5	4	24	0	166	0	5	4	20	13	178	87	5	3
stockcycle	72	0	2522	0	0	0	96	0	2325	0	0	0	48	3	2638	150	0	0
st_test8	17	2	12	2	13	15	22	0	12	0	13	18	17	7	17	14	13	15
st_testgr1	10	0	51	2	10	10	10	0	51	0	10	10	10	2	57	10	10	10
st_testgr3	12	0	47	0	12	12	14	0	50	0	12	12	12	2	60	12	12	12
super1	1129	215	115163	181	341	609	1442	0	110139	0	341	631	1069	306	111473	15828	340	638
super2	1125	199	104966	127	356	609	1444	0	107089	0	355	628	1041	276	104621	11532	356	628
super3	1146	186	130325	131	354	602	1454	0	126445	0	353	637	1070	277	114880	19081	353	606
super3t	853	159	95314	131	256	512	1026	0	85543	0	256	498	804	244	106848	13693	256	532
supplychain	12	1	82	0	4	6	12	0	82	0	4	6	8	4	81	23	4	6
supplychainp1_022020	60	0	13257	0	24	38	80	0	15852	0	24	37	40	3	15417	1173	24	36
supplychainp1_030510	25	0	989	0	10	10	30	0	988	0	10	10	17	7	1329	447	10	12
supplychainp1_053050	130	0	82449	0	50	59	160	0	68027	0	50	76	114	41	123255	55473	50	95
supplychainr1_053050	95	1	189541	0	3	33	160	0	230721	0	3	41	43	29	291416	65292	3	33
syn05h	16	0	69	0	6	11	12	0	72	0	6	10	18	14	60	23	6	14
syn05m02m	10	0	121	0	4	4	12	0	121	0	4	4	6	8	174	92	4	5
syn05m03m	13	0	138	0	5	9	18	0	145	0	5	9	9	11	250	150	4	7
syn05m04m	20	1	179	1	6	9	24	0	139	0	6	7	8	8	355	191	6	8
syn10h	29	7	162	8	1	19	36	0	156	0	1	18	20	18	234	149	1	19
syn10m02m	21	1	254	2	0	7	24	0	253	0	0	8	10	13	398	265	0	9
syn10m03m	31	0	630	0	0	15	36	0	724	0	0	15	18	18	831	488	0	16
syn10m04m	44	1	860	3	0	13	48	0	838	0	0	12	22	23	1415	769	0	18
syn10m	7	0	48	0	6	6	12	0	78	0	6	8	6	2	70	23	6	6
syn15m02m	39	2	738	1	8	12	48	0	722	0	8	16	22	14	1238	532	8	15
syn15m03m	57	4	2157	3	17	23	72	0	2151	0	17	22	32	16	3332	1264	17	22
syn15m04m	78	5	2400	4	2	29	96	0	1991	0	2	26	42	9	2969	803	2	27
syn15m	19	0	138	0	4	7	22	0	138	0	4	7	10	7	247	137	4	7
syn20m02m	49	3	682	2	0	14	56	0	675	0	0	18	31	17	1026	441	0	23
syn20m03m	71	5	836	6	0	23	84	0	938	0	0	21	35	26	2098	1657	0	17
syn20m04m	95	5	1792	6	0	30	112	0	1947	0	0	25	57	28	4517	2687	0	28
syn20m	24	1	295	2	1	5	28	0	295	0	6	5	14	9	374	203	1	7
syn30m02m	58	4	709	6	6	22	80	0	632	0	5	22	38	12	1047	449	6	29
syn30m03m	91	3	1141	5	5	37	120	0	1196	0	10	36	58	27	2463	1747	5	37
syn30m04m	131	11	1356	10	40	40	160	0	1316	0	3	47	76	28	1973	1072	10	54
syn30m	29	2	296	3	3	12	42	0	283	0	3	12	21	9	352	172	3	15
syn40m02m	81	7	850	8	0	36	112	0	827	0	0	34	56	20	1820	1180	0	36
syn40m03m	141	6	1458	6	0	47	168	0	1423	0	0	52	82	41	3637	2716	0	55
syn40m04m	192	13	2158	18	0	51	224	0	2210	0	0	53	111	35	6552	5087	0	69
syn40m	39	2	453	3	0	17	56	0	471	0	0	18	31	9	728	312	0	18
synheat	78	22	275	18	1	24	120	0	358	0	1	28	33	38	302	302	1	24

Table A.6 continued

Instance	greedy (base)						greedy+filt. off						greedy+filt. aggr.					
	lp	lp$_{\text{filt}}$	iter	iter$_{\text{filt}}$	b$_{\text{obbt}}$	lvb	lp	lp$_{\text{filt}}$	iter	iter$_{\text{filt}}$	b$_{\text{obbt}}$	lvb	lp	lp$_{\text{filt}}$	iter	iter$_{\text{filt}}$	b$_{\text{obbt}}$	lvb
synthes1	3	2	9	0	0	1	4	0	9	0	0	1	1	5	11	11	0	1
synthes2	6	2	33	1	2	4	8	0	33	0	2	3	3	6	44	44	2	3
synthes3	10	2	59	1	0	1	12	0	47	0	0	2	6	10	59	51	0	1
tanksize	26	1	104	2	7	16	34	0	104	0	7	16	19	6	170	73	7	15
tln12	207	2	7240	2	38	64	288	0	7882	0	38	71	108	11	2463	2037	38	60
tln4	30	0	149	0	9	13	40	0	179	0	9	10	15	8	152	81	9	10
tln5	33	1	310	1	15	18	60	0	345	0	15	17	24	8	262	74	15	16
tln6	55	2	384	5	19	22	84	0	386	0	19	21	41	5	323	102	19	21
tln7	89	0	836	0	5	21	112	0	830	0	5	22	41	18	479	337	5	23
tloss	51	0	227	0	21	23	84	0	229	0	21	22	39	7	211	76	21	23
tls12	230	12	1893	1	24	109	436	0	1616	0	24	160	207	22	1774	685	24	88
tls2	9	2	56	2	0	2	12	0	72	0	0	2	7	6	111	57	0	0
tls4	58	13	438	15	15	29	74	0	505	0	15	30	42	31	685	524	15	35
tls5	65	14	1037	54	20	36	86	0	992	0	20	40	62	44	1717	1346	20	45
tls6	88	21	1264	84	20	32	120	0	1306	0	20	40	74	47	1870	1509	20	40
tls7	120	19	2222	23	70	78	182	0	3446	0	70	84	122	43	2344	1054	70	78
tltr	42	3	444	0	6	27	54	0	406	0	6	17	36	25	633	341	6	29
transswitch0009r	80	4	459	5	0	4	94	0	484	0	0	4	18	11	373	299	0	2
transswitch0014r	181	20	1739	23	1	37	206	0	1749	0	1	41	71	69	1688	1508	1	40
transswitch0030r	366	55	4041	18	0	111	396	0	3876	0	0	102	150	116	2649	2233	0	118
transswitch0039r	432	20	5755	30	0	8	500	0	5887	0	0	6	117	37	5388	3292	0	7
transswitch0057r	649	56	7112	57	0	134	736	0	6913	0	0	145	309	238	6837	5665	0	147
transswitch0118r	1551	132	19203	100	0	349	1830	0	19339	0	0	333	732	547	22161	18175	0	337
transswitch0300r	3491	306	39201	235	0	539	4230	0	40442	0	0	572	1358	923	45501	36712	0	565
transswitch2383wpr	25117	1227	179768	623	0	1533	28728	0	181489	0	0	1435	5583	2424	133530	111048	0	1419
transswitch2736spr	32113	1602	236137	699	0	2020	38460	0	241232	0	0	1894	8214	2965	201263	141018	0	1907
tspn05	34	1	707	2	17	15	54	0	870	0	17	15	34	8	925	434	17	15
tspn08	74	2	2001	1	27	34	108	0	1851	0	27	34	34	12	2453	1390	27	34
tspn10	146	1	11546	4	128	133	188	0	12298	0	128	133	147	13	10649	1229	128	133
tspn12	206	4	21790	14	172	182	256	0	23316	0	172	181	195	15	21119	2163	172	181
tspn15	283	11	19453	19	129	142	346	0	20179	0	129	142	150	23	16233	6490	129	144
unitcommit1	300	18	14905	131	1	30	480	0	14792	0	1	42	198	86	18784	8960	1	20
unitcommit2	500	129	855155	0	0	101	812	0	779984	0	0	91	296	235	287186	77177	0	82
util	11	0	62	0	9	9	12	0	62	0	9	9	10	7	73	26	9	9
waste	246	28	2444	26	2	85	450	0	2291	0	23	112	176	104	4444	3722	17	65
water3	88	8	1444	9	23	43	120	0	1506	0	43	43	60	30	1412	568	23	45
watercontamination0202	521	16	57183	81	50	67	1180	0	55465	0	50	96	257	121	40882	10293	50	66
watercontamination0202r	129	45	823	15	0	15	188	0	851	0	0	51	85	64	1330	1108	0	0
watercontamination0303	940	88	32783	415	90	179	2446	0	30429	0	90	285	267	163	21338	9296	90	166
watercontamination0303r	201	44	2978	16	0	19	370	0	2909	0	0	35	65	92	1681	1681	0	0
waternd1	34	3	364	2	0	15	54	0	495	0	0	12	20	13	660	351	0	13
waternd2	112	12	5086	10	0	63	184	0	4403	0	0	61	75	45	3807	1400	0	62
waterno2_01	44	14	561	18	21	33	58	0	583	0	21	37	44	32	913	548	21	34
waterno2_02	100	38	1740	28	13	54	116	0	1891	0	13	58	83	72	3469	2576	13	45
waterno2_03	143	61	3787	88	15	76	174	0	3596	0	15	87	146	130	5788	3976	15	72
waterno2_04	174	64	4771	40	20	83	232	0	4777	0	20	113	165	129	7018	4010	20	83

Table A.6 continued

Instance	greedy (base)						greedy+filt. off						greedy+filt. aggr.					
	lp	lp$_{\text{filt}}$	iter	iter$_{\text{filt}}$	b$_{\text{obbt}}$	lvb	lp	lp$_{\text{filt}}$	iter	iter$_{\text{filt}}$	b$_{\text{obbt}}$	lvb	lp	lp$_{\text{filt}}$	iter	iter$_{\text{filt}}$	b$_{\text{obbt}}$	lvb
waterno2.06	271	119	8017	86	17	139	348	0	7748	0	17	170	255	206	12669	8452	17	139
waterno2.09	406	151	13282	128	28	218	522	0	11547	0	28	259	388	308	21903	13450	28	197
waterno2.12	525	205	17886	194	43	290	696	0	18277	0	43	341	513	403	34596	22624	43	288
waterno2.18	814	324	26333	222	60	412	1044	0	26533	0	60	505	762	610	69999	53301	60	401
waterno2.24	1116	416	38669	310	72	583	1392	0	39617	0	72	677	1097	852	88973	64398	72	596
watertreatnd_conc	59	9	426	1	5	25	110	0	513	0	5	17	29	21	404	252	5	17
watertreatnd_flow	65	16	1794	12	7	48	110	0	1850	0	7	61	61	31	2910	1067	7	46
waterx	101	24	2267	2	2	17	148	0	2267	0	2	17	93	68	1920	822	2	14

Table A.7: Detailed results comparing filtering strategies for OBBT at the root node on test set GO. See Table 6.1 for aggregated results.

lp — number of LPs solved during OBBT
lp_{filt} — number of LPs solved during OBBT's filtering
iter — number of LP iterations during OBBT
$iter_{filt}$ — number of LP iterations of OBBT's filtering LPs
b_{obbt} — number of bounds tightened by OBBT
lvb — number of lVBs found by OBBT

Instance	greedy (base)						greedy+filt. off						greedy+filt. aggr.					
	lp	lp_{filt}	iter	$iter_{filt}$	b_{obbt}	lvb	lp	lp_{filt}	iter	$iter_{filt}$	b_{obbt}	lvb	lp	lp_{filt}	iter	$iter_{filt}$	b_{obbt}	lvb
alkylation	15	3	87	3	9	10	16	0	87	0	9	10	14	7	116	34	9	10
alkyl	20	0	93	0	13	15	26	0	93	0	13	15	19	6	128	39	13	15
arki0003	4357	1237	16100	566	350	1228	8968	0	20052	0	350	1505	2491	1753	354189	40125	350	2199
arki0004	4983	130	24091	223	0	2235	6474	0	25368	0	0	2323	3905	1736	67760	54431	0	1598
arki0015	2917	626	515010	365	394	1895	3968	0	509218	0	394	1894	2459	794	491857	18746	394	1935
arki0018	19668	2682	191737	2682	329	15021	39216	0	202065	0	329	16808	17053	6997	146956	32702	329	15313
arki0019	1547	1	4858	0	0	708	2038	0	4817	0	0	716	1020	9	5748	2087	0	718
arki0020	3771	1	11474	1	0	1734	5046	0	11817	0	0	1731	2523	7	13565	4379	0	1732
bayes2.10	89	44	638	35	7	62	128	0	609	0	7	66	71	40	690	198	7	67
bayes2.20	79	32	933	18	7	73	128	0	1015	0	7	81	75	33	1164	271	7	75
bayes2.30	94	44	1072	17	7	88	128	0	1178	0	7	90	88	46	1219	161	7	87
bayes2.50	87	24	1627	11	7	75	128	0	1644	0	7	76	83	26	1874	180	7	70
bearing	23	7	233	2	4	16	34	0	240	0	4	19	19	8	328	40	4	16
btest14	95	7	2082	6	16	56	166	0	2269	0	16	58	99	32	2410	506	16	69
camshape100	160	0	483	0	61	98	200	0	483	0	61	99	119	34	584	385	61	108
camshape200	316	0	973	0	113	171	400	0	973	0	113	172	214	68	1337	852	113	184
camshape400	626	0	2030	0	218	335	800	0	2030	0	218	336	470	186	2738	1762	218	432
camshape800	1244	0	4046	0	414	695	1600	0	4046	0	414	696	935	381	5269	3440	414	852
catmix100	501	0	0	0	0	0	600	0	0	0	0	0	400	6	0	0	0	0
catmix200	1001	0	0	0	0	0	1200	0	0	0	0	0	800	6	0	0	0	0
catmix400	2001	0	0	0	0	0	2400	0	0	0	0	0	1600	6	0	0	0	0
catmix800	4001	0	0	0	0	0	4800	0	0	0	0	0	3200	6	0	0	0	0
chain100	507	32	15186	0	68	302	800	0	18932	0	68	302	382	132	21156	11127	68	302
chain200	1010	92	25427	0	155	603	1600	0	30929	0	155	603	682	261	42156	22309	155	603
chain50	263	22	8439	7	51	154	400	0	9836	0	51	154	215	48	20961	9319	51	154
chance	9	0	15	0	8	8	10	0	15	0	8	8	9	4	29	14	8	8
chem	53	6	237	5	13	39	62	0	230	0	13	39	46	15	307	111	13	40
elec200	20995	262	143948	12073	0	0	41000	0	167275	0	0	0	77	82	38811	38811	0	0
elec25	438	34	5647	408	0	0	750	0	4838	0	0	0	23	29	804	804	0	0
elec50	1487	95	20714	3756	0	0	2750	0	20525	0	0	0	58	64	2589	2589	0	0
etamac	56	3	186	9	8	17	88	0	247	0	8	17	41	11	250	134	8	16
ex14.1.1	4	0	30	0	4	4	4	0	30	0	4	4	4	4	49	31	4	4
ex14.1.2	10	0	85	0	4	7	10	0	85	0	4	7	9	6	150	47	4	7
ex14.1.3	4	0	25	0	2	4	4	0	25	0	2	4	4	4	29	8	2	4
ex14.1.5	21	2	42	0	16	1	22	0	42	0	16	1	19	8	45	27	16	1
ex14.1.6	14	0	32	0	4	10	14	0	32	0	4	10	9	6	48	34	4	9

Table A.7 continued

Instance	greedy (base)						greedy+filt. off						greedy+filt. aggr.					
	lp	lp$_{filt}$	iter	iter$_{filt}$	b$_{obbt}$	lvb	lp	lp$_{filt}$	iter	iter$_{filt}$	b$_{obbt}$	lvb	lp	lp$_{filt}$	iter	iter$_{filt}$	b$_{obbt}$	lvb
ex14.1.7	76	23	743	29	2	17	84	0	591	0	2	14	64	56	746	421	2	16
ex14.1.8	12	0	41	0	4	9	12	0	41	0	4	9	12	8	56	24	4	9
ex14.1.9	2	0	5	0	0	1	6	0	6	0	0	2	1	4	5	5	0	1
ex14.2.4	82	0	45	0	6	7	86	0	45	0	6	7	81	6	112	62	6	7
ex14.2.7	89	0	47	0	6	4	90	0	56	0	0	4	85	3	48	8	0	4
ex14.2.9	23	0	74	0	13	18	28	0	74	0	13	18	19	5	102	42	13	18
ex2.1.10	22	0	37	0	21	22	40	0	37	0	21	22	22	4	50	12	21	22
ex2.1.1	6	0	5	0	0	1	10	0	5	0	0	1	2	3	6	4	0	1
ex2.1.7	26	0	249	0	20	20	40	0	244	0	20	20	20	5	253	28	20	20
ex2.1.9	19	0	67	0	10	10	20	0	67	0	10	10	20	2	60	0	10	10
ex3.1.1	13	0	35	0	7	3	16	0	35	0	7	3	11	3	38	16	7	3
ex3.1.2	7	0	15	0	7	7	10	0	15	0	7	7	7	2	21	4	7	7
ex3.1.4	3	0	5	0	0	1	6	0	5	0	0	1	1	3	6	4	0	1
ex3pb	8	0	52	0	0	2	10	0	57	0	0	4	7	5	81	45	0	1
ex4.1.1	2	0	16	0	1	2	2	0	16	0	1	2	2	3	20	20	1	2
ex4.1.2	1	0	0	0	0	0	2	0	0	0	0	0	0	2	0	0	0	0
ex4.1.3	2	0	8	0	1	1	2	0	8	0	1	1	1	2	19	19	1	1
ex4.1.4	2	0	16	0	0	2	2	0	16	0	0	2	2	4	17	17	0	2
ex4.1.5	3	0	9	0	0	0	4	0	9	0	0	0	2	3	9	1	0	0
ex4.1.6	2	0	11	0	1	1	2	0	11	0	1	1	1	2	7	7	1	1
ex4.1.7	2	0	20	0	2	2	2	0	20	0	2	2	2	2	28	16	2	2
ex4.1.9	2	0	5	0	2	2	6	0	5	0	0	2	2	3	4	4	0	2
ex5.2.2.case1	5	0	14	0	3	3	6	0	14	0	3	3	5	5	26	14	3	4
ex5.2.2.case2	5	0	12	0	1	4	6	0	12	0	1	4	4	4	23	12	1	4
ex5.2.2.case3	5	0	14	0	3	3	6	0	14	0	3	5	5	5	26	14	3	5
ex5.2.4	9	0	32	0	0	1	10	0	32	0	0	3	5	8	42	28	0	3
ex5.2.5	27	7	159	4	0	7	54	0	146	0	0	15	13	19	76	66	0	10
ex5.3.2	13	3	37	2	2	6	16	0	35	0	2	6	10	12	37	33	2	6
ex5.3.3	57	10	324	9	1	29	98	0	392	0	1	44	51	45	411	335	1	27
ex5.4.2	12	0	46	0	6	7	16	0	46	0	6	7	10	3	56	16	6	8
ex5.4.3	14	1	60	1	8	6	20	0	64	0	8	6	11	7	80	52	8	5
ex5.4.4	24	3	148	5	6	4	36	0	131	0	6	6	23	18	222	153	6	9
ex6.1.1	19	4	40	0	4	19	28	0	40	0	4	20	20	15	44	25	4	18
ex6.1.2	9	0	11	0	2	7	10	0	11	0	2	7	8	5	25	11	2	7
ex6.1.4	13	0	65	0	3	10	18	0	65	0	3	10	16	2	102	38	3	9
ex6.2.10	90	28	1173	7	9	37	106	0	1173	0	9	43	77	58	148963	148213	9	45
ex6.2.11	37	1	179	0	6	19	56	0	179	0	6	23	23	18	232	101	6	21
ex6.2.12	42	15	154	0	0	31	60	0	154	0	0	31	35	36	163	105	0	27
ex6.2.13	62	13	231	15	0	28	82	0	246	0	0	29	31	30	168	133	0	27
ex6.2.5	64	38	491	54	0	45	90	0	525	0	0	48	61	55	739	421	0	46
ex6.2.6	27	0	195	0	19	22	36	0	118	0	19	23	27	13	176	84	19	27
ex6.2.7	73	11	1054	21	0	31	96	0	1042	0	0	30	50	23	852	381	0	27
ex6.2.8	28	0	197	0	20	25	32	0	197	0	20	25	30	13	189	66	20	26
ex6.2.9	68	24	660	28	6	33	76	0	476	0	6	40	53	42	495	257	6	38
ex7.2.2	10	0	49	0	3	7	12	0	49	0	3	7	9	3	73	26	3	7
ex7.2.3	27	5	233	1	4	16	36	0	293	0	4	17	22	25	249	161	4	12

Table A.7 continued

Instance	greedy (base)						greedy+filt. off						greedy+filt. aggr.					
	lp	lp$_{filt}$	iter	iter$_{filt}$	b$_{obbt}$	lvb	lp	lp$_{filt}$	iter	iter$_{filt}$	b$_{obbt}$	lvb	lp	lp$_{filt}$	iter	iter$_{filt}$	b$_{obbt}$	lvb
ex7.2.4	19	1	169	0	5	11	24	0	169	0	5	11	13	9	265	110	5	11
ex7.3.1	19	0	31	0	10	10	22	0	32	0	10	10	19	2	32	12	10	10
ex7.3.2	10	0	8	0	7	7	10	0	8	0	7	7	11	6	13	3	7	8
ex7.3.3	5	0	4	0	4	4	6	0	5	0	4	4	6	7	6	3	4	6
ex7.3.4	37	2	24	1	15	15	40	0	24	0	15	15	42	15	39	25	15	20
ex7.3.5	36	3	49	2	4	4	46	0	58	0	4	4	36	14	75	31	4	6
ex7.3.6	3	0	3	0	3	3	4	0	3	0	3	3	3	4	11	8	3	3
ex8.1.3	15	0	29	0	0	0	16	0	29	0	0	0	12	5	29	2	0	0
ex8.1.4	4	0	1	0	0	0	4	0	1	0	0	0	4	2	1	0	0	0
ex8.1.5	4	0	0	0	0	0	4	0	0	0	0	0	4	2	0	0	0	0
ex8.1.6	10	0	40	0	2	2	10	0	40	0	2	2	9	2	19	2	2	2
ex8.1.7	10	0	48	0	21	26	12	0	57	0	21	26	3	6	89	54	21	26
ex8.2.1b	74	0	246	0	21	26	114	0	240	0	21	26	33	8	271	110	21	26
ex8.2.2b	6074	93	30636	93	40	40	15044	0	30889	0	40	40	44	15	32262	9112	40	40
ex8.2.3b	13482	0	37854	0	77	85	31282	0	37353	0	77	85	86	51	31186	9403	77	85
ex8.2.4b	82	0	193	0	35	38	122	0	191	0	35	38	41	9	202	84	35	38
ex8.2.5b	2080	3	17797	9	52	54	5068	0	17752	0	52	54	59	15	19166	4765	52	54
ex8.3.11	129	8	261	9	7	40	280	0	328	0	7	42	47	44	168	188	7	41
ex8.3.12	98	6	231	6	7	38	210	0	237	0	7	42	35	38	391	154	7	33
ex8.3.13	149	26	314	20	7	65	290	0	372	0	7	80	94	61	389	160	7	70
ex8.3.14	183	26	583	25	7	52	330	0	536	0	7	64	103	69	233	367	7	47
ex8.3.1	127	10	237	9	7	38	280	0	311	0	7	47	65	34	180	130	7	46
ex8.3.5	88	4	233	5	7	33	180	0	204	0	7	36	37	37	241	170	7	32
ex8.3.7	123	23	259	30	7	48	214	0	253	0	7	48	63	67	303	221	7	44
ex8.3.8	100	4	339	9	7	37	210	0	329	0	7	47	44	26	303	176	7	32
ex8.3.9	70	3	166	1	6	38	136	0	162	0	6	36	39	40	109	92	6	35
ex8.4.1	22	0	205	0	6	7	42	0	240	0	6	8	12	12	364	207	6	8
ex8.4.2	54	25	472	28	1	31	86	0	468	0	1	31	45	47	545	519	1	31
ex8.4.4	43	2	327	0	17	22	48	0	377	0	17	22	33	9	384	88	17	23
ex8.4.5	25	1	183	0	3	3	30	0	183	0	3	4	6	12	382	280	3	4
ex8.4.6	57	3	113	2	1	3	92	0	209	0	1	3	24	17	136	134	1	3
ex8.4.7	110	10	5196	4	48	93	184	0	6344	0	48	94	110	31	7430	1324	48	100
ex8.4.8_bnd	202	31	5245	8	71	159	274	0	4929	0	71	177	202	64	5143	1187	71	167
ex8.5.1	22	4	38	0	3	8	24	0	38	0	3	8	20	11	51	28	3	8
ex8.5.2	25	1	72	0	8	8	26	0	72	0	8	8	22	5	104	29	8	6
ex8.5.4	15	1	67	1	2	6	18	0	67	0	2	6	13	12	69	44	2	6
ex8.5.5	18	2	63	2	3	4	20	0	63	0	3	4	15	11	72	32	3	5
ex8.5.6	28	3	17	1	11	13	32	0	16	0	13	13	26	9	42	19	11	13
ex8.6.1	160	3	57	7	8	8	228	0	59	0	8	8	95	6	238	224	8	8
ex9.2.2	3	0	5	0	2	3	4	0	5	0	2	3	3	4	10	5	2	3
ex9.2.4	4	0	16	0	0	3	4	0	16	0	0	3	0	4	18	18	0	0
ex9.2.5	4	0	18	0	1	2	4	0	18	0	1	3	4	6	21	17	1	3
ex9.2.6	6	0	20	0	2	3	8	0	20	0	2	2	2	6	20	20	0	2
ex9.2.7	3	0	11	0	1	3	4	0	11	0	1	3	3	3	10	5	2	3
glider100	3394	598	3318	199	1	701	4198	0	3884	0	1	701	3094	901	4542	2989	1	601
glider200	6794	1198	6777	399	1	1401	8398	0	7693	0	1	1401	6194	1801	9066	5986	1	1201

Table A.7 continued

Instance	greedy (base)						greedy+filt. off						greedy+filt. aggr.					
	lp	lp$_{\text{filt}}$	iter	iter$_{\text{filt}}$	b$_{\text{obbt}}$	lvb	lp	lp$_{\text{filt}}$	iter	iter$_{\text{filt}}$	b$_{\text{obbt}}$	lvb	lp	lp$_{\text{filt}}$	iter	iter$_{\text{filt}}$	b$_{\text{obbt}}$	lvb
glider400	13594	2398	12991	799	1	2801	16798	0	15201	0	1	2801	12394	3601	18776	11980	1	2401
glider50	1694	298	1713	99	1	351	2098	0	1906	0	1	351	1544	451	2301	1489	1	301
gsg_0001	100	23	854	52	50	90	120	0	1119	0	50	90	92	28	1105	313	50	90
hhfair	28	3	35	2	6	10	42	0	42	0	6	10	24	12	35	20	6	10
himmel11	7	0	11	0	7	7	10	0	11	0	7	7	7	2	19	4	7	7
himmel16	18	0	91	0	0	0	18	0	91	0	7	0	6	8	76	62	7	0
house	8	0	9	0	6	8	10	0	9	0	6	8	8	8	14	5	6	8
hs62	9	0	22	0	8	9	12	0	31	0	8	9	9	4	40	18	8	9
hybriddynamic_fixedcc	46	0	70	0	26	27	64	0	74	0	26	29	43	7	83	18	26	27
hybriddynamic_varcc	107	3	1353	2	37	40	146	0	1368	0	37	40	84	21	1795	550	37	40
infeas1	2847	31	191666	18	2631	2710	2948	0	192980	0	2631	2710	2869	56	108394	12208	2631	2708
jbearing100	5001	0	2500	0	0	0	10000	0	2500	0	0	0	5000	5	2501	1	0	0
jbearing25	1251	0	625	0	0	0	2500	0	625	0	0	0	1250	5	626	1	0	0
jbearing50	2501	0	1250	0	0	0	5000	0	1250	0	0	0	2500	5	1251	1	0	0
jbearing75	3751	0	1875	0	0	0	7500	0	1875	0	0	0	3750	5	1876	1	0	0
kall_circles_c6a	25	3	94	2	7	9	32	0	110	0	7	13	16	15	68	68	7	14
kall_circles_c6b	25	3	107	2	7	10	32	0	99	0	7	9	18	18	61	61	7	16
kall_circles_c6c	27	1	98	0	8	11	36	0	95	0	8	11	19	15	58	58	8	17
kall_circles_c7a	25	1	109	1	3	10	36	0	126	0	3	10	17	7	188	72	3	12
kall_circles_c8a	29	3	151	7	9	11	40	0	162	0	9	10	22	9	161	75	9	14
kall_circlespolygons_c1p12	28	4	164	4	2	2	30	0	173	0	2	2	24	24	168	118	2	2
kall_circlespolygons_c1p13	27	3	163	2	2	2	30	0	173	0	2	2	20	21	183	134	2	2
kall_circlespolygons_c1p5a	100	39	700	59	1	1	124	0	741	0	1	2	76	71	893	607	1	2
kall_circlespolygons_c1p5b	587	197	5343	244	0	1	694	0	6254	0	0	1	449	418	5121	3353	0	1
kall_circlespolygons_c1p6a	876	307	8928	428	2	1	988	0	9905	0	2	3	677	645	7785	5393	0	1
kall_circlesrectangles_c1r12	33	4	129	4	2	2	34	0	120	0	2	3	23	22	147	111	2	2
kall_circlesrectangles_c1r13	31	4	133	3	0	4	34	0	157	0	3	3	21	24	186	133	0	0
kall_circlesrectangles_c6r1	154	42	1016	51	0	5	168	0	1151	0	6	4	119	110	942	707	0	3
kall_circlesrectangles_c6r29	313	77	2181	78	0	5	344	0	2625	0	0	2	257	228	2316	1712	0	5
kall_circlesrectangles_c6r39	512	142	3781	152	0	6	556	0	4353	0	0	4	438	404	4062	2924	0	5
kall_congruentcircles_c31	11	2	26	2	7	8	16	0	27	0	7	8	8	4	15	12	7	7
kall_congruentcircles_c32	11	1	39	1	1	2	16	0	40	0	1	5	6	5	44	19	1	4
kall_congruentcircles_c41	7	0	9	0	3	1	8	0	9	0	3	1	6	6	12	7	3	3
kall_congruentcircles_c42	14	4	42	2	2	8	20	0	42	0	2	6	11	10	60	33	2	8
kall_congruentcircles_c51	19	5	47	5	5	8	24	0	61	0	5	7	20	21	88	59	5	14
kall_congruentcircles_c52	15	4	53	4	3	3	24	0	58	0	3	4	13	14	68	41	3	8
kall_congruentcircles_c61	22	5	64	5	6	10	28	0	60	0	6	6	16	14	50	40	6	15
kall_congruentcircles_c62	15	2	73	6	4	2	28	0	73	0	4	6	15	15	105	62	4	9
kall_congruentcircles_c63	20	2	86	1	9	9	28	0	80	0	9	9	18	10	88	32	9	13
kall_congruentcircles_c71	26	5	109	5	0	7	32	0	132	0	0	9	17	12	149	51	0	14
kall_congruentcircles_c72	21	4	105	8	5	4	32	0	100	0	5	4	17	16	96	54	5	11
kall_diffcircles_10	29	2	78	3	1	4	38	0	85	0	1	5	9	6	93	55	1	3
kall_diffcircles_5a	17	2	46	0	5	12	24	0	48	0	5	10	13	10	65	42	5	11
kall_diffcircles_5b	18	0	49	3	6	9	24	0	43	0	6	8	10	8	44	36	6	9
kall_diffcircles_6	18	2	35	3	2	4	28	0	45	0	2	6	5	9	41	36	2	4
kall_diffcircles_7	20	3	54	3	2	6	32	0	61	0	2	8	7	11	54	50	2	7

Table A.7 continued

Instance	greedy (base)						greedy+filt. off						greedy+filt. aggr.					
	lp	lp_{filt}	iter	$iter_{filt}$	b_{obbr}	lvb	lp	lp_{filt}	iter	$iter_{filt}$	b_{obbr}	lvb	lp	lp_{filt}	iter	$iter_{filt}$	b_{obbr}	lvb
kall_difficircles.8	21	0	73	0	8	8	30	0	75	0	8	8	11	4	58	45	8	8
kall_difficircles.9	24	2	67	2	1	3	34	0	66	0	1	4	11	8	82	46	1	6
knp3-12	64	30	426	174	0	0	72	0	416	0	0	0	1	7	250	250	0	0
knp4-24	187	1	2548	1	0	0	192	0	2548	0	0	0	7	13	827	827	0	0
knp5-40	395	1	7223	148	0	0	400	0	7223	0	0	0	116	121	4065	4065	0	0
knp5-41	401	6	7873	0	0	0	410	0	8456	0	0	0	38	45	3228	3228	0	0
knp5-42	412	1	9755	3	0	0	420	0	9755	0	0	0	25	31	3466	3466	0	0
knp5-43	424	0	9888	0	0	0	430	0	9552	0	0	0	21	28	3162	3162	0	0
knp5-44	434	0	12542	0	0	0	440	0	12542	0	0	0	49	56	4503	4503	0	0
launch	85	1	170	2	49	62	140	0	161	0	49	74	69	15	157	81	49	59
least	21	32	58	31	0	6	28	0	60	0	0	6	10	12	55	46	0	6
like	697	0	661	0	10	1	790	0	1051	0	10	91	667	37	716	92	10	1
linear	30	0	75	0	0	16	40	0	75	0	0	16	27	4	71	24	0	16
mathopt5-7	2	0	8	0	0	1	2	0	8	0	0	1	1	2	15	15	0	1
mathopt5-8	2	0	7	0	0	1	2	0	7	0	0	1	1	2	18	18	0	1
maxmin	148	39	709	95	66	10	204	0	653	0	66	8	180	102	1370	957	66	11
methanol100	7590	0	199	0	0	0	7594	0	199	0	0	0	7589	5	199	0	0	0
methanol50	3966	0	224	0	0	0	3970	0	224	0	0	0	3965	5	190	0	0	0
minlphi	18	5	44	1	7	5	20	0	53	0	7	5	18	9	73	33	7	5
oaer	4	0	13	0	2	4	4	0	13	0	2	5	3	2	16	10	2	3
pindyck	174	26	2862	0	112	124	190	0	2862	0	112	124	174	34	3178	416	112	124
pinene50	1226	0	1233	0	0	2	1230	0	1233	0	0	0	1225	3	1233	0	0	3
pointpack04	14	0	27	0	2	4	16	0	36	0	2	4	8	5	29	18	2	3
pointpack06	21	3	38	1	1	8	24	0	65	0	1	10	7	10	67	55	1	4
pointpack08	25	6	61	3	0	9	32	0	62	0	0	9	8	14	52	52	0	7
pointpack10	31	13	62	9	0	12	40	0	61	0	0	13	11	17	85	85	0	9
pointpack12	40	19	97	16	0	13	48	0	79	0	0	17	15	21	81	81	0	13
pointpack14	45	22	107	23	0	19	56	0	116	0	0	17	17	23	93	93	0	16
portfol_buyin	13	1	73	3	7	10	18	0	59	0	7	11	12	10	140	67	7	12
powerflow0009r	86	3	315	6	0	2	94	0	310	0	0	23	38	14	472	329	0	2
powerflow0014r	148	14	1198	1	1	19	206	0	1505	0	1	6	55	42	1658	1278	1	18
powerflow0030r	291	2	1334	3	0	7	396	0	1376	0	0	4	90	17	1578	1154	0	7
powerflow0039r	441	2	1720	1	0	4	480	0	1740	0	0	4	151	28	2191	1760	0	4
powerflow0057r	522	36	4962	11	0	40	736	0	5748	0	0	46	142	116	7759	6756	0	37
powerflow0118r	1277	54	17492	21	0	105	1826	0	17656	0	0	114	352	228	27827	23735	0	92
powerflow0300r	3101	166	39841	47	0	257	4146	0	41902	0	0	278	1085	463	57435	41895	0	229
powerflow2383wpt	23076	280	74025	168	0	538	26676	0	77489	0	0	539	4481	625	79578	65937	0	493
powerflow2736spr	27638	553	70238	283	0	911	31780	0	71843	0	0	920	8540	927	89057	72114	0	855
prob06	3	0	6	0	2	2	4	0	6	0	2	2	7	5	7	7	2	2
process	16	1	101	0	8	10	16	0	101	0	8	10	15	5	131	45	8	10
procsyn	32	0	88	0	1	0	40	0	92	0	1	0	12	7	68	48	1	0
prolog	9	0	50	0	4	4	12	0	52	0	4	4	6	5	86	45	4	4
qp3	145	1	258	0	114	131	200	0	254	0	114	129	147	8	251	13	114	133
rocket100	1114	108	2782	192	1	104	1800	0	2637	0	1	107	1133	640	3272	2507	1	395
rocket200	2443	439	4550	205	1	432	3600	0	4826	0	1	323	2249	1256	6527	5036	1	790
rocket400	4748	745	10075	948	1	714	7200	0	9231	0	1	658	4516	2523	12995	10186	1	1586

Table A.7 continued

Instance	greedy (base)						greedy+filt. off						greedy+filt. aggr.					
	lp	lp_{flt}	iter	$iter_{flt}$	b_{obbt}	lvb	lp	lp_{flt}	iter	$iter_{flt}$	b_{obbt}	lvb	lp	lp_{flt}	iter	$iter_{flt}$	b_{obbt}	lvb
rocket50	560	54	1359	55	1	50	900	0	1381	0	1	50	550	307	1617	1242	1	195
shiporig	21	0	19	0	0	0	30	0	17	0	0	0	20	5	17	0	0	0
st_bpaf1a	14	0	19	0	13	13	20	0	22	0	13	13	14	2	24	4	13	13
st_bsj2	5	0	9	0	4	4	6	0	9	0	4	4	5	2	13	4	4	4
st_bsj4	7	0	6	0	6	6	12	0	6	0	6	6	6	4	9	5	6	6
st_e03	16	2	126	1	8	10	16	0	126	0	8	10	15	5	170	58	8	10
st_e04	7	1	83	1	2	5	10	0	83	0	2	5	6	9	120	70	2	5
st_e05	8	2	10	0	5	5	8	0	10	0	5	5	8	6	16	7	5	5
st_e07	5	0	17	0	1	3	6	0	17	0	1	3	4	4	25	11	1	2
st_e08	4	0	7	0	4	4	4	0	7	0	4	4	4	2	9	1	4	4
st_e09	4	1	2	0	4	4	4	0	2	0	4	4	4	4	4	2	4	4
st_e11	4	0	2	0	0	0	4	0	2	0	0	0	4	3	3	1	0	0
st_e16	16	0	72	0	9	6	20	0	59	0	9	6	14	6	95	36	9	6
st_e19	4	0	17	0	0	0	4	0	17	0	0	0	0	4	17	17	0	0
st_e22	4	0	3	0	3	3	4	0	3	0	3	3	4	2	5	2	3	3
st_e24	3	0	2	0	3	3	4	0	4	0	3	3	3	4	5	4	3	3
st_e25	8	0	18	0	6	2	8	0	18	0	6	2	7	4	18	8	6	2
st_e26	4	1	2	0	2	4	4	0	2	0	2	4	4	5	2	2	2	4
st_e28	7	0	11	0	7	7	10	0	11	0	7	7	7	2	19	4	7	7
st_e30	7	0	29	0	2	2	10	0	32	0	2	2	5	4	38	21	2	2
st_e33	5	0	14	0	1	3	6	0	14	0	1	3	4	4	27	11	1	3
st_e41	17	1	52	0	11	15	24	0	52	0	11	18	16	10	69	38	11	15
st_fp7a	28	0	235	0	20	20	40	0	236	0	20	20	20	5	253	24	20	20
st_fp7b	24	0	224	0	20	20	40	0	227	0	20	20	20	5	245	24	20	20
st_fp7c	28	0	235	0	20	20	40	0	236	0	20	20	20	5	242	26	20	20
st_fp7d	28	0	235	0	20	20	40	0	236	0	20	20	20	5	249	24	20	20
st_fp7e	26	0	249	0	20	20	40	0	244	0	20	20	20	5	253	28	20	20
st_glmp_fp1	4	0	5	0	4	2	4	0	5	0	4	2	4	2	10	4	4	2
st_glmp_fp2	3	0	12	0	3	2	4	0	14	0	3	2	3	4	16	6	3	2
st_glmp_fp3	4	0	13	0	3	0	4	0	13	0	3	0	4	4	23	13	3	0
st_glmp_kk90	3	0	5	0	3	3	4	0	5	0	3	3	3	4	8	4	3	3
st_glmp_kk92	4	0	18	0	2	3	4	0	18	0	2	3	4	2	22	6	2	3
st_glmp_kky	3	0	6	0	3	3	4	0	8	0	3	3	3	2	8	4	3	3
st_glmp_ss1	3	0	7	0	2	3	4	0	11	0	2	3	3	2	7	5	2	3
st_glmp_ss2	3	0	5	0	3	3	4	0	5	0	3	3	3	2	6	3	3	3
st_iqpbk1	15	0	99	0	2	6	16	0	99	0	2	6	7	7	71	36	11	7
st_iqpbk2	16	0	66	0	2	7	16	0	66	0	2	7	7	5	75	41	20	7
st_jcbpaf2	16	1	57	1	12	12	20	0	53	0	12	12	16	6	79	30	12	12
st_pan1	5	0	6	0	1	4	6	0	6	0	1	4	4	4	12	5	1	4
st_ph14	4	0	4	0	4	4	6	0	4	0	4	4	4	2	6	2	4	4
st_ph15	7	0	13	0	4	4	8	0	13	0	4	4	6	2	19	5	4	4
st_ph1	7	0	10	0	4	6	12	0	10	0	4	7	6	4	13	7	4	6
st_ph20	3	0	7	0	3	2	4	0	7	0	3	2	3	4	19	11	3	2
st_ph2	7	0	10	0	4	6	12	0	10	0	4	7	6	3	16	7	4	6
st_ph3	5	0	7	0	5	5	8	0	7	0	5	6	5	2	10	3	5	5
st_phex	4	0	4	0	4	2	4	0	4	0	4	2	4	2	10	6	4	2

Table A.7 continued

Instance	greedy (base)						greedy+filt. off						greedy+filt. aggr.					
	lp	lp$_{filt}$	iter	iter$_{filt}$	b$_{obbt}$	lvb	lp	lp$_{filt}$	iter	iter$_{filt}$	b$_{obbt}$	lvb	lp	lp$_{filt}$	iter	iter$_{filt}$	b$_{obbt}$	lvb
st-qpc-m0	3	0	3	0	2	0	4	0	3	0	2	0	3	7	3	0	2	1
st-qpc-m1	9	0	33	0	6	6	10	0	33	0	6	6	8	2	32	7	6	6
st-qpc-m3a	13	0	98	0	1	6	20	0	101	0	1	6	10	4	84	26	1	10
st-qpc-m3b	17	6	148	36	7	9	20	0	128	0	7	9	10	3	160	88	7	9
st-qpk1	3	0	4	0	2	0	4	0	4	0	2	0	4	8	5	1	2	0
st-qpk2	10	1	27	0	0	6	12	0	19	0	0	6	6	3	44	13	0	6
st-qpk3	21	1	80	0	0	11	22	0	88	0	0	11	11	3	115	23	0	11
st_rv1	12	0	25	0	10	12	20	0	23	0	10	12	12	5	35	12	10	12
st_rv2	24	0	130	0	21	21	40	0	127	0	21	21	23	4	152	19	21	21
st_rv3	28	1	174	0	20	23	40	0	174	0	20	23	26	5	226	40	20	23
st_rv7	37	0	294	0	29	32	60	0	291	0	29	32	35	5	310	24	29	32
st_rv8	47	0	637	0	39	43	80	0	631	0	39	43	44	6	719	63	39	43
st_rv9	58	0	632	0	42	50	100	0	641	0	42	50	51	4	663	48	42	49
st_z	6	0	10	0	5	4	6	0	10	0	5	4	5	2	11	0	5	4
supplychain1_020306	15	0	262	0	7	8	18	0	262	0	7	8	13	12	358	243	7	12
sym05m	4	0	27	0	3	4	6	0	27	0	3	4	4	2	44	17	3	4
tricp	71	0	108	0	20	20	108	0	115	0	20	20	20	7	125	59	20	20
wall	16	0	1	0	0	0	16	0	1	0	0	0	16	2	54	17	0	0
wastewater02m1	13	0	67	0	9	11	20	0	74	0	9	12	13	5	54	17	9	12
wastewater02m2	15	3	95	0	1	15	20	0	91	0	1	16	15	15	148	107	1	15
wastewater04m1	18	4	109	0	6	15	24	0	94	0	6	16	20	12	116	46	6	18
wastewater04m2	23	5	97	2	10	18	24	0	95	0	10	18	30	19	195	82	10	20
wastewater05m1	26	2	199	0	3	13	48	0	169	0	3	15	24	16	239	106	3	16
wastewater05m2	43	9	1089	27	6	30	48	0	1152	0	6	33	45	42	1507	960	6	37
wastewater11m1	75	3	1750	5	58	65	140	0	1873	0	58	69	69	12	1483	181	58	67
wastewater11m2	89	16	3081	6	6	66	140	0	3138	0	6	77	90	90	2125	1264	6	81
wastewater12m1	133	3	6463	14	112	117	260	0	6684	0	112	126	124	10	6945	577	112	120
wastewater12m2	154	21	6559	2	10	131	260	0	5345	0	10	129	149	139	5015	2544	10	134
wastewater13m1	276	4	12608	12	246	258	540	0	14539	0	246	266	265	16	8607	1171	246	262
wastewater13m2	282	10	20279	0	13	228	540	0	26621	0	13	244	301	253	15320	8019	13	285
wastewater14m1	51	3	971	7	30	42	90	0	917	0	30	44	44	13	1091	267	30	44
wastewater14m2	51	6	3228	1	13	40	90	0	2181	0	13	69	52	34	2784	810	13	49
wastewater15m1	28	3	381	7	12	22	48	0	443	0	12	26	22	12	372	131	12	22
wastewater15m2	37	6	1232	7	13	33	48	0	1261	0	13	43	38	23	1574	550	13	36
waterund01	41	10	495	21	14	18	54	0	503	0	14	19	29	21	666	393	14	23
waterund08	98	21	2877	37	34	50	124	0	2794	0	34	49	59	26	3009	830	34	53
waterund11	54	9	904	8	20	27	78	0	803	0	20	30	36	21	817	430	20	28
waterund14	140	19	3434	25	42	67	160	0	3844	0	41	64	83	45	3986	1257	41	68
waterund17	78	10	1601	22	25	42	96	0	1504	0	25	41	57	34	1812	619	25	47
waterund18	62	12	876	12	21	31	78	0	884	0	21	32	42	22	976	390	21	34
waterund22	128	32	2698	64	32	49	160	0	2737	0	32	52	71	33	2091	742	32	51
waterund25	103	15	4021	23	34	47	140	0	4016	0	34	48	70	26	4886	1163	34	50
waterund27	321	91	61193	198	105	148	400	0	57524	0	105	144	205	64	49595	5702	105	148
waterund28	950	57	923371	48	254	488	1320	0	906130	0	255	524	658	194	768548	32680	254	489
waterund32	741	94	150635	159	180	360	1160	0	151030	0	180	381	567	107	91337	15635	182	377
waterund36	334	84	54394	252	101	133	432	0	52570	0	101	133	193	53	54552	8883	101	143

Table A.7 continued

Instance	greedy (base)						greedy+filt. off						greedy+filt. aggr.					
	lp	lp_{filt}	iter	$iter_{filt}$	b_{obbt}	lvb	lp	lp_{filt}	iter	$iter_{filt}$	b_{obbt}	lvb	lp	lp_{filt}	iter	$iter_{filt}$	b_{obbt}	lvb
weapons	182	37	2143	38	0	69	310	0	2177	0	0	78	157	57	3017	1032	0	78

Table A.8: Detailed results comparing ordering strategies for OBBT at the root node on test set INT. Results for strategy "greedy (base)" are listed in Table A.6. See Table 6.1 for aggregated results.

lp — number of LPs solved during OBBT
lp_{filt} — number of LPs solved during OBBT's filtering
iter — number of LP iterations during OBBT
$iter_{filt}$ — number of LP iterations of OBBT's filtering LPs
b_{obbt} — number of bounds tightened by OBBT
lvb — number of LVBs found by OBBT

Instance	random						min-max						greedy (base)						reverse greedy					
	lp	lp_{filt}	iter	$iter_{filt}$	b_{obbt}	lvb	lp	lp_{filt}	iter	$iter_{filt}$	b_{obbt}	lvb	lp	lp_{filt}	iter	$iter_{filt}$	b_{obbt}	lvb	lp	lp_{filt}	iter	$iter_{filt}$	b_{obbt}	lvb
4stufen	54	9	1130	12	8	28	49	11	1018	7	8	30	54	9	649	3	8	31	52	14	1244	9	8	30
alan	5	0	39	0	5	5	5	0	23	0	5	5	6	0	33	0	5	5	6	0	33	0	5	5
batch0812_nc	77	14	1310	5	40	55	78	13	1535	6	40	55	76	9	1723	1	40	54	79	20	1430	11	40	55
batch0812	49	4	964	3	34	43	51	3	1050	4	34	43	51	2	672	0	34	44	49	5	1020	4	34	42
batchdes	15	3	161	1	9	12	15	2	146	0	9	12	15	0	104	0	9	11	15	3	224	1	9	12
batch_nc	54	14	2145	4	20	39	57	11	2417	3	20	39	56	10	1947	3	20	36	53	14	2031	3	20	38
batch	34	1	791	0	16	24	33	2	787	0	16	23	36	0	638	0	16	27	33	1	1155	0	16	23
batchs101006m	68	4	2116	2	19	24	74	2	1962	0	19	27	86	1	1710	1	19	32	71	7	2306	1	19	30
batchs121208m	96	11	9279	1	23	39	98	9	9415	1	23	41	105	0	7776	0	23	39	96	14	10556	1	23	35
batchs151208m	89	12	3306	1	23	42	90	10	3030	0	23	46	108	0	3504	0	23	48	94	19	3724	0	23	45
batchs201210m	96	14	4703	0	23	48	93	13	3934	2	23	51	119	2	4635	0	23	47	87	14	5128	6	23	45
bchoco05	80	17	2586	50	37	75	80	14	2712	39	37	74	84	6	1848	12	37	80	83	19	3136	53	37	77
bchoco06	143	59	8138	245	6	120	144	63	7499	401	2	118	142	42	6622	90	6	115	138	59	6667	130	6	104
blend029	33	6	377	16	2	16	38	8	509	17	2	19	36	5	469	4	2	22	35	7	520	20	2	15
blend146	93	26	2119	150	0	45	91	23	2851	127	0	53	108	13	3729	53	0	54	98	27	3508	183	0	45
blend480	130	36	4294	56	0	83	128	31	3553	37	0	87	142	26	5335	78	0	88	135	34	4444	33	0	68
blend531	104	23	2217	56	0	65	105	23	2365	33	0	59	118	15	2311	48	0	57	110	29	2843	43	0	57
blend718	86	15	2101	61	1	49	80	12	2012	50	1	46	90	6	2087	20	1	43	91	21	2355	61	1	39
blend721	95	19	2709	20	0	57	94	25	2347	27	0	56	104	17	2903	95	0	58	104	31	2794	52	0	47
blend852	129	33	3203	42	0	71	125	33	2711	47	0	62	145	14	3981	55	0	72	130	29	3204	52	0	70
carton7	40	11	1414	64	4	12	38	9	1740	50	0	15	39	6	1644	2	4	16	40	13	1888	69	4	13
carton9	57	18	2317	24	11	18	60	21	1927	23	11	19	60	7	1945	19	11	21	62	24	2870	71	11	18
casctanks	431	101	49363	75	185	347	439	114	42794	74	185	357	452	53	22032	22	185	352	489	102	53955	67	185	354
cecil_13	101	11	21454	0	43	78	109	15	9790	3	43	79	108	2	3864	2	43	79	93	15	18557	1	43	77
chp_partload	1487	384	357337	169	399	982	1488	380	338716	208	399	984	1561	297	146966	137	399	995	1501	408	423902	218	399	998
clay0203h	258	148	9376	133	0	229	262	140	12132	88	0	227	264	140	3895	100	0	235	268	145	18737	112	0	223
clay0203m	9	0	85	0	0	1	9	0	105	0	0	0	12	0	135	0	0	0	9	1	91	1	0	0
clay0204h	357	177	23796	110	0	319	353	176	19114	186	0	315	357	176	5427	121	0	321	353	174	26741	117	0	312
clay0204m	12	0	180	0	0	1	14	0	148	0	0	1	15	0	158	0	0	1	9	0	171	0	0	1
clay0205h	478	317	26540	205	0	359	490	317	20862	219	0	397	489	267	16457	162	0	430	516	312	41173	149	0	403
clay0205m	18	0	448	0	0	1	18	0	381	0	0	1	19	0	394	0	0	3	19	0	600	0	0	0
clay0303h	426	277	10015	246	0	313	439	288	9339	227	0	316	454	249	4629	16	0	332	424	264	10831	115	0	324
clay0303m	9	0	134	0	0	0	10	0	159	0	0	1	12	0	174	0	0	0	10	0	217	0	0	0
clay0304h	567	366	12543	128	0	422	548	336	12907	176	0	361	589	323	9242	82	0	425	566	341	16607	320	0	443
clay0304m	14	0	225	0	2	2	14	1	235	0	0	1	16	0	236	0	0	0	12	0	223	0	0	1
clay0305h	716	460	19102	197	0	548	712	464	17464	364	0	536	757	413	15207	269	0	558	722	463	18244	339	0	565

Table A.8 continued

Instance	random						min-max						greedy (base)						reverse greedy					
	lp	lp_{filt}	$iter$	$iter_{filt}$	b_{obbt}	lvb	lp	lp_{filt}	$iter$	$iter_{filt}$	b_{obbt}	lvb	lp	lp_{filt}	$iter$	$iter_{filt}$	b_{obbt}	lvb	lp	lp_{filt}	$iter$	$iter_{filt}$	b_{obbt}	lvb
clay0305m	18	0	485		0	1	18		406		0	0	20		288	1	0	0	17	0	522	0	0	1
crudeoil_lee1.05	90	39	5967	159	34	48	86	37	5518	82	33	50	91	34	3787	70	32	50	92	42	6786	108	33	52
crudeoil_lee1.06	113	65	9892	175	22	59	115	67	9291	204	21	55	117	48	8147	150	21	59	107	57	11720	145	22	52
crudeoil_lee1.07	121	63	14835	141	25	73	132	76	12646	206	25	80	135	50	10008	116	25	72	131	74	15285	183	24	80
crudeoil_lee1.08	163	97	20275	297	28	87	154	87	20567	235	30	87	176	69	12957	140	28	95	162	93	21835	268	31	90
crudeoil_lee1.09	181	101	27229	318	36	97	181	98	26518	291	34	109	192	83	18964	201	36	104	180	102	30227	204	35	105
crudeoil_lee1.10	194	105	33570	472	44	109	193	113	31059	456	41	117	195	64	18502	161	42	115	192	104	32176	331	44	117
crudeoil_lee2.05	148	66	29494	101	39	86	148	68	27636	159	39	83	158	48	20051	86	39	86	143	65	30231	112	39	87
crudeoil_lee2.06	200	92	49423	195	55	119	211	102	51343	191	55	118	212	64	30629	130	55	120	201	96	52271	245	55	116
crudeoil_lee2.07	239	111	75143	577	64	137	241	109	81824	458	64	137	246	81	46873	261	64	139	236	112	81417	459	64	142
crudeoil_lee2.08	284	138	103947	473	74	176	280	129	102900	336	74	166	304	122	62279	373	74	172	292	149	106808	543	74	167
crudeoil_lee2.09	318	147	156297	452	88	199	314	142	163736	517	88	198	342	117	99020	303	88	210	325	159	163760	703	88	190
crudeoil_lee2.10	370	185	192113	775	97	212	365	181	184391	1038	97	221	378	123	110685	434	97	209	366	177	190908	909	97	218
crudeoil_lee3.05	287	126	24112	298	120	186	283	118	27882	279	120	195	289	91	18775	172	120	181	289	118	28354	252	120	195
crudeoil_lee3.06	384	161	90551	401	173	253	392	169	86420	401	173	261	385	125	61367	385	173	256	375	154	92173	340	173	258
crudeoil_lee3.07	470	180	103216	436	226	338	464	179	103795	463	226	325	473	146	65455	388	226	340	472	193	93411	646	226	333
crudeoil_lee3.08	549	185	239891	602	287	384	535	187	225806	522	287	383	556	157	154527	420	287	400	543	201	226734	599	287	363
crudeoil_lee3.09	633	229	309283	763	341	466	639	236	300295	575	341	463	653	177	199420	465	341	472	650	248	307962	1039	341	458
crudeoil_lee3.10	729	265	418387	939	397	512	715	250	423729	949	397	520	743	212	264413	897	397	550	731	272	413710	1175	397	524
crudeoil_lee4.05	188	62	42158	208	75	127	186	75	40577	89	75	126	194	49	22919	137	75	126	185	62	38900	125	75	136
crudeoil_lee4.06	240	79	83813	284	100	168	244	81	92723	199	100	171	247	55	54350	102	100	170	245	83	87453	231	100	173
crudeoil_lee4.07	296	106	108887	559	117	203	288	99	124704	249	117	192	311	85	74047	180	117	204	286	98	116459	279	117	199
crudeoil_lee4.08	332	123	150556	518	126	219	328	114	153321	330	126	222	351	82	95270	146	126	228	330	122	155091	347	126	208
crudeoil_lee4.09	401	159	209233	394	150	269	398	155	213011	563	150	272	417	102	132726	268	150	264	404	162	224760	622	150	271
crudeoil_lee4.10	408	140	252133		157	286	406	150	227715	428	157	287	430	107	157458	330	157	295	409	149	235760	501	157	281
crudeoil_li01	103	15	3404	20	36	36	100	8	3257	6	36	37	95	6	2553	8	36	32	98	16	4062	21	36	35
crudeoil_li02	31	7	2871	1	13	26	30	8	2636	2	13	25	32	6	1550	2	13	28	31	9	3064	3	13	26
crudeoil_li03	238	69	14974	130	21	30	242	56	14280	76	21	31	254	31	12160	13	21	33	226	85	14994	169	21	28
crudeoil_li05	202	35	8888	514	13	20	212	34	8826	206	13	20	236	31	7467	229	13	21	189	47	10114	234	13	21
crudeoil_li06	246	64	13636	72	14	23	250	64	11239	94	14	23	255	25	11183	16	14	25	233	75	13159	114	14	30
crudeoil_li11	240	59	16730	75	19	19	237	60	15701	57	19	25	257	26	14927	16	19	21	240	78	16610	176	19	21
crudeoil_li21	235	69	350065	3338	17	13	249	77	30150	91	17	16	261	31	31300	362	17	11	234	85	32682	1116	17	18
csched1a	14	1	86	0	3	7	15	1	123	1	3	7	16	1	104	0	3	7	11	1	109	1	3	7
csched1	15	2	155	1	1	6	14	1	149	1	1	5	15	1	130	1	1	7	10	0	181	0	1	5
csched2a	112	62	4445	31	10	61	107	59	6793	31	10	58	125	28	6817	12	10	70	111	66	6769	45	10	62
csched2	152	84	1563	71	1	64	151	79	1577	65	1	59	152	64	1306	55	1	63	162	93	1333	83	1	64
densitymod	315	0	191478	0	315	315	315	0	188197	0	315	315	316	0	184553	0	315	315	315	0	186328	0	315	315
du-opt5	12	0	229	0	4	5	14	0	231	1	4	4	17	0	240	0	4	5	9	1	232	1	4	5
du-opt	18	0	224	0	4	5	12	0	212	0	4	4	17	0	184	4	4	5	11	0	241	0	4	5
edgecross10-060	44	0	3113	2	0	0	42	0	3291	0	0	0	52	4	2082	4	0	0	54	3	7045	0	0	0
edgecross14-039	143	2	16262	7	2	0	126	14	11717	23	0	0	161	4	7492	0	0	0	151	3	23834	127	0	0
elf	12	3	209	7	7	6	12	1	120	3	0	4	11	2	286	0	5	5	11	3	259	8	0	5
eniplac	69	15	432	21	21	18	67	11	431	19	21	21	65	2	369	5	21	23	70	21	405	23	21	20
enpro48pb	36	0	1174	10	10	14	37	0	1282	0	14	14	42	0	1204	5	14	14	32	0	1441	0	14	14
enpro56pb	35	1	2052	8	8	16	35	1	2044	1	15	15	39	0	1667	0	8	17	30	1	2470	1	5	18
ethanolh	149	46	5997	46	5	74	157	56	3953	56	5	73	159	30	2729	28	5	73	145	54	6916	49	5	61

Table A.8 continued

Instance	random						min-max						greedy (base)						reverse greedy					
	lp	lp_{filt}	iter	$iter_{filt}$	b_{obbr}	lvb	lp	lp_{filt}	iter	$iter_{filt}$	b_{obbr}	lvb	lp	lp_{filt}	iter	$iter_{filt}$	b_{obbr}	lvb	lp	lp_{filt}	iter	$iter_{filt}$	b_{obbr}	lvb
ethanolm	64	16	1130	14	7	34	74	16	906	10	7	36	81	10	700	9	7	37	66	21	1171	16	7	39
ex1224	8	0	90	0	4	7	8	0	128	0	4	6	8	0	67	0	4	6	8	1	71	0	4	6
ex1233	71	20	262	43	16	27	70	16	319	32	16	26	75	1	191	0	16	26	67	22	226	44	16	27
ex1243	43	5	81	6	6	5	42	3	58	5	0	4	42	4	85	4	0	5	42	3	70	7	0	4
ex1244	59	5	260	6	2	14	59	3	294	0	2	9	59	3	200	4	2	12	59	5	200	7	2	11
ex1252a	24	10	250	3	1	12	26	6	272	5	1	9	32	14	209	0	1	17	36	22	325	25	1	13
ex1252	36	22	274	18	3	18	30	16	326	11	3	16	32	17	500	7	3	16	41	30	471	32	3	21
ex1263a	21	1	135	1	12	14	22	2	97	1	12	17	22	1	112	20	12	16	22	2	103	3	12	16
ex1263	27	5	353	16	11	16	27	5	323	11	11	14	26	1	362	1	11	15	24	3	291	3	11	14
ex1264a	21	0	106	9	0	10	22	0	86	9	0	9	23	2	90	1	9	9	21	1	105	1	9	11
ex1264	21	2	220	9	8	13	32	0	232	1	8	13	24	1	198	0	8	13	22	1	228	3	8	15
ex1265a	33	2	282	2	12	17	32	4	249	2	12	15	40	0	231	0	12	16	33	3	171	3	12	16
ex1265	32	5	189	4	6	17	44	3	175	2	6	16	36	3	251	2	6	15	30	3	211	4	6	14
ex1266a	44	3	202	2	21	21	42	1	721	7	21	21	52	0	223	0	21	25	42	3	195	2	21	22
ex1266	44	4	617	17	19	26	9	1	155	0	19	29	55	1	770	1	19	26	41	3	621	8	19	26
ex4	8	1	205	0	4	5	19	10	87	13	0	5	8	0	238	0	4	5	7	1	256	0	2	5
fac1	22	9	101	10	0	10	45	16	172	20	0	8	23	11	70	14	4	8	18	13	75	13	13	7
fac2	41	14	168	24	0	23	60	24	388	29	0	31	48	21	156	25	0	37	45	16	172	20	0	31
fac3	59	17	413	14	0	29	359	64	12653	69	0	36	69	17	338	16	0	36	61	25	371	24	0	37
feedtray	354	75	14256	32	241	112	359	64	12653	69	241	131	372	57	8258	71	241	115	336	75	22129	35	241	109
fin2bb	84	0	7436	0	61	61	84	0	4756	0	61	61	84	0	2975	61	61	61	84	0	3636	61	61	61
flay02h	3	0	68	0	2	2	3	0	74	0	2	2	3	0	74	0	2	2	4	0	82	0	2	2
flay02m	4	0	107	0	2	2	4	0	75	0	2	2	4	0	61	0	2	3	4	0	124	0	3	2
flay03h	6	0	316	0	3	4	6	0	300	0	3	4	6	0	213	0	3	4	6	0	357	0	3	4
flay03m	6	0	106	0	3	3	6	0	122	0	3	3	6	0	72	0	3	3	6	0	168	0	3	3
flay04h	8	0	277	0	4	1	8	0	254	0	4	1	8	0	234	0	4	3	8	0	317	0	4	1
flay04m	8	1	158	0	0	4	8	0	111	0	0	4	8	0	80	0	1	4	7	0	144	0	1	4
flay05h	10	0	1334	0	0	5	10	0	1366	0	0	6	10	0	1071	0	0	6	10	0	1653	0	0	6
flay05m	9	0	184	0	0	5	9	0	193	0	0	5	9	0	103	0	0	5	9	0	212	0	0	5
flay06h	12	0	2070	0	0	4	12	0	2064	0	0	5	12	0	1776	0	0	5	12	0	2609	0	0	5
flay06m	12	1	310	1	0	1	12	0	251	0	0	5	12	0	253	0	0	5	11	0	394	0	0	5
fo7_2	21	1	142	1	0	14	24	1	121	1	0	1	27	4	101	4	0	2	19	2	118	3	0	1
fo7.ar2_1	18	0	145	1	0	14	19	1	129	1	0	14	20	0	98	0	0	14	16	1	151	1	0	14
fo7.ar25_1	16	0	177	0	0	4	18	0	109	0	0	4	19	0	149	0	0	4	14	0	149	0	0	4
fo7.ar3_1	15	0	92	0	0	1	17	0	100	0	0	1	19	0	103	0	0	1	15	2	107	1	0	1
fo7.ar4_1	14	0	108	0	0	5	19	1	100	0	0	5	17	0	94	0	0	5	15	2	128	0	0	5
fo7.ar5_1	19	1	159	0	0	6	23	0	171	0	0	6	23	1	107	0	0	6	18	2	176	2	0	7
fo7	17	1	317	2	0	16	18	0	263	0	0	16	20	0	183	2	0	16	16	1	220	1	0	16
fo8.ar2_1	19	0	108	0	0	7	22	0	126	0	0	7	22	0	99	0	0	7	17	0	177	0	4	7
fo8.ar25_1	22	0	119	0	0	4	24	0	164	0	0	4	25	0	123	0	0	7	22	0	215	0	0	7
fo8.ar3_1	17	0	136	0	0	8	19	0	164	0	0	8	18	0	106	0	0	4	16	0	198	0	0	4
fo8.ar4_1	20	5	100	0	0	6	22	5	128	0	0	6	18	0	114	0	0	7	21	3	211	0	0	8
fo8.ar5_1	21	2	180	0	0	1	24	1	163	0	0	1	25	2	122	0	0	6	19	0	186	0	0	6
fo8	22	0	346	0	0	18	23	0	325	0	0	18	26	0	240	2	0	1	21	0	331	0	0	1
fo9.ar2_1	22	0	115	0	0	18	24	0	119	0	0	7	25	0	58	0	0	18	19	0	151	0	0	18
fo9.ar25_1	23	0	115	0	0	7	23	0	127	0	0	7	24	0	104	0	0	7	21	0	208	0	0	7

Table A.8 continued

Instance	random						min–max						greedy (base)						reverse greedy					
	lp	lp_{filt}	iter	$iter_{filt}$	b_{obbt}	lvb	lp	lp_{filt}	iter	$iter_{filt}$	b_{obbt}	lvb	lp	lp_{filt}	iter	$iter_{filt}$	b_{obbt}	lvb	lp	lp_{filt}	iter	$iter_{filt}$	b_{obbt}	lvb
fo9_ar3_1	18	0	251	0	0	4	23	0	186	0	0	4	25	0	160	0	0	4	19	0	225	0	0	4
fo9_ar4_1	24	0	243	0	0	7	24	0	200	0	0	7	24	1	178	0	0	7	21	3	260	1	0	7
fo9_ar5_1	23	1	222	0	0	8	28	1	205	1	0	8	30	2	153	0	0	8	24	5	279	3	0	8
fo9	19	0	256	0	0	0	24	5	311	4	0	0	31	1	266	0	0	0	18	1	365	1	0	0
fuel	10	2	231	0	4	9	10	1	263	0	4	9	11	1	192	1	4	8	11	3	276	3	4	9
fuzzy	120	15	10377	7	80	84	119	22	8657	8	80	84	123	19	5363	5	80	83	116	15	8858	6	80	84
gams01	216	25	271251	23	118	143	212	27	233527	24	118	138	236	8	120856	13	118	141	216	27	296185	22	118	142
gasprod_sarawak01	37	9	547	2	6	32	39	11	635	2	6	34	37	11	511	4	6	32	37	13	657	8	6	34
gasprod_sarawak16	566	193	52080	91	113	479	565	166	40639	58	113	486	572	149	30238	70	113	494	564	213	48142	88	113	494
gasprod_sarawak81	2846	1123	607509	620	337	2432	2882	1026	428958	523	337	2420	2899	813	265274	608	337	2443	2819	1227	453014	543	337	2393
gastrans	57	5	616	3	35	34	56	5	774	3	35	34	59	2	589	3	35	32	55	6	760	0	35	32
gear2	7	4	46	3	0	4	7	2	57	1	0	4	7	4	44	0	0	4	7	1	73	4	0	4
gear3	6	2	22	0	1	5	7	2	20	0	1	5	6	2	17	1	1	5	8	1	37	0	1	5
gear4	6	2	48	1	2	2	6	2	46	1	2	2	6	2	30	0	2	2	6	1	48	0	2	2
gear	6	1	22	0	1	5	7	2	20	0	1	5	6	2	17	0	1	5	8	1	37	0	1	5
genpooling_lee1	27	6	115	10	0	6	26	4	164	8	0	5	26	3	130	7	0	3	26	6	149	22	0	4
genpooling_lee2	32	5	109	21	0	8	31	4	126	5	0	8	29	3	88	4	0	9	30	9	140	25	0	10
genpooling_meyer04	31	5	1824	43	0	0	31	5	1224	28	0	10	34	1	1231	11	0	0	33	7	1402	45	0	7
genpooling_meyer10	128	19	3150	49	0	39	131	16	3481	41	0	36	135	5	6396	19	0	49	128	15	3970	29	0	53
genpooling_meyer15	252	43	4446	113	0	99	268	45	4378	118	0	96	270	26	4818	92	0	106	260	48	4301	110	0	113
ghg_1veh	55	16	656	12	3	21	56	14	568	9	3	19	52	14	457	7	3	19	54	20	795	15	3	20
ghg_2veh	117	43	3685	32	27	58	118	44	3380	29	27	58	113	44	1918	21	27	59	123	40	4627	27	27	60
ghg_3veh	196	74	10064	49	46	104	192	73	7425	80	46	108	191	65	4903	45	46	107	199	72	11585	31	46	111
gkocis	4	0	19	2	2	3	4	0	12	0	2	3	4	0	19	0	2	3	4	0	26	0	2	3
heatexch_gen1	115	55	636	64	0	60	115	49	713	83	1	56	115	32	610	42	2	58	110	47	892	94	2	59
heatexch_spec1	95	47	450	59	1	30	97	45	379	61	1	27	94	31	373	40	1	21	93	50	372	53	1	26
heatexch_spec2	87	35	745	48	10	40	84	30	847	52	10	44	84	16	458	26	10	39	69	25	608	42	10	45
heatexch_spec3	424	139	2560	221	0	86	373	134	2156	202	0	101	442	86	1643	210	0	89	408	177	2268	294	0	93
heatexch_trigen	103	39	1232	33	16	57	96	36	1249	36	16	56	112	31	944	24	16	59	92	39	1163	32	16	59
hybriddynamic_var	37	1	849	1	22	25	40	1	428	0	22	25	52	1	381	1	22	25	35	5	948	7	22	25
hydroenergy1	133	58	1756	55	13	78	132	57	1579	47	13	80	154	13	1271	14	13	78	135	50	1873	52	13	73
hydroenergy2	229	82	3240	41	13	114	223	84	2931	49	13	115	284	33	2135	23	13	108	237	81	3490	53	13	111
hydroenergy3	436	198	8333	132	13	270	429	195	8446	85	13	285	480	89	5007	57	13	274	472	195	8871	134	13	279
jit1	8	0	84	0	6	6	8	0	56	0	6	6	8	0	67	0	6	6	8	0	43	0	6	6
kport20	152	44	679	31	8	31	148	45	669	41	8	33	172	27	718	42	8	32	142	45	733	35	8	33
kport40	284	102	5042	113	8	93	287	96	4274	101	8	94	310	46	3692	34	8	94	282	94	4271	81	0	99
lip	8	1	81	0	8	7	9	0	71	0	8	7	8	0	103	0	8	7	8	0	92	0	8	7
m3	10	1	37	0	0	2	9	0	21	0	0	2	9	1	27	1	0	2	9	0	56	0	0	2
m6	14	0	177	0	0	3	17	0	169	0	0	3	18	1	103	0	0	3	17	3	184	2	0	3
m7_ar2_1	18	1	113	1	0	4	18	1	83	1	0	4	14	1	84	0	0	4	15	0	93	0	0	4
m7_ar25_1	16	3	110	1	0	3	18	1	121	1	0	3	18	1	99	0	0	3	18	2	138	0	0	3
m7_ar3_1	22	4	150	0	0	6	23	2	127	1	0	6	22	2	124	0	0	6	21	6	211	0	0	6
m7_ar4_1	20	4	216	0	0	7	19	3	206	0	0	7	19	2	153	0	0	7	20	7	224	2	0	7
m7_ar5_1	25	4	193	0	0	5	23	4	215	2	0	5	23	3	124	0	0	5	25	7	292	2	0	5
m7	18	0	131	0	0	1	17	0	129	0	0	1	16	0	139	0	0	1	16	0	142	0	0	1
meanvarxsc	9	2	44	0	0	7	8	0	61	0	0	7	10	0	46	0	0	8	9	0	66	0	0	8

Table A.8 continued

Instance	random						min-max						greedy (base)						reverse greedy					
	lp	lp_{filt}	iter	$iter_{filt}$	b_{obbt}	lvb	lp	lp_{filt}	iter	$iter_{filt}$	b_{obbt}	lvb	lp	lp_{filt}	iter	$iter_{filt}$	b_{obbt}	lvb	lp	lp_{filt}	iter	$iter_{filt}$	b_{obbt}	lvb
milinfract	416	0	4394	4	0	1	431	0	4401	4	0	1	494	0	3759	0	0	1	393	0	4504	0	0	1
minlphix	24	3	68	4	6	9	23	3	62	4	6	9	23	4	36	5	6	8	24	3	88	5	6	10
multiplants.mtg1c	49	15	16991	14	19	24	53	16	13501	6	19	21	48	5	13566	3	19	27	51	15	17934	11	19	20
multiplants.mtg5	65	10	2671	5	33	49	65	8	2216	2	33	48	70	5	1654	6	33	44	68	13	3281	14	33	51
multiplants.stg1b	151	73	33250	171	8	36	130	49	35611	162	8	30	215	70	32291	109	8	37	155	71	35078	90	8	19
multiplants.stg1c	91	29	29701	41	0	15	88	29	39740	85	0	16	126	19	10159	29	33	41	128	64	21437	101	9	14
multiplants.stg6	147	87	80288	195	9	29	143	44	50059	233	8	40	217	42	48181	113	9	87	188	91	75262	251	9	35
ndcc12	613	33	28770	19	97	317	601	27	22051	6	97	331	836	5	19748	4	97	319	612	36	15637	25	97	325
ndcc12persp	130	1	2818	0	0	73	129	0	1962	0	4	77	134	1	2002	0	4	76	129	0	2701	0	4	68
ndcc13	624	21	32094	8	210	408	607	13	23716	3	210	403	837	1	20898	0	210	405	604	11	14640	5	210	403
ndcc13persp	127	0	1901	0	72	90	127	0	1667	0	72	84	129	1	2067	0	72	86	126	1	1732	3	72	86
ndcc14	844	38	62684	17	82	446	841	41	44662	18	82	457	1138	4	37164	4	82	462	826	23	26092	3	82	427
ndcc14persp	159	4	4083	1	2	96	157	4	3092	1	2	88	160	1	3123	1	2	88	160	2	4411	1	2	85
ndcc15	643	16	64087	6	87	354	648	15	46671	9	87	381	880	1	46475	0	87	361	639	6	27119	4	87	328
ndcc16	1047	39	83759	13	60	532	1050	49	51773	16	60	543	1430	1	51437	0	60	535	1039	22	34194	8	60	460
ndcc16persp	177	3	4952	0	0	70	175	4	4168	0	0	64	177	0	4552	0	0	58	176	0	5348	0	0	58
netmod_dol1	8	1	22	1	0	2	7	0	19	0	0	2	10	4	18	0	0	4	6	0	22	0	0	4
netmod_dol2	10	4	5288	5	0	1	9	3	3856	4	0	1	10	4	3547	5	0	2	8	0	9094	0	0	3
netmod_kar1	6	1	13	1	0	2	6	1	12	1	0	2	8	2	14	2	0	1	5	1	13	1	0	2
netmod_kar2	6	1	13	1	0	2	6	1	12	1	0	2	8	2	14	2	0	1	5	1	13	1	0	2
no7_ar2_1	19	0	130	0	0	14	24	0	139	0	0	14	20	0	113	0	0	14	16	0	170	0	0	14
no7_ar25_1	16	0	178	0	0	4	19	0	143	0	0	4	19	0	114	0	0	4	18	0	205	0	0	4
no7_ar3_1	19	1	158	0	0	1	21	0	176	0	0	1	27	0	137	0	0	1	18	2	200	3	0	1
no7_ar4_1	18	2	196	0	0	6	20	2	162	0	0	6	23	0	144	0	0	6	18	3	171	0	0	6
no7_ar5_1	24	3	211	2	0	6	27	2	250	0	0	6	22	2	188	0	0	7	22	3	281	0	0	6
nous1	50	22	300	21	6	32	50	17	352	12	0	33	54	10	285	13	6	34	51	24	304	23	6	33
nous2	60	29	387	26	7	37	55	26	344	19	7	41	61	19	381	20	7	37	55	25	391	20	7	34
nuclear104	2228	40	2142	17	0	29	2232	40	2303	17	0	29	2234	39	1764	16	0	29	2227	40	2680	17	0	29
nuclear14a	593	195	370944	85466	10	376	592	187	239133	79491	10	383	618	162	202427	89805	10	373	565	175	201612	53606	10	376
nuclear14b	610	113	140665	3054	203	375	618	112	78886	2044	203	365	613	95	66415	4155	203	368	605	113	107550	3248	203	377
nuclear14	417	0	657	0	0	0	417	0	611	0	0	0	417	0	325	0	0	0	416	1	861	1	0	0
nuclear25a	574	185	85266	11054	7	374	578	181	65180	12629	7	377	597	148	237521	137860	7	372	569	187	219967	138747	7	371
nuclear25b	759	174	166037	7933	215	417	765	175	113147	6249	215	411	763	155	104529	2182	215	409	769	190	167498	9644	215	421
nuclear25	433	1	651	1	0	0	433	1	682	1	0	0	435	0	386	1	0	0	432	1	790	1	0	0
nuclear49a	1256	401	324252	60214	8	794	1242	388	225194	62799	8	787	1390	361	94723	36372	8	809	1217	395	266001	41856	8	783
nuclear49b	1664	390	971746	27547	458	939	1662	392	618448	45659	458	919	1665	311	295948	9365	458	916	1658	390	767076	50106	458	913
nuclear49	949	11	2241	6	0	8	947	10	1489	5	0	7	951	10	673	5	0	0	948	11	2576	6	0	7
nuclearva	189	2	220	2	0	0	188	2	259	2	0	0	189	2	132	2	0	0	188	2	260	2	0	0
nuclearvb	188	1	168	1	0	0	187	1	199	1	0	0	188	1	121	1	0	0	187	1	248	1	0	0
nuclearvc	188	1	168	1	0	0	187	1	197	1	0	0	188	1	121	1	0	0	187	1	248	1	0	0
nuclearvd	188	2	148	3	0	0	187	1	264	1	0	0	188	1	135	1	0	0	188	2	262	2	0	0
nuclearve	188	2	148	3	0	0	187	1	264	1	0	0	188	1	135	1	0	0	188	2	262	2	0	0
nuclearvf	188	2	148	3	0	0	187	1	264	1	0	0	188	1	135	1	0	0	188	2	262	2	0	0
nvs01	14	1	138	0	3	5	12	0	161	0	3	5	11	1	110	1	3	5	13	4	207	4	3	6
nvs02	5	0	21	0	0	0	5	0	28	0	0	0	5	0	21	0	0	0	7	0	54	3	0	0
nvs05	23	2	136	1	2	6	24	3	149	2	2	7	25	1	111	0	2	7	27	6	125	5	2	7

Table A.8 continued

Instance	random						min–max						greedy (base)						reverse greedy					
	lp	lp_filt	iter	iter_filt	b_obbt	lvb	lp	lp_filt	iter	iter_filt	b_obbt	lvb	lp	lp_filt	iter	iter_filt	b_obbt	lvb	lp	lp_filt	iter	iter_filt	b_obbt	lvb
nvs06	8	1	29	1	5	7	9	1	30	0	5	7	9	0	27	0	5	7	8	1	27	0	5	7
nvs08	4	1	48	0	1	1	5	0	19	0	1	1	5	0	36	0	1	1	4	0	31	0	1	1
nvs10	4	0	10	0	4	4	4	0	7	0	4	4	4	0	5	0	4	4	4	0	7	0	4	4
nvs11	6	0	41	0	6	6	6	0	49	0	6	6	6	0	36	0	6	6	6	0	41	0	6	6
nvs12	8	0	51	0	8	8	8	0	49	0	8	8	8	0	51	0	8	8	8	0	63	0	8	8
nvs13	9	0	72	0	8	8	9	0	65	0	8	8	10	0	53	0	8	8	8	0	69	0	8	8
nvs15	4	0	20	0	2	2	4	0	18	0	2	2	4	0	20	0	2	2	4	0	22	0	2	2
nvs16	10	2	53	2	2	1	12	2	49	2	0	0	13	4	42	0	0	0	12	4	45	5	0	0
nvs17	11	0	102	0	11	11	11	0	117	0	11	11	14	0	112	0	11	11	11	0	132	0	11	11
nvs18	10	0	96	0	10	10	10	0	82	0	10	10	12	0	90	0	10	10	10	0	108	0	10	10
nvs19	13	0	189	0	12	12	13	0	193	0	12	12	16	0	176	0	12	12	12	0	200	0	12	12
nvs20	40	2	1977	2	33	33	37	2	2076	2	33	33	43	0	1660	0	33	33	37	3	1780	3	33	33
nvs21	5	1	5	0	0	2	5	1	5	0	2	2	5	0	5	0	0	2	6	2	5	0	0	2
nvs23	15	0	238	0	13	13	15	0	227	0	13	13	17	0	249	0	13	13	14	0	260	0	13	13
nvs24	17	0	301	1	12	12	17	0	330	0	12	12	20	0	349	0	12	12	16	2	322	0	12	12
o7_2	21	1	108	0	0	0	22	0	129	0	0	0	23	0	118	0	0	0	18	0	143	1	0	0
o7_ar2_1	17	0	208	0	0	14	22	0	319	0	0	14	20	0	173	0	0	14	15	0	289	0	0	14
o7_ar25_1	19	1	133	1	0	4	20	0	198	0	0	4	26	0	143	0	0	4	25	4	261	0	0	4
o7_ar3_1	21	0	205	0	0	1	19	0	198	0	0	1	27	0	145	0	0	1	18	0	174	0	0	1
o7_ar4_1	18	2	190	0	0	6	22	4	186	1	0	0	23	0	182	0	0	6	18	3	168	0	0	6
o7_ar5_1	23	4	225	0	6	6	25	1	210	2	4	4	25	2	117	0	0	6	21	3	215	2	6	6
o7	16	0	105	0	0	0	16	0	111	0	0	0	21	0	83	0	0	0	16	6	137	0	0	0
o8_ar4_1	21	3	251	0	0	8	24	3	266	0	0	8	22	1	232	0	0	8	21	6	247	0	0	8
o9_ar4_1	25	2	222	0	0	7	28	4	318	0	0	7	27	0	211	0	0	6	27	4	275	0	0	7
oil2	1078	1	67858	1	839	1073	1082	1	47390	1	839	1073	1081	0	10787	0	839	1073	1083	1	16682	1	839	1073
oil	1360	126	63281	63	863	639	1367	134	34025	46	863	610	1370	104	11555	52	863	616	1366	138	22809	55	863	748
ortez	51	16	1535	39	8	34	49	12	1223	35	8	34	51	12	922	21	8	32	51	18	1519	49	8	33
pooling-epa1	125	52	6299	32	28	66	123	52	6882	45	28	63	128	57	5547	34	28	67	126	49	7939	46	28	55
pooling-epa2	183	72	8491	76	32	90	190	89	9670	123	32	91	201	74	8024	86	32	95	199	90	11410	105	32	91
pooling-epa3	581	200	115046	260	96	261	593	220	128257	284	96	262	620	194	91900	235	96	283	581	228	127675	395	96	251
portfol_card	17	4	53	1	10	17	17	5	38	0	10	17	17	0	25	0	10	17	17	5	69	0	10	17
portfol_classical050_1	87	0	785	0	59	61	98	0	893	0	59	61	77	0	536	0	59	61	88	0	841	0	59	61
portfol_classical200_2	354	0	4217	0	263	271	400	0	4506	0	263	271	355	0	3041	0	263	271	323	0	4398	0	263	271
portfol_robust050_34	177	0	2053	0	168	172	184	0	2102	0	168	172	178	0	1453	0	168	172	174	0	2371	0	168	172
portfol_robust100_09	354	0	3940	0	334	337	365	0	3803	0	334	337	358	0	3493	0	334	337	337	0	4810	0	334	337
portfol_robust200_03	727	0	7991	0	703	708	758	0	8534	0	703	708	752	0	6259	0	703	708	708	0	9329	0	703	708
portfol_shortfall050_68	200	11	829	2	188	200	200	12	967	3	188	200	200	11	605	0	188	200	200	12	1012	4	188	200
portfol_shortfall100_04	404	0	2563	0	404	404	404	0	2599	0	404	404	404	0	2237	0	404	404	404	0	2832	0	404	404
portfol_shortfall200_05	804	0	6276	0	804	804	804	0	6746	0	804	804	804	0	5178	0	804	804	804	0	7062	0	804	804
primary	117	18	3140	11	47	69	118	21	2668	17	47	64	117	9	1664	2	47	68	117	19	2480	12	47	64
procsel	2	0	6	0	2	2	2	0	6	0	2	2	2	0	6	0	2	2	2	0	6	0	2	2
product2	243	7	6129	3	157	211	238	8	6524	0	157	212	255	2	2123	1	157	211	238	6	5950	12	157	213
product	183	51	3333	28	0	51	191	45	2997	58	0	40	215	31	2819	20	0	32	178	57	2663	39	0	56
qapw	450	0	25320	0	450	450	450	0	25427	0	450	450	450	0	22954	0	450	450	450	0	31382	0	450	450
ravempb	32	1	1472	0	3	10	39	1	1591	0	3	10	41	0	976	0	3	12	40	7	1930	4	3	10
risk2bpb	6	0	383	0	3	3	6	0	328	0	3	3	6	0	285	0	3	3	6	0	273	0	3	3

Table A.8 continued

instance	random lp	lp_filt	iter	iter_filt	b_obbt	lvb	min-max lp	lp_filt	iter	iter_filt	b_obbt	lvb	greedy (base) lp	lp_filt	iter	iter_filt	b_obbt	lvb	reverse greedy lp	lp_filt	iter	iter_filt	b_obbt	lvb
routingdelay_bigm	3940	73	440	4	46	171	3941	65	443	2	46	155	3941	37	426	8	46	174	3932	118	440	5	46	177
routingdelay-proj	4075	68	367	0	0	127	4081	53	377	0	0	97	4080	20	351	1	0	135	4073	122	345	2	0	139
rsyn0805h	17	0	29	2	3	5	16	5	29	0	3	5	18	4	31	2	3	17	15	6	25	0	3	5
rsyn0805m02h	26	4	563	2	0	18	26	5	394	2	4	18	27	0	351	0	2	4	26	1	664	2	0	17
rsyn0805m02m	8	1	124	7	0	5	10	4	145	3	0	6	11	0	116	2	0	11	7	4	133	1	1	6
rsyn0805m03m	16	5	282	3	0	9	18	6	185	7	0	7	18	1	269	1	0	13	11	0	284	4	0	9
rsyn0805m04m	21	5	459	3	0	12	21	5	606	5	0	14	23	1	499	1	1	2	22	3	683	0	0	14
rsyn0805m	5	0	61	0	0	2	5	0	41	0	0	2	5	0	43	0	0	2	4	3	32	0	0	2
rsyn0810m02m	18	4	311	5	0	8	16	7	256	1	0	9	22	2	159	0	0	9	17	14	303	5	0	9
rsyn0810m03m	26	5	554	7	0	12	28	7	636	11	0	9	33	2	470	3	0	16	30	0	673	14	0	12
rsyn0810m04m	37	7	863	9	0	18	38	10	1042	10	0	15	45	4	872	5	0	19	38	10	953	13	0	18
rsyn0810m	10	0	83	0	0	3	8	0	71	0	0	5	9	0	62	0	0	3	8	19	63	0	0	3
rsyn0815m02m	33	7	731	5	0	17	32	7	790	8	0	17	39	3	589	3	0	15	32	23	503	13	0	12
rsyn0815m03m	48	11	2441	14	2	18	49	17	2161	25	2	18	61	1	2026	0	0	21	52	14	2210	25	2	17
rsyn0815m04m	66	15	2961	20	3	27	70	18	3221	30	3	26	83	1	2744	3	3	29	65	15	3409	35	3	27
rsyn0815m	19	4	216	5	0	9	16	6	216	7	0	5	19	2	139	3	0	4	15	34	219	8	0	5
rsyn0820m02m	42	12	1099	12	0	18	40	11	1069	17	2	18	50	4	862	1	0	21	43	5	1106	32	3	19
rsyn0820m03m	57	12	2074	11	0	25	56	20	1944	26	0	22	68	4	1476	10	0	24	57	9	1809	29	2	24
rsyn0820m04m	84	20	2433	36	3	28	83	20	2288	25	3	28	102	1	1917	1	0	29	88	24	2614	57	3	29
rsyn0820m	21	3	196	3	0	9	25	7	258	9	0	5	26	4	190	1	0	7	21	37	291	5	0	7
rsyn0830m02m	54	13	1146	12	0	25	55	11	1126	10	3	31	62	4	830	3	0	25	54	8	1414	11	3	27
rsyn0830m03m	92	26	1653	53	0	34	89	28	1880	55	0	35	98	8	1428	9	0	34	82	24	1634	50	0	31
rsyn0830m04m	117	31	3951	38	4	52	111	26	4412	41	4	50	136	7	3120	7	4	53	118	27	2787	42	4	53
rsyn0830m	25	2	398	1	0	10	24	2	329	2	0	14	29	0	288	0	0	13	27	39	332	8	0	14
rsyn0840m02m	75	16	1320	22	0	32	78	21	1356	33	0	33	91	3	1115	8	0	36	74	9	1214	26	0	37
rsyn0840m03m	120	26	3420	44	1	55	122	29	2951	30	0	64	145	7	2006	15	1	51	112	1	2535	28	0	62
rsyn0840m04m	150	30	5005	35	1	65	152	32	5461	27	3	64	190	0	3313	6	3	53	145	37	4392	42	3	59
rsyn0840m	37	0	543	0	0	14	37	5	551	5	1	11	42	0	366	0	0	15	36	9	512	15	1	16
sep1	10	0	69	0	6	9	10	0	67	0	6	8	10	0	61	0	6	8	10	1	75	1	6	7
sepasequ.complex	570	147	22152	118	53	222	602	164	23069	234	53	219	653	101	12396	102	53	210	535	173	16099	181	53	222
sepasequ.convent	667	220	110455	13421	258	387	678	212	115685	14890	258	398	743	208	116229	12022	258	386	667	218	99377	13783	258	404
sfacloc1.2.80	147	67	1515	182	8	28	154	55	1497	139	8	36	162	15	1415	13	8	42	140	72	1479	122	8	35
sfacloc1.2.90	146	52	1552	85	2	27	146	39	1660	51	8	30	152	24	1441	28	8	23	130	66	1527	71	0	23
sfacloc1.2.95	148	55	2089	83	2	16	147	54	2051	77	2	17	153	31	1828	61	2	16	140	80	1762	126	2	20
sfacloc1.3.80	220	99	2980	169	0	48	222	98	2584	172	0	57	232	34	2585	38	0	53	223	125	2293	199	0	60
sfacloc1.3.90	204	78	4031	116	0	45	208	99	3944	144	0	44	229	46	3472	87	0	35	215	113	3556	159	0	38
sfacloc1.3.95	203	90	2687	141	1	37	216	99	2557	175	1	44	225	56	2068	90	1	40	225	131	2189	179	1	44
sfacloc1.4.80	297	130	3700	144	0	85	293	113	3400	142	0	86	313	36	4018	33	0	102	283	145	3245	179	0	100
sfacloc1.4.90	272	98	4888	121	0	51	265	121	4795	121	0	53	293	33	4758	37	0	58	265	125	4420	135	0	48
sfacloc1.4.95	303	156	3767	226	0	58	299	147	3900	191	0	49	313	73	3261	114	0	49	285	158	3543	232	0	51
sfacloc2.2.80	115	28	2366	30	0	38	106	31	1890	31	0	33	125	11	1903	10	0	40	105	38	1914	36	0	31
sfacloc2.2.90	104	30	1650	24	0	31	107	54	1711	14	0	36	121	6	1613	8	0	35	96	25	1792	33	0	31
sfacloc2.2.95	55	7	1358	4	0	21	57	12	1160	17	0	20	63	6	1059	4	0	18	55	15	1262	16	0	19
sfacloc2.3.80	60	0	1516	0	0	51	60	0	1583	0	0	50	60	0	1463	0	0	51	60	0	1401	0	0	51
sfacloc2.3.90	60	0	1020	0	0	48	60	0	1194	0	0	49	60	0	1087	0	0	48	60	0	1079	0	0	50
sfacloc2.3.95	32	0	829	0	0	26	32	0	779	0	0	24	32	0	996	0	0	23	32	0	666	0	0	26

Table A.8 continued

Instance	random						min-max						greedy (base)						reverse greedy					
	lp	lp$_{filt}$	iter	iter$_{filt}$	b$_{obbr}$	lvb	lp	lp$_{filt}$	iter	iter$_{filt}$	b$_{obbr}$	lvb	lp	lp$_{filt}$	iter	iter$_{filt}$	b$_{obbr}$	lvb	lp	lp$_{filt}$	iter	iter$_{filt}$	b$_{obbr}$	lvb
sfacloc2_4_80	75	0	3264	0	0	66	75	0	3154	0	0	67	75	0	2130	0	0	64	75	0	3033	0	0	68
sfacloc2_4_90	75	0	1737	0	0	62	75	0	1785	0	0	67	75	0	1531	0	0	69	75	0	1564	0	0	66
sfacloc2_4_95	40	0	1331	0	0	35	40	0	1327	0	0	33	40	0	1107	0	0	33	40	0	1709	0	0	35
slay04h	13	2	593	1	4	9	15	2	675	4	4	8	14	1	636	5	4	9	14	2	756	1	0	8
slay04m	12	3	325	0	3	4	12	2	331	5	3	4	16	1	231	1	3	4	12	1	297	3	3	4
slay05h	18	1	1050	4	5	11	19	1	989	0	5	11	18	0	973	0	5	11	19	5	1248	1	5	10
slay05m	14	0	346	0	0	0	15	1	404	1	0	0	19	0	390	0	0	0	17	2	475	3	0	0
slay06h	23	1	2221	2	6	18	23	6	2068	9	6	17	24	1	1912	1	6	18	24	3	2637	1	6	19
slay06m	19	2	477	0	0	0	15	2	457	3	0	0	22	0	508	0	3	0	14	0	634	0	0	0
slay07h	27	4	3032	2	3	18	26	4	3687	0	3	19	28	1	3397	0	0	20	25	6	3235	3	3	19
slay07m	21	5	843	2	4	0	23	7	768	5	4	0	27	0	653	0	3	0	20	6	1131	2	1	1
slay08h	31	5	3323	1	4	25	28	4	3150	7	4	23	32	1	2707	1	4	27	30	9	4201	8	4	25
slay08m	26	1	1094	2	2	0	25	7	1318	5	5	0	32	0	1191	1	0	1	28	2	1713	6	5	0
slay09h	34	0	5370	0	5	28	34	8	6526	4	5	29	36	1	5282	0	5	28	33	10	5984	7	5	26
slay09m	28	1	1702	1	0	0	26	2	2141	2	0	0	30	0	1419	0	0	0	29	4	2521	5	0	0
slay10h	38	6	8052	1	5	32	37	4	7136	2	5	34	40	2	5700	0	5	35	38	7	8806	0	5	34
slay10m	32	4	1979	1	5	0	29	1	2168	1	0	0	38	1	1797	1	0	0	31	1	2947	1	0	0
smallinvDAXr1b010-011	11	0	77	0	8	8	12	0	72	0	8	8	13	0	87	0	8	8	10	0	80	0	8	8
smallinvDAXr1b020-022	13	0	71	0	8	8	13	0	79	0	8	8	15	0	76	0	8	8	11	0	86	0	8	8
smallinvDAXr1b050-055	14	0	108	0	12	12	14	0	96	0	12	12	14	0	88	0	12	12	14	0	113	0	12	12
smallinvDAXr1b100-110	14	0	106	0	13	13	15	0	105	0	13	13	15	0	106	0	13	13	14	0	117	0	13	13
smallinvDAXr1b150-165	31	0	132	0	26	28	32	0	130	0	26	28	34	0	131	0	26	28	30	0	127	0	26	28
smallinvDAXr1b200-220	31	0	137	0	26	28	32	0	133	0	26	28	34	0	131	0	26	28	29	0	129	0	26	28
smallinvDAXr2b010-011	10	0	90	0	7	7	10	0	81	0	7	7	14	0	88	0	7	7	8	0	79	0	7	7
smallinvDAXr2b020-022	12	0	82	0	9	9	14	0	89	0	9	9	15	0	92	0	9	9	12	0	96	0	9	9
smallinvDAXr2b050-055	31	0	159	0	27	27	32	0	152	0	27	27	34	0	160	0	27	27	30	0	165	0	27	27
smallinvDAXr2b100-110	13	0	86	0	12	12	13	0	87	0	12	12	14	0	93	0	12	12	13	0	123	0	12	12
smallinvDAXr2b150-165	15	0	115	0	13	13	15	0	113	0	13	13	15	0	85	0	13	13	15	0	122	0	13	13
smallinvDAXr2b200-220	14	0	105	0	13	13	14	0	100	0	13	13	14	0	92	0	13	13	14	0	113	0	13	13
smallinvDAXr3b010-011	11	0	92	0	7	7	11	0	81	0	7	7	15	0	81	0	7	7	11	0	86	0	7	7
smallinvDAXr3b020-022	14	0	119	0	10	10	14	0	103	0	10	10	14	0	100	0	10	10	14	0	117	0	10	10
smallinvDAXr3b050-055	14	0	136	0	13	13	15	0	116	0	13	13	15	0	95	0	13	13	14	0	133	0	13	13
smallinvDAXr3b100-110	14	0	128	0	13	13	14	0	110	0	13	13	15	0	110	0	13	13	14	0	135	0	13	13
smallinvDAXr3b150-165	14	0	111	0	13	13	14	0	115	0	13	13	14	0	101	0	13	13	14	0	122	0	13	13
smallinvDAXr3b200-220	15	0	118	0	13	13	15	0	108	0	13	13	16	0	99	0	13	13	15	0	118	0	13	13
smallinvDAXr4b010-011	12	0	86	0	7	7	11	0	85	0	7	7	14	0	105	0	7	7	12	0	96	0	7	7
smallinvDAXr4b020-022	13	0	113	0	11	11	14	0	127	0	11	11	15	0	113	0	11	11	13	0	121	0	11	11
smallinvDAXr4b050-055	14	0	128	0	13	13	14	0	134	0	13	13	15	0	136	0	13	13	14	0	147	0	13	13
smallinvDAXr4b100-110	14	0	123	0	13	13	14	0	121	0	13	13	14	0	109	0	13	13	14	0	127	0	13	13
smallinvDAXr4b150-165	13	0	108	0	13	13	13	0	110	0	13	13	13	0	97	0	13	13	13	0	113	0	13	13
smallinvDAXr4b200-220	15	0	113	0	13	13	15	0	111	0	13	13	15	0	102	0	13	13	15	0	137	0	13	13
smallinvDAXr5b010-011	12	0	93	0	8	8	12	0	93	0	8	8	13	0	101	0	8	8	11	0	91	0	8	8
smallinvDAXr5b020-022	13	0	107	0	11	11	12	0	88	0	11	11	15	0	128	0	11	11	12	0	118	0	11	11
smallinvDAXr5b050-055	14	0	97	0	13	13	14	0	99	0	13	13	14	0	99	0	13	13	14	0	114	0	13	13
smallinvDAXr5b100-110	14	0	132	0	9	9	14	0	122	0	9	9	16	0	99	0	9	9	12	0	133	0	9	9
smallinvDAXr5b150-165	14	0	102	0	13	13	14	0	105	0	13	13	15	0	98	0	13	13	14	0	120	0	13	13

Table A.8 continued

Instance	random						min-max						greedy (base)						reverse greedy					
	lp	lp$_{filt}$	iter	iter$_{filt}$	b$_{obbt}$	lvb	lp	lp$_{filt}$	iter	iter$_{filt}$	b$_{obbt}$	lvb	lp	lp$_{filt}$	iter	iter$_{filt}$	b$_{obbt}$	lvb	lp	lp$_{filt}$	iter	iter$_{filt}$	b$_{obbt}$	lvb
smallinvDAXr5b200-220	15	0	110	0	13	13	15	0	115	0	13	13	15	0	98	0	13	13	15	0	114	0	13	13
smallinvSNPr1b010-011	100	0	1037	0	68	72	100	0	1068	0	68	72	114	0	963	0	68	72	99	0	885	0	68	72
smallinvSNPr1b020-022	101	0	1275	0	67	79	101	0	1195	0	67	79	112	0	1045	0	67	79	99	0	1041	0	67	79
smallinvSNPr1b050-055	101	0	1407	0	39	81	101	0	1242	0	39	81	107	0	1101	0	39	81	99	0	1195	0	39	81
smallinvSNPr1b100-110	101	0	1655	0	26	79	101	0	1509	0	26	79	111	0	1232	0	26	79	99	0	1282	0	26	79
smallinvSNPr1b150-165	101	0	1352	0	31	84	101	0	1231	0	31	84	112	0	1176	0	31	84	99	0	1104	0	31	84
smallinvSNPr1b200-220	101	0	1430	0	33	84	101	0	1367	0	33	84	111	0	1175	0	33	84	99	0	1150	0	33	84
smallinvSNPr2b010-011	100	0	818	0	75	76	101	0	826	0	75	76	113	0	781	0	75	76	100	0	806	0	75	76
smallinvSNPr2b020-022	101	0	1255	0	47	63	101	0	1204	0	47	63	111	0	1000	0	47	63	101	0	1053	0	47	63
smallinvSNPr2b050-055	101	0	1595	0	36	82	101	0	1389	0	36	82	110	0	1237	0	36	82	100	0	1265	0	36	82
smallinvSNPr2b100-110	101	0	1470	0	30	81	101	0	1374	0	30	81	110	0	1106	0	30	81	100	0	1198	0	30	81
smallinvSNPr2b150-165	101	0	1763	0	22	85	101	0	1541	0	22	85	109	0	1371	0	22	85	100	0	1356	0	22	85
smallinvSNPr2b200-220	101	0	1579	0	28	84	101	0	1565	0	28	84	112	0	1338	0	28	84	100	0	1359	0	28	84
smallinvSNPr3b010-011	100	0	928	0	68	69	100	0	954	0	68	69	109	0	755	0	68	69	100	0	781	0	68	69
smallinvSNPr3b020-022	101	0	906	0	61	72	100	0	950	0	61	72	109	0	839	0	61	72	99	0	828	0	61	72
smallinvSNPr3b050-055	101	0	1181	0	26	73	100	0	1161	0	26	73	110	0	997	0	26	73	100	0	969	0	26	73
smallinvSNPr3b100-110	101	0	1461	0	32	87	100	0	1472	0	32	87	110	0	1268	0	32	87	100	0	1217	0	32	87
smallinvSNPr3b150-165	102	0	1404	0	21	85	100	0	1393	0	21	85	110	0	1151	0	21	85	100	0	1260	0	21	85
smallinvSNPr3b200-220	101	0	1388	0	13	78	101	0	1399	0	13	78	114	0	1143	0	13	78	100	0	1138	0	13	78
smallinvSNPr4b010-011	101	0	891	0	56	57	101	0	855	0	56	57	110	0	719	0	56	57	98	0	787	0	56	57
smallinvSNPr4b020-022	101	0	1192	0	43	61	101	0	1070	0	43	61	110	0	876	0	43	61	98	0	815	0	43	61
smallinvSNPr4b050-055	101	0	1270	0	35	80	101	0	1226	0	35	80	110	0	1062	0	35	80	98	0	1083	0	35	80
smallinvSNPr4b100-110	101	0	1593	0	18	82	102	0	1400	1	18	82	110	0	1197	0	18	82	99	0	1174	0	18	82
smallinvSNPr4b150-165	101	0	1366	0	18	88	101	0	1380	0	18	88	109	0	1024	0	18	88	98	0	1098	0	18	88
smallinvSNPr4b200-220	101	0	1706	0	20	89	100	0	1629	0	20	89	110	0	1308	0	20	89	98	0	1279	0	20	89
smallinvSNPr5b010-011	99	0	748	0	66	67	99	0	730	0	66	67	110	0	620	0	66	67	98	0	662	0	66	67
smallinvSNPr5b020-022	100	0	1074	0	44	60	100	0	1111	0	44	60	110	0	798	0	44	60	98	0	857	0	44	60
smallinvSNPr5b050-055	100	0	887	0	12	65	100	0	1028	0	12	65	108	0	747	0	12	65	98	0	694	0	12	65
smallinvSNPr5b100-110	101	0	987	0	35	89	101	0	938	0	35	89	109	0	709	0	35	89	98	0	720	0	35	89
smallinvSNPr5b150-165	100	0	1062	0	15	90	100	0	1038	0	15	90	110	0	883	0	15	90	98	0	792	0	15	90
smallinvSNPr5b200-220	100	0	1229	0	34	92	101	0	1302	0	34	92	111	0	1044	0	34	92	98	0	995	0	34	92
space25a	64	23	79	2	0	31	62	22	68	1	0	30	62	19	81	128	0	28	67	26	71	26	0	30
space25	43	330	89	263	1	1	43	318	77	251	0	1	44	112	80	0	0	0	42	336	125	1	0	2
space960	2089	0	547345	0	12	0	2092	0	639268	0	12	0	2494	0	804391	0	12	0	2149	0	734197	287	12	0
spectra2	31	0	857	0	26	26	31	0	773	14	26	26	36	8	817	14	26	26	30	16	794	0	26	26
sporttournament14	75	10	501	14	2	2	66	11	427	0	2	2	118	2	496	0	2	2	56	2	533	27	2	2
spring	23	2	197	0	7	12	23	2	249	0	7	12	23	2	117	3	7	12	23	2	309	0	7	12
squfl010-025	290	72	18950	93	0	166	249	28	20060	30	0	171	449	57	14415	3	0	160	248	40	16503	49	0	185
squfl010-025persp	564	66	123215	70	267	444	612	94	118245	123	267	430	705	5	89169	65	267	424	527	59	133231	62	267	437
squfl010-040	410	45	53820	25	0	265	401	40	52032	16	0	273	726	78	38553	2	0	273	392	28	56904	8	0	278
squfl010-040persp	913	84	516367	132	508	724	949	82	478019	157	508	712	1214	136	325905	105	508	712	854	81	526724	105	508	726
squfl010-080	694	186	39984	39	0	477	716	201	40205	82	0	471	1391	78	36181	15	0	493	753	241	42969	115	0	479
squfl010-080persp	1917	451	942637	452	800	1347	1933	460	941096	510	800	1331	2809	136	610351	144	800	1327	1746	438	923404	481	800	1326
squfl015-060	855	136	148368	20	0	640	878	134	141633	52	0	637	1718	713	86294	5	0	577	852	110	145907	13	0	623
squfl015-060persp	2313	835	2031133	2210	900	1561	2506	969	2078376	2365	900	1563	2909	713	1081371	1657	900	1391	2072	753	2101965	2271	900	1518
squfl015-080	1069	381	136465	162	0	815	1098	391	128955	215	0	802	2236	53	93262	32	0	752	1066	316	134552	91	0	829

Table A.8 continued

Instance	random						min-max						greedy (base)						reverse greedy					
	lp	lp_{filt}	iter	$iter_{filt}$	b_{obbr}	lvb	lp	lp_{filt}	iter	$iter_{filt}$	b_{obbr}	lvb	lp	lp_{filt}	iter	$iter_{filt}$	b_{obbr}	lvb	lp	lp_{filt}	iter	$iter_{filt}$	b_{obbr}	lvb
squfl015-080persp	3085	543	2272852	634	1200	2058	3179	572	2193950	710	1200	2035	3860	322	1377896	380	1200	2008	2949	548	2247949	671	1200	2096
squfl020-040	809	81	110149	30	0	572	844	108	103812	69	0	560	1551	28	74615	6	0	492	841	127	113882	69	0	545
squfl020-040persp	1964	279	2079652	366	1109	1498	2008	308	1865504	460	1109	1479	2265	252	881474	438	1109	1459	1873	267	2137165	337	1109	1504
squfl020-050	1061	266	243097	119	0	727	1023	186	232174	67	0	687	1906	54	138898	8	0	639	1062	245	248620	107	0	725
squfl020-050persp	2349	273	3777672	422	1277	1909	2373	261	3272307	361	1277	1891	2877	192	1529480	250	1277	1896	2175	211	3974302	213	1277	1924
squfl020-150persp	7873	1072	6440077	1503	3000	5230	8309	1086	6405817	1614	3000	5176	9480	1279	4623861	2121	3000	5102	6847	767	5515301	998	3000	5227
squfl025-025	647	29	91617	22	0	405	724	114	94988	103	0	409	1188	2	59358	3	0	361	669	69	82316	51	0	432
squfl025-025persp	1426	164	1125663	330	919	1204	1465	170	988580	355	919	1186	1628	132	616521	276	919	1165	1360	140	1119077	267	919	1201
squfl025-030	826	87	120114	77	0	477	792	58	121324	51	0	508	1482	13	76202	11	0	441	820	83	116761	78	0	525
squfl025-030persp	1716	280	1624054	653	788	1379	1786	314	1368587	527	788	1367	1953	240	734069	418	788	1278	1631	229	1604636	359	788	1391
squfl025-040	1069	199	351241	103	0	674	1070	167	300238	91	0	698	1939	24	199743	9	0	644	1026	146	342754	58	0	676
squfl025-040persp	2457	475	2379379	665	1353	1839	2489	473	1987106	624	1353	1851	2791	302	1024115	419	1353	1845	2435	526	2520972	631	1353	1860
squfl030-100persp	8868	3224	4451696	5660	3000	5037	8903	3213	4799764	5946	3000	4906	8672	2314	3607947	4100	3000	5138	7613	2435	4198360	4813	3000	4866
squfl030-150persp	13733	4167	7279649	6351	4500	7468	13606	3934	7091007	5717	4500	7536	14300	3119	6118433	4888	4500	7839	11661	2844	6301574	4253	4500	7526
squfl040-080	3644	1402	1146707	961	0	2129	3333	945	1069174	578	0	2074	5885	159	568778	94	0	1989	3489	1142	1045691	745	0	2112
squfl040-080persp	9083	3814	3095437	6997	3200	5128	8982	3714	3116829	7314	3200	5117	9821	3614	2989762	6934	3200	5337	8798	3826	2975984	5985	3200	5249
sssd08-04	14	4	181	1	4	10	13	6	160	2	4	10	14	3	147	3	4	10	13	7	167	3	4	10
sssd08-04persp	28	11	298	20	4	12	27	12	280	14	4	14	27	9	341	19	4	15	25	12	248	17	4	15
sssd12-05	18	6	308	1	4	13	18	9	281	4	4	13	18	4	294	0	0	13	18	11	259	6	5	13
sssd12-05persp	32	10	513	20	5	17	34	14	474	33	5	14	36	11	512	23	5	20	33	16	401	26	5	21
sssd15-04	16	6	319	1	0	12	16	6	364	5	0	12	16	5	293	1	0	12	16	10	317	4	0	12
sssd15-04persp	28	14	439	29	4	12	25	11	390	15	4	16	31	10	424	23	4	14	28	15	391	21	4	16
sssd15-06	19	5	467	1	0	14	20	4	450	0	0	14	20	4	503	0	0	14	20	11	460	6	0	14
sssd15-06persp	36	8	742	9	6	22	36	11	608	14	6	21	39	9	705	19	6	22	35	14	599	16	6	25
sssd15-08	26	7	769	3	0	18	26	8	771	5	8	18	26	9	652	1	8	18	26	13	693	8	8	18
sssd15-08persp	45	6	1258	8	8	23	48	14	1061	22	8	30	49	12	925	17	8	24	45	16	855	8	8	33
sssd16-07	24	9	515	3	7	17	24	6	518	1	7	16	24	6	484	1	8	17	24	14	438	7	7	17
sssd16-07persp	45	14	837	19	7	23	46	20	647	25	7	27	48	6	774	22	7	25	46	22	622	25	7	26
sssd18-06	20	6	514	2	6	14	20	9	540	6	6	13	20	6	561	2	6	14	20	11	462	6	6	14
sssd18-06persp	37	16	647	21	6	21	37	12	708	20	6	19	40	13	737	19	6	19	37	16	669	20	6	20
sssd18-08	28	10	937	4	8	20	28	11	804	4	8	20	28	7	840	0	8	20	28	15	820	8	8	20
sssd18-08persp	50	13	1330	19	8	31	51	19	977	29	8	33	54	17	1136	31	8	33	52	23	1125	23	8	31
sssd20-04	14	8	237	4	0	10	14	5	250	1	0	10	14	5	283	4	0	10	14	8	275	4	0	10
sssd20-04persp	27	11	400	12	4	16	24	10	363	12	8	14	27	7	449	17	4	12	27	13	420	23	4	16
sssd20-08	26	10	706	4	8	18	26	9	567	3	8	18	26	7	675	2	8	18	26	15	590	8	8	18
sssd20-08persp	47	17	947	28	8	26	47	13	968	15	8	22	52	12	915	28	8	29	50	23	718	25	8	30
sssd22-08	28	12	813	11	8	20	28	10	843	9	8	20	28	7	789	1	8	20	28	16	770	8	8	20
sssd22-08persp	47	16	1086	34	8	26	46	13	1122	22	8	29	55	17	1029	35	8	26	51	24	975	32	8	29
sssd25-04	14	5	370	2	4	10	14	5	397	3	4	10	14	4	399	1	4	10	14	8	362	8	4	10
sssd25-04persp	26	10	510	15	4	13	28	13	520	20	4	15	28	8	501	17	4	14	20	9	430	19	4	14
sssd25-08	26	8	754	3	8	18	26	11	680	8	8	18	26	6	814	2	8	18	26	14	735	11	8	18
sssd25-08persp	48	11	1107	15	8	29	48	14	1125	19	8	25	51	13	1097	25	8	28	50	23	968	25	8	30
st.e29	8	0	90	0	4	7	8	0	128	0	2	6	8	0	67	0	4	6	8	1	71	0	4	6
st.e31	7	0	26	0	2	2	7	0	41	0	2	2	7	0	28	0	2	2	6	0	26	0	2	2
st.e32	56	4	514	4	17	23	54	7	551	5	17	23	63	2	373	2	17	23	56	5	587	3	17	23
st.e36	13	2	101	1	4	3	13	3	142	1	4	3	13	3	111	1	4	3	13	4	99	1	4	3

Table A.8 continued

Instance	random						min-max						greedy (base)						reverse greedy					
	lp	lp$_{filt}$	iter	iter$_{filt}$	b$_{obbt}$	lvb	lp	lp$_{filt}$	iter	iter$_{filt}$	b$_{obbt}$	lvb	lp	lp$_{filt}$	iter	iter$_{filt}$	b$_{obbt}$	lvb	lp	lp$_{filt}$	iter	iter$_{filt}$	b$_{obbt}$	lvb
st_e38	10	0	43	3	8	7	10	0	53	3	8	7	10	1	47	0	8	7	10	1	60	0	8	7
st_e40	23	6	225	3	5	4	24	10	169	3	5	3	23	4	166	2	5	4	23	10	184	2	5	1
stockcycle	49	0	1947	1	0	0	50	0	2205	0	0	0	72	0	2522	0	0	0	48	0	2782	3	0	0
st_test8	17	2	24	1	13	15	17	2	18	1	13	15	17	2	12	2	13	15	17	3	31	3	13	15
st_testgr1	10	0	47	0	10	10	10	0	55	0	10	10	10	0	51	0	10	10	10	0	55	0	10	10
st_testgr3	12	0	57	0	12	12	12	0	56	0	12	12	12	0	12	0	12	12	12	0	65	0	12	12
super1	1081	256	232483	170	341	611	1057	269	220784	168	340	595	1129	215	115163	181	341	609	1094	301	221829	234	340	593
super2	1081	257	245036	208	355	603	1089	266	209761	177	356	607	1125	199	104966	127	356	609	1084	281	220676	184	355	604
super3	1105	284	250326	176	353	603	1102	279	213212	207	354	602	1146	186	130325	131	354	602	1095	300	257448	185	353	588
super3t	793	204	192428	134	256	504	796	210	157683	154	256	499	853	159	95314	131	256	512	805	213	197118	134	256	500
supplychain	12	2	83	1	4	6	12	2	97	0	4	6	12	1	82	0	4	6	12	2	93	2	4	6
supplychainp1_022020	60	0	14327	0	24	37	60	0	19196	0	24	35	60	0	13257	0	24	38	60	0	13231	0	24	37
supplychainp1_030510	25	0	879	0	10	10	25	0	867	0	10	10	25	0	989	0	10	10	24	1	847	1	10	10
supplychainp1_053050	122	3	88505	4	50	63	130	5	78188	13	50	63	130	0	82449	0	50	59	80	0	59891	6	50	63
supplychainr1_053050	48	6	218824	8	3	26	45	3	205098	1	6	23	95	1	189541	0	3	33	34	4	222844	3	3	20
syn05h	16	4	70	1	6	9	16	3	55	1	6	10	16	0	69	0	6	11	16	5	68	0	6	10
syn05m02m	7	0	120	1	4	5	8	0	123	1	4	5	10	0	121	0	4	4	7	0	128	0	4	5
syn05m03m	13	1	174	1	5	9	12	1	218	1	5	9	13	0	138	0	5	9	10	0	138	0	5	8
syn05m04m	17	1	401	1	6	8	16	1	268	1	6	12	20	1	179	1	6	9	15	0	419	0	6	10
syn10h	29	7	209	7	1	15	29	8	187	11	1	14	29	7	162	8	1	19	28	6	226	2	2	15
syn10m02m	18	3	323	4	0	10	18	8	241	5	2	12	21	7	254	2	0	7	19	5	227	8	0	8
syn10m03m	29	5	931	3	0	18	27	4	848	5	0	14	31	1	630	0	0	15	27	5	623	6	0	12
syn10m04m	34	3	1245	3	0	19	34	4	1300	2	0	18	44	0	860	3	0	13	33	9	919	9	0	16
syn10m	7	0	46	0	6	6	7	0	44	0	6	6	7	0	48	0	6	6	7	0	47	0	6	7
syn15m02m	34	8	831	7	8	15	33	6	878	5	8	15	39	2	738	1	8	12	30	4	796	4	8	15
syn15m03m	48	8	2509	13	17	21	48	0	2862	9	17	21	57	4	2157	3	17	23	47	7	2229	7	17	26
syn15m04m	69	14	2978	13	2	29	63	11	2812	9	2	29	78	5	2400	2	2	29	66	17	2247	21	2	32
syn15m	17	3	187	3	4	7	14	1	175	1	4	6	19	0	138	0	4	7	16	5	160	7	4	8
syn20m02m	39	10	737	11	0	14	41	13	846	14	0	14	49	3	682	0	4	14	41	15	715	15	4	14
syn20m03m	64	21	1232	33	0	19	63	19	1191	27	0	23	71	5	836	6	0	23	57	19	1062	26	0	22
syn20m04m	83	22	2457	21	0	35	78	22	2648	25	0	23	95	5	1792	6	0	30	76	28	2437	31	1	34
syn20m	22	5	286	6	6	5	19	2	311	3	6	8	24	4	295	2	1	5	18	2	312	3	1	7
syn30m02m	49	14	925	16	6	29	50	10	780	12	6	22	58	4	709	6	6	22	48	10	896	16	6	22
syn30m03m	77	16	1819	35	5	35	76	22	1875	47	5	35	91	3	1141	5	5	37	78	25	1372	35	5	31
syn30m04m	113	26	2269	42	10	52	111	37	2041	51	10	48	131	11	1356	10	6	40	109	33	1722	65	10	46
syn30m	24	4	342	6	3	12	26	6	416	6	3	14	29	2	296	3	2	12	28	6	325	8	3	12
syn40m02m	69	13	1383	10	0	38	66	13	1384	16	0	36	81	7	850	8	0	36	70	19	1050	23	0	33
syn40m03m	113	24	1984	29	0	57	112	28	1790	30	0	52	141	7	1458	6	0	47	112	31	1685	30	0	59
syn40m04m	159	36	3556	62	0	68	150	43	3443	57	0	71	192	13	2158	18	0	51	148	41	2728	59	0	63
syn40m	34	5	621	7	0	19	33	7	552	11	0	15	39	2	453	3	1	17	35	10	489	14	0	10
synheat	79	31	387	30	1	28	87	41	317	40	0	25	78	22	275	18	1	24	77	41	311	46	1	26
synthes1	3	0	13	0	1	1	3	1	10	1	0	1	3	0	9	0	0	0	4	1	10	1	0	1
synthes2	6	1	18	1	2	3	5	1	17	1	2	3	6	2	33	1	2	4	4	1	37	1	2	2
synthes3	10	1	76	0	3	3	7	1	43	0	2	2	10	2	59	0	1	2	7	2	53	2	2	2
tanksize	27	4	177	8	7	15	25	2	120	4	7	16	26	1	104	2	7	16	23	2	166	1	7	16
tln12	148	3	4243	6	38	59	150	4	4350	4	38	63	207	2	7240	2	38	64	164	26	1514	83	38	62

Table A.8 continued

Instance	random						min–max						greedy (base)						reverse greedy					
	lp	lp$_{filt}$	iter	iter$_{filt}$	b$_{obbr}$	lvb	lp	lp$_{filt}$	iter	iter$_{filt}$	b$_{obbr}$	lvb	lp	lp$_{filt}$	iter	iter$_{filt}$	b$_{obbr}$	lvb	lp	lp$_{filt}$	iter	iter$_{filt}$	b$_{obbr}$	lvb
tln4	26	3	163	3	9	10	24	2	150	2	2	12	30	0	149	0	9	13	23	3	174	3	9	10
tln5	28	1	245	1	15	16	27	1	295	1	15	18	33	1	310	1	15	18	28	3	302	2	9	16
tln6	47	5	304	11	19	25	44	5	365	10	19	22	55	2	384	5	19	22	44	6	333	11	19	22
tln7	61	1	525	4	5	22	58	4	593	4	5	20	89	0	836	0	5	21	60	4	542	5	5	16
tloss	44	4	230	3	21	24	44	4	187	7	21	21	51	0	227	1	21	23	42	4	191	4	21	22
tls12	229	12	1142	3	24	103	229	14	1295	7	24	106	230	12	1893	24	24	109	229	14	2092	6	24	97
tls2	9	2	88	2	0	2	9	1	57	1	0	1	9	2	56	2	0	2	7	0	81	0	0	1
tls4	56	16	385	21	15	28	50	11	467	18	15	27	58	13	438	15	15	29	49	12	552	22	15	31
tls5	66	17	1076	36	20	37	64	15	959	58	20	34	65	14	1037	54	20	36	61	15	1084	91	20	37
tls6	83	15	1407	91	20	33	80	12	1067	89	20	32	88	21	1264	84	20	32	86	19	1184	101	20	38
tls7	121	26	1844	26	70	78	123	24	2553	36	70	79	120	19	2222	23	70	78	121	25	1589	21	70	83
tltr	36	7	550	5	6	16	33	5	525	4	6	17	42	3	444	0	6	27	33	5	585	3	6	23
transswitch0009r	64	6	574	14	6	3	66	8	585	10	4	4	80	4	459	5	0	4	60	10	431	12	2	3
transswitch0014r	149	27	2107	37	1	40	158	37	1802	43	1	42	181	20	1739	23	1	37	148	37	2611	60	1	36
transswitch0030r	322	59	4913	15	0	122	325	57	4559	16	0	125	366	55	4041	18	0	111	319	69	5978	26	0	105
transswitch0039r	358	27	6066	33	0	6	360	114	6085	33	0	7	432	20	5755	30	0	8	346	27	7814	38	0	8
transswitch0057r	588	116	8566	159	0	129	560	114	6705	150	0	130	649	56	7112	57	0	134	556	128	10067	173	0	131
transswitch0118r	1344	262	22221	277	0	329	1327	238	21731	238	0	308	1551	132	19203	100	0	349	1279	282	27930	293	0	326
transswitch0300r	3030	483	47650	415	0	544	2976	484	45604	429	0	528	3491	306	39201	235	0	539	2803	526	55430	470	0	538
transswitch2383wpr	20478	1792	212206	1215	0	1498	20621	1697	214110	1126	0	1523	25117	1227	179768	623	0	1533	18994	2060	260358	1384	0	1502
transswitch2736spr	26853	2226	280462	1163	0	2059	26944	2201	290641	1204	0	2024	32113	1602	236137	699	0	2020	24563	2672	335567	1668	0	1969
tspn05	35	4	781	4	17	15	36	6	775	4	17	15	34	2	707	2	17	15	35	4	1264	4	17	15
tspn08	71	4	1678	4	27	34	71	6	1896	5	27	34	74	2	2001	2	27	34	68	2	3662	2	27	34
tspn10	148	2	14757	12	128	133	150	7	12121	9	128	133	146	1	11546	1	128	133	144	2	18926	17	128	133
tspn12	211	11	27560	12	172	181	210	10	23758	18	172	181	206	4	21790	4	172	182	216	20	34942	29	172	181
tspn15	257	32	27389	50	129	142	281	33	25056	41	129	142	283	11	19453	14	129	142	240	35	34519	42	129	142
unitcommit1	246	44	20230	50	1	25	263	54	15191	55	1	23	300	18	14905	19	1	30	214	51	15400	1	1	30
unitcommit2	422	169	1117364	218	0	80	459	153	1403553	183	0	86	500	129	855155	131	0	101	442	203	1319684	243	0	89
util	10	0	65	0	9	9	10	0	55	0	9	9	10	0	62	0	9	9	10	1	74	9	9	9
waste	248	38	3329	55	2	94	240	36	2908	36	2	99	246	28	2444	26	2	85	245	39	3131	31	2	88
watercontamination0202	439	132	130903	1036	50	64	452	149	123455	910	50	66	521	16	57183	81	50	67	431	132	82242	1118	50	62
watercontamination0202r	128	46	1145	17	0	21	130	47	1001	20	0	4	129	45	823	15	0	15	132	50	929	12	0	0
watercontamination0303	837	220	52651	1685	90	178	864	216	53317	1774	90	177	940	88	32783	415	90	179	811	279	35224	1792	90	180
watercontamination0303r	189	39	5652	5	0	25	194	30	4108	2	0	19	201	44	2978	16	0	15	166	11	4220	0	0	0
watermd1	28	5	463	4	0	13	27	9	468	9	0	13	34	3	364	2	0	15	31	8	629	14	0	11
watermd2	103	32	4455	28	0	59	107	34	5241	53	21	60	112	12	5086	10	0	63	100	34	5718	37	0	62
waterno2.01	46	17	1339	27	21	34	44	14	1092	27	21	33	44	14	561	18	21	33	44	16	1319	36	21	30
waterno2.02	98	56	3246	47	13	48	98	55	2871	38	13	52	100	38	1740	28	13	54	99	63	3773	76	13	52
waterno2.03	144	86	5643	130	15	72	142	85	4691	164	15	73	143	61	3787	88	15	76	138	85	5921	208	15	73
waterno2.04	170	95	7388	89	20	82	173	99	8557	153	20	85	174	64	4771	40	20	83	168	100	8331	100	20	85
waterno2.06	259	152	14046	123	17	130	267	163	13398	205	17	126	271	119	8017	86	17	139	258	165	13099	119	17	140
waterno2.09	402	221	24408	181	28	216	399	223	22416	143	28	221	406	151	13282	128	28	218	397	241	20470	160	28	215
waterno2.12	532	304	31068	225	43	273	527	301	31638	177	43	284	525	205	17886	194	43	290	526	328	29869	244	43	274
waterno2.18	801	456	52333	344	60	426	802	460	49503	280	60	432	814	324	26333	222	60	412	787	489	47066	417	60	435
waterno2.24	1105	620	70969	444	72	576	1090	621	64758	489	72	586	1116	416	38669	310	72	583	1095	680	68732	558	72	601
watersym1	110	18	6045	3	86	92	114	24	5601	3	86	89	116	12	5259	1	86	90	113	23	5332	9	86	87

Table A.8 continued

instance	random						min–max						greedy (base)						reverse greedy					
	lp	lp_{filt}	iter	$iter_{filt}$	b_{obbt}	lvb	lp	lp_{filt}	iter	$iter_{filt}$	b_{obbt}	lvb	lp	lp_{filt}	iter	$iter_{filt}$	b_{obbt}	lvb	lp	lp_{filt}	iter	$iter_{filt}$	b_{obbt}	lvb
watertreatnd_conc	59	12	384	10	5	23	54	11	398	8	5	24	59	9	426	1	5	25	50	14	284	8	5	22
watertreatnd_flow	64	21	2003	35	7	47	64	22	1692	45	7	46	65	16	1794	12	7	48	64	22	2048	42	7	48
waterx	91	29	3788	16	2	15	89	35	3602	10	2	20	101	24	2267	2	2	17	89	44	4537	33	2	16
waterz	76	27	839	44	2	29	77	29	722	32	2	21	84	15	825	19	2	28	78	37	881	50	2	26

Table A.9: Detailed results comparing ordering strategies for OBBT at the root node on test set GO. Results for strategy "greedy (base)" are listed in Table A.7. See Table 6.1 for aggregated results.

lp — number of LPs solved during OBBT
lp_{filt} — number of LPs solved during OBBT's filtering
$iter$ — number of LP iterations during OBBT
$iter_{filt}$ — number of LP iterations of OBBT's filtering LPs
b_{obbt} — number of bounds tightened by OBBT
lvb — number of LVBs found by OBBT

Instance	random lp	lp_{filt}	$iter$	$iter_{filt}$	b_{obbt}	lvb	min-max lp	lp_{filt}	$iter$	$iter_{filt}$	b_{obbt}	lvb	greedy (base) lp	lp_{filt}	$iter$	$iter_{filt}$	b_{obbt}	lvb	reverse greedy lp	lp_{filt}	$iter$	$iter_{filt}$	b_{obbt}	lvb
alkylation	15	2	104	0	9	11	15	1	106	0	9	11	15	3	87	3	9	10	15	3	98	4	9	11
alkyl	20		108	1	13	15	21	2	105	1	13	15	20	0	93	0	13	15	20	3	109	2	13	15
arki0015	2883	979	1539968	498	394	1940	2947	990	1042378	959	394	1948	2917	626	515010	365	394	1895	3083	1132	1517341	2752	394	1862
arki0018	20130	4894	1390320	1305	329	14039	17949	4746	1326615	1199	329	14039	19668	2682	191737	2682	329	15021	23942	3543	1466425	6	329	14173
arki0019	1274	169	3709	364	0	727	1233	212	3609	508	0	741	1547	0	4858	1	0	708	1060	71	1768	138	0	554
arki0020	3464	410	8552	856	0	1792	3327	497	7655	1010	0	1788	3771	1	11474	1	0	1734	2346	627	7004	1618	0	1755
bayes2.10	82	38	944	14	7	63	81	40	818	26	7	60	89	44	638	35	7	62	82	49	845	21	7	61
bayes2.20	83	39	1486	11	7	74	84	44	1379	11	7	74	79	32	933	18	7	73	82	45	1509	19	7	75
bayes2.30	96	46	1956	15	7	87	98	52	1885	22	7	88	94	44	1072	17	7	88	98	56	1741	17	7	88
bayes2.50	84	27	3237	17	7	77	84	27	3233	6	7	78	87	24	1627	11	7	75	86	31	3379	35	7	72
bearing	23	8	310	9	4	16	21	8	380	8	4	15	23	7	233	2	4	16	22	6	418	2	4	13
btest14	91	11	5888	2	16	55	87	2	6287	0	16	56	95	7	2082	6	16	56	86	14	7850	28	16	54
camshape100	129	43	2893	1	61	117	131	46	783	1	61	120	160	0	483	0	61	98	128	44	4028	0	61	120
camshape200	246	96	7634	2	113	223	249	95	2193	1	113	227	316	0	973	0	113	171	240	94	12777	2	113	227
camshape400	468	201	24414	2	218	447	479	199	7484	6	218	444	626	0	2030	0	218	335	473	200	42185	1	218	447
camshape800	930	407	87721	10	414	874	936	401	22068	18	414	864	1244	0	4046	0	414	695	946	411	164359	11	414	875
catmix100	501	0	0	0	0	0	501	0	0	0	0	0	501	0	0	0	0	0	501	0	0	0	0	0
catmix200	1001	0	0	0	0	0	1001	0	0	0	0	0	1001	0	0	0	0	0	1001	0	0	0	0	0
catmix400	2001	0	0	0	0	0	2001	0	0	0	0	0	2001	0	0	0	0	0	2001	0	0	0	0	0
catmix800	4001	0	0	0	0	0	4001	0	0	0	0	0	4001	0	0	0	0	0	4001	0	0	0	0	0
chain100	447	109	59201	0	68	302	486	100	40307	0	68	302	507	32	15186	0	68	302	388	115	45107	0	68	302
chain200	904	219	151272	2	155	603	971	206	104730	1	155	603	1010	92	25427	0	155	603	838	220	138530	0	155	603
chain50	240	55	25460	14	51	154	236	46	17228	8	51	154	263	22	8439	7	51	154	228	62	32074	13	51	154
chance	9	0	22	2	8	8	9	4	23	0	8	8	9	0	15	0	8	8	8	5	21	8	8	8
chem	50	4	451	2	13	39	50	4	321	1	13	40	53	6	237	5	13	39	49	5	476	1	13	40
elec200	20893	412	99162	12909	0	0	20782	492	47218	14781	0	0	20995	262	143948	12073	0	0	20677	625	20739	7597	0	0
elec25	423	48	5097	650	0	0	424	72	3889	928	0	0	438	34	5647	408	0	0	395	71	1306	485	0	0
elec50	1481	104	17299	3256	0	0	1454	123	12398	3566	0	0	1487	95	20714	3756	0	0	1423	106	3998	1009	0	0
etamac	53	4	181	12	8	17	57	4	200	12	8	17	56	3	186	9	8	17	53	4	211	12	8	17
ex14.1.1	4	0	38	0	4	4	4	0	35	0	4	4	4	0	30	0	4	4	4	0	46	0	4	4
ex14.1.2	10	0	108	0	4	7	10	1	118	0	4	7	10	0	85	0	4	7	10	1	119	0	4	7
ex14.1.3	4	2	27	0	2	1	4	0	25	0	2	1	4	0	25	0	2	1	4	0	13	0	2	1
ex14.1.5	20	2	68	0	16	7	21	2	44	0	16	7	21	2	42	0	16	7	22	3	56	1	16	4
ex14.1.6	13	1	56	2	4	1	12	2	49	2	4	1	14	0	32	0	4	1	13	3	94	0	4	1
ex14.1.7	72	35	1171	52	2	15	72	36	831	46	2	14	76	23	743	29	2	17	74	37	1127	39	2	9
ex14.1.8	12	0	40	0	4	9	11	0	46	0	4	9	12	0	41	0	4	9	11	0	73	0	4	9

Table A.9 continued

Instance	random						min–max						greedy (base)						reverse greedy					
	lp	lp$_{\text{filt}}$	iter	iter$_{\text{filt}}$	b$_{\text{obbt}}$	lvb	lp	lp$_{\text{filt}}$	iter	iter$_{\text{filt}}$	b$_{\text{obbt}}$	lvb	lp	lp$_{\text{filt}}$	iter	iter$_{\text{filt}}$	b$_{\text{obbt}}$	lvb	lp	lp$_{\text{filt}}$	iter	iter$_{\text{filt}}$	b$_{\text{obbt}}$	lvb
ex14.1.9	2	0	4	0	0	0	2	0	4	0	0	0	2	0	5	0	0	0	2	0	4	0	0	0
ex14.2.4	83	0	56	0	6	7	83	0	82	0	6	7	82	0	45	0	6	7	82	0	92	0	6	7
ex14.2.7	86	0	33	0	0	4	87	0	32	0	0	4	89	0	47	0	0	4	85	0	40	0	0	0
ex14.2.9	24	0	256	0	13	18	26	1	258	1	13	18	23	0	74	0	13	18	23	0	258	0	13	18
ex2.1.10	23	0	47	0	21	22	23	0	51	0	21	22	22	0	37	0	21	22	22	0	44	0	21	22
ex2.1.1	6	0	5	0	0	1	6	0	6	0	0	1	6	0	5	0	0	1	5	0	4	0	0	0
ex2.1.7	22	0	244	0	20	20	22	1	228	1	20	20	26	0	249	0	20	20	20	0	227	0	20	20
ex2.1.9	11	0	44	1	10	10	12	1	48	1	10	10	19	0	67	0	10	10	10	0	48	0	10	10
ex3.1.1	12	1	39	0	0	3	13	0	38	0	7	2	13	0	35	0	7	3	12	0	40	0	7	3
ex3.1.2	7	0	13	1	7	7	7	0	13	0	7	7	7	0	15	0	7	7	7	0	15	0	7	3
ex3.1.4	2	0	5	0	0	1	3	0	4	0	0	1	3	0	5	0	0	1	3	0	4	0	0	1
ex3pb	8	0	82	0	0	2	8	0	57	0	0	2	8	0	52	0	0	2	8	1	98	1	0	2
ex4.1.1	2	0	20	0	1	2	2	0	20	0	1	2	2	0	16	0	1	2	2	0	20	0	1	2
ex4.1.2	1	0	0	0	0	0	1	0	0	0	0	0	1	0	0	0	0	0	1	0	0	0	0	0
ex4.1.3	2	0	19	0	1	1	2	0	19	0	1	1	2	0	8	0	1	1	2	0	19	0	1	1
ex4.1.4	2	0	17	0	0	2	2	0	17	0	0	2	2	0	16	0	0	2	2	0	17	0	1	2
ex4.1.5	3	0	9	0	0	1	3	0	9	0	0	0	3	0	9	0	0	0	4	0	13	0	0	1
ex4.1.6	2	0	7	0	1	1	2	0	7	0	1	1	2	0	11	0	1	1	2	0	7	0	0	1
ex4.1.7	2	0	16	0	2	2	2	0	16	0	2	2	2	0	20	0	2	2	2	0	16	0	2	2
ex4.1.9	2	0	4	0	0	2	2	0	4	0	0	2	2	0	5	0	0	2	2	0	4	0	2	2
ex5.2.2_case1	5	0	20	0	3	4	5	0	19	0	3	4	5	0	14	0	3	3	4	0	17	0	3	4
ex5.2.2_case2	5	1	16	0	1	4	5	1	15	0	1	3	5	0	12	0	1	3	5	2	13	1	1	3
ex5.2.2_case3	5	0	21	0	3	4	5	1	20	0	3	5	5	0	14	0	3	4	4	0	18	0	3	4
ex5.2.4	9	2	27	2	0	3	7	0	23	2	0	2	9	0	32	0	0	3	6	0	31	0	0	3
ex5.2.5	24	4	133	3	2	12	27	8	129	9	2	6	27	4	159	4	2	7	27	6	124	5	0	9
ex5.3.2	12	6	40	9	0	3	13	5	33	4	1	7	13	9	37	2	2	6	14	5	46	8	1	5
ex5.3.3	58	29	346	79	1	31	59	28	351	55	1	25	57	10	324	9	1	29	57	22	447	29	2	30
ex5.4.2	13	2	49	2	6	8	13	0	42	0	6	8	12	0	46	0	6	7	11	0	43	0	6	7
ex5.4.3	13	1	58	1	8	6	14	2	53	0	8	7	14	1	60	5	8	6	15	4	64	5	8	4
ex5.4.4	22	4	146	3	6	6	24	6	136	9	6	8	24	3	148	0	6	4	23	4	151	12	6	6
ex6.1.2	10	1	25	0	2	7	10	0	26	0	2	7	9	0	11	0	2	7	9	2	15	1	2	7
ex6.1.4	14	1	88	0	3	10	15	1	97	0	3	8	13	0	65	0	3	10	14	3	94	3	3	9
ex6.2.10	83	38	1663	6	9	45	88	34	1530	17	9	36	90	28	1173	7	9	37	71	33	1579	11	9	50
ex6.2.11	36	4	272	1	6	19	35	5	349	2	6	18	37	1	179	9	6	19	34	3	249	2	6	16
ex6.2.12	45	13	206	2	0	25	47	18	149	2	0	25	42	15	154	0	0	31	44	16	158	5	0	27
ex6.2.13	64	21	181	23	0	28	60	26	196	25	0	27	62	13	231	15	0	28	59	22	317	19	0	29
ex6.2.5	68	46	460	35	0	44	66	43	970	43	0	46	64	38	491	54	0	45	59	40	509	30	0	43
ex6.2.6	26	0	304	0	19	22	28	0	157	60	19	22	27	0	195	0	19	22	27	0	370	0	19	22
ex6.2.7	71	21	1874	61	0	28	68	20	1752	60	0	27	73	11	1054	21	0	31	74	19	2559	27	19	29
ex6.2.8	25	0	350	0	20	24	26	0	310	0	20	24	28	0	197	0	20	25	27	1	410	1	20	24
ex6.2.9	61	27	687	34	0	37	60	24	873	339	6	38	68	24	660	28	6	33	57	29	482	29	6	35
ex7.2.2	9	0	68	0	3	7	10	0	65	0	3	7	10	0	49	0	3	7	9	0	88	0	6	7
ex7.2.3	26	9	173	11	4	12	27	5	153	1	4	13	27	5	233	1	4	16	26	10	225	4	4	11
ex7.2.4	17	1	184	0	5	11	16	2	186	5	5	10	19	1	169	0	5	11	18	2	211	5	5	10
ex7.3.1	19	0	76	0	10	10	19	0	49	0	10	10	19	0	31	0	10	10	19	0	50	0	10	10
ex7.3.2	10	0	9	0	7	7	10	0	7	0	7	7	10	0	8	0	7	7	10	0	7	0	7	7

Table A.9 continued

Instance	random lp	lp_filt	iter	iter_filt	b_obbt	lvb	min-max lp	lp_filt	iter	iter_filt	b_obbt	lvb	greedy (base) lp	lp_filt	iter	iter_filt	b_obbt	lvb	reverse greedy lp	lp_filt	iter	iter_filt	b_obbt	lvb
ex7.3.3	5	0	4	0	4	4	5	0	3	0	4	4	5	0	4	0	4	4	5	0	3	0	4	4
ex7.3.4	37	2	23	0	15	15	37	2	29	1	15	15	37	1	24	1	15	15	37	2	28	0	15	15
ex7.3.5	35	3	53	1	4	4	36	3	47	0	4	4	36	0	49	0	4	4	35	3	53	2	4	4
ex7.3.6	3	0	3	0	3	3	3	0	6	0	3	3	3	0	3	0	3	3	3	0	14	0	3	3
ex8.1.3	15	0	36	0	0	3	15	0	28	0	0	3	15	0	29	0	0	3	15	0	51	1	0	3
ex8.1.4	4	0	1	0	0	0	4	0	1	0	0	0	4	0	1	0	0	0	4	0	1	0	0	0
ex8.1.5	4	0	0	0	0	0	4	0	0	0	0	0	4	0	0	0	0	0	4	0	0	0	0	0
ex8.1.6	9	0	47	0	0	0	10	0	37	0	0	0	10	0	40	0	0	0	10	0	51	0	0	0
ex8.1.7	8	0	53	0	2	2	10	0	49	0	2	2	6	0	48	0	2	2	6	0	70	0	2	2
ex8.2.1b	71	0	342	0	21	26	74	0	298	104	21	26	70	0	246	0	21	26	70	0	372	0	21	26
ex8.2.2b	6332	88	40680	88	40	40	6074	93	29662	104	40	40	6074	93	30636	93	40	40	7317	62	33567	62	40	40
ex8.2.3b	13259	0	83246	0	77	85	13482	0	54682	0	77	85	13482	0	37854	0	77	85	13421	3	64821	3	77	85
ex8.2.4b	82	3	296	3	35	38	82	3	245	1	35	38	82	3	193	3	35	38	78	1	338	0	35	38
ex8.2.5b	2113	3	27002	3	52	54	2080	3	22645	1	52	54	2080	3	17797	9	52	54	2597	0	26910	1	52	54
ex8.3.11	126	18	246	15	7	34	129	8	239	14	7	35	129	6	261	6	7	38	118	21	235	17	7	32
ex8.3.12	98	13	158	17	7	30	98	6	195	15	7	34	98	20	231	20	7	65	95	7	182	8	7	26
ex8.3.13	142	26	341	20	7	68	127	26	345	37	7	66	127	9	314	9	7	38	133	46	331	44	7	64
ex8.3.1	122	13	223	16	7	37	88	10	264	16	7	33	88	5	237	5	7	33	127	15	262	14	7	30
ex8.3.5	85	11	228	12	7	34	123	4	228	26	7	39	123	30	233	30	7	48	90	14	231	13	7	34
ex8.3.7	117	24	203	32	7	48	100	23	174	36	7	45	100	9	259	9	7	37	125	39	212	48	7	48
ex8.3.8	101	19	282	29	7	35	70	4	333	27	7	35	70	1	339	1	7	38	107	22	343	32	7	30
ex8.3.9	72	12	127	9	7	36	22	3	145	11	7	37	22	28	166	28	7	31	67	15	166	13	7	39
ex8.4.1	21	12	249	0	6	7	54	0	242	0	6	7	54	2	205	2	6	7	22	0	284	0	6	7
ex8.4.2	54	30	542	20	1	27	43	25	534	22	1	33	43	0	472	0	1	31	66	41	883	34	1	27
ex8.4.4	39	4	425	3	17	22	25	2	464	2	17	22	25	2	327	2	17	22	37	6	573	4	17	22
ex8.4.5	19	3	260	1	3	3	57	1	203	0	3	3	57	1	183	3	3	3	17	3	214	3	3	3
ex8.4.6	51	5	154	4	1	2	110	3	197	13	1	2	110	3	113	2	1	3	50	8	179	12	1	2
ex8.4.7	105	12	7119	4	48	93	202	10	5500	6	48	94	202	4	5196	8	48	93	111	22	12595	2	48	94
ex8.4.8.bnd	184	45	14545	13	71	158	202	31	13982	15	71	160	202	8	5245	8	71	159	190	50	19312	19	71	158
ex8.5.1	20	2	44	0	3	7	22	4	51	0	3	9	22	0	38	0	3	8	19	3	27	1	3	8
ex8.5.2	21	1	96	0	8	8	25	1	92	2	8	6	25	3	72	3	8	8	21	3	76	3	8	6
ex8.5.4	13	1	54	1	2	6	15	3	60	1	2	4	15	1	67	1	2	6	14	3	51	2	2	6
ex8.5.6	27	2	23	1	11	13	28	0	25	0	11	13	28	1	17	1	11	13	26	2	39	0	11	13
ex8.6.1	159	1	58	6	8	8	160	3	54	9	8	8	160	7	57	7	8	8	152	4	54	9	8	8
ex9.2.2	4	1	5	0	2	3	3	0	5	0	2	3	3	0	5	0	2	3	4	1	6	0	2	3
ex9.2.4	3	0	11	0	0	0	4	0	13	1	0	0	4	0	16	0	0	0	3	0	10	0	0	0
ex9.2.5	4	0	16	1	1	2	6	1	18	1	1	2	6	1	18	1	1	3	3	1	30	1	1	3
ex9.2.6	4	0	20	0	0	2	6	0	13	1	0	2	4	0	20	0	0	2	3	0	11	0	0	2
ex9.2.7	3	0	11	0	2	3	3	0	11	0	2	3	3	0	11	0	2	3	3	0	17	0	2	3
glider100	3363	674	6553	319	1	603	3394	598	4771	162	1	701	3394	199	3318	199	1	701	3181	625	5072	145	1	606
glider200	6712	1348	11449	1005	1	1208	6794	1198	9216	324	1	1401	6794	399	6777	399	1	1401	6372	1246	10043	364	1	1203
glider400	13460	2699	21717	1517	1	2403	13594	2398	20306	643	1	2801	12991	799	12991	799	1	2801	12742	2506	20239	919	1	2403
glider50	1673	324	2753	135	1	303	1694	298	2396	79	1	351	1694	99	1713	99	1	351	1592	308	2493	95	1	304
gsg_0001	104	30	2071	73	50	90	100	23	2517	68	50	90	854	52	854	52	50	90	108	90	3209	79	50	90
hhfair	25	5	28	1	6	11	28	3	27	1	6	10	28	2	35	2	6	10	26	5	22	0	6	13
himmel11	7	0	11	0	7	7	7	0	11	0	7	7	7	0	11	0	7	7	7	0	14	0	7	7

Table A.9 continued

Instance	random						min–max						greedy (base)						reverse greedy					
	lp	lp_{filt}	iter	$iter_{filt}$	b_{obbt}	lvb	lp	lp_{filt}	iter	$iter_{filt}$	b_{obbt}	lvb	lp	lp_{filt}	iter	$iter_{filt}$	b_{obbt}	lvb	lp	lp_{filt}	iter	$iter_{filt}$	b_{obbt}	lvb
himmel16	15	3	85	5	0	0	16	3	88	6	0	0	18	0	91	0	0	0	13	0	104	0	0	0
house	9	0	14	0	6	8	8	0	8	0	6	8	8	0	9	0	6	8	8	1	6	0	6	8
hs62	9	0	28	0	8	9	9	0	20	0	8	9	9	0	22	0	8	9	9	1	28	0	8	9
hybriddynamic_fixedcc	46	1	110	10	26	27	45	1	76	0	26	27	46	0	70	0	26	27	44	1	118	24	26	27
hybriddynamic_varcc	102	55	2770	19	37	40	94	2	2452	1	37	40	107	3	1353	18	37	40	98	7	2834	2	37	40
infeas1	2830	55	3307309	2631	2631	2707	2839	53	805051	12	2631	2709	2847	31	191666	18	2631	2710	2846	55	7056532	13	2631	2709
jbearing100	5001	0	2500	0	0	0	5001	0	2500	0	0	0	5001	0	2500	0	0	0	5001	0	2501	0	0	0
jbearing25	1251	0	625	0	0	0	1251	0	625	0	0	0	1251	0	625	0	0	0	1251	0	626	0	0	0
jbearing50	2501	0	1250	0	0	0	2501	0	1250	0	0	0	2501	0	1250	0	0	0	2501	0	1251	0	0	0
jbearing75	3751	0	1875	0	0	0	3751	0	1875	0	0	0	3751	0	1875	0	0	0	3751	0	1876	0	0	0
kall_circles_c6a	22	2	62	1	7	12	23	3	78	3	7	13	25	3	94	2	7	9	21	3	55	3	7	9
kall_circles_c6b	21	2	43	1	7	11	23	3	81	3	7	11	25	3	107	2	7	10	22	3	46	1	7	12
kall_circles_c6c	26	4	76	2	8	16	25	4	87	4	8	11	27	1	98	0	8	11	24	1	67	0	8	14
kall_circles_c7a	24	3	163	1	9	13	25	3	204	2	3	12	25	1	109	1	3	10	25	4	184	1	3	11
kall_circles_c8a	28	3	225	10	9	12	31	2	222	9	9	11	29	3	151	7	9	11	27	5	183	11	9	12
kall_circlespolygons_c1p12	27	7	142	2	2	2	29	4	181	2	2	2	28	4	164	4	2	2	28	5	190	0	2	2
kall_circlespolygons_c1p13	27	7	179	2	2	2	27	5	191	6	2	2	28	3	163	2	2	2	27	5	258	0	2	2
kall_circlespolygons_c1p5a	93	22	882	20	1	1	99	31	840	70	1	1	100	39	700	59	1	1	96	13	1199	5	1	1
kall_circlespolygons_c1p5b	576	112	7315	97	0	1	592	89	6661	140	0	1	587	197	5343	244	0	1	579	92	10195	76	0	0
kall_circlespolygons_c1p6a	835	210	11086	184	0	0	881	133	9950	144	0	0	876	307	8928	428	0	0	857	126	15048	98	0	0
kall_circlesrectangles_c1r12	32	4	145	5	2	2	30	6	115	7	2	2	33	4	129	4	2	2	31	6	195	2	2	2
kall_circlesrectangles_c1r13	32	8	184	7	0	1	32	6	186	6	0	1	31	4	133	3	0	1	32	3	201	4	0	1
kall_circlesrectangles_c6r1	140	25	1134	25	0	6	142	29	1183	20	0	6	154	42	1016	51	0	2	139	19	1460	6	0	2
kall_circlesrectangles_c6r29	297	62	2650	61	0	4	309	57	2435	67	0	4	313	77	2181	78	0	5	304	37	3492	27	0	3
kall_circlesrectangles_c6r39	485	84	4927	64	0	4	506	95	4504	133	0	7	512	142	3781	152	0	6	493	68	6169	36	0	6
kall_congruentcircles_c31	11	2	27	2	7	8	11	2	30	2	7	8	11	2	26	2	7	8	11	3	21	3	7	8
kall_congruentcircles_c32	11	0	35	0	1	3	10	0	26	0	1	3	11	1	39	1	1	2	10	3	42	1	1	4
kall_congruentcircles_c41	7	0	11	0	3	1	7	1	11	1	3	1	7	0	9	0	3	1	7	1	13	0	3	3
kall_congruentcircles_c42	14	5	65	4	2	4	14	3	50	4	2	4	14	5	42	5	2	8	13	3	57	3	2	5
kall_congruentcircles_c51	19	5	53	6	5	8	19	5	57	5	5	8	19	4	47	5	5	8	19	3	45	4	5	8
kall_congruentcircles_c52	15	3	83	4	3	5	16	5	70	6	3	7	15	4	53	4	3	3	15	6	79	8	3	5
kall_congruentcircles_c61	21	4	65	3	6	12	22	7	61	8	6	12	22	5	64	6	6	10	21	6	49	4	6	10
kall_congruentcircles_c62	19	5	87	6	4	5	20	8	91	21	4	1	15	5	73	6	4	2	18	7	112	8	4	5
kall_congruentcircles_c63	20	2	123	1	9	10	21	2	63	1	9	13	20	2	86	1	9	9	20	3	120	3	9	10
kall_congruentcircles_c71	24	4	182	4	0	13	24	6	142	4	0	13	26	5	109	9	0	7	24	8	156	5	0	12
kall_congruentcircles_c72	22	7	147	6	5	5	22	6	115	5	5	5	21	5	105	8	5	4	21	8	129	9	5	2
kall_difficircles_10	25	2	99	2	1	5	23	1	85	0	1	5	29	2	78	3	1	4	25	1	121	1	1	6
kall_difficircles_5a	15	2	45	2	5	10	15	2	44	2	5	12	17	2	46	1	5	12	15	2	59	2	5	10
kall_difficircles_5b	16	0	49	0	6	10	17	2	48	6	6	9	18	0	49	0	6	9	15	2	38	1	1	8
kall_difficircles_6	15	1	28	1	2	3	14	0	48	0	2	7	18	2	35	3	2	4	16	1	51	0	6	8
kall_difficircles_7	17	1	38	0	2	8	21	2	51	2	2	3	20	3	54	3	2	6	18	1	72	1	2	5
kall_difficircles_8	18	0	40	0	8	8	15	0	34	0	8	8	21	0	73	0	8	8	17	1	50	1	8	8
kall_difficircles_9	21	0	95	0	1	5	20	0	75	0	1	0	24	2	67	2	1	0	21	1	87	1	1	4
knp3-12	58	18	587	68	0	0	58	35	526	220	0	0	64	30	426	174	0	0	51	17	575	91	0	0
knp4-24	128	5	3499	48	0	0	130	4	3626	10	0	0	187	1	2548	1	0	0	140	6	5594	20	0	0
knp5-40	258	3	12162	4	0	0	273	9	13009	218	0	0	395	1	7223	1	0	0	249	20	14436	312	0	0

Table A.9 continued

Instance	random						min-max						greedy (base)						reverse greedy					
	lp	lp_{filt}	iter	$iter_{filt}$	b_{obbt}	lvb	lp	lp_{filt}	iter	$iter_{filt}$	b_{obbt}	lvb	lp	lp_{filt}	iter	$iter_{filt}$	b_{obbt}	lvb	lp	lp_{filt}	iter	$iter_{filt}$	b_{obbt}	lvb
knp5-41	269	11	12676	73	0	0	266	16	13655	394	0	0	401	6	7873	148	0	0	245	31	14948	746	0	0
knp5-42	285	3	16648	14	0	0	291	5	17439	204	0	0	412	1	9755	3	0	0	309	20	24269	674	0	0
knp5-43	281	2	16310	86	0	0	192	9	11075	51	0	0	424	0	9888	0	0	0	206	12	14736	254	0	0
knp5-44	222	7	16463	55	0	0	301	3	21749	4	0	0	434	0	12542	0	0	0	222	7	23891	162	0	0
launch	80	10	416	12	49	59	83	8	257	16	49	57	85	4	170	2	49	62	76	11	193	5	49	59
least	18	3	50	0	0	6	16	2	53	1	0	2	21	3	58	0	0	6	15	2	49	1	0	5
like	697	32	769	31	0	1	697	31	645	30	0	2	697	32	661	31	0	1	757	91	991	0	0	92
linear	26	0	127	0	10	16	23	0	86	0	10	16	30	0	75	0	10	16	25	0	160	0	10	16
mathopt5.7	2	0	15	0	0	1	2	0	15	0	0	1	2	0	8	0	0	1	2	0	15	0	0	1
mathopt5.8	2	0	18	0	0	1	2	0	18	0	0	1	2	0	7	0	0	1	2	0	18	0	0	1
maxmin	150	48	886	195	66	6	154	54	909	146	66	8	148	39	709	95	66	10	174	22	815	46	66	11
methanol100	7590	0	1336	0	0	0	7590	0	1526	0	0	0	7590	0	199	0	0	0	7590	0	2170	0	0	0
methanol50	3966	0	794	0	0	0	3966	0	976	0	0	0	3966	0	224	0	0	5	3966	0	961	0	0	0
minlphi	19	3	63	1	7	5	18	4	61	0	7	4	18	5	44	1	7	5	19	3	88	1	7	5
oaer	4	0	12	0	2	4	4	0	7	0	2	4	4	0	13	1	2	4	4	2	6	2	2	4
pindyck	174	26	10199	0	112	124	174	26	7222	0	112	124	174	26	2862	0	112	124	174	26	8538	0	112	124
pineme50	1226	0	1233	0	0	0	1226	0	1233	0	0	0	1226	0	1233	0	0	0	1226	0	1233	0	0	0
pointpack04	12	5	25	5	2	5	13	3	31	4	2	4	14	0	27	0	2	4	12	4	32	4	2	4
pointpack06	18	7	39	6	1	5	20	5	53	6	1	4	21	3	38	1	1	8	16	5	67	3	1	5
pointpack08	22	3	75	1	0	6	22	11	59	5	0	6	25	13	61	3	1	9	23	5	90	3	0	10
pointpack10	29	10	86	9	0	8	26	12	70	6	0	8	31	13	62	9	0	12	33	18	126	29	0	6
pointpack12	34	17	95	16	0	14	36	20	94	10	0	15	40	19	97	16	0	13	36	17	109	19	0	12
pointpack14	40	17	105	10	0	18	41	21	105	16	0	16	45	22	107	23	0	19	39	22	94	24	0	9
portfol.buyin	12	0	69	0	7	12	13	2	84	2	7	10	13	1	73	2	7	10	11	1	98	0	7	11
powerflow0009r	64	16	350	1	0	2	63	3	374	3	0	2	86	3	315	3	0	19	56	4	509	4	0	2
powerflow0014r	124	16	1600	22	1	17	129	11	1616	2	1	16	148	14	1198	6	1	19	108	17	1523	7	1	16
powerflow0030r	218	4	1488	3	0	6	216	2	1525	1	0	8	291	2	1334	3	0	7	203	5	1937	2	0	8
powerflow0039r	339	0	2218	0	3	3	332	3	1969	3	0	8	441	2	1720	1	0	4	273	4	2526	4	0	3
powerflow0057r	457	47	6459	40	0	42	437	44	6288	22	0	41	522	36	4962	11	0	40	413	60	7746	39	0	36
powerflow0118r	1041	91	19254	34	0	96	1038	77	19913	30	0	87	1277	54	17492	21	0	105	976	95	25292	39	0	95
powerflow0300r	2498	229	45803	95	0	238	2487	218	44621	87	0	243	3101	166	39841	47	0	257	2345	249	55851	129	0	238
powerflow2383wpr	17314	305	85369	201	0	502	17140	273	92481	156	0	526	23076	280	74025	168	0	538	13706	385	119174	268	0	494
powerflow2736spr	20515	556	84288	356	0	872	20498	460	92504	263	0	888	27638	553	70238	283	0	911	16556	744	105515	575	0	850
prob06	2	2	6	0	2	2	16	2	6	0	2	2	3	0	6	0	2	2	16	0	2	0	2	2
process	16	2	119	0	8	10	28	1	164	1	8	10	16	1	101	0	8	10	29	0	195	7	8	10
procsyn	23	0	51	0	1	0	8	0	56	1	1	0	32	0	88	0	1	0	7	0	87	2	1	0
prolog	8	1	60	0	1	4	8	0	61	0	1	4	9	0	50	0	4	4	7	0	42	0	1	4
qp3	147	1	350	1	114	129	149	1	351	1	114	133	145	1	258	0	114	131	146	1	366	0	114	129
rocket100	1143	530	4368	428	1	283	1327	409	2030	257	1	320	1114	108	2782	192	1	104	1047	466	2642	140	1	369
rocket200	2338	1018	8855	601	1	703	2338	710	5161	824	1	431	2443	439	4550	205	1	432	2025	945	5486	297	1	744
rocket400	4450	2266	12444	1194	1	1313	5623	1934	7578	1102	1	1292	4748	745	10075	948	1	714	4196	2054	10893	697	1	1495
rocket50	611	237	2375	155	1	150	656	197	1028	127	1	144	560	54	1359	55	1	50	508	215	1296	60	1	181
shiporig	21	0	16	0	0	0	21	0	15	0	0	0	21	0	19	0	0	0	24	2	29	5	2	0
st.bpafla	13	0	19	0	13	13	14	0	24	0	13	13	14	0	19	0	13	13	13	0	11	0	13	13
st.bsj2	5	0	12	0	4	4	5	0	8	0	4	4	5	0	9	0	4	4	5	0	7	0	4	3
st.bsj4	7	0	9	0	6	6	7	0	8	0	6	6	7	0	6	0	6	6	7	0	7	0	6	6

Table A.9 continued

instance	random						min–max						greedy (base)						reverse greedy					
	lp	lp$_{filt}$	iter	iter$_{filt}$	b$_{obbt}$	lvb	lp	lp$_{filt}$	iter	iter$_{filt}$	b$_{obbt}$	lvb	lp	lp$_{filt}$	iter	iter$_{filt}$	b$_{obbt}$	lvb	lp	lp$_{filt}$	iter	iter$_{filt}$	b$_{obbt}$	lvb
st_e03	16	3	167	1	8	10	16	2	146	1	8	10	16	2	126	1	8	10	15	2	212	0	8	10
st_e04	7	1	64	1	2	5	7	2	69	2	2	5	7	1	83	1	2	5	9	3	53	2	2	5
st_e05	8	1	11	0	5	5	8	1	11	0	5	5	8	2	10	0	5	5	8	1	15	1	5	5
st_e07	5	0	16	0	1	3	5	0	12	0	1	3	5	0	17	0	1	3	6	1	21	1	1	3
st_e08	4	0	8	4	4	4	4	0	7	0	4	4	4	0	7	0	4	4	4	0	10	0	4	4
st_e09	4	1	3	0	4	4	4	0	2	0	4	4	4	1	2	0	4	4	4	1	4	0	4	4
st_e11	4	1	4	0	0	0	4	0	3	0	0	0	4	0	2	0	0	0	4	1	3	1	0	0
st_e16	15	1	53	0	9	7	14	1	57	1	9	7	16	0	72	0	9	6	15	1	85	0	9	6
st_e19	4	0	15	1	0	0	4	1	17	1	0	0	4	0	17	0	0	0	4	0	15	1	0	0
st_e22	4	0	5	0	3	3	4	0	4	0	3	3	4	0	3	0	3	3	4	0	6	0	3	3
st_e24	3	0	3	0	3	3	3	0	3	0	3	3	3	0	3	0	3	3	3	0	3	0	3	3
st_e25	7	0	13	0	6	4	3	0	16	0	6	2	8	1	18	0	6	2	7	1	22	0	6	2
st_e26	4	1	3	0	2	4	4	1	2	0	2	4	4	0	2	0	2	4	4	1	3	1	2	4
st_e28	7	0	11	0	7	7	7	0	11	0	7	7	7	0	11	0	7	7	7	0	14	0	7	7
st_e30	7	0	25	0	2	7	7	0	36	0	2	7	7	0	29	0	2	7	7	0	28	0	2	7
st_e33	5	0	14	0	1	3	5	0	10	0	1	4	5	1	14	0	1	3	5	0	19	0	1	3
st_e41	17	3	59	3	11	15	16	4	67	5	11	15	17	0	52	0	11	15	16	3	105	3	11	15
st_fp7a	21	0	225	0	20	20	22	0	239	0	20	20	28	0	235	0	20	20	20	0	220	0	20	20
st_fp7b	21	0	235	0	20	20	22	0	235	0	20	20	24	0	224	0	20	20	20	0	228	0	20	20
st_fp7c	21	0	225	0	20	20	22	0	239	0	20	20	28	0	235	0	20	20	20	0	220	0	20	20
st_fp7d	21	0	225	0	20	20	22	0	239	0	20	20	28	0	235	0	20	20	20	0	220	0	20	20
st_fp7e	22	0	244	0	20	20	22	0	228	0	20	20	26	0	249	0	20	20	20	0	227	0	20	20
st_glmp-fp1	4	0	13	0	4	2	4	0	12	0	4	2	4	0	5	0	4	2	4	0	13	0	4	2
st_glmp-fp2	3	0	10	0	3	2	3	0	10	0	3	2	3	0	12	0	3	2	3	0	10	0	3	2
st_glmp-fp3	4	0	35	0	3	0	4	0	27	0	3	0	4	0	13	0	3	0	4	0	35	0	3	0
st_glmp-kk90	4	0	3	0	2	3	3	0	3	0	2	3	3	0	5	0	3	3	4	0	3	0	3	3
st_glmp-kk92	3	0	26	0	3	3	3	0	19	0	3	3	3	0	18	0	3	3	3	0	26	0	3	3
st_glmp-kky	3	0	4	0	3	3	3	0	4	0	3	3	3	0	6	0	3	3	3	0	4	0	3	3
st_glmp-ss1	3	0	7	0	3	3	3	0	7	0	3	2	3	0	7	0	2	2	3	0	4	0	2	2
st_glmp-ss2	3	0	3	0	3	3	3	0	3	0	3	3	3	0	5	0	2	3	3	0	8	0	3	3
st_iqpbk1	16	0	175	0	2	7	15	1	122	0	2	7	15	0	99	0	2	6	15	0	228	0	2	7
st_iqpbk2	16	0	151	0	7	7	16	0	119	0	7	7	16	0	66	0	2	7	14	0	185	0	2	7
st_jcbpaf2	13	0	52	0	12	12	13	0	77	0	12	12	16	1	57	1	12	12	12	0	75	0	12	12
st_pan1	5	0	13	0	1	4	5	0	10	1	1	4	5	0	6	0	1	4	5	1	13	0	1	4
st_ph14	4	0	4	0	4	4	4	0	4	0	4	4	4	0	4	0	4	4	4	0	4	0	4	4
st_ph15	7	0	23	0	4	6	7	0	19	0	4	6	7	0	13	0	4	4	6	0	14	0	4	6
st_ph1	7	0	13	0	4	5	7	0	14	0	3	4	7	0	10	0	4	6	8	0	16	0	3	2
st_ph20	3	0	9	0	4	5	3	0	7	0	4	6	3	0	7	0	3	2	3	0	7	0	4	6
st_ph2	7	0	13	0	5	5	7	0	14	0	5	2	7	0	10	0	4	6	8	0	15	0	5	5
st_ph3	5	0	7	0	2	5	5	0	8	0	2	6	5	0	7	0	5	5	5	0	10	0	4	5
st_phex	4	0	7	0	6	2	4	0	6	0	6	5	4	0	4	0	4	2	4	0	8	0	2	2
st_qpc-m0	3	0	3	0	1	0	3	0	3	0	1	2	3	0	3	0	2	0	4	0	5	0	2	0
st_qpc-m1	8	0	33	0	7	6	8	0	41	2	7	6	9	0	33	0	6	6	7	0	35	0	6	6
st_qpc-m3a	11	0	158	48	7	8	13	1	118	48	7	7	13	0	98	0	1	6	14	3	120	0	1	5
st_qpc-m3b	17	6	160	0	10	10	17	6	164	0	7	10	17	6	148	36	7	6	11	3	107	3	7	9
st_qpk1	3	0	4	0	2	0	3	0	4	0	2	0	3	0	4	0	2	0	4	0	7	0	2	0

Table A.9 continued

Instance	random						min-max						greedy (base)						reverse greedy					
	lp	lp$_{filt}$	iter	iter$_{filt}$	b$_{obbt}$	lvb	lp	lp$_{filt}$	iter	iter$_{filt}$	b$_{obbt}$	lvb	lp	lp$_{filt}$	iter	iter$_{filt}$	b$_{obbt}$	lvb	lp	lp$_{filt}$	iter	iter$_{filt}$	b$_{obbt}$	lvb
st_qpk2	7	0	27	0	0	6	7	0	15	0	0	6	10	1	27	0	0	6	6	0	25	0	0	6
st_qpk3	12	0	64	0	0	11	12	0	90	0	0	11	21	1	80	0	0	11	12	1	88	0	0	11
st_rv1	12	0	35	0	10	12	12	0	33	0	10	12	12	0	25	0	10	12	12	0	31	0	10	12
st_rv2	23	0	148	0	21	21	22	0	157	0	21	21	24	1	130	0	21	21	21	0	139	0	21	21
st_rv3	25	0	223	0	20	23	24	1	226	0	20	23	28	0	174	0	20	23	24	1	220	0	20	23
st_rv7	35	0	329	0	29	32	33	0	333	0	29	32	37	0	294	0	29	32	33	0	344	0	29	32
st_rv8	44	0	697	0	39	43	43	0	668	0	39	43	47	0	637	0	39	43	43	0	763	0	39	43
st_rv9	52	0	690	0	42	50	53	0	776	0	42	50	58	0	632	0	42	50	51	0	718	0	42	50
st_z	5	0	12	0	0	4	15	0	8	0	5	4	6	0	10	0	5	4	5	0	8	0	5	4
supplychainp1_020306	15	0	255	0	5	7	15	0	264	0	7	7	15	0	262	0	7	8	15	0	225	0	7	7
syn05m	4	0	29	0	3	7	4	0	31	0	3	4	4	0	27	0	3	4	4	0	21	0	3	4
tricp	65	1	255	1	20	20	62	1	203	1	20	20	71	0	108	0	20	20	54	0	457	0	20	20
wall	16	0	1	0	0	0	16	0	1	0	0	0	16	0	1	0	0	0	16	0	1	0	0	0
wastewater02m1	12	1	36	0	9	12	12	0	45	0	9	12	13	0	67	0	9	11	12	1	46	0	9	12
wastewater02m2	15	6	110	0	1	13	15	6	78	0	1	14	15	3	95	0	1	15	15	8	66	0	1	15
wastewater04m1	18	5	61	1	6	13	18	6	88	6	6	16	18	4	109	0	6	15	18	6	72	4	6	14
wastewater04m2	23	7	108	3	10	16	23	6	110	1	10	19	23	5	97	2	10	18	23	6	139	6	10	17
wastewater05m1	25	5	124	4	3	12	26	4	123	2	3	14	26	2	199	0	3	13	25	4	98	2	3	12
wastewater05m2	43	18	1329	22	6	32	42	20	1185	12	6	33	43	9	1089	27	6	30	44	24	1170	36	6	33
wastewater11m1	74	4	1517	12	58	65	74	5	1703	21	58	64	75	3	1750	5	58	74	74	4	1579	22	58	65
wastewater11m2	88	25	3232	7	6	68	89	25	2253	9	6	70	89	16	3081	6	6	66	89	39	2267	48	6	74
wastewater12m1	135	4	3669	6	112	118	135	3	3162	3	112	120	133	3	6463	4	112	117	135	5	3028	6	112	121
wastewater12m2	153	38	4630	48	10	130	155	36	7079	18	10	120	154	21	6559	14	10	131	154	56	5080	89	10	133
wastewater13m1	277	7	10020	28	246	255	277	7	9872	12	246	253	276	4	12608	2	246	258	275	5	9531	14	246	257
wastewater13m2	284	38	16925	26	13	226	283	32	15941	48	13	222	282	10	20279	12	13	228	280	63	10043	136	13	261
wastewater14m1	50	6	978	24	30	42	52	7	932	13	30	42	51	3	971	10	30	42	50	5	902	9	30	44
wastewater14m2	51	11	2766	13	13	47	51	10	2406	8	13	43	51	6	3228	7	13	40	51	13	2553	8	13	48
wastewater15m1	28	3	276	0	12	21	28	3	309	0	12	22	28	3	381	1	12	22	29	7	378	23	12	21
wastewater15m2	37	11	1168	4	13	35	38	12	1061	7	13	34	37	6	1232	7	13	33	37	13	1373	13	13	32
waterund01	37	11	440	16	14	18	40	16	499	31	14	20	41	10	495	21	14	18	39	17	663	30	14	15
waterund08	88	23	4013	40	34	52	87	27	4045	36	34	54	98	21	2877	37	34	50	88	31	3751	55	34	45
waterund11	52	13	918	17	20	27	54	16	836	14	20	30	54	9	904	8	20	27	50	14	1128	14	20	24
waterund14	124	45	4303	84	41	62	134	54	4162	152	42	60	140	19	3434	25	42	67	112	45	4406	80	42	60
waterund17	68	26	1647	24	25	41	72	24	1648	35	25	40	78	10	1601	22	25	42	72	35	1580	52	25	41
waterund18	52	14	814	14	21	28	53	15	1000	20	21	27	62	12	876	12	21	31	51	20	1086	51	21	26
waterund22	120	46	2412	58	32	53	124	50	2361	100	32	54	128	32	2698	64	32	49	116	50	2717	78	32	51
waterund25	89	28	3789	47	34	44	101	34	4154	57	34	41	103	15	4021	23	34	47	96	36	5075	68	34	47
waterund27	289	107	57259	286	105	144	305	121	61276	293	105	141	321	91	61193	198	105	148	302	126	59765	496	105	140
waterund28	751	175	1103181	117	254	484	748	186	1121567	279	254	481	950	57	923371	48	254	488	759	193	1065053	147	254	494
waterund32	668	140	179095	184	180	374	663	150	147944	206	180	367	741	94	150635	159	180	360	675	128	180056	184	180	336
waterund36	289	103	57248	101	101	129	284	97	46912	279	101	135	334	84	54394	252	101	133	291	101	53513	210	101	133
weapons	183	41	2933	58	0	80	169	32	2777	20	0	78	182	37	2143	38	0	69	166	37	2503	28	0	67

Table A.10: Detailed results comparing SCIP without OBBT, SCIP with OBBT only, and SCIP with OBBT and LVB propagation on test set INT. See Table 6.2 for aggregated results.

n — number of branch-and-bound nodes
t — total running time
t_{obbt} — time used by OBBT
lp — LPs solved by OBBT
b_{obbt} — bounds tightened by OBBT
lvb — LVBs found by OBBT
t_{lvb} — time used by LVB propagation
b_{lvb} — bounds tightened by LVB propagation

Instance	SCIP plain		SCIP+OBBT					SCIP+OBBT+LVB						
	n	t	n	t	t_{obbt}	lp	b_{obbt}	n	t	t_{obbt}	t_{lvb}	b_{obbt}	b_{lvb}	lvb
alan	3	0.1	4	0.1	0.00	34	8	4	0.0	0.00	0.00	8	0	9
batch	8	0.2	16	0.3	0.01	259	16	16	0.2	0.02	0.01	16	72	24
batch0812	4	0.3	4	0.4	0.03	449	82	3	0.5	0.02	0.00	74	30	85
batch0812_nc	911	2.2	792	1.9	0.13	717	77	592	2.0	0.15	0.00	69	906	111
batchdes	1	0.1	1	0.0	0.00	55	9	1	0.1	0.01	0.00	9	7	14
batchs101006m	1631	9.8	2008	10.3	0.14	608	19	1584	9.2	0.14	0.04	19	963	25
batchs121208m	3852	17.6	3378	14.8	0.59	2629	23	3094	13.0	0.59	0.01	23	1897	48
batchs151208m	3932	28.2	2797	19.2	0.47	4439	35	5151	31.1	0.54	0.02	35	2036	75
batchs201210m	11277	60.8	6267	33.4	0.73	6477	33	7189	39.4	0.73	0.02	33	3593	78
blend029	29046	12.0	21241	9.6	0.02	305	2	33150	13.1	0.02	0.06	2	3974	18
blend721	1073801	717.4	1536191	1013.7	0.22	1739	0	580639	375.0	0.21	0.95	0	25950	61
carton7	595797	459.1	499975	367.8	0.14	2031	4	318913	229.8	0.13	0.53	4	2761	11
clay0203h	102	2.3	626	3.5	0.44	3469	0	2991	5.7	0.60	0.01	0	4472	233
clay0203m	20	0.2	20	0.2	0.01	128	0	20	0.2	0.01	0.01	0	0	0
clay0204h	1172	13.4	1172	13.6	0.75	3342	0	1021	11.9	0.75	0.00	0	4326	328
clay0204m	1086	1.2	1143	1.1	0.01	176	0	1145	1.2	0.01	0.00	0	1	1
clay0205h	12475	103.2	14493	112.0	1.31	6806	0	13605	131.5	1.51	0.10	0	49273	381
clay0205m	8107	3.9	8635	4.2	0.01	255	0	8717	4.0	0.01	0.01	0	0	1
clay0303h	395	6.8	395	7.1	0.46	4162	0	395	7.4	0.45	0.01	0	890	315
clay0303m	23	0.3	23	0.3	0.00	133	0	23	0.3	0.01	0.00	0	2	1
clay0304h	30328	412.8	26806	411.2	1.90	16058	0	18802	275.4	2.00	0.07	0	54488	414
clay0304m	535	1.0	610	1.1	0.01	195	0	652	0.9	0.01	0.00	0	3	4
clay0305h	9917	201.0	9415	216.8	3.73	22875	0	12022	276.9	3.67	0.06	0	5733	532
clay0305m	9493	5.3	9527	5.2	0.02	280	0	9537	5.3	0.01	0.00	0	0	1
crudeoil_lee1.05	8	1.1	12	1.2	0.24	1613	34	3	1.3	0.39	0.00	34	3	52
crudeoil_lee1.06	43	2.0	42	2.8	0.52	2310	21	45	2.9	0.63	0.00	21	53	64
crudeoil_lee1.07	56	4.0	40	5.5	0.75	3235	25	46	5.5	0.67	0.00	25	59	78
crudeoil_lee1.08	176	7.8	153	8.8	1.06	3416	32	161	9.0	1.06	0.00	32	175	96
crudeoil_lee1.09	52	7.3	33	8.3	1.39	4531	35	35	9.0	1.42	0.01	35	87	100
crudeoil_lee1.10	66	8.7	165	17.2	2.30	6101	43	165	18.6	2.46	0.02	43	257	117
crudeoil_lee2.05	12	2.0	12	4.9	2.53	11746	92	18	4.6	2.50	0.00	103	25	248
crudeoil_lee2.06	157	10.0	250	14.1	3.27	10239	55	257	14.1	3.55	0.00	55	337	119
crudeoil_lee2.07	192	11.0	330	17.8	4.95	10292	63	301	19.3	4.99	0.00	63	242	141
crudeoil_lee2.08	543	22.9	345	26.9	7.45	13794	74	455	27.9	7.69	0.02	74	229	174

Table A.10 continued

Instance	SCIP plain		SCIP +OBBT					SCIP +OBBT+LVB						
	n	t	n	t	t_{obbt}	lp	b_{obbt}	n	t	t_{obbt}	t_{lvb}	b_{obbt}	b_{lvb}	lvb
crudeoil_lee2.09	517	28.6	730	39.6	13.14	17313	86	666	39.5	13.07	0.02	86	827	199
crudeoil_lee2.10	685	35.4	596	40.5	13.87	16925	97	775	52.4	13.83	0.01	97	483	229
crudeoil_lee3.05	409961	506.2	335361	430.6	1.48	5369	120	535451	692.0	1.50	7.57	120	1453824	192
crudeoil_lee4.05	43	4.0	66	6.4	2.23	8398	75	99	6.9	2.35	0.00	75	47	134
crudeoil_lee4.06	83	10.2	5	16.5	5.33	7834	100	5	15.7	5.13	0.00	100	3	165
crudeoil_lee4.07	88	9.8	195	26.8	9.24	13056	116	412	28.4	9.26	0.01	116	729	201
crudeoil_lee4.08	91	19.8	218	36.2	10.85	8190	109	303	33.6	10.15	0.01	109	190	213
crudeoil_lee4.09	128	24.9	83	42.5	16.56	13255	113	101	44.8	16.52	0.00	113	98	235
crudeoil_lee4.10	451	43.5	243	58.2	22.67	20971	156	383	59.1	23.02	0.00	156	491	294
crudeoil_li06	92305	966.2	51531	526.2	1.67	5289	14	44145	512.5	1.67	0.02	14	5949	24
du-opt	1808	2.8	1364	2.1	0.01	181	4	1323	2.6	0.00	0.02	4	295	5
du-opt5	64	0.5	66	0.5	0.00	163	4	57	0.5	0.00	0.01	4	35	5
edgecross10-060	42081	79.0	42081	79.1	0.09	1368	0	42081	79.6	0.09	0.02	0	0	0
edgecross14-039	28597	505.2	28579	508.6	0.42	5040	0	28579	505.2	0.42	0.04	0	0	0
elf	477	0.8	477	1.0	0.01	283	0	350	0.8	0.00	0.00	0	283	6
eniplac	232	1.2	232	1.1	0.01	397	0	232	1.2	0.01	0.00	0	54	21
enpro48pb	20	1.0	34	1.1	0.10	1127	32	32	1.2	0.14	0.01	33	25	44
enpro56pb	1318	3.0	546	2.1	0.08	547	8	563	2.3	0.09	0.01	8	93	14
ex1224	4	0.1	1	0.2	0.00	76	4	1	0.1	0.00	0.00	4	5	6
ex1243	131	1.1	131	1.1	0.00	57	0	135	1.1	0.00	0.00	0	81	15
ex1244	4154	4.4	3473	5.6	0.01	228	2	3306	5.8	0.01	0.01	2	855	10
ex1263	276	0.7	467	0.8	0.02	401	12	371	0.6	0.03	0.00	12	335	30
ex1263a	67	0.1	728	0.9	0.00	53	12	188	0.3	0.00	0.00	12	178	15
ex1264	253	0.6	153	0.4	0.00	154	8	103	0.2	0.00	0.00	8	112	14
ex1264a	66	0.1	137	0.1	0.00	61	9	145	0.2	0.01	0.01	9	118	12
ex1265	98	0.3	224	0.7	0.01	190	6	294	0.9	0.01	0.00	6	167	17
ex1265a	40	0.1	78	0.2	0.01	109	12	74	0.1	0.00	0.00	12	57	14
ex1266	339	1.0	289	1.2	0.08	1340	23	99	1.0	0.08	0.03	23	145	49
ex1266a	5	0.1	121	0.2	0.00	115	21	153	0.5	0.00	0.00	21	138	23
ex4	14	0.7	14	0.7	0.03	287	8	11	0.7	0.02	0.00	7	11	19
fac1	2	0.1	2	0.1	0.00	149	0	2	0.1	0.00	0.00	0	0	20
fac3	9	0.3	9	0.3	0.01	75	0	9	0.3	0.01	0.00	0	0	36
fin2bb	5000	230.5	18473	637.1	0.21	1215	61	5570	238.1	0.18	0.00	61	403	61
flay02h	7	0.1	7	0.1	0.00	156	2	7	0.1	0.02	0.00	2	3	2
flay02m	7	0.1	7	0.1	0.00	62	2	7	0.1	0.00	0.00	2	1	2
flay03h	111	0.9	103	0.7	0.04	512	3	103	0.8	0.03	0.00	3	2	4
flay03m	103	0.2	107	0.2	0.00	71	3	107	0.2	0.00	0.00	3	68	3
flay04h	2196	4.1	2335	4.1	0.02	256	3	2335	4.3	0.03	0.00	4	0	0
flay04m	2363	1.8	2385	2.0	0.00	106	0	2385	2.0	0.01	0.00	0	0	4
flay05h	106721	191.2	106443	191.0	0.06	631	0	99444	182.7	0.06	0.39	0	933	6
flay05m	95853	54.6	98881	55.3	0.01	118	0	98107	55.8	0.00	0.38	0	10317	5
flay06m	3609871	1793.5	3596393	1777.4	0.01	207	0	3814781	1878.0	0.01	9.60	0	2272908	6
fo7	168006	85.0	207840	98.1	0.01	194	0	207840	98.9	0.02	0.35	0	0	1
fo7_2	55500	24.5	53727	24.4	0.00	93	0	53727	24.5	0.00	0.03	0	0	4
fo7_ar25_1	36189	18.1	42734	19.1	0.00	116	0	42734	19.3	0.01	0.10	0	0	4
fo7_ar2_1	113648	51.2	142494	61.6	0.01	100	0	142494	62.1	0.00	0.22	0	0	14

Table A.10 continued

Instance	SCIP plain		SCIP +OBBT					SCIP +OBBT+LVB						
	n	t	n	t	t_{obbt}	lp	b_{obbt}	n	t	t_{obbt}	t_{lvb}	b_{obbt}	b_{lvb}	lvb
fo7.ar3.1	52706	23.9	51514	22.4	0.01	92	0	51514	22.4	0.01	0.10	0	0	1
fo7.ar4.1	62643	31.3	62404	30.6	0.01	88	0	62404	30.6	0.01	0.12	0	0	5
fo7.ar5.1	35545	20.2	36539	23.4	0.01	120	0	36539	23.5	0.00	0.06	0	0	6
fo8	211923	120.9	220300	136.2	0.02	162	0	210539	124.4	0.01	0.26	0	1010	1
fo8.ar25.1	614176	249.1	605817	250.2	0.01	121	0	605817	242.8	0.00	0.85	0	0	7
fo8.ar2.1	443504	184.0	275619	113.0	0.01	94	0	275619	113.9	0.01	0.33	0	0	16
fo8.ar3.1	198579	106.4	162891	83.8	0.01	98	0	162891	83.1	0.01	0.19	0	0	4
fo8.ar4.1	180458	87.9	134526	70.6	0.00	139	0	134526	71.4	0.01	0.20	0	0	11
fo8.ar5.1	75792	41.9	110941	59.5	0.01	98	0	110941	59.5	0.01	0.18	0	0	6
fo9	2358400	1243.6	2078934	1127.7	0.02	242	0	2078934	1131.4	0.03	1.79	0	0	0
fo9.ar2.1	3667495	1263.0	3653351	1256.3	0.01	93	0	3653351	1261.5	0.01	4.12	0	0	18
fo9.ar3.1	330999	171.7	524507	242.0	0.01	168	0	524507	244.5	0.01	0.57	0	0	4
fo9.ar4.1	1909806	1097.2	1102537	581.4	0.02	161	0	1102537	574.0	0.02	1.36	0	0	7
fo9.ar5.1	1013002	640.5	775573	442.9	0.01	113	0	775573	444.2	0.02	1.18	0	0	8
fuel	1	0.1	1	0.0	0.00	83	4	1	0.0	0.00	0.00	4	3	9
gasprod_sarawak01	652	1.5	28	1.6	0.05	529	10	16	1.3	0.06	0.00	10	27	64
gastrans	112	0.7	6	0.3	0.05	931	46	6	0.2	0.07	0.00	46	1	59
gear	1	0.0	1	0.0	0.00	26	1	1	0.0	0.00	0.00	1	0	5
gear2	1	0.0	1	0.1	0.00	32	0	1	0.1	0.00	0.00	0	0	4
gear3	1	0.0	1	0.0	0.00	26	1	1	0.0	0.00	0.00	1	0	5
genpooling_lee1	4928	6.3	4928	6.4	0.00	126	0	5199	6.9	0.01	0.01	0	749	5
genpooling_lee2	155441	123.7	155441	128.2	0.00	120	0	86231	94.7	0.01	0.25	0	16751	9
ghg_1veh	7481	38.5	9901	45.3	0.01	315	3	10451	52.4	0.02	0.04	3	4804	20
gkocis	4	0.1	3	0.1	0.00	20	2	3	0.1	0.00	0.00	2	0	4
hybriddynamic.var	9011	8.2	3391	3.2	0.01	179	22	3131	3.1	0.02	0.00	22	1946	25
jit1	10	0.0	11	0.0	0.00	68	6	9	0.0	0.00	0.00	6	12	6
johnall	771	179.6	7	435.7	188.62	9799	5158	7	450.3	196.16	0.01	8	1355	5259
kport20	187383	159.3	310503	274.6	0.02	816	8	268805	270.5	0.03	1.45	8	482559	35
lip	5	0.2	1	0.2	0.00	8	8	1	0.2	0.00	0.00	8	0	7
m3	19	0.2	19	0.2	0.01	35	0	19	0.2	0.00	0.00	0	0	2
m6	3817	2.4	3817	2.4	0.01	120	0	3817	2.4	0.01	0.00	0	0	3
m7	4499	4.3	4499	4.5	0.01	103	0	4499	4.7	0.01	0.01	0	0	1
m7.ar25.1	1689	2.0	1674	1.8	0.00	89	0	1674	1.9	0.01	0.04	0	0	3
m7.ar2.1	11338	6.6	10546	6.0	0.01	90	0	10546	6.1	0.01	0.01	0	0	4
m7.ar3.1	7729	4.8	7729	4.8	0.01	110	0	8349	5.8	0.02	0.01	0	33	6
m7.ar4.1	2679	3.7	2643	3.8	0.01	167	0	2643	3.7	0.02	0.02	0	0	7
m7.ar5.1	4616	5.0	10723	9.2	0.01	135	0	10723	9.5	0.01	0.01	0	0	5
meanvarxsc	6	0.2	5	0.2	0.00	70	14	5	0.2	0.00	0.00	14	1	26
netmod_dol1	52477	1813.5	56979	1900.8	0.01	12	0	57905	1914.9	0.01	0.06	0	419	4
netmod_dol2	209	71.0	209	71.8	0.70	5698	0	259	74.6	0.82	0.01	0	217	4
netmod_kar1	336	3.5	336	3.3	0.00	6	0	336	3.9	0.00	0.00	0	1	1
netmod_kar2	336	3.6	336	3.3	0.00	6	0	336	3.5	0.00	0.00	0	1	1
no7.ar25.1	246825	120.2	250005	116.7	0.01	125	0	250005	116.8	0.01	0.34	0	0	4
no7.ar2.1	68666	36.9	68395	34.4	0.01	113	0	68395	36.2	0.01	0.09	0	0	14
no7.ar3.1	613840	259.6	514058	221.1	0.00	135	0	514058	222.3	0.01	0.53	0	0	1
no7.ar4.1	279454	138.4	305424	152.5	0.01	140	0	305424	154.4	0.01	0.32	0	0	7

Table A.10 continued

Instance	SCIP plain		SCIP +OBBT					SCIP +OBBT+LVB						
	n	t	n	t	t_{obbt}	lp	b_{obbt}	n	t	t_{obbt}	t_{lvb}	b_{obbt}	b_{lvb}	lvb
no7_ar5_1	115855	68.7	114540	62.1	0.01	156	0	114540	62.9	0.02	0.21	0	0	6
nous2	8651	6.7	6161	5.4	0.01	187	7	13231	10.7	0.01	0.03	7	4956	44
nvs01	9	0.1	8	0.1	0.00	114	3	9	0.1	0.01	0.00	3	4	5
nvs02	4	0.0	4	0.0	0.00	26	0	4	0.0	0.01	0.00	0	0	0
nvs06	17	0.1	11	0.1	0.00	35	5	11	0.1	0.00	0.00	5	4	8
nvs08	3	0.1	5	0.1	0.00	34	1	5	0.1	0.00	0.00	1	2	1
nvs09	17365951	3600.0	1	0.1	0.01	244	1	1	0.1	0.02	0.00	1	0	7
nvs10	1	0.0	1	0.0	0.00	8	4	1	0.0	0.00	0.00	4	0	4
nvs11	5	0.1	1	0.0	0.00	23	6	1	0.0	0.00	0.00	6	0	6
nvs12	5	0.1	5	0.1	0.00	33	8	5	0.0	0.00	0.00	8	0	8
nvs13	21	0.1	11	0.1	0.01	30	8	11	0.1	0.00	0.00	8	4	8
nvs15	6	0.0	6	0.0	0.00	25	2	6	0.0	0.00	0.00	2	0	2
nvs16	6	0.1	6	0.1	0.00	62	0	6	0.1	0.00	0.00	0	0	0
nvs17	45	0.2	47	0.2	0.00	42	11	43	0.2	0.00	0.00	11	55	11
nvs18	33	0.1	31	0.2	0.00	34	10	33	0.1	0.01	0.00	10	22	10
nvs19	105	0.4	99	0.3	0.00	59	12	81	0.4	0.01	0.00	12	97	12
nvs21	255	0.5	255	0.6	0.00	5	0	255	0.5	0.00	0.00	0	91	2
nvs23	117	0.6	109	0.6	0.00	50	13	101	0.5	0.01	0.00	13	67	13
nvs24	139	0.7	144	0.5	0.01	75	12	138	0.6	0.01	0.00	12	97	12
o7	4697648	1971.1	4752796	2020.9	0.00	95	0	4752796	2022.9	0.00	4.08	0	0	0
o7_2	1556312	625.6	1399982	560.8	0.01	129	0	1399982	560.8	0.01	1.05	0	0	0
o7_ar25_1	735468	333.7	811159	364.8	0.02	124	0	811159	366.3	0.01	1.30	0	0	4
o7_ar2_1	448027	206.7	398089	185.1	0.01	161	0	398089	185.8	0.02	0.65	0	0	14
o7_ar3_1	1549801	690.3	1425964	634.3	0.01	148	0	1425964	621.2	0.01	1.67	0	0	1
o7_ar4_1	2494812	1214.3	2284976	1072.1	0.01	131	0	2284976	1077.5	0.02	3.48	0	0	6
o7_ar5_1	736393	335.6	688302	340.6	0.01	131	0	688302	337.1	0.01	0.84	0	0	6
oil2	1140	11.3	2	5.0	2.19	3096	839	4	0.4	2.17	0.01	839	8	1163
ortez	9	0.5	4	0.4	0.05	477	27	5	0.3	0.03	0.00	8	13	36
portfol_card	8	0.3	6	0.3	0.00	130	18	5	0.4	0.00	0.00	18	15	32
portfol_classical050_1	21882	73.9	20437	67.7	0.01	192	59	19195	66.5	0.03	0.22	59	33442	61
portfol_robust050_34	272	8.7	310	6.4	0.13	658	314	324	6.4	0.09	0.00	314	322	348
portfol_robust100_09	3913	69.5	4237	73.0	0.31	570	334	4655	77.6	0.25	0.33	334	31939	338
portfol_shortfall050_68	1609	12.8	1085	10.9	0.05	277	188	1062	10.8	0.02	0.03	188	11355	200
procsel	1	0.0	1	0.0	0.00	8	2	1	0.0	0.00	0.00	2	0	2
product	2681	16.6	2681	17.3	0.22	5002	2	2331	16.2	0.22	0.00	0	148	27
product2	362	9.6	2	3.4	0.14	1665	157	2	3.4	0.15	0.00	157	4	214
ravempb	31	1.2	19	0.8	0.08	1064	3	19	0.8	0.08	0.00	3	6	11
risk2bpb	7	0.3	6	0.3	0.01	226	3	6	0.3	0.01	0.00	3	0	3
routingdelay_bigm	1041210	3600.1	4699	75.0	0.86	4288	138	145491	528.3	0.94	0.27	138	107747	360
rsyn0805h	7926888	3600.0	14	0.8	0.00	53	3	20	1.1	0.01	0.00	3	0	5
rsyn0805m	645	1.0	645	1.2	0.00	40	0	645	1.1	0.01	0.00	0	16	2
rsyn0805m02h	1375	3.3	1392	3.4	0.07	517	4	1277	3.5	0.07	0.03	4	108	20
rsyn0805m02m	7680	9.0	7680	9.0	0.01	93	0	7680	9.2	0.00	0.00	0	1520	5
rsyn0805m03m	7421	18.9	6839	16.9	0.04	601	0	6839	17.2	0.04	0.03	0	289	22
rsyn0805m04m	4483	17.3	4043	16.6	0.04	402	0	4043	16.4	0.04	0.01	0	230	15
rsyn0810h	108	0.8	90	0.7	0.02	265	5	90	0.7	0.02	0.00	5	19	14

Table A.10 continued

Instance	SCIP plain		SCIP +OBBT					SCIP +OBBT+LVB						
	n	t	n	t	t_{obbt}	lp	b_{obbt}	n	t	t_{obbt}	t_{lvb}	b_{obbt}	b_{lvb}	lvb
rsyn0810m	921	1.6	921	1.6	0.01	96	0	921	1.4	0.01	0.00	0	32	5
rsyn0810m02m	34147	39.8	34147	40.0	0.02	225	0	33663	37.5	0.01	0.15	0	17398	10
rsyn0810m03m	30636	56.2	30636	56.5	0.02	398	0	30796	57.8	0.02	0.04	0	1000	10
rsyn0810m04m	44515	104.0	60082	135.2	0.12	993	0	59924	129.2	0.11	0.05	0	206	34
rsyn0815m	383	0.9	383	1.0	0.01	170	0	383	1.0	0.01	0.00	0	6	8
rsyn0815m02m	54903	61.6	45159	54.5	0.05	794	0	45900	55.6	0.04	0.09	0	10889	13
rsyn0815m03m	52362	75.8	55757	76.9	0.36	2905	2	53327	77.7	0.35	0.16	2	8092	42
rsyn0815m04m	1893032	3329.5	2060091	3479.4	0.32	3067	3	1959755	3252.5	0.32	3.18	3	410497	31
rsyn0820m	3125	3.8	3125	3.7	0.02	122	0	3125	3.8	0.01	0.00	0	356	7
rsyn0820m02m	56626	82.3	56626	81.8	0.05	634	0	49906	76.1	0.05	0.08	0	9073	18
rsyn0830m	2310	3.8	2310	4.0	0.03	254	0	2310	4.2	0.02	0.00	0	140	11
rsyn0830m02m	141891	216.4	148153	227.4	0.07	806	3	146332	235.0	0.07	0.27	3	6072	29
rsyn0840m	2269	4.4	2097	4.2	0.03	391	1	2097	4.0	0.03	0.02	1	91	17
rsyn0840m02m	185484	314.6	211288	392.9	0.08	1124	0	202303	373.3	0.08	0.35	0	49444	34
sep1	37	0.6	21	0.5	0.01	33	6	19	0.5	0.00	0.00	6	1	8
sepasequ_convent	504	5.0	2000	24.9	7.57	26534	239	3197	24.1	7.56	0.12	239	433	378
sfacloc2_2_80	748	5.8	872	5.8	0.11	1103	0	872	5.6	0.11	0.00	0	487	40
sfacloc2_2_90	409	1.6	409	1.5	0.06	1131	0	409	1.5	0.09	0.00	0	496	41
sfacloc2_2_95	195	0.6	195	0.7	0.04	635	0	195	0.7	0.04	0.00	0	146	19
sfacloc2_3_90	230215	130.7	229749	131.9	0.08	737	0	256360	144.9	0.10	0.42	0	45632	54
sfacloc2_3_95	15839	10.4	15839	10.5	0.06	217	0	15573	10.4	0.05	0.04	0	3940	24
sfacloc2_4_90	1613184	1163.6	1698010	1219.0	0.11	521	0	1655381	1193.6	0.11	3.04	0	518857	69
sfacloc2_4_95	103694	61.7	108256	66.7	0.10	354	0	104671	63.4	0.10	0.13	0	20550	33
slay04h	52	0.8	55	1.0	0.05	758	4	55	0.9	0.04	0.00	4	0	9
slay04m	38	0.9	40	0.7	0.01	227	3	42	0.7	0.00	0.00	3	23	4
slay05h	342	2.7	289	3.4	0.11	1495	5	279	2.9	0.10	0.00	5	2	19
slay05m	33	1.0	33	0.9	0.01	344	0	33	0.8	0.01	0.00	0	0	0
slay06h	1533	8.8	1337	7.8	0.22	3868	6	679	7.0	0.20	0.01	6	15	24
slay06m	75	1.7	73	1.7	0.01	345	0	73	1.7	0.00	0.00	0	0	0
slay07h	8863	46.6	2841	27.3	0.45	7008	3	4884	33.6	0.43	0.01	3	165	22
slay07m	358	2.4	236	2.0	0.02	560	0	236	2.0	0.02	0.00	0	2	1
slay08h	4740	42.9	5087	43.6	0.52	7008	4	4270	35.1	0.39	0.00	4	353	34
slay08m	667	6.6	667	6.8	0.04	620	0	667	6.6	0.03	0.00	0	0	0
slay09h	8091	77.2	280805	1179.7	1.03	12596	5	10346	127.0	1.04	0.07	5	414	33
slay09m	5063	24.6	3972	22.2	0.06	879	0	3972	21.4	0.05	0.02	0	0	1
slay10m	30497	134.3	55465	225.1	0.04	1412	0	55465	221.1	0.05	0.03	0	0	0
smallinvDAXr1b010-011	5119	0.9	4343	0.9	0.00	51	8	1096	0.5	0.00	0.02	8	417	8
smallinvDAXr1b020-022	96	0.2	110	0.2	0.00	53	8	99	0.2	0.00	0.00	8	97	8
smallinvDAXr1b050-055	280	0.4	236	0.4	0.00	46	13	181	0.3	0.00	0.00	13	101	13
smallinvDAXr1b100-110	9302382	3332.0	678	0.8	0.00	49	14	582	0.8	0.00	0.02	14	285	14
smallinvDAXr1b150-165	1020	0.9	1326	1.2	0.00	50	26	1326	1.1	0.00	0.00	26	49	28
smallinvDAXr1b200-220	780	1.0	311	0.8	0.01	50	26	311	0.9	0.00	0.02	26	44	28
smallinvDAXr2b010-011	5095	1.1	9838	1.7	0.00	54	7	2434	0.8	0.00	0.02	7	613	7
smallinvDAXr2b020-022	110	0.2	107	0.3	0.01	51	9	101	0.3	0.00	0.00	9	28	9
smallinvDAXr2b050-055	212	0.6	193	0.5	0.00	53	25	293	0.7	0.00	0.00	25	56	25
smallinvDAXr2b100-110	2843962	958.9	512	0.7	0.00	45	13	462	0.5	0.01	0.01	13	119	13

Table A.10 continued

Instance	SCIP plain		SCIP +OBBT					SCIP +OBBT+LVB						
	n	t	n	t	t_{obbt}	lp	b_{obbt}	n	t	t_{obbt}	t_{lvb}	b_{obbt}	b_{lvb}	lvb
smallinvDAXr2b150-165	2907046	3600.0	3550365	3600.0	0.01	54	13	2401412	864.8	0.00	7.33	13	176575	13
smallinvDAXr2b200-220	111	0.3	281	0.6	0.00	39	13	317	0.5	0.01	0.00	13	25	13
smallinvDAXr3b010-011	4881	1.0	9575	1.3	0.02	51	7	5194	0.9	0.00	0.01	7	1241	7
smallinvDAXr3b020-022	32273	4.7	41816	6.0	0.00	52	10	4326	1.1	0.00	0.00	10	405	10
smallinvDAXr3b050-055	227	0.3	262	0.4	0.00	52	13	202	0.2	0.00	0.00	13	52	13
smallinvDAXr3b100-110	329	0.7	492	0.6	0.01	57	13	472	0.6	0.01	0.00	13	61	13
smallinvDAXr3b150-165	9352132	1299.6	993502	100.8	0.00	46	13	65682	11.8	0.00	0.15	13	8205	13
smallinvDAXr4b010-011	5346	1.2	9210	1.5	0.00	51	7	3308	0.8	0.00	0.00	7	624	7
smallinvDAXr4b020-022	31409	4.7	28331	4.3	0.00	56	11	1149	0.6	0.00	0.04	11	218	11
smallinvDAXr4b050-055	218	0.3	289	0.6	0.00	63	13	232	0.4	0.00	0.00	13	70	13
smallinvDAXr4b100-110	3913031	702.9	475322	124.1	0.00	39	13	12150	3.0	0.00	0.05	13	3151	13
smallinvDAXr4b150-165	5952282	2040.1	10212	2.5	0.01	28	13	872	0.7	0.00	0.00	13	68	13
smallinvDAXr5b010-011	5874	1.1	4706	1.1	0.00	72	8	132	0.2	0.00	0.00	8	36	8
smallinvDAXr5b020-022	75	0.2	101	0.2	0.00	48	11	90	0.2	0.00	0.00	11	71	11
smallinvDAXr5b050-055	257	0.4	228	0.3	0.00	45	13	224	0.3	0.00	0.00	13	83	13
smallinvDAXr5b100-110	7000991	3600.0	12248804	3600.1	0.01	46	9	438	0.6	0.00	0.00	9	71	9
smallinvDAXr5b150-165	14209156	3600.0	16272742	1776.2	0.00	42	13	1955490	333.3	0.00	3.39	13	124188	13
smallinvDAXr5b200-220	43347891	3600.0	8592721	3600.0	0.00	48	13	4114605	905.6	0.00	6.64	13	220065	13
smallinvSNPr1b010-011	794	6.5	806	7.1	0.01	362	68	735	5.9	0.01	0.04	68	139	72
smallinvSNPr1b020-022	28178	117.9	16605	72.3	0.01	335	67	26358	112.4	0.02	0.43	67	388	79
smallinvSNPr1b050-055	710662	3600.0	681172	3600.0	0.01	236	39	610118	3016.5	0.01	9.91	39	1309	81
smallinvSNPr2b010-011	511	5.3	564	5.7	0.01	346	75	695	5.8	0.01	0.04	75	516	76
smallinvSNPr2b020-022	4382	22.1	5993	29.3	0.01	240	47	2995	17.5	0.02	0.08	47	405	63
smallinvSNPr2b050-055	48883	232.3	194470	878.8	0.02	342	36	175781	809.0	0.02	3.25	36	758	82
smallinvSNPr3b010-011	297	3.7	402	3.5	0.01	290	68	317	3.7	0.01	0.00	68	451	69
smallinvSNPr3b020-022	930	7.8	824	8.5	0.02	281	61	1929	12.6	0.01	0.04	61	925	72
smallinvSNPr3b050-055	17318	83.4	10035	51.3	0.01	304	26	10144	50.7	0.01	0.10	26	537	73
smallinvSNPr3b100-110	208403	1012.1	559743	2855.6	0.02	274	32	114885	531.0	0.01	2.03	32	3920	87
smallinvSNPr4b010-011	294	2.7	263	3.0	0.01	404	56	427	3.1	0.02	0.01	56	106	57
smallinvSNPr4b020-022	436	5.5	441	5.2	0.01	329	43	659	6.2	0.02	0.03	43	214	61
smallinvSNPr4b050-055	1575	14.1	2127	14.2	0.02	328	35	1800	13.9	0.01	0.06	35	569	80
smallinvSNPr4b100-110	58741	271.8	18046	83.4	0.02	273	18	7022	35.5	0.01	0.19	18	1398	82
smallinvSNPr4b150-165	49636	242.9	48629	228.1	0.01	293	18	106884	461.7	0.03	1.54	18	13691	88
smallinvSNPr4b200-220	50809	232.6	80304	384.0	0.01	437	20	160943	682.5	0.02	2.33	20	24679	89
smallinvSNPr5b010-011	212	2.4	155	1.9	0.00	271	66	217	2.3	0.00	0.01	66	119	67
smallinvSNPr5b020-022	314	3.8	230	3.6	0.02	290	44	243	3.2	0.01	0.05	44	241	60
smallinvSNPr5b050-055	998	10.3	964	8.7	0.01	287	12	831	9.7	0.01	0.06	12	501	65
smallinvSNPr5b100-110	4308	28.6	4856	32.2	0.02	314	35	3065	20.6	0.01	0.07	35	1199	89
smallinvSNPr5b150-165	8152	41.4	4326	26.1	0.01	334	15	12756	72.8	0.01	0.27	15	3458	90
smallinvSNPr5b200-220	24188	117.0	29683	156.6	0.01	254	34	264735	1240.0	0.01	4.70	34	48933	92
spectra2	23	0.3	33	1.3	0.03	227	26	33	1.3	0.03	0.00	26	0	26
spring	36	0.3	23	0.2	0.00	137	17	22	0.4	0.01	0.00	17	28	24
squfl010-025	706	2294.3	710	2424.7	1.25	1082	0	710	2400.8	1.25	0.01	0	2427	163
squfl010-025persp	10	2.1	8	12.1	9.88	7480	578	6	11.2	9.18	0.02	394	480	479
squfl010-040persp	6	2.4	8	19.6	17.13	12668	450	6	18.0	15.53	0.01	476	124	807
squfl010-080persp	133	24.4	133	61.3	36.31	28039	422	133	59.1	35.10	0.01	424	307	535

Table A.10 continued

Instance	SCIP plain		SCIP +OBBT					SCIP +OBBT+LVB						
	n	t	n	t	t_{obbt}	lp	b_{obbt}	n	t	t_{obbt}	t_{lvb}	b_{obbt}	b_{lvb}	lvb
squfl015-060persp	215	36.8	211	82.9	43.23	18428	243	211	82.5	42.88	0.01	243	243	243
squfl015-080persp	541	77.5	537	129.5	45.02	33311	265	537	129.6	45.15	0.01	265	0	265
squfl020-040persp	39	10.9	28	125.3	115.42	37136	986	28	127.8	117.38	0.16	986	1237	1405
squfl020-050persp	90	21.7	83	265.1	247.38	64516	1285	83	262.8	245.42	0.17	1285	2728	1875
squfl020-150persp	6355	2083.7	6261	2597.6	516.17	112533	1458	5989	2559.0	523.54	0.23	1458	3551	1467
squfl025-025persp	18	7.3	18	82.2	74.35	34997	1274	25	70.0	61.96	0.20	1022	307	1115
squfl025-030persp	35	12.8	45	58.9	45.58	18678	746	39	58.6	45.94	0.06	746	333	760
squfl025-040persp	271	31.0	291	127.7	101.39	22712	997	299	129.1	101.28	0.17	997	1700	1334
sssd08-04	34140	9.7	31928	9.7	0.01	78	4	34932	10.7	0.00	0.23	4	48271	10
sssd08-04persp	56521	13.8	55164	13.3	0.01	264	4	49989	12.5	0.01	0.20	4	10294	16
st_e29	4	0.1	1	0.1	0.00	76	4	1	0.2	0.00	0.00	4	5	6
st_e31	451	1.6	381	1.5	0.00	18	2	381	1.6	0.00	0.00	2	6	2
st_e32	2713	5.1	1467	3.5	0.01	280	17	1596	3.4	0.01	0.01	17	3232	23
st_e35	18551	22.9	13571	15.8	0.06	521	18	6021	8.5	0.04	0.03	18	4165	44
st_e36	1135	1.1	833	1.2	0.00	105	4	833	1.1	0.00	0.00	4	0	3
st_e38	11	0.2	11	0.2	0.00	28	8	11	0.2	0.00	0.00	8	3	7
st_e40	22	0.1	15	0.1	0.00	98	5	15	0.1	0.00	0.00	5	5	3
st_test8	1	0.0	1	0.0	0.00	27	13	1	0.0	0.00	0.00	13	0	15
st_testgr1	38	0.1	27	0.0	0.00	40	10	22	0.0	0.00	0.00	10	18	10
st_testgr3	12	0.1	4	0.0	0.00	38	15	4	0.0	0.00	0.00	15	2	16
stockcycle	29590	150.7	29590	150.1	0.04	198	0	29590	149.9	0.04	0.04	0	0	0
supplychain	141	0.5	121	0.5	0.00	30	4	143	0.5	0.00	0.00	4	21	6
supplychainp1_022020	2531189	3600.1	1165327	2225.4	1.90	1213	24	157251	537.7	1.90	2.87	24	8399	36
supplychainp1_030510	2211	4.6	347	2.0	0.11	462	10	347	2.0	0.07	0.01	10	2	12
supplychainr1_022020	116764	124.9	1804365	3600.0	1.51	9978	32	2379500	3600.0	1.50	17.82	32	95834	64
syn05h	355927222	3600.0	4	0.1	0.00	46	6	4	0.1	0.00	0.00	6	0	14
syn05m02m	3	0.2	3	0.3	0.00	139	18	3	0.2	0.00	0.00	18	4	19
syn05m03m	4	0.4	4	0.3	0.02	225	23	2	0.2	0.02	0.00	17	6	22
syn05m04m	3	0.4	4	0.2	0.01	242	34	4	0.5	0.01	0.00	34	7	36
syn10h	1	0.2	1	0.2	0.01	156	1	1	0.2	0.01	0.00	1	0	19
syn10m	1	0.1	1	0.1	0.00	29	6	1	0.1	0.00	0.00	6	0	6
syn10m02h	13	0.5	5	0.6	0.06	829	4	5	0.5	0.05	0.00	4	0	50
syn10m02m	13	0.4	13	0.4	0.03	568	14	13	0.4	0.04	0.00	14	3	34
syn10m03h	34	0.9	34	0.9	0.04	717	3	35	0.8	0.03	0.00	3	5	41
syn10m03m	54	0.7	40	0.7	0.06	805	13	54	0.8	0.05	0.00	13	1	39
syn10m04m	18	0.8	18	0.9	0.11	1149	18	18	0.8	0.11	0.00	16	3	39
syn15h	10	0.4	2	0.3	0.03	351	7	2	0.3	0.02	0.00	7	0	25
syn15m	11	0.3	11	0.2	0.02	277	13	11	0.3	0.01	0.00	13	5	17
syn15m02h	63	0.8	63	0.9	0.06	559	22	63	0.8	0.06	0.00	22	23	58
syn15m02m	66	0.6	60	0.6	0.07	869	28	35	0.6	0.10	0.00	28	24	38
syn15m03m	186	1.0	169	1.0	0.21	2214	49	170	0.9	0.14	0.01	46	161	60
syn15m04m	615	2.0	577	2.1	0.10	844	2	577	2.2	0.10	0.00	2	168	27
syn20m	70	0.5	58	0.7	0.03	375	13	58	0.7	0.02	0.00	13	10	21
syn20m02h	447	2.6	85	1.2	0.12	1870	17	75	1.3	0.16	0.00	17	41	164
syn20m02m	345	1.5	345	1.6	0.06	462	0	345	1.6	0.06	0.00	0	63	23
syn20m03m	1585	3.5	1547	3.7	0.16	2814	10	1581	3.9	0.23	0.01	10	202	47

Table A.10 continued

Instance	SCIP plain		SCIP + OBBT					SCIP + OBBT + LVB						
	n	t	n	t	t_{obbt}	lp	b_{obbt}	n	t	t_{obbt}	t_{lvb}	b_{obbt}	b_{lvb}	lvb
syn20m04h	18316	87.7	16520	80.5	0.33	2194	20	18816	91.0	0.32	0.22	20	21502	160
syn20m04m	10791	12.9	11147	14.7	0.41	4399	7	10905	13.8	0.36	0.06	7	2606	66
syn30m	15	0.4	14	0.4	0.01	190	3	14	0.3	0.01	0.00	3	0	15
syn30m02m	397	2.4	423	1.8	0.02	482	6	423	2.1	0.06	0.00	6	80	29
syn30m03m	1318	4.3	1334	4.9	0.10	1791	5	1328	5.0	0.09	0.00	5	477	37
syn30m04m	75167	191.0	72732	189.3	0.07	1126	10	72261	180.6	0.09	0.12	10	27054	54
syn40m	110	0.7	110	0.6	0.05	339	0	110	0.8	0.03	0.00	0	8	18
syn40m02m	5100	9.3	5388	9.1	0.07	1227	0	5235	8.9	0.07	0.01	0	2115	36
syn40m03m	492145	798.0	464931	655.7	0.16	2773	0	484423	689.7	0.19	1.01	0	253290	55
syn40m04m	1639582	3335.2	1711034	3309.7	0.43	5180	0	1623399	3319.2	0.41	3.83	0	726420	69
synthes1	5	0.1	5	0.1	0.00	11	0	5	0.1	0.00	0.00	0	0	1
synthes2	4	0.1	4	0.1	0.00	61	8	4	0.1	0.00	0.00	8	0	9
synthes3	3	0.1	3	0.1	0.00	53	0	3	0.1	0.00	0.00	0	0	1
tanksize	2211	1.9	2805	2.5	0.00	92	7	4381	2.9	0.00	0.03	7	2101	15
tln4	3487	1.7	4385	1.6	0.00	93	9	2370	1.4	0.00	0.00	9	1151	10
tln5	623529	304.4	4107464	1953.6	0.01	94	15	875200	447.0	0.00	0.00	15	555519	16
tloss	65	0.2	49	0.1	0.00	113	21	68	0.1	0.00	0.01	21	89	23
tls2	10	0.1	10	0.1	0.00	63	0	10	0.2	0.00	0.00	0	0	0
tls4	23124	23.1	21975	24.4	0.02	363	15	20011	24.3	0.01	0.06	15	18952	35
tltr	5	0.4	8	0.4	0.02	608	6	8	0.5	0.02	0.00	6	12	38
tspn05	19941	53.9	14401	32.6	0.01	453	17	15451	37.2	0.02	0.05	17	4883	15
unicommit1	2	1.6	4	3.3	1.66	11757	309	4	3.7	1.76	0.01	305	2	419
util	6	0.4	17	0.3	0.00	51	9	119	0.4	0.00	0.00	9	366	9
watercontamination0202	35	429.0	35	1299.8	1.81	10432	50	35	1291.4	1.69	0.00	50	17	66
watercontamination0202r	6	651.3	6	1224.4	0.07	1142	0	6	1225.9	0.08	0.00	0	0	0
waterrnd1	84391	68.1	84391	69.3	0.02	367	21	54171	48.2	0.01	0.04	0	15217	13
waterno2.01	21	0.8	11	0.3	0.03	569	21	11	0.4	0.04	0.00	21	1	34
waterno2.02	261	5.8	181	7.0	0.25	2336	13	191	7.2	0.26	0.00	13	70	45
waterno2.03	66813	195.3	58551	164.5	0.42	3707	15	65964	182.6	0.45	0.00	15	21243	72
watertreatnd_conc	11301	11.6	16361	16.5	0.01	268	5	7221	9.9	0.01	0.32	5	2260	17
watertreatnd_flow	51391	61.1	154361	181.7	0.09	947	2	203903	223.4	0.07	0.29	2	6432	42

Table A.11: Detailed results comparing SCIP without OBBT, SCIP with OBBT only, and SCIP with OBBT and LVB propagation on test set GO. See Table 6.2 for aggregated results.

n — number of branch-and-bound nodes
t — total running time
t_{obbt} — time used by OBBT
lp — LPs solved by OBBT
b_{obbt} — bounds tightened by OBBT
lvb — LVBs found by OBBT
t_{lvb} — time used by LVB propagation
b_{lvb} — bounds tightened by LVB propagation

Instance	SCIP plain		SCIP+OBBT					SCIP+OBBT+LVB						
	n	t	n	t	t_{obbt}	lp	b_{obbt}	n	t	t_{obbt}	t_{lvb}	b_{obbt}	b_{lvb}	lvb
alkyl	101	0.3	48	0.3	0.01	57	13	48	0.3	0.00	0.00	13	1	15
bearing	7361	10.3	7051	11.2	0.01	52	4	8404	15.7	0.00	0.11	4	2831	16
chance	11	0.1	7	0.1	0.00	23	8	7	0.1	0.00	0.00	8	2	8
chem	1370091	3429.6	1897422	3600.0	0.01	150	13	1313341	2543.8	0.01	11.19	13	2439604	40
ex14.1.1	61	0.4	82	0.4	0.00	35	4	82	0.4	0.00	0.00	4	0	4
ex14.1.2	94	0.6	94	0.6	0.00	55	4	94	0.6	0.00	0.00	4	0	7
ex14.1.3	1	0.1	1	0.1	0.00	11	2	1	0.1	0.00	0.00	2	0	4
ex14.1.5	387	0.5	260	0.4	0.00	43	16	260	0.4	0.00	0.00	16	0	1
ex14.1.6	55	0.3	38	0.3	0.00	41	4	38	0.3	0.00	0.00	4	0	9
ex14.1.7	111551	125.8	36724	111.0	0.03	505	2	339431	338.0	0.03	0.33	2	1520	15
ex14.1.8	1	0.1	1	0.1	0.00	34	4	1	0.1	0.00	0.00	4	0	9
ex2.1.1	17	0.2	17	0.1	0.00	6	0	17	0.1	0.00	0.00	0	8	1
ex2.1.10	9	0.1	5	0.1	0.00	34	21	5	0.1	0.00	0.00	21	0	22
ex2.1.7	5415	1.5	7877	1.8	0.00	24	6	7710	1.7	0.00	0.00	6	6550	6
ex2.1.9	3697	3.8	3601	3.4	0.00	20	10	3681	3.6	0.00	0.00	10	283	10
ex3.1.1	2631	0.8	2901	1.0	0.00	27	7	2901	1.0	0.00	0.00	7	0	3
ex3.1.2	3	0.1	1	0.1	0.00	11	7	1	0.1	0.00	0.00	7	0	7
ex3.1.4	29	0.2	29	0.3	0.00	5	0	29	0.3	0.00	0.00	0	1	1
ex3pb	4	0.3	3	0.2	0.00	64	5	3	0.2	0.00	0.00	5	1	6
ex4.1.1	31	0.3	31	0.3	0.00	21	1	31	0.3	0.00	0.00	1	0	2
ex4.1.2	42	0.3	42	0.3	0.00	0	0	42	0.3	0.00	0.00	0	0	0
ex4.1.3	29	0.2	35	0.3	0.00	20	1	35	0.2	0.00	0.00	1	0	1
ex4.1.4	45	0.3	45	0.2	0.00	17	0	45	0.3	0.00	0.00	0	0	2
ex4.1.6	69	0.3	69	0.3	0.00	8	1	69	0.3	0.00	0.00	1	0	1
ex4.1.7	27	0.2	29	0.3	0.00	18	2	29	0.2	0.00	0.00	2	0	2
ex4.1.9	35	0.2	35	0.2	0.00	5	0	35	0.2	0.00	0.00	0	0	2
ex5.2.2.case1	65	0.3	181	0.3	0.00	18	3	181	0.3	0.00	0.00	3	24	4
ex5.2.2.case2	47	0.2	129	0.2	0.00	16	1	419	0.4	0.00	0.00	1	66	4
ex5.2.2.case3	41	0.1	77	0.3	0.00	18	3	85	0.3	0.00	0.00	3	14	5
ex5.2.4	591	0.6	571	0.7	0.00	31	0	571	0.6	0.00	0.00	0	11	3
ex5.3.2	901	0.5	869	0.7	0.00	35	2	876	0.7	0.00	0.01	2	181	6
ex5.4.2	771	0.5	511	0.4	0.00	26	6	511	0.4	0.00	0.00	6	30	8
ex5.4.3	39	0.3	16	0.2	0.00	62	8	16	0.2	0.00	0.00	8	0	5
ex5.4.4	2757336	3600.0	1021821	1084.7	0.01	166	6	869231	1046.1	0.00	2.61	6	27736	9

Table A.11 continued

Instance	SCIP plain		SCIP +OBBT					SCIP +OBBT+LVB						
	n	t	n	t	t_{obbt}	lp	b_{obbt}	n	t	t_{obbt}	t_{lvb}	b_{obbt}	b_{lvb}	lvb
ex6.1.1	45101	26.8	36381	22.2	0.00	35	4	39081	24.1	0.00	0.12	4	2501	18
ex6.1.2	100	0.4	96	0.4	0.00	19	2	81	0.4	0.00	0.00	2	10	7
ex6.1.3	332351	317.1	1251081	1216.3	0.00	56	3	575561	464.5	0.00	1.18	3	3876	19
ex6.1.4	802	1.0	1240	1.1	0.00	54	3	1511	1.0	0.00	0.00	3	168	9
ex6.2.14	954461	1464.2	758871	1520.0	0.02	255	6	715921	3600.0	0.03	1.17	6	4916	32
ex6.2.6	31911	29.1	34501	28.3	0.00	106	19	17981	16.9	0.00	0.20	19	128652	27
ex7.2.2	101	0.4	71	0.3	0.00	35	3	71	0.3	0.00	0.00	3	14	7
ex7.2.4	7351	6.5	7361	6.2	0.00	120	5	7921	6.6	0.00	0.01	5	815	11
ex7.3.1	5	0.1	1	0.1	0.00	31	10		0.1	0.00	0.00	10	0	10
ex7.3.2	9	0.1	3	0.1	0.00	17	7	3	0.1	0.00	0.00	7	0	8
ex7.3.3	27	0.2	23	0.1	0.00	7	4	23	0.2	0.00	0.00	4	4	6
ex7.3.6	7	0.1	7	0.1	0.00	11	3	7	0.1	0.01	0.00	3	0	3
ex8.1.6	1	0.1	1	0.1	0.00	11	0	1	0.1	0.00	0.00	0	0	0
ex8.1.7	106	0.8	111	0.7	0.00	57	2	57	0.7	0.00	0.00	2	18	2
ex8.2.3b	208623	3600.0	1	42.7	22.80	9489	77	1	40.6	22.22	0.01	77	1	85
ex8.2.4b	12713279	3600.0	39691	11.7	0.01	125	35	39691	12.0	0.00	0.06	35	1	38
ex8.4.1	17511	268.3	15761	266.7	0.00	218	6	16701	269.3	0.01	0.16	6	2431	8
ex8.4.4	35101	65.0	29441	56.0	0.01	118	17	26991	56.2	0.02	0.09	17	31560	23
ex8.4.5	131	2.0	175	2.3	0.00	283	3	211	2.2	0.01	0.00	3	8	4
ex8.4.6	460501	2783.9	518431	3133.3	0.00	150	1	482731	2921.2	0.00	1.66	1	5250	3
ex9.2.2	3	0.1	3	0.1	0.00	7	2	3	0.0	0.00	0.00	2	17	3
ex9.2.4	3	0.1	3	0.1	0.00	18	0	3	0.1	0.00	0.00	0	0	0
ex9.2.5	5	0.1	5	0.1	0.00	18	1	5	0.1	0.00	0.00	1	0	3
ex9.2.6	7	0.1	7	0.1	0.00	20	0	7	0.1	0.00	0.00	0	3	2
ex9.2.7	5	0.1	5	0.1	0.00	7	2	5	0.1	0.00	0.00	2	0	3
gsg_0001	733501	901.5	151361	338.2	0.04	216	50	66601	61.9	0.04	0.23	50	10391	90
himmel11	3	0.0	1	0.0	0.00	11	7	1	0.0	0.00	0.00	7	0	7
himmel16						66	0					0	0	0
house	2755	2.4	2709	2.4	0.01	11	6	2709	2.4	0.00	0.00	6	5	8
hs62	40001599	3600.0	701	1.2	0.00	27	8	691	1.0	0.00	0.01	8	491	9
hybriddynamic.fixedcc	861	1.3		0.1	0.00	59	26	1	0.0	0.03	0.00	26	0	27
hybriddynamic.varcc	1	0.1	18035	15.6	0.03	625	37	16221	14.2	0.00	0.04	37	32	40
kall_circles.c6a	9781	9.1	413896	143.4	0.00	75	7	356091	122.8	0.00	0.59	7	88640	14
kall_circles.c6b	346931	121.7	152473	50.8	0.00	68	7	150795	55.0	0.00	0.32	7	53639	16
kall_circles.c7a	183845	66.0	280786	91.2	0.00	128	2	206724	74.0	0.01	0.42	7	83590	12
kall_circles.c8a	265491	84.6	4637793	1790.6	0.00	92	9	874238	373.4	0.01	1.56	9	373611	14
kall_congruentcircles.c31	1321535	527.7	109	0.4	0.00	19	7	109	0.4	0.00	0.00	7	3	7
kall_congruentcircles.c32	149	0.4	199	0.4	0.00	23	1	197	0.3	0.00	0.00	1	25	4
kall_congruentcircles.c41	215	0.3	11	0.3	0.00	10	3	11	0.3	0.00	0.00	3	0	3
kall_congruentcircles.c42	11	0.3	141	0.3	0.00	39	2	158	0.4	0.00	0.00	2	417	3
kall_congruentcircles.c51	194	0.4	9406	3.3	0.00	61	5	9662	3.5	0.00	0.08	5	4523	8
kall_congruentcircles.c52	7191	2.9	959	1.0	0.00	45	3	995	0.9	0.00	0.00	3	627	14
kall_congruentcircles.c61	1247	0.8	54788	17.8	0.00	47	6	64917	21.4	0.00	0.10	6	28855	8
kall_congruentcircles.c62	53881	17.1	3461	1.8	0.00	67	4	3055	1.7	0.00	0.00	4	3201	15
kall_congruentcircles.c63	4000	1.8	2155	1.4	0.00	44	9	1993	1.3	0.01	0.01	9	4598	9
kall_congruentcircles.c71	326242	113.3	326242	112.8	0.00	60	0	648184	223.7	0.00	1.31	0	212731	14

Table A.11 continued

Instance	SCIP plain		SCIP +OBBT					SCIP +OBBT+lVB						
	n	t	n	t	t_{obbt}	lp	b_{obbt}	n	t	t_{obbt}	t_{lvb}	b_{obbt}	b_{lvb}	lvb
kall_congruentcircles_c72	18849	6.9	17809	6.7	0.00	60	5	17631	6.6	0.01	0.05	5	9435	11
kall_difficircles_10	11986760	3600.0	3021901	862.9	0.01	62	1	2170571	622.5	0.00	2.84	1	13361	3
kall_difficircles_5a	2251	1.0	2031	1.0	0.00	50	5	2461	1.0	0.00	0.01	5	1156	11
kall_difficircles_5b	14051	3.6	20551	4.6	0.00	43	6	15941	3.9	0.00	0.02	6	5460	9
kall_difficircles_6	24151	6.7	25431	5.9	0.00	38	2	14161	3.9	0.00	0.03	2	86	4
kall_difficircles_7	72461	14.9	28187	7.9	0.01	52	2	27374	7.5	0.00	0.01	8	270	7
kall_difficircles_8	5773801	1344.1	5621451	1287.7	0.01	56	8	5621451	1286.7	0.00	6.38	1	1	8
kall_difficircles_9	1053901	259.2	12988887	3600.0	0.01	53	1	13756590	3600.0	0.00	15.29	1	13901	6
linear	1	2.8	1	2.7	0.00	51	10	1	2.8	0.00	0.00	10	0	16
mathopt5_7	33	0.2	33	0.2	0.00	16	0	33	0.2	0.00	0.00	0	0	1
mathopt5_8	33	0.3	33	0.2	0.00	19	0	33	0.2	0.00	0.00	0	0	1
oaer	2	0.1		0.0	0.00	13	2		0.0	0.00	0.00	0	0	3
pointpack04	13	0.2	5	0.1	0.01	25	2	5	0.1	0.00	0.00	2	1	3
pointpack06	10161	3.6	6263	2.3	0.00	56	1	6515	2.3	0.00	0.03	1	833	4
pointpack08	456455	175.3	456455	174.8	0.01	52	0	405110	156.9	0.00	0.83	0	195310	7
portfol_buyin	4	0.1	4	0.2	0.01	93	13	4	0.2	0.00	0.00	13	0	20
prob06	1	0.0	1	0.0	0.00	8	2	1	0.0	0.00	0.00	2	0	2
prob07	139911	269.3	24081	51.1	0.01	151	48	30061	73.6	0.01	0.29	48	24072	52
process	100	0.4	131	0.5	0.00	59	8	131	0.5	0.00	0.00	8	0	10
procsyn	11	3.1	1	3.2	0.00	60	1	1	3.2	0.00	0.00	1	0	0
st_bpaf1a	11	0.1	3	0.0	0.00	21	13	1	0.1	0.00	0.00	13	100	13
st_bsj2	11	0.1	7	0.1	0.00	9	4	7	0.1	0.00	0.00	4	2	4
st_bsj4	3	0.0	3	0.0	0.00	11	6	3	0.0	0.00	0.00	6	0	6
st_e04	17	0.2	21	0.2	0.00	73	2	21	0.2	0.00	0.00	6	4	5
st_e05	64	0.3	66	0.2	0.00	12	5	66	0.2	0.00	0.00	5	0	5
st_e07	89	0.3	85	0.3	0.00	15	1	85	0.3	0.00	0.00	1	0	2
st_e08	1	0.0	1	0.0	0.00	5	4	1	0.0	0.00	0.00	4	0	4
st_e09	11	0.1	11	0.1	0.00	6	4	11	0.1	0.00	0.00	4	0	4
st_e11	6	0.2	6	0.2	0.00	4	0	6	0.2	0.00	0.00	0	0	0
st_e16	35	0.3	23	0.3	0.00	49	9	21	0.2	0.01	0.00	9	7	6
st_e19	145	0.3	145	0.3	0.00	17	0	145	0.3	0.00	0.00	0	0	0
st_e22	1	0.0	1	0.0	0.00	6	3	1	0.0	0.00	0.00	3	0	3
st_e24	3	0.0	3	0.0	0.00	7	3	3	0.0	0.00	0.00	3	0	3
st_e25	15	0.1	11	0.1	0.00	15	6	11	0.1	0.00	0.00	6	0	2
st_e26	1	0.0	1	0.0	0.00	4	2	1	0.0	0.01	0.00	2	0	4
st_e28	3	0.0	1	0.0	0.00	11	7	1	0.0	0.00	0.00	7	0	7
st_e30	97	0.4	97	0.4	0.00	25	2	97	0.4	0.00	0.00	2	2	2
st_e33	89	0.3	85	0.3	0.00	15	1	87	0.3	0.00	0.00	1	12	3
st_e41	13	0.2	11	0.2	0.00	50	11	11	0.2	0.00	0.00	11	12	15
st_fp7a	3815	1.2	213	0.3	0.00	47	19	227	0.4	0.02	0.00	19	381	19
st_fp7b	1885	0.9	1667	0.8	0.00	21	7	1320	0.7	0.00	0.00	7	1141	7
st_fp7c	3347	1.1	857	0.6	0.00	26	10	901	0.6	0.00	0.00	10	1655	10
st_fp7d	3777	1.1	507	0.4	0.01	51	19	413	0.5	0.01	0.00	19	804	19
st_fp7e	5415	1.4	7877	1.9	0.00	24	6	7710	1.8	0.00	0.04	6	6550	6
st_glmp_fp1	5	0.1		0.0	0.00	8	4		0.0	0.01	0.00	4	0	2
st_glmp_fp2	9	0.1	5	0.1	0.00	9	3	5	0.1	0.00	0.00	3	0	2

Table A.11 continued

Instance	SCIP plain		SCIP +OBBT					SCIP +OBBT+LVB						
	n	t	n	t	t_{obbt}	lp	b_{obbt}	t_{obbt}	t_{lvb}	t	n	b_{obbt}	b_{lvb}	lvb
st_glmp_fp3	3	0.0	3	0.0	0.00	17	3	0.00	0.00	0.0	3	3	0	0
st_glmp_kk90	9	0.0	7	0.0	0.00	7	3	0.00	0.00	0.1	7	3	0	3
st_glmp_kk92	3	0.0	1	0.0	0.00	10	2	0.00	0.00	0.0	1	2	0	3
st_glmp_kky	1	0.0	1	0.0	0.00	7	3	0.00	0.00	0.0	1	3	0	3
st_glmp_ss1	21	0.1	21	0.1	0.00	7	2	0.00	0.00	0.1	21	2	2	2
st_glmp_ss2	1	0.0	1	0.0	0.00	6	3	0.00	0.00	0.0	1	3	0	3
st_iqpbk1	37	0.3	31	0.3	0.00	42	2	0.00	0.00	0.3	31	2	0	3
st_iqpbk2	39	0.3	29	0.3	0.00	48	2	0.00	0.00	0.3	29	2	0	7
st_jcbpaf2	105	0.2	7	0.1	0.00	45	12	0.00	0.00	0.1	7	12	1	12
st_pan1	7	0.1	7	0.1	0.00	9	1	0.00	0.00	0.1	7	1	2	4
st_ph1	3	0.0	1	0.0	0.00	12	4	0.00	0.00	0.0	1	4	0	6
st_ph14	3	0.0	1	0.0	0.00	6	4	0.00	0.00	0.0	1	4	0	4
st_ph15	11	0.1	5	0.1	0.00	11	4	0.00	0.00	0.1	5	4	40	6
st_ph2	1	0.1	1	0.1	0.00	13	3	0.00	0.00	0.1	1	3	0	2
st_ph20	5	0.0	3	0.0	0.00	14	5	0.00	0.00	0.0	3	5	0	5
st_ph3	3	0.0	1	0.0	0.00	8	4	0.00	0.00	0.0	1	4	0	2
st_phex	1	0.0	1	0.0	0.00	10	0	0.00	0.00	0.0	1	0	0	0
st_qpc-m0	15	0.1	15	0.1	0.00	0	6	0.00	0.00	0.1	15	6	0	6
st_qpc-m1	23	0.1	9	0.1	0.00	15	1	0.00	0.00	0.1	7	1	1	10
st_qpc-m3a	171	0.4	181	0.4	0.00	35	7	0.00	0.00	0.4	181	7	33	9
st_qpc-m3b	9	0.1	7	0.1	0.00	98	7	0.00	0.00	0.1	7	7	1	0
st_qpk1	13	0.1	13	0.1	0.00	0	0	0.00	0.00	0.1	13	0	0	6
st_qpk2	35	0.3	35	0.3	0.00	19	0	0.00	0.00	0.3	35	0	39	11
st_qpk3	234	0.2	234	0.3	0.00	34	0	0.00	0.00	0.3	234	0	472	12
st_rv1	19	0.1	17	0.1	0.00	24	10	0.00	0.00	0.1	17	10	8	21
st_rv2	63	0.4	70	0.3	0.00	42	21	0.00	0.00	0.4	69	21	49	21
st_rv3	341	0.4	301	0.4	0.00	66	20	0.00	0.00	0.4	297	20	86	23
st_rv7	297	0.4	255	0.4	0.00	70	28	0.00	0.00	0.4	255	28	111	31
st_rv8	573	0.6	463	0.5	0.01	82	27	0.01	0.01	0.5	321	27	120	31
st_rv9	1361	1.1	1931	1.0	0.00	88	31	0.01	0.02	0.9	1691	31	2886	37
st_z	11	0.1	7	0.1	0.00	5	5	0.00	0.00	0.1	7	5	0	4
supplychainp1_020306	21	0.7	25	0.9	0.02	276	16	0.02	0.00	0.8	31	16	131	34
supplychain1_020306	14	0.3	14	0.6	0.01	8	4	0.00	0.00	0.5	14	4	104	11
supplychain1_030510	4	0.1	4	0.2	0.04	463	22	0.04	0.00	0.2	4	22	0	28
syn05m	3	0.2	2	0.1	0.01	21	3	0.00	0.00	0.1	2	3	0	0
syn05m03h	3	0.4	4	0.5	0.01	241	33	0.01	0.00	0.5	4	33	37	44
wastewater02m1	95	0.3	151	0.3	0.00	29	9	0.00	0.00	0.3	121	9	59	12
wastewater02m2	127	0.6	137	0.7	0.01	113	1	0.01	0.00	0.6	139	1	75	15
wastewater04m1	111	0.4	1141	0.6	0.00	61	6	0.00	0.00	0.4	151	6	52	18
wastewater04m2	108	0.7	69	0.7	0.01	99	10	0.01	0.00	0.7	69	10	53	20
wastewater05m1	478	3600.0	7591	5.0	0.00	122	3	0.01	0.04	4.7	6191	3	9730	16

A.2.2 Experiments with New Branching Rules

Table A.12: Detailed results comparing the performance of three new branching rules for MINLP in SCIP. Values where SCIP hit the time limit of one hour are printed in italics. See Section 7.4.4 for aggregated results.

n — total number of branch-and-bound nodes
t — total running time (in seconds)

Instance	DEFAULT		VARTYPE		COVER		NLSCORE	
	n	t	n	t	n	t	n	t
autocorr_bern20-10	411810	645.2	411810	642.3	415439	657.3	389593	635.0
autocorr_bern20-15	488941	2022.2	488941	2024.8	491792	2001.3	487196	2053.4
batch0812_nc	3227	3.6	3935	3.4	3095	3.0	3227	3.4
batch0812	3	0.3	3	0.2	3	0.2	3	0.2
carton7	476801	356.6	441597	320.5	–	–	450078	382.4
clay0203h	102	2.3	102	2.1	173	3.2	116	2.2
clay0204h	1241	13.3	1241	13.2	1619	16.6	1367	15.1
clay0205h	13315	100.9	13381	101.8	16382	176.3	10398	64.7
clay0303h	427	7.0	424	7.1	315	3.2	393	6.3
clay0304h	*286634*	*3600.0*	53345	953.8	11458	108.5	9266	104.1
clay0305h	29042	410.5	29650	419.5	164649	3165.9	56382	1095.5
crudeoil_lee2_05	22	2.3	22	2.2	23	3.6	25	2.0
du-opt5	64	0.4	64	0.4	–	–	64	0.4
du-opt	1808	2.4	1808	2.5	–	–	1808	2.4
eniplac	231	1.2	231	1.1	261	1.0	231	1.1
ex1263a	67	0.0	67	0.1	–	–	67	0.0
ex1264a	66	0.1	66	0.1	–	–	103	0.1
ex1265a	40	0.1	40	0.1	136	0.4	182	0.3
ex1266a	5	0.0	5	0.0	165	0.5	126	0.2
fac3	9	0.2	9	0.2	7	0.2	9	0.2
gasprod_sarawak01	220	1.0	255	1.1	720	2.7	1010	1.8
gastrans	82	0.7	162	0.8	78	0.6	168	0.7
gear3	4241	1.7	4241	1.6	–	–	4241	1.8
gear4	6954	2.3	6954	2.2	–	–	6903	2.4
gear	4241	1.7	4241	1.7	–	–	4241	1.7
graphpart_2g-0044-1601	74	0.7	74	0.6	90	0.7	74	0.4
graphpart_2g-0055-0062	1241	7.5	1241	8.0	1533	9.7	1241	7.5
graphpart_2g-0066-0066	9761	148.9	9761	148.7	6803	94.5	9761	147.3
graphpart_3g-0234-0234	1401	18.2	1401	18.1	1472	17.6	1401	18.2
graphpart_3g-0244-0244	4070	145.4	4070	144.6	3488	143.5	8486	291.4
graphpart_3g-0333-0333	4164	32.4	4164	31.9	4439	32.3	3528	25.7
graphpart_3g-0334-0334	40839	667.1	40839	662.1	45190	958.1	40839	665.4
jit1	10	0.0	10	0.0	–	–	10	0.0
kport20	219503	189.0	223456	189.3	223524	211.0	240354	205.4
lip	9	0.1	9	0.1	–	–	9	0.1
nvs01	11	0.0	11	0.0	–	–	11	0.0
nvs02	4	0.0	4	0.0	–	–	4	0.0
nvs06	11	0.0	11	0.0	–	–	11	0.0
nvs08	3	0.0	3	0.0	–	–	3	0.0
nvs11	5	0.0	5	0.0	–	–	5	0.0
nvs12	5	0.0	5	0.0	–	–	5	0.0
nvs13	21	0.0	21	0.0	–	–	21	0.0
nvs15	6	0.0	6	0.0	6	0.0	6	0.0
nvs16	6	0.0	6	0.0	–	–	6	0.0
nvs17	49	0.3	49	0.3	–	–	49	0.3
nvs18	33	0.1	33	0.1	–	–	33	0.1
nvs19	105	0.3	105	0.3	–	–	105	0.3
nvs20	3289	1.4	195	0.5	–	–	3289	1.6
nvs21	255	0.4	255	0.6	–	–	17	0.1
nvs23	119	0.6	119	0.6	–	–	119	0.6
nvs24	139	0.7	139	0.6	–	–	139	0.7
rsyn0805m02h	1383	3.1	1383	3.1	1195	3.1	1080	3.1
rsyn0805m02m	7680	8.6	7680	8.6	6740	8.5	6344	7.3
rsyn0805m03m	4363	11.4	4363	11.6	4450	11.3	4850	13.3
rsyn0805m04h	448	2.3	448	2.4	243	1.9	12	1.3
rsyn0805m04m	4483	16.2	4483	16.1	4165	15.3	8112	23.0
rsyn0810h	7	0.4	7	0.4	7	0.5	7	0.4
rsyn0810m02h	13127	20.9	13171	20.6	23171	29.5	1867	7.6
rsyn0810m02m	34147	38.7	34147	38.8	27040	32.1	33514	42.3
rsyn0810m03h	1242020	2319.0	1162287	2256.8	1107080	2347.1	765989	1627.1
rsyn0810m03m	30636	54.8	30636	54.6	27882	50.1	22643	40.9

Table A.12 continued

Instance	DEFAULT		VARTYPE		COVER		NLSCORE	
	n	t	n	t	n	t	n	t
rsyn0810m04h	1074	11.1	1072	11.0	1555	13.6	54	4.5
rsyn0810m04m	37968	102.3	37968	102.1	37064	99.3	35328	85.1
rsyn0810m	921	1.3	921	1.4	926	1.3	971	1.4
rsyn0815h	43	0.6	43	0.6	57	0.9	45	0.5
rsyn0815m02h	105342	190.5	106278	191.1	66563	112.6	19447	49.9
rsyn0815m02m	56124	67.8	56124	67.8	59589	62.3	543776	381.9
rsyn0815m03h	100619	313.2	112547	340.0	97200	298.7	4084	26.4
rsyn0815m04m	2189407	3366.5	2189407	3333.1	2331047	3456.9	*2239882*	*3600.0*
rsyn0815m	391	1.1	391	0.9	372	0.7	333	0.8
rsyn0820h	833	4.9	833	5.0	930	4.7	720	4.5
rsyn0820m02h	1127	12.0	1127	11.5	1226	11.2	43	5.9
rsyn0820m02m	47303	68.3	47303	68.7	64909	82.2	51771	67.9
rsyn0820m	3098	3.7	3098	3.7	3598	3.1	3361	4.0
rsyn0830h	790	5.0	790	5.1	311	4.1	1181	6.5
rsyn0830m02h	33876	93.3	33876	91.4	16370	41.9	1394	12.5
rsyn0830m02m	119840	185.1	119840	184.0	132212	195.0	157961	227.0
rsyn0830m03h	86524	405.8	86990	403.7	5131	44.4	1260	20.6
rsyn0830m04h	*394698*	*3600.0*	*442726*	*3600.0*	*450610*	*3600.0*	44611	208.7
rsyn0830m	2310	3.6	2310	3.6	2178	3.5	2145	5.0
rsyn0840h	404	6.7	404	6.8	304	6.3	769	7.0
rsyn0840m02h	545471	1181.1	545471	1173.9	16580	71.1	1720	21.4
rsyn0840m02m	158693	273.5	158693	270.9	241047	358.3	228938	368.9
rsyn0840m03h	*877140*	*3600.0*	*934565*	*3600.0*	*929762*	*3600.0*	46130	313.1
rsyn0840m04h	*285253*	*3600.0*	*309757*	*3600.0*	*317990*	*3600.0*	225195	2244.1
rsyn0840m	2412	3.2	2412	3.1	2522	3.2	2196	3.4
sepasequ_convent	*4625521*	*3600.0*	*3266734*	*3600.0*	*3939955*	*3600.0*	504854	1046.0
smallinvDAXr1b010-011	69	0.2	69	0.2	—	—	69	0.2
smallinvDAXr1b020-022	67	0.2	67	0.2	—	—	67	0.3
smallinvDAXr1b050-055	290	0.5	290	0.5	—	—	290	0.5
smallinvDAXr1b100-110	407	0.7	407	0.6	—	—	407	0.7
smallinvDAXr1b150-165	1037	1.0	1037	1.0	—	—	1037	0.8
smallinvDAXr1b200-220	967	1.0	967	0.9	—	—	967	1.0
smallinvDAXr2b010-011	79	0.4	79	0.4	—	—	79	0.4
smallinvDAXr2b020-022	79	0.3	79	0.3	—	—	79	0.3
smallinvDAXr2b050-055	212	0.5	212	0.6	—	—	212	0.5
smallinvDAXr2b100-110	345	0.4	345	0.4	—	—	345	0.4
smallinvDAXr2b150-165	848	0.8	848	0.8	—	—	848	0.9
smallinvDAXr2b200-220	339	0.6	339	0.6	—	—	339	0.6
smallinvDAXr3b010-011	75	0.2	75	0.2	—	—	75	0.2
smallinvDAXr3b020-022	55	0.1	55	0.1	—	—	55	0.1
smallinvDAXr3b050-055	270	0.5	270	0.4	—	—	270	0.5
smallinvDAXr3b100-110	457	0.7	457	0.6	—	—	457	0.6
smallinvDAXr3b150-165	966	0.9	966	0.9	—	—	966	0.8
smallinvDAXr3b200-220	337	0.6	337	0.5	—	—	337	0.5
smallinvDAXr4b010-011	99	0.3	99	0.3	—	—	99	0.3
smallinvDAXr4b020-022	64	0.2	64	0.2	—	—	64	0.2
smallinvDAXr4b050-055	260	0.4	260	0.4	—	—	260	0.3
smallinvDAXr4b100-110	441	0.5	441	0.5	—	—	441	0.5
smallinvDAXr4b150-165	735	0.6	735	0.6	—	—	735	0.7
smallinvDAXr4b200-220	535	0.7	535	0.8	—	—	535	0.6
smallinvDAXr5b010-011	97	0.3	97	0.3	—	—	97	0.3
smallinvDAXr5b020-022	59	0.2	59	0.2	—	—	59	0.2
smallinvDAXr5b050-055	187	0.4	187	0.4	—	—	187	0.4
smallinvDAXr5b100-110	301	0.5	301	0.5	—	—	301	0.5
smallinvDAXr5b150-165	769	0.7	769	0.7	—	—	769	0.7
smallinvDAXr5b200-220	399	0.6	399	0.5	—	—	399	0.5
smallinvSNPr1b010-011	794	6.4	794	6.3	—	—	794	6.4
smallinvSNPr1b020-022	28178	117.0	28178	123.1	—	—	28178	117.0
smallinvSNPr2b010-011	511	5.4	511	5.2	—	—	511	5.2
smallinvSNPr2b020-022	4382	21.7	4382	21.9	—	—	4382	22.4
smallinvSNPr2b050-055	48883	232.0	48883	230.0	—	—	48883	229.9
smallinvSNPr3b010-011	297	3.4	297	3.5	—	—	297	3.5
smallinvSNPr3b020-022	930	7.9	930	7.8	—	—	930	7.6
smallinvSNPr3b050-055	17318	83.9	17318	82.4	—	—	17318	82.2
smallinvSNPr3b100-110	208403	1024.9	208403	1012.5	—	—	208403	1007.8
smallinvSNPr4b010-011	294	2.5	294	2.4	—	—	294	2.4
smallinvSNPr4b020-022	436	5.2	436	5.3	—	—	436	5.2
smallinvSNPr4b050-055	1575	14.0	1575	13.8	—	—	1575	13.7
smallinvSNPr4b100-110	58741	269.1	58741	268.2	—	—	58741	271.9
smallinvSNPr4b150-165	49636	241.8	49636	240.5	—	—	49636	241.0
smallinvSNPr4b200-220	50809	231.7	50809	232.2	—	—	50809	230.9
smallinvSNPr5b010-011	212	2.4	212	2.2	—	—	212	2.5

Table A.12 continued

Instance	DEFAULT n	DEFAULT t	VARTYPE n	VARTYPE t	COVER n	COVER t	NLSCORE n	NLSCORE t
smallinvSNPr5b020-022	314	3.5	314	3.5	–	–	314	3.4
smallinvSNPr5b050-055	998	10.0	998	10.0	–	–	998	9.9
smallinvSNPr5b100-110	4308	28.9	4308	28.3	–	–	4308	28.7
smallinvSNPr5b150-165	8152	41.2	8152	41.6	–	–	8152	41.0
smallinvSNPr5b200-220	24188	116.6	24188	117.7	–	–	24188	116.9
spring	41	0.3	41	0.3	32	0.3	41	0.3
squfl010-025persp	12	1.4	12	1.5	–	–	12	1.4
squfl010-040persp	5	2.4	5	2.8	–	–	5	2.3
squfl010-080persp	121	23.8	121	24.4	–	–	121	23.5
squfl015-060persp	197	35.6	197	35.6	–	–	197	35.2
squfl015-080persp	407	64.5	407	63.9	–	–	407	63.6
squfl020-040persp	9	7.7	9	7.7	–	–	9	7.9
squfl020-050persp	110	48.6	110	51.1	–	–	110	51.4
squfl020-150persp	6473	2324.7	6473	2306.0	–	–	6473	2304.1
squfl025-025persp	19	6.0	19	6.0	–	–	19	6.0
squfl025-030persp	190	22.1	190	22.3	–	–	190	22.5
squfl025-040persp	397	30.3	397	30.5	–	–	397	30.2
sssd08-04persp	58817	14.0	56521	14.1	–	–	58392	13.7
st_e32	919	2.4	953	2.8	–	–	919	2.2
st_e36	1638	1.3	1149	1.1	–	–	1638	1.2
st_e38	13	0.1	19	0.1	–	–	13	0.0
st_e40	22	0.1	22	0.1	–	–	22	0.0
st_testgr1	21	0.0	21	0.0	–	–	21	0.0
st_testgr3	9	0.0	9	0.0	–	–	9	0.0
supplychainp1_020306	*13592932*	*3600.0*	29211	11.1	*13905589*	*3600.0*	305	0.9
supplychainp1_022020	790341	2258.6	124984	894.1	790341	2257.6	779131	1963.2
supplychainp1_030510	12249	9.9	7942568	3177.8	12249	9.6	2213344	776.3
syn05m02h	3	0.1	3	0.1	3	0.1	3	0.1
syn05m03h	3	0.1	3	0.1	3	0.2	3	0.1
syn05m03m	1	0.1	1	0.1	1	0.1	1	0.1
syn05m04h	3	0.2	3	0.2	3	0.2	3	0.2
syn10h	3	0.1	3	0.1	3	0.1	1	0.1
syn10m02h	3	0.2	3	0.2	3	0.2	3	0.2
syn10m02m	13	0.3	13	0.2	13	0.3	13	0.3
syn10m03h	5	0.5	5	0.2	5	0.5	5	0.6
syn10m03m	47	0.5	47	0.5	43	0.5	45	0.5
syn10m04h	19	0.7	19	0.7	17	0.7	17	0.7
syn10m04m	31	0.6	31	0.7	25	0.6	31	0.6
syn15h	12	0.2	12	0.2	12	0.2	12	0.2
syn15m02h	3	0.4	3	0.4	3	0.4	3	0.3
syn15m02m	66	0.5	66	0.5	51	0.4	66	0.4
syn15m03h	21	0.7	21	0.8	21	0.7	11	0.8
syn15m03m	178	0.7	178	0.9	117	0.7	181	0.7
syn15m04h	23	1.8	23	1.7	23	1.7	23	1.8
syn15m04m	615	1.8	615	1.8	601	2.3	593	2.2
syn15m	11	0.2	11	0.2	11	0.2	11	0.1
syn20h	23	0.3	23	0.3	25	0.4	24	0.3
syn20m02h	19	0.8	19	0.6	19	0.7	19	0.8
syn20m02m	288	1.4	288	1.3	395	1.4	288	1.1
syn20m03h	31	1.2	31	1.2	29	1.0	36	1.1
syn20m03m	1603	3.4	1603	3.4	2526	4.7	1479	3.5
syn20m04h	31	1.3	31	1.5	33	1.4	27	1.4
syn20m04m	9539	12.3	9539	12.0	11085	14.7	8161	10.3
syn20m	61	0.3	61	0.3	51	0.5	51	0.3
syn30h	1205	2.2	1205	2.2	1295	2.3	1173	2.2
syn30m02h	54719	85.3	55376	84.4	45154	65.4	43987	67.3
syn30m02m	331	1.9	331	1.8	493	2.1	392	2.2
syn30m03m	1286	4.3	1286	4.1	4325	8.9	1517	5.8
syn30m04m	74612	181.8	74612	181.9	99888	220.9	69329	169.6
syn30m	15	0.2	15	0.3	15	0.3	21	0.2
syn40h	5684	7.2	5684	7.2	4559	6.8	4526	6.8
syn40m02h	377	4.1	377	4.2	362	4.1	393	3.6
syn40m02m	5168	8.1	5168	8.1	6378	9.2	6388	11.3
syn40m03m	522959	810.9	522959	807.6	486066	670.5	584735	937.1
syn40m04h	*11845*	*3600.0*	*12466*	*3600.0*	*12298*	*3600.0*	*12412*	*3600.0*
syn40m04m	690673	1728.1	690673	1721.1	1047299	2347.7	845522	2018.0
syn40m	110	0.6	110	0.6	117	1.0	110	0.6
tanksize	2878	2.1	2316	1.9	–	–	2878	2.0
tln4	3487	1.9	3487	1.8	–	–	9349	3.9
tln5	493683	231.3	484029	233.8	–	–	106078	52.7
tloss	65	0.1	65	0.1	59	0.1	154	0.2
tls2	10	0.1	10	0.1	–	–	10	0.0

Table A.12 continued

Instance	DEFAULT		VARTYPE		COVER		NLSCORE	
	n	t	n	t	n	t	n	t
tltr	6	0.3	6	0.3	9	0.3	7	0.3
tspn05	*3702620*	*3600.0*	389855	636.7	731542	1139.4	912087	1242.4
unitcommit1	3	2.1	3	2.2	3	2.1	3	2.1